Lecture Notes in Computer Science 15523

Founding Editors

Gerhard Goos
Juris Hartmanis

Editorial Board Members

Elisa Bertino, *Purdue University, West Lafayette, IN, USA*
Wen Gao, *Peking University, Beijing, China*
Bernhard Steffen ⓘ, *TU Dortmund University, Dortmund, Germany*
Moti Yung ⓘ, *Columbia University, New York, NY, USA*

The series Lecture Notes in Computer Science (LNCS), including its subseries Lecture Notes in Artificial Intelligence (LNAI) and Lecture Notes in Bioinformatics (LNBI), has established itself as a medium for the publication of new developments in computer science and information technology research, teaching, and education.

LNCS enjoys close cooperation with the computer science R & D community, the series counts many renowned academics among its volume editors and paper authors, and collaborates with prestigious societies. Its mission is to serve this international community by providing an invaluable service, mainly focused on the publication of conference and workshop proceedings and postproceedings. LNCS commenced publication in 1973.

Ichiro Ide · Ioannis Kompatsiaris ·
Changsheng Xu · Keiji Yanai · Wei-Ta Chu ·
Naoko Nitta · Michael Riegler ·
Toshihiko Yamasaki
Editors

MultiMedia Modeling

31st International Conference on Multimedia Modeling, MMM 2025
Nara, Japan, January 8–10, 2025
Proceedings, Part IV

Editors
Ichiro Ide
Nagoya University
Nagoya, Japan

Changsheng Xu
Chinese Academy of Sciences
Beijing, China

Wei-Ta Chu
National Cheng Kung University
Tainan City, Taiwan

Michael Riegler
Simula
Oslo, Norway

Ioannis Kompatsiaris
Centre of Research and Technology
Thermi, Greece

Keiji Yanai
The University of Electro-Communications
Tokyo, Japan

Naoko Nitta
Mukogawa Women's University
Nishinomiya, Japan

Toshihiko Yamasaki
The University of Tokyo
Tokyo, Japan

ISSN 0302-9743 ISSN 1611-3349 (electronic)
Lecture Notes in Computer Science
ISBN 978-981-96-2070-8 ISBN 978-981-96-2071-5 (eBook)
https://doi.org/10.1007/978-981-96-2071-5

© The Editor(s) (if applicable) and The Author(s), under exclusive license
to Springer Nature Singapore Pte Ltd. 2025
Chapter "Temporal Closeness for Enhanced Cross-Modal Retrieval of Sensor and Image Data" is licensed under the terms of the Creative Commons Attribution 4.0 International License (http://creativecommons.org/licenses/by/4.0/). For further details see license information in the chapter.

This work is subject to copyright. All rights are solely and exclusively licensed by the Publisher, whether the whole or part of the material is concerned, specifically the rights of translation, reprinting, reuse of illustrations, recitation, broadcasting, reproduction on microfilms or in any other physical way, and transmission or information storage and retrieval, electronic adaptation, computer software, or by similar or dissimilar methodology now known or hereafter developed.
The use of general descriptive names, registered names, trademarks, service marks, etc. in this publication does not imply, even in the absence of a specific statement, that such names are exempt from the relevant protective laws and regulations and therefore free for general use.
The publisher, the authors and the editors are safe to assume that the advice and information in this book are believed to be true and accurate at the date of publication. Neither the publisher nor the authors or the editors give a warranty, expressed or implied, with respect to the material contained herein or for any errors or omissions that may have been made. The publisher remains neutral with regard to jurisdictional claims in published maps and institutional affiliations.

This Springer imprint is published by the registered company Springer Nature Singapore Pte Ltd.
The registered company address is: 152 Beach Road, #21-01/04 Gateway East, Singapore 189721, Singapore

If disposing of this product, please recycle the paper.

Preface

It is with great pleasure that we welcome you to the 31st International Conference on Multimedia Modeling (MMM 2025), held from January 8 to 10, 2025, in the historic city of Nara, Japan. This conference continued its tradition of bringing together experts, researchers, and practitioners from around the globe to present and discuss the latest advancements in multimedia technologies. Over the past three decades, MMM has grown to become a leading venue for multimedia research, reflecting the rapid evolution of this field. The topics covered this year span cutting-edge areas such as multimedia content analysis, retrieval, interaction, and applications powered by AI and machine learning. The convergence of these technologies is shaping the future of multimedia, and MMM 2025 aimed to showcase the innovation driving this progress.

The program featured keynote talks by world-renowned experts, a diverse range of technical research presentations, interactive poster and demonstration presentations, and the Video Browser Showdown competition. These sessions not only highlighted the technical achievements of participants but also fostered collaboration and sparked new ideas among attendees.

We were honored to feature three keynote presentations by Nancy F. Chen from A*STAR, Singapore, Kiyoharu Aizawa from The University of Tokyo, Japan, and Andrei Bursuc from valeo.ai, France.

This year's program received an impressive number of submissions, reflecting the vibrancy and growth of the community. From the 277 Regular Paper submissions, 122 high-quality papers were selected, demonstrating rigorous research across a range of topics from multimedia contents analysis to cutting-edge approaches and AI applications. In addition, 24 out of 30 Demonstration Papers were chosen, providing an interactive experience where researchers could showcase their work in action. The Video Browser Showdown, one of the highlights of the conference, featured 17 systems selected for this year's competition, which delivered exciting demonstrations of state-of-the-art multimedia retrieval technologies.

Special Sessions (SS) played an integral role in the conference, fostering discussions on focused topics within the multimedia community. This year, we accepted five Special Sessions, each addressing a critical area in multimedia research:

- ExpertSUM: Expert-Level Text Summarization from Fine-Grained Multimedia Analytics
- MLLMA: Multimodal Large Language Models and Applications
- Multimedia Research in Robotics
- SpIMA: Spatial Intelligence in Multimedia Analytics
- Simulating Edge Computing and Multimodal AI: A Benchmark for Real-World Applications

We received a total of 23 submissions for the five Special Sessions, and selected 13 papers for presentation.

We are immensely grateful to the authors, program committee, reviewers, and sponsors who dedicated their time and effort to make this conference a success. Their contributions shaped a rich and exciting program, and we are confident that MMM 2025 was a valuable experience for all participants.

Nara, with its cultural heritage and serene surroundings, offers a unique setting for intellectual exchange and networking. In particular, the conference venue is situated in the center of Nara Park, where many deer roam about. We hope you took the time to explore this beautiful park while engaging with fellow researchers and professionals from across the multimedia community.

We hope MMM 2025 inspired new collaborations and ideas, and look forward to seeing the future impact of the work presented here.

January 2025

Ichiro Ide
Ioannis Kompatsiaris
Changsheng Xu
Keiji Yanai
Wei-Ta Chu
Naoko Nitta
Michael Riegler
Toshihiko Yamasaki

Organization

Honorary Chairs

Kiyoharu Aizawa	The University of Tokyo, Japan
Noboru Babaguchi	Fukui University of Technology, Japan/Osaka University, Japan

General Chairs

Ichiro Ide	Nagoya University, Japan
Ioannis Kompatsiaris	Centre for Research & Technology Hellas, Greece
Changsheng Xu	Chinese Academy of Sciences, China
Keiji Yanai	The University of Electro-Communications, Japan

Program Committee Chairs

Wei-Ta Chu	National Cheng Kung University, Taiwan
Naoko Nitta	Mukogawa Women's University, Japan
Michael Riegler	Simula, Norway
Toshihiko Yamasaki	The University of Tokyo, Japan

Program Coordinators (Community Direction Chairs)

Chong-Wah Ngo	Singapore Management University, Singapore
Shin'ichi Satoh	National Institute of Informatics, Japan

Demonstration Chairs

Min-Chun Hu	National Tsinghua University, Taiwan
Wolfgang Hürst	Utrecht University, The Netherlands
Marie Katsurai	Doshisha University, Japan
Taishi Sawabe	Nara Institute of Science and Technology, Japan

Publicity Chairs

Huynh Thi Thanh Binh	Hanoi University of Science and Technology, Vietnam
Claudio Gennaro	ISTI-CNR, Italy
Xirong Li	Renmin University of China, China
Yoko Yamakata	The University of Tokyo, Japan

Web Chairs

Takahiro Komamizu	Nagoya University, Japan
Xueting Wang	CyberAgent, Japan

Publication Chairs

Marc A. Kastner	Hiroshima City University, Japan
Qiang Ma	Kyoto Institute of Technology, Japan

Financial Chair

Go Irie	Tokyo University of Science, Japan

Registration Chairs

Andreu Girbau-Xalabarder	Denso IT Laboratory, Japan
Yusuke Matsui	The University of Tokyo, Japan

Local Arrangements Chairs

Takuya Funatomi	Nara Institute of Science and Technology, Japan
Yasutomo Kawanishi	RIKEN, Japan
Kazuya Kitano	Nara Institute of Science and Technology, Japan

Video Browser Showdown Chairs

Werner Bailer	Joanneum Research, Austria
Cathal Gurrin	Dublin City University, Ireland
Jakub Lokoč	Charles University, Czech Republic
Klaus Schöffmann	Klagenfurt University, Austria

Steering Committee

Kiyoharu Aizawa	The University of Tokyo, Japan
Phoebe Chen	La Trobe University, Australia
Wen-Huang Cheng	National Taiwan University, Taiwan
Tat-Seng Chua	National University of Singapore, Singapore
Peng Cui	Tsinghua University, China
Duc-Tien Dang-Nguyen	University of Bergen, Norway
Cathal Gurrin	Dublin City University, Ireland
Richang Hong	Hefei University of Technology, China
Benoit Huet	Median Technologies Inc., France
Björn Þór Jónsson	Reykjavik University, Iceland
Guo-Jun Qi	University of Central Florida, USA
Klaus Schöffmann	Klagenfurt University, Austria

Special Session Organizers

ExpertSUM: Expert-Level Text Summarization from Fine-Grained Multimedia Analytics

Takahiro Komamizu	Nagoya University, Japan
Satoshi Yamazaki	NEC Corp., Japan

MLLMA: Multimodal Large Language Models and Applications

Avinash Anand	Indraprastha Institute of Information Technology, Delhi, India
Yaman Kumar	Adobe Systems, USA
Rajiv Ratn Shah	Indraprastha Institute of Information Technology, Delhi, India
Astha Verma	National University of Singapore, Singapore

Multimedia Research in Robotics

Yasutomo Kawanishi	RIKEN, Japan
Takashi Minato	RIKEN, Japan
Yutaka Nakamura	RIKEN, Japan
Wataru Sato	RIKEN, Japan
Koichiro Yoshino	Institute of Science Tokyo, Japan

SpIMA: Spatial Intelligence in Multimedia Analytics

Demir Begüm	TU Berlin, Germany
Ilias Gialampoukidis	Centre for Research & Technology Hellas, Greece
Björn Þór Jónsson	Reykjavik University, Iceland
Ioannis Papoutsis	National Technical University of Athens, Greece
Maria Pegia	Centre for Research & Technology Hellas, Greece
Stefanos Vrochidis	Centre for Research & Technology Hellas, Greece

Simulating Edge Computing and Multimodal AI: A Benchmark for Real-World Applications

Minh-Son Dao	National Institute of Information and Communications Technology, Japan
Sadanori Ito	National Institute of Information and Communications Technology, Japan
Takamasa Mizoi	National Institute of Information and Communications Technology, Japan
Do-Van Nguyen	National Institute of Information and Communications Technology, Japan
Anh-Khoa Tran	National Institute of Information and Communications Technology, Japan
Koji Zettsu	National Institute of Information and Communications Technology, Japan

Program Committee

Alan Smeaton	Dublin City University, Ireland
Albert Ali Salah	Utrecht University, The Netherlands
Alex Falcon	University of Udine, Italy
Atsushi Hashimoto	OMRON SINIC X Corp., Japan

Atsuyuki Miyai	The University of Tokyo, Japan
Avinash Anand	Indraprastha Institute of Information Technology, Delhi, India
Boyun Li	Sichuan University, China
Cathal Gurrin	Dublin City University, Ireland
Chen Fu	The University of Tokyo, Japan
Cheng-Hsin Hsu	National Tsing Hua University, Taiwan
Cheng-Kang Tan	National Cheng Kung University, Taiwan
Chenyang Lyu	Mohamed bin Zayed University of Artificial Intelligence, United Arab Emirates
Chi Zhang	The University of Tokyo, Japan
Chieh-Yu Pan	National Cheng Kung University, Taiwan
Chih-Wei Lin	Fujian Agriculture and Forestry University, China
Christian Timmerer	Klagenfurt University, Austria
Chuck-Jee Chau	The Chinese University of Hong Kong, China
Chutisant Kerdvibulvech	National Institute of Development Administration, Thailand
Fan Zhang	Communication University of Zhejiang, China
Fangzhou Luo	McMaster University, Canada
Fei Wu	Nanjing University of Posts and Telecommunications, China
Feng Li	University of Science and Technology of China, China
Fumito Shinmura	Shizuoka University, Japan
George Awad	National Institute of Standards and Technology, USA
Go Irie	Tokyo University of Science, Japan
Guo-Shiang Lin	National Chin-Yi University of Technology, Taiwan
Haodian Wang	University of Science and Technology of China, China
Haruya Kyutoku	Aichi University of Technology, Japan
Hisashi Miyamori	Kyoto Sangyo University, Japan
Hsuan-Yu Fan	National Cheng Kung University, Taiwan
Hui Cui	Qilu University of Technology, China
Huy Quang Ung	KDDI Research, Inc., Japan
Ilias Gialampoukidis	Centre for Research & Technology Hellas, Greece
Itthisak Phueaksri	RIKEN, Japan
Jaime Boanerjes Fernandez Roblero	Dublin City University, Ireland
Jakub Lokoc	Charles University, Czech Republic
Jeonghun Baek	The University of Tokyo, Japan
Jiafeng Mao	CyberAgent, Japan

Jian Hou	Dongguan University of Technology, China
Jing Liu	Capital Medical University, China
John See	Heriot-Watt University Malaysia, Malaysia
Jun Li	Nanjing Normal University, China
Kaede Shiohara	The University of Tokyo, Japan
Kazu Mishiba	Tottori University, Japan
Kazuaki Nakamura	Tokyo University of Science, Japan
Kazuya Kitano	Nara Institute of Science and Technology, Japan
Keisuke Doman	Chukyo University, Japan
Keith Curtis	Technological University of the Shannon, Ireland
Kimiaki Shirahama	Doshisha University, Japan
Koichi Shinoda	Institute of Science Tokyo, Japan
Koichiro Yoshino	Institute of Science Tokyo, Japan
Kosetsu Tsukuda	National Institute of Advanced Industrial Science and Technology, Japan
Ladislav Peska	Charles University, Czech Republic
Lang Huang	The University of Tokyo, Japan
Li-Wei Kang	National Taiwan Normal University, Taiwan
Liang Wang	Beijing University of Technology, China
Ling Xiao	The University of Tokyo, Japan
Linlin Shen	Shenzhen University, China
Luca Rossetto	Dublin City University, Ireland
Luwei Zhang	The University of Tokyo, Japan
Manisha Saini	BML Munjal University, India
Maorong Wang	The University of Tokyo, Japan
Marc A. Kastner	Hiroshima City University, Japan
Marcel Worring	University of Amsterdam, The Netherlands
Maria Pegia	Centre for Research & Technology Hellas, Greece
Mark Quinlan	University College London, UK
Masahiro Toyoura	University of Yamanashi, Japan
Masakazu Iwamura	Osaka Metropolitan University, Japan
Michihiro Kuroki	The University of Tokyo, Japan
Minh-Son Dao	National Institute of Information and Communications Technology, Japan
Mohamed Saleem Nazmudeen	Universiti Teknologi Brunei, Brunei Darussalam
Na Qi	Beijing University of Technology, China
Naoto Inoue	CyberAgent, Japan
Naye Ji	Communication University of Zhejiang, China
Nicolas Michel	Gustave Eiffel University, France
Norifumi Kawabata	Computational Imaging Lab, Japan
Palaiahnakote Shivakumara	University of Salford, UK
Patrick Ramos	Osaka University, Japan

Pengcheng Zhao	Hefei University of Technology, China
Pravin Nagar	Dolby Laboratories, India
Pål Halvorsen	SimulaMet, Norway
Qian Cao	Renmin University of China, China
Qiong Chang	Institute of Science Tokyo, Japan
Raj Shivprakash Jaiswal	Indraprastha Institute of Information Technology, Delhi, India
Rajiv Ratn Shah	Indraprastha Institute of Information Technology, Delhi, India
Ren Togo	Hokkaido University, Japan
Ryan Ramos	Osaka University, Japan
Ryosuke Furuta	The University of Tokyo, Japan
Ryosuke Yamanishi	Kansai University, Japan
Sarah Fernandes Pinto Fachada	Université Libre de Bruxelles, Belgium
Satoshi Kosugi	Institute of Science Tokyo, Japan
Satoshi Yamazaki	NEC Corp., Japan
Shaodong Li	Guangxi University, China
Shengzhou Yi	The University of Tokyo, Japan
Shih-Wei Sun	Taipei National University of the Arts, Taiwan
Shijie Hao	Hefei University of Technology, China
Shintami Chusnul Hidayati	Institut Teknologi Sepuluh Nopember, Indonesia
Shogo Okada	Japan Advanced Institute of Science and Technology, Japan
Shuhei Yamamoto	University of Tsukuba, Japan
Shuntaro Masuda	The University of Tokyo, Japan
Soichiro Kumano	The University of Tokyo, Japan
Sourav Mishra	theAstate Inc., Japan
Steven Hicks	SimulaMet, Norway
Supatta Viriyavisuthisakul	Panyapiwat Institute of Management, Thailand
Taishi Sawabe	Nara Institute of Science and Technology, Japan
Takahiro Komamizu	Nagoya University, Japan
Takashi Minato	RIKEN, Japan
Takashi Shibata	NEC Corp., Japan
Takayuki Nakatsuka	National Institute of Advanced Industrial Science and Technology, Japan
Tao Peng	Soochow University, China
Thitirat Siriborvornratanakul	National Institute of Development Administration, Thailand
Thu Nguyen	SimulaMet, Norway
Tien-Dung Mai	University of Information Technology, VNU HCM, Vietnam
Tomohiro Fujita	Doshisha University, Japan

Tor-Arne Schmidt Nordmo	UiT The Arctic University of Norway, Norway
Toru Tamaki	Nagoya Institute of Technology, Japan
Tran Khoa	National Institute of Information and Communications Technology, Japan
Vajira Thambawitta	SimulaMet, Norway
Vijay John	RIKEN, Japan
Wataru Ohyama	Tokyo Denki University, Japan
Wei-Jun Chen	Carl Zeiss AG, Germany
Weifeng Liu	China University of Petroleum, China
Wenbin Gan	National Institute of Information and Communications Technology, Japan
Xiangling Ding	Hunan University of Science and Technology, China
Xiao Luo	University of California, Los Angeles, USA
Xiujuan Zheng	Sichuan University, China
Xueting Wang	CyberAgent, Inc., Japan
Yasutomo Kawanishi	RIKEN, Japan
Yi-Hsuan Lu	National Cheng Kung University, Taiwan
Yingling Chen	National Defense University, Taiwan
Yiwei Zhang	theAstate Inc., Japan
Yoko Yamakata	The University of Tokyo, Japan
Yongpan Sheng	Southwest University, China
Yongqing Sun	Nihon University, Japan
Yoshiyuki Shoji	Shizuoka University, Japan
Yu Chun Chen	National Central University, Taiwan
Yu Liu	Tongji University, China
Yu-Hsiang Chen	National Cheng Kung University, Taiwan
Yuan Lin	Kristiania University College, Norway
Yuan Zeng	Shenzhen Technology University, China
Yuanzhi Cai	Commonwealth Scientific and Industrial Research Organisation, Australia
Yue Zhao	Shandong University, China
Yuen Peng Loh	Multimedia University, Malaysia
Yulan Su	Institute of Information Engineering, Chinese Academy of Sciences, China
Yusuke Matsui	The University of Tokyo, Japan
Yuta Nakashima	Osaka University, Japan
Yutaka Nakamura	RIKEN, Japan
Zepu Yi	Huazhong University of Science and Technology, China
Zerun Wang	The University of Tokyo, Japan
Zhenping Xie	Jiangnan University, China

Zhenqiang Li	Zhengzhou University, China
Zhenzhen Quan	Shandong University, China
Zhi Liu	North China University of Technology, China
Zhibin Zhang	Inner Mongolia University, China
Zhicheng Du	Tsinghua University, China
Zichun Zhong	Wayne State University, USA

Demonstration Program Committee

Hung-Kuo Chu	National Tsing Hua University, Taiwan
Kimiaki Shirahama	Doshisha University, Japan
Marie Katsurai	Doshisha University, Japan
Min-Chun Hu	National Tsing Hua University, Taiwan
Taishi Sawabe	Nara Institute of Science and Technology, Japan
Tomoya Sawada	Doshisha University, Japan
Tse-Yu Pan	National Taiwan University of Science and Technology, Taiwan
Wolfgang Hürst	Utrecht University, The Netherlands

VBS Program Committee

Bao Le Hoang	Dublin City University, Ireland
Cathal Gurrin	Dublin City University, Ireland
Ivana Sixtova	Charles University, Czech Republic
Jakub Lokoc	Charles University, Czech Republic
Jürgen Primus	Klagenfurt University, Austria
Klaus Schöffmann	Klagenfurt University, Austria
Linh Tran	Dublin City University, Ireland
Luca Rossetto	Dublin City University, Ireland
Ly Duyen Tran	Dublin City University, Ireland
Mario Leopold	Klagenfurt University, Austria
Martin Winter	Joanneum Research, Austria
Stefanie Onsori-Wechtitsch	Joanneum Research, Austria
Thang-Long Nguyen-Ho	Dublin City University, Ireland
Thu Nguyen	SimulaMet, Norway
Vajira Thambawitta	SimulaMet, Norway
Van-Tu Ninh	Dublin City University, Ireland
Werner Bailer	Joanneum Research, Austria

Contents

Regular Papers

SES-Net: Multi-dimensional Spot-Edge-Surface Network for Nuclei Segmentation 3
    Congjian Lu, Shuwang Zhou, Ke Shan, Hongkuan Zhang, and Zhaoyang Liu

Skin-Adapter: Fine-Grained Skin-Color Preservation for Text-to-Image Generation 16
    Zhuowei Chen, Mengqi Huang, Nan Chen, and Zhendong Mao

Small Tunes Transformer: Exploring Macro and Micro-level Hierarchies for Skeleton-Conditioned Melody Generation 30
    Yishan Lv, Jing Luo, Boyuan Ju, and Xinyu Yang

SMG-Diff: Adversarial Attack Method Based on Semantic Mask-Guided Diffusion 44
    Yongliang Zhang and Jing Liu

SPLGAN-TTS: Learning Semantic and Prosody to Enhance the Text-to-Speech Quality of Lightweight GAN Models 58
    Ding-Chi Chang, Shiou-Chi Li, and Jen-Wei Huang

SSCDUF: Spatial-Spectral Correlation Transformer Based on Deep Unfolding Framework for Hyperspectral Image Reconstruction 71
    Hui Zhao, Na Qi, Qing Zhu, and Xiumin Lin

SSDL: Sensor-to-Skeleton Diffusion Model with Lipschitz Regularization for Human Activity Recognition 85
    Nikhil Sharma, Changchang Sun, Zhenghao Zhao, Anne Hee Hiong Ngu, Hugo Latapie, and Yan Yan

Structural Information-Guided Fine-Grained Texture Image Inpainting 100
    Zhiyi Fang, Yi Qian, and Xiyue Dai

Style Separation and Content Recovery for Generalizable Sketch Re-identification and a New Benchmark 114
    Lingyi Lu, Xin Xu, and Xiao Wang

Synchronization and Calibration of Video Sequences Acquired Using
Multiple Plenoptic 2.0 Cameras 128
   Daniele Bonatto, Sarah Fachada, Jaime Sancho, Eduardo Juarez,
   Gauthier Lafruit, and Mehrdad Teratani

Target-Oriented Dynamic Denosing Curriculum Learning for Multimodel
Stance Detection 141
   Zihao Suo and Shanliang Pan

TDM: Temporally-Consistent Diffusion Model for All-in-One Real-World
Video Restoration 155
   Yizhou Li, Zihua Liu, Yusuke Monno, and Masatoshi Okutomi

Temporal Closeness for Enhanced Cross-Modal Retrieval of Sensor
and Image Data 170
   Shuhei Yamamoto and Noriko Kando

The Right to an Explanation Under the GDPR and the AI Act 184
   Bjørn Aslak Juliussen

Toward Appearance-Based Autonomous Landing Site Identification
for Multirotor Drones in Unstructured Environments 198
   Joshua Springer, Gylf Þór Guðmundsson, and Marcel Kyas

Towards Inclusive Education: Multimodal Classification of Textbook
Images for Accessibility 212
   Saumya Yadav, Élise Lincker, Caroline Huron, Stéphanie Martin,
   Camille Guinaudeau, Shin'ichi Satoh, and Jainendra Shukla

Towards Visual Storytelling by Understanding Narrative Context Through
Scene-Graphs 226
   Itthisak Phueaksri, Marc A. Kastner, Yasutomo Kawanishi,
   Takahiro Komamizu, and Ichiro Ide

TPS-YOLO: The Efficient Tiny Person Detection Network Based
on Improved YOLOv8 and Model Pruning 240
   Li Yao, Qianni Huang, and Yan Wan

Uncertainty-Guided Joint Semi-supervised Segmentation and Registration
of Cardiac Images 253
   Junjian Chen and Xuan Yang

Understanding the Roles of Visual Modality in Multimodal Dialogue:
An Empirical Study 268
   Qian Cao, Ruihua Song, and Xu Chen

Vision-Language Pretraining for Variable-Shot Image Classification 283
   *Sotirios Papadopoulos, Konstantinos Ioannidis, Stefanos Vrochidis,
Ioannis Kompatsiaris, and Ioannis Patras*

Visual Anomaly Detection on Topological Connectivity Under Improved
YOLOv8 298
   Yu Li and Zhenping Xie

Wavelet Integrated Convolutional Neural Network for ECG Signal
Denoising 311
   Takamasa Terada and Masahiro Toyoura

WavFusion: Towards Wav2vec 2.0 Multimodal Speech Emotion
Recognition 325
   Feng Li, Jiusong Luo, and Wanjun Xia

Zero-Shot Sketch-Based Image Retrieval with Hybrid Information Fusion
and Sample Relationship Modeling 337
   Weijie Wu, Jun Li, Zhijian Wu, and Jianhua Xu

**Special Session: ExpertSUM: Special Session on Expert-Level Text
Summarization from Fine-Grained Multimedia Analytics**

CalorieVoL: Integrating Volumetric Context Into Multimodal Large
Language Models for Image-Based Calorie Estimation 353
   Hikaru Tanabe and Keiji Yanai

Can Masking Background and Object Reduce Static Bias for Zero-Shot
Action Recognition? 366
   Takumi Fukuzawa, Kensho Hara, Hirokatsu Kataoka, and Toru Tamaki

**Special Session: MLLMA: Special Session on Multimodal Large
Language Models and Applications**

Enhanced Anomaly Detection in 3D Motion Through Language-Inspired
Occlusion-Aware Modeling 383
   *Su Li, Liang Wang, Jianye Wang, Ziheng Zhang, Junjun Zhang,
and Lei Zhang*

Evaluating VQA Models' Consistency in the Scientific Domain 398
   Khanh-An C. Quan, Camille Guinaudeau, and Shin'ichi Satoh

Image2Text2Image: A Novel Framework for Label-Free Evaluation
of Image-to-Text Generation with Text-to-Image Diffusion Models 413
   *Jia-Hong Huang, Hongyi Zhu, Yixian Shen, Stevan Rudinac,
and Evangelos Kanoulas*

Quantifying Image-Adjective Associations by Leveraging Large-Scale
Pretrained Models 428
   *Chihaya Matsuhira, Marc A. Kastner, Takahiro Komamizu,
Takatsugu Hirayama, and Ichiro Ide*

TACST: Time-Aware Transformer for Robust Speech Emotion Recognition 442
   Wei Wei, Bingkun Zhang, and Yibing Wang

TS-MEFM: A New Multimodal Speech Emotion Recognition Network
Based on Speech and Text Fusion 454
   Wei Wei, Bingkun Zhang, and Yibing Wang

Author Index 469

Regular Papers

SES-Net: Multi-dimensional Spot-Edge-Surface Network for Nuclei Segmentation

Congjian Lu[1,2], Shuwang Zhou[2,3(✉)], Ke Shan[2], Hongkuan Zhang[2], and Zhaoyang Liu[2(✉)]

[1] School of Mathematics and Statistics, Qilu University of Technology (Shandong Academy of Sciences), Jinan, China
[2] Shandong Artificial Intelligence Institute, Qilu University of Technology (Shandong Academy of Sciences), Jinan 250014, Shandong, China
{zhoushw,liuzhyang}@qlu.edu.cn
[3] College of Computer Science and Engineering, Shandong University of Science and Technology, Qingdao 266590, Shandong, China

Abstract. On account of the dense distribution and blurred boundaries of nuclei in pathological images, nuclei instance segmentation remains a challenging task. Existing methods often fail to fully utilize information across points, lines, and regions. To fulfill the research gap, this paper proposes a novel multi-dimensional nuclei instance segmentation network (SES-Net). First, we introduce a three-branch network structure that includes a keypoint branch for precise localization of nuclei, an edge branch for describing the shape and contours of nuclei, and a mask branch for comprehensive characterization of nuclei regions. Next, we present a region optimization module (ROM) that effectively integrates keypoint and edge features to distinguish overlapping instances and refine segmentation masks. Finally, we describe a post-processing method that combines the ROM output with nuclei mask images to produce the final instance segmentation results. Extensive experiments validate the effectiveness of each component in SES-Net and demonstrate its promising performance on publicly available MoNuSeg, CoNSep and Kumar datasets.

Keywords: Nuclei Segmentation · Three-branch network · Medical Imaging

1 Introduction

Within the scope of digital pathology, the task of [1–4] nuclei image segmentation on histopathology images is fundamental for morphological quantification and tumor grading assessments. The outcome of nuclei segmentation can provide essential visual information about nuclei such as size, shape, and color, which are crucial for understanding cellular morphology. Therefore, this task is of significant importance. However, nuclei segmentation has posed considerable challenges due to complexities such as variations in nuclei size, blurred nuclear boundaries, uneven staining, cell clustering and overlapping cells [5]. For example, as discussed in Schmidt [6], unlike objects in natural images [7–9], the nuclei are often overlapping, which makes it difficult to distinguish adjacent nuclei and lacks robustness for instance segmentation methods designed for natural images such as Mask-RCNN [7] and PA-Net [10]. Another major challenge is the blur edge between the nuclei contacts, which increases the difficulty of nuclei segmentation. Numerous methods [1–5] have been attempted to address the aforementioned challenges, particularly focusing on the encoder-decoder neural structure, with the U-Net [11] model being a prime example. Due to the limitations of U-Net in effectively separating closely positioned objects in complex histological images, the literature has introduced various approaches that focus on three main aspects, namely, designing proper network architectures, introducing auxiliary task learning and proposing optimized segmentation strategies. Unfortunately, these models frequently lack spatial awareness, which could enhance instance segmentation of overlapping nuclei-especially when cancers are further advanced.

The "from point to line to surface" approach in everyday life is a systematic and hierarchical method of analysis commonly used to describe and understand complex structures. In biomedical image analysis, particularly in the description and segmentation of cell nuclei features, this approach helps us gradually build a complete and accurate model. Based on this idea, we propose a novel multidimensional spot-edge-surface network (SES-Net), a three-branch network structure based on U-Net [11], consisting of a cell nucleus keypoint branch, a boundary branch, and a mask branch. This structure uses point, line, and surface features to achieve a three-dimensional description of the nuclei. Furthermore, we introduce a region optimization module for overlapping cell nuclei instances. Guided by the keypoints and boundaries, this module distinguishes and optimizes overlapping cell nuclei. By using post-processing strategies to correct segmentation errors, we achieve the final segmentation results. This method, which refines from local details to the overall structure, shows significant advantages in the accuracy and robustness of nuclei instance segmentation compared to traditional methods.

The main contributions of this work are three-fold:

1. We introduce a three-branch network structure encompassing keypoint detection, boundary detection, and mask prediction branches. This structure enhances boundary information supervision and significantly improves nuclear segmentation performance.
2. We present a region optimization module that improves the accuracy of predicted locations by combining the keypoints and boundaries of each nuclear instance.
3. We propose a post-processing strategy that leverages multi-dimensional features to effectively correct mis-segmentation and missed segmentation predictions.

2 Related Work

Current nucleus instance segmentation methods can be broadly categorized into top-down and bottom-up approaches.

2.1 Top-Down Methods

Top-down methods, such as Mask-RCNN [7], have made great strides in the segmentation of natural. However, their application in nucleus segmentation is limited [5,12,13], primarily due to two reasons: First, on the data side, histopathological images contain many severely overlapping nuclei. As a result, a single bounding box proposal often includes multiple nuclei with indistinct boundaries, making it difficult for the network to optimize. This issue arises because the bounding box may enclose several nuclei, which complicates the segmentation task and reduces the accuracy of individual nucleus identification. Second, on the model side, top-down methods typically predict segmentation masks at a fixed resolution (e.g., 28×28) and then resample these masks to match the size of their corresponding bounding boxes. This resampling process can introduce quantization errors [12], posing challenges for accurately segmenting the boundaries of nuclei. The fixed resolution of the masks can lead to loss of detail and inaccuracies in the final segmentation output.

2.2 Bottom-Up Methods

Bottom-up methods first regress various types of nuclear proxies and then employ fine-grained post-processing to assign pixels to individual instances. Due to their accuracy, bottom-up methods are widely used in nucleus instance segmentation. These methods [1–5] typically involve designing robust network architectures, incorporating auxiliary task learning, and proposing optimized segmentation strategies. Specifically, DIST [4] transforms the nucleus segmentation problem into distance map regression by predicting the distance from each pixel to its associated nucleus center. Along a predetermined set of directions, StarDist [14] and its extension CPP-Net [15] anticipate the distances between each foreground pixel and the boundary of its corresponding instance. CIA-Net [1] enhances information richness for nucleus segmentation by adding additional contour supervision. In order to improve segmentation efficiency, Triple U-Net [16] creates a parallel feature aggregation network that combines features from RGB and hematoxylin pictures. CD-Net [17] utilizes centripetal direction to describe spatial relationships within the nucleus, effectively identifying boundaries. By estimating the vertical and horizontal distances between nucleus pixels and the centroid, Hover-Net [18] is the first model to simultaneously propose nucleus segmentation and categorization. Schmidt [6] proposed detecting star-convex polygons instead of bounding boxes to locate nuclei, offering new solutions for segmenting overlapping and clustered nuclei.

The representation of local to global properties of nucleus pixels needs to be reconsidered, despite the fact that the previously discussed techniques have enhanced segmentation performance. This is because segmenting overlapping and crowded nuclei in histopathology pictures can be challenging. In this study, we propose SES-Net to address this bottleneck.

3 Method

In this section, we propose a multi-dimensional nuclei instance segmentation network (-SES-Net-) to accurately segment nuclei instances. The model workflow is shown in Fig. 1. In general, we aim to learn the key point features, edge features and nuclear features of the instance. Specifically, our method first generates three feature results: a keypoint heatmap, a nuclear edge map and a nuclear mask map. Next, Gaussian filtering is applied to local peaks in the keypoint heatmap to locate the key points of nuclei, with each instance corresponding to one nucleus, as shown in Fig. 2. Then, a region optimization algorithm is used to further refine and expand the results. The nuclear keypoints are treated as seed points, and the nuclei boundaries are used as constraints. Finally, post-processing operations are performed between the optimized results and the nuclei mask to obtain the final segmentation results.

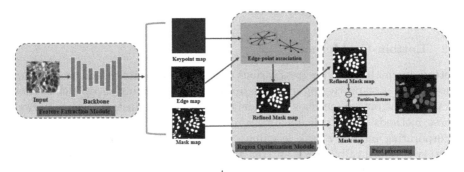

Fig. 1. The overall architecture for the proposed Spot-Edge-Surface network with three dimensions (SES-Net).

3.1 Feature Extraction Module

Traditional feature extraction backbone networks, such as ResNet [19] and VGGNet [20], because they sequentially encode the input image into high-level, low-resolution representations. They fail to capture strong and representative features for small target [12]. Moreover, there are differences in the scale of the nuclei of different tissues, and a significant number of nuclei are extremely small (smaller than 8×8 pixels), which is not ideal for detecting and segmenting tiny nuclei.

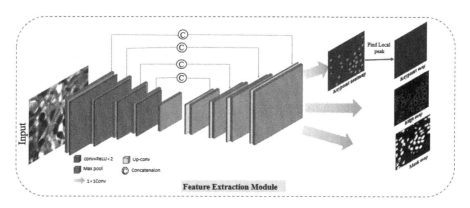

Fig. 2. Feature Extraction Module. The keypoint branch generates semantic keypoint heatmaps to locate the center of the cell nuclei, the boundary branch generates the cell nucleus boundaries, and the mask branch generates semantic segmentation maps.

We use the high-resolution network U-Net [11] as the backbone to solve this issue. U-Net, with its encoder-decoder structure, maintains high-resolution features, as seen in Fig. 2. The decoder portion of U-Net reconstructs a higher resolution image by upsampling and combining features from the appropriate encoder layers, while the encoder portion gradually downsamples the input image to gather contextual information. Using U-Net allows us to obtain detailed feature maps for further prediction. Subsequently, we utilize 1×1 convolutions to split U-Net [11] into three tasks: predicting nuclear keypoint heatmaps, nuclear boundary contour maps, and nuclear masks. The 1×1 convolutions effectively separate these tasks, ensuring each one benefits from high-resolution feature maps. For the obtained keypoint heatmaps, we apply local Gaussian filtering to produce keypoint maps. Specifically, Gaussian filtering smooths the keypoint heatmaps to more accurately locate the center of the nucleus. This step is crucial for reducing noise and enhancing the precision of keypoint localization. The synergy of these three tasks significantly improves the accuracy and robustness of nuclear detection and segmentation.

3.2 Keypoint Detection

The keypoint feature map is transformed from the keypoint heatmap, and the key heatmap generation process is described below. For an input image I with width W and height H, respectively, and given the corresponding ground truth heat map Y, our goal is to predict a low-resolution keypoint heat map. The predicted output $\hat{Y}$ is a tensor with values in the range $[0, 1]$ and has the shape $\left(\frac{W}{R} \times \frac{H}{R} \times C\right)$, where R is the downsampling factor and C is the number of keypoint types. The number of can be $C = 1$ for nuclear detection alone or greater than 1 if both nuclear detection and classification are needed. In this context, a prediction $\hat{Y}_{xyc} = 1$ indicates the detection of a keypoint, while $\hat{Y}_{xyc} =$

0 denotes the background. Here, x and y represent the spatial coordinates, and c denotes the class.

We train the keypoint prediction branch, using the Gaussian kernel without normalization $\exp\left(-\frac{(x-\tilde{p}_x)^2+(y-\tilde{p}_y)^2}{2\sigma_\rho^2}\right)$. Among them, σ_ρ^2 is the adaptive standard deviation of object size [21]. The training objective is a penalized reduced pixel logistic regression with focus loss [22]:

$$L_{\text{keypoint}} = -\frac{1}{N_{\text{pos}}^k} \sum_{xyc} \begin{cases} (1-\hat{Y}_{xyc})^\alpha \log \hat{Y}_{xyc}, & \text{if } Y_{xyc}=1 \\ (1-Y_{xyc})^\beta (\hat{Y}_{xyc})^\alpha \log(1-\hat{Y}_{xyc}), & \text{otherwise} \end{cases} \quad (1)$$

3.3 Region Optimization Module

In star-convex [6], only predicting distance through keypoint features might not be sufficient due to the significant variation in the size of nuclei, which means that pixel points might lack the contextual information necessary for accurate distance prediction. To address this issue, we propose a region optimization module (ROM). This module utilizes more precise boundary pixels and keypoint pixels to output an optimized mask image.

Through the region optimization Algorithm 1, the nucleus keypoints are first used as seed points q, and the 8 neighborhood points $N(q)$ of this point are obtained. For each neighborhood point $N(q)$, region growing is performed through boundary features and extended to the nuclear boundary region constrained by the nuclear boundary. In this way, each key point can be extended to the corresponding nuclear boundary to derive a continuous nuclei mask.

The ROM algorithm can effectively utilize richer contextual information, using point features for locating nuclei and distinguishing dense nuclei, and using edge features for shape description, thereby improving the accuracy and robustness of overlapping nuclear segmentation tasks.

3.4 Loss Function

Three branches make up the network model's output: keypoint, edge and mask. The loss function, represented by L_{total}, has the following definition:

$$L_{\text{total}} = \omega_1 L_{\text{keypoint}} + \omega_2 L_{\text{edge}} + \omega_3 L_{\text{mask}}. \quad (2)$$

Here, L_{keypoint}, L_{edge} and L_{mask} represent the loss of mask, edge and keypoints in the three branches, respectively. Each weight coefficient is represented by ω_i. For each of L_{edge} and L_{mask}, we compute the loss by summing the standard binary cross-entropy loss and the Dice loss.

$$\text{CE} = -\sum_{(r,c)}^{(H,W)} \left[P_{G(r,c)} \log P_S(r,c) + (1-P_{G(r,c)}) \log(1-P_S(r,c)) \right]. \quad (3)$$

Algorithm 1. Region Optimization

1: **Input:** The keypoint map $P = \{p_1, p_2, \ldots, p_n\}$, The edge maps $C = \{c_1, c_2, \ldots, c_n\}$
2: **Output:** Refined Mask Prediction M
3: Initialize a mask image M with the same size as the edge maps and pixel values set to $(0, 0, 0)$
4: **for** each i, center in heat map p **do**
5:    $Q = \{p_i\} \leftarrow i$, center
6:    **while** not $Q.empty()$ **do**
7:       $q \leftarrow Q.get()$
8:       $N(q) \leftarrow$ Get neighboring points(q)
9:       **for** each n in $N(q)$ **do**
10:          **if** n is within contour and $M(n) = (0, 0, 0)$ **then**
11:             $M(n) \leftarrow (i+1, i+1, i+1)$
12:             $Q.put(n)$
13:          **end if**
14:       **end for**
15:    **end while**
16: **end for**
17: **return** M

$$\text{Dice} = 1 - \sum_{(r,c)}^{(H,W)} \left(\frac{2 \times P_S(r,c) \times P_G(r,c)}{P_S(r,c)^2 + P_G(r,c)^2} \right). \quad (4)$$

$$L_{\text{edge}} = L_{\text{mask}} = \text{CE} + \text{Dice}. \quad (5)$$

The image's width is indicated by W, and its height by H. The image's pixel represented by (r, c), $P_G(r, c)$ denotes the ground truth label for that pixel, and $P_S(r, c)$ signifies the model's predicted probability that the pixel is a positive label. For the loss of key points, we follow Eqn. 1 to compute the loss key points. The model can learn to accurately predict the position of the nuclei, which is helpful to improve the accuracy of nuclei localization.

3.5 Post Processing

After the improvement, due to the unclei results of the boundary branch output, there are still discontinuous boundaries in morphology after the boundary and key points are fed to ROM, and the over-segmentation problem occurs on the basis of distinguishing each instance. To solve this problem, we propose a new post-processing method. Specifically, we start by subtracting the ROM output from the partition mask and instantiating each disconnected region to remove redundant partitions. Then, we use connected components to represent overlapping and isolated cores. In addition, by removing small targets with an area of less than 20 pixels. Finally, the process is completed by propagating the instance results outward until the foreground information is entirely filled. This method makes full use of the complementarity between ROM output and segment mask, and successfully simplifies and optimizes the small case recovery unit.

Algorithm 2. Instance Segmentation Processing

1: **Input:** Refined mask map M; Mask map N
2: **Output:** $Final\_mask$
3: $Mask\_Init \leftarrow (N - M)$
4: $CC \leftarrow$ Remove_small_regions($Mask\_Init$)
5: **for** each CC_i in CC **do**
6:    $kernel \leftarrow \begin{bmatrix} 1 & 1 & 1 \\ 1 & 1 & 1 \\ 1 & 1 & 1 \end{bmatrix}$
7:    $Final\_mask[CC\_mask] \leftarrow$ Dilate_mask(CC_i)
8: **end for**
9: **return** $Final\_mask$

---

4 Experiments

We conducted extensive experiments to evaluate the effectiveness of our proposed SES-Net on nuclei segmentation both quantitatively and qualitatively. Section 4.1 details the dataset, evaluation metrics and implementation specifics. Section 4.2 presents both quantitative and qualitative comparisons with existing methods. Additionally, Sect. 4.3 includes ablation studies to assess the efficacy of the SES-Net structure.

4.1 Experiment Settings

Datasets. (1) MoNuSeg [23]: The dataset comprises histopathological tissue slices from Triple-Negative Breast Cancer (TNBC) and Castration-Resistant Prostate Cancer (CRPC). Released by Kumar et al. in 2017, this medical imaging dataset is intended to aid research in tumor segmentation. The training set contains 30 images from 7 organs, with annotations for 21,623 individual cell nuclei. The test set includes 14 images from the same 7 organs, though two of these organs are not present in the training set. Each image is in color with dimensions of 1000×1000 pixels.

(2) CoNSep [18]: The dataset comprises 41 H&E stained images of colorectal adenocarcinoma (CRA). These images were collected using a 40x scanner by the University Hospitals Coventry and Warwickshire. For segmentation tasks, we employed the dataset referenced in the literature, which includes 27 images for training and 14 images for testing.

(3) Kumar: This dataset comprises 30 H&E stained histology images, each with dimensions of 1000×1000 pixels. These images originate from 7 different organs, showcasing diverse histological staining features. The images, collected with a 40x scanner by the National Institutes of Health (NIH), are used for segmentation tasks as described in the literature. The dataset comprises 16 training images and 14 test images. High performance on this dataset demonstrates strong generalization capability.

Evaluation Metrics. To evaluate the overall segmentation performance of the proposed SES-Net, we employ four metrics as described in [23]: F1-score (F1), average Dice coefficient (Dice), Panoptic Quality (PQ) and Aggregated Jaccard Index (AJI).

Implementation Details. We implemented these models using PyTorch 3.7 on an NVIDIA Tesla V100 system. During training, we maintained the original sizes of all datasets and employed data augmentation techniques, including random horizontal and vertical flips, as well as random rotations. We used stochastic gradient descent (SGD) with a learning rate of 0.00001 and an adaptive learning rate decay strategy for optimization.In the training of the keypoint branch, the default downsampling factor R is 4. The number of keypoint types C is 1, only for nuclear segmentation. If two Gaussian of the same class overlap, we take the maximum value of the element. In the loss function, ω_i represents a set of weight coefficients, which are all set to 1 in the experiments.

For the MoNuSeg dataset, the input size was 1000×1000 with a batch size of 12; for the Kumar dataset, the input size was 1000×1000 with a batch size of 8; and for the CoNSep dataset, the input size was 512×512 with a batch size of 12. We saved checkpoints every 20 epochs during the 200 epochs of training. Additionally, we employed specific methods in the post-processing stage to address discontinuities in nuclear segmentation.

4.2 Experimental Results

In this section, we perform a quantitative and qualitative comparison of our SES-Net method against other nuclei segmentation techniques. As detailed in Table 1, SES-Net achieved the highest performance in nuclei segmentation tasks. Specifically, in the MoNuSeg dataset, our proposed SES-Net achieved Dice of 0.837 and an AJI of 0.672 and PQ of 0.648. In the CoNSep dataset, SES-Net achieved Dice

Table 1. Quantiative evaluation. Comparison of our model with other models on the MoNuSeg, CoNSeP and Kumar datasets.

Model	MoNuSeg			CoNSeP			Kumar		
	Dice	AJI	PQ	Dice	AJI	PQ	Dice	AJI	PQ
U-Net [11]	0.758	0.556	0.478	0.724	0.482	0.328	0.758	0.556	0.478
Mask-R-CNN [7]	0.760	0.548	0.509	0.740	0.474	0.460	0.760	0.546	0.509
DITS [4]	0.786	0.560	0.443	0.804	0.502	0.398	0.789	0.559	0.443
DCAN [3]	0.793	0.525	0.492	0.733	0.289	0.256	0.791	0.556	0.478
Micro-Net [24]	0.797	0.560	0.519	0.794	0.527	0.449	0.797	0.560	0.519
HoVer-Net [18]	0.826	0.618	0.597	0.853	0.571	0.547	0.811	0.618	0.597
HARU-Net [25]	**0.838**	0.656	0.642	0.856	0.554	0.518	0.825	0.613	0.572
SES-Net	0.837	**0.672**	**0.648**	**0.865**	**0.584**	**0.551**	**0.829**	**0.636**	**0.581**

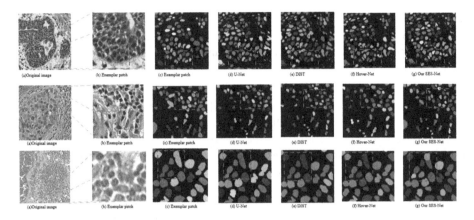

Fig. 3. Qualitative evaluation. The segmentation results for the MoNuSeg (top), Kumar (middle) and CoNSeP (bottom) datasets are visualized. Panels include: (a) Original image; (b) Sample patch; (c) Ground truth; Predictions from (d) U-Net [11]; (e) DIST [4]; (f) Hover-Net [11]; and (g) our proposed SES-Net. Different instances are distinguished by various colors. Red rectangles are used to highlight specific areas for clearer comparison.

of 0.865 and an AJI of 0.584 and PQ of 0.551. In the Kumar dataset, SES-Net achieved Dice of 0.829 and an AJI of 0.636 and PQ of 0.581. In the MoNuSeg dataset, HARU-Net[23] achieves a slightly higher Dice score compared to SES-Net, primarily due to HARU-Net's adoption of a complex network architecture and deeper network layers, which enhances segmentation accuracy. In contrast, SES-Net is designed to be lightweight, with fewer model parameters and a simpler network structure. Despite its lower model complexity and parameter count, SES-Net still manages to achieve Dice scores comparable to HARU-Net. Furthermore, SES-Net demonstrates superior performance compared to other methods across the MoNuSeg, CoNSep and Kumar datasets.

Qualitative Evaluation. We conducted detailed qualitative visual analyses on the MoNuSeg, CoNSep and Kumar datasets. As shown in Fig. 3, SES-Net consistently outperforms all other methods in the nuclear instance segmentation task. Specifically, the red rectangles in Fig. 3 vividly demonstrate the unique capability of our method to effectively separate nuclear pixels from the background and other segmented cluster instances. These findings not only confirm the superior performance of our method through quantitative metrics but also provide qualitative evidence in handling complex tissue images.

4.3 Ablation Studies

To assess the performance of each module in SES-Net, we carried out a series of ablation experiments using the MoNuSeg dataset, with U-Net serving as the baseline model.

Table 2. Performance on MoNuSeg dataset.

Mask	Edge	Point	ROM	MoNuSeg		
				F1	Dice	AJI
✓				0.792	0.751	0.554
✓	✓			0.851	0.809	0.556
✓	✓	✓		0.860	0.812	0.591
✓	✓	✓	✓	**0.874**	**0.837**	**0.672**

Effectiveness of Edge Branch. We studied whether adding boundary branches would have a positive impact on segmentation performance. These boundary branches are integrated into the network structure and perform a simple mask addition fusion operation on the output results. From the results of the first and second rows in Table 2, it can be observed that the F1 and Dice scores have improved by 5.9% and 5.8%, respectively.

Effectiveness of Keypoint Branch. To validate the effectiveness of keypoint branches, we incorporated them into the model, integrating keypoints into the segmentation results. From the results of the second and third rows in Table 2, it can be observed that the F1, Dice, and AJI scores improved by 0.9%, 0.3%, and 3.5%, respectively. The findings indicate that introducing point supervision strategies alongside boundary supervision strategies in SES-Net helps the model learn more effective features.

Effectiveness of ROM. Furthermore, to validate the effectiveness of the Region of Interest Module (ROM) in interacting with different supervisions, we compared segmentation performance with and without ROM. Results from the third and fourth rows of Table 2 on the MoNuSeg dataset show improvements in F1, Dice, and AJI scores by 1.4%, 2.5%, and 8.1%, respectively. Thus, ROM ensures the fusion between boundaries and keypoints, leveraging the strengths of each branch to enhance and optimize the final cell nuclei masks, thereby achieving more precise cell nuclei segmentation results.

5 Conclusion

In this paper, we propose a network structure called SES-Net, which combines three dimensions of point, line and surface for nuclear segmentation from WSI images. This represents a novel paradigm and perspective for comprehensive extraction and analysis of nuclear image features. In addition, we have introduced a regional optimization module. This module utilizes point features and boundary features to effectively separate the issues of contact and overlap between cell nuclei. Finally, we propose a more reasonable post-processing method that refines nuclear instance boundaries using features from different tasks. Extensive experiments and ablation studies on three datasets demonstrate the superiority of our proposed network.

Acknowledgement. This study was funded by the Taishan Scholar Young Expert Program (No.tsqn201909137), the Key Research and Development Project of Shandong Province (No.2023CXGC010113) and the Jinan 20 New Colleges and Universities Project (No.202228068).

Disclosure of Interests. The authors have no competing interests to declare that are relevant to the content of this article.

References

1. Zhou, Y., Onder, O.F., Dou, Q., Tsougenis, E., Chen, H., Heng, P.-A.: CIA-Net: robust nuclei instance segmentation with contour-aware information aggregation. In: Chung, A.C.S., Gee, J.C., Yushkevich, P.A., Bao, S. (eds.) IPMI 2019. LNCS, vol. 11492, pp. 682–693. Springer, Cham (2019). https://doi.org/10.1007/978-3-030-20351-1_53
2. Chen, H., Qi, X., Lequan, Yu., Dou, Q., Qin, J., Heng, P.-A.: DCAN: deep contour-aware networks for object instance segmentation from histology images. Med. Image Anal. **36**, 135–146 (2017)
3. Oda, H., et al.: BESNet: boundary-enhanced segmentation of cells in histopathological images. In: Frangi, A.F., Schnabel, J.A., Davatzikos, C., Alberola-López, C., Fichtinger, G. (eds.) MICCAI 2018. LNCS, vol. 11071, pp. 228–236. Springer, Cham (2018). https://doi.org/10.1007/978-3-030-00934-2_26
4. Naylor, P., Laé, M., Reyal, F., Walter, T.: Segmentation of nuclei in histopathology images by deep regression of the distance map. IEEE Trans. Med. Imaging **38**(2), 448–459 (2018)
5. Vu, QD., et al.: Methods for segmentation and classification of digital microscopy tissue images. Front. Bioeng. Biotechnol **7**, 433738 (2019)
6. Schmidt, U., Weigert, M., Broaddus, C., Myers, G.: Cell detection with star-convex polygons. In: Frangi, A.F., Schnabel, J.A., Davatzikos, C., Alberola-López, C., Fichtinger, G. (eds.) MICCAI 2018. LNCS, vol. 11071, pp. 265–273. Springer, Cham (2018). https://doi.org/10.1007/978-3-030-00934-2_30
7. Beibei, X., et al.: Automated cattle counting using mask R-CNN in quadcopter vision system. Comput. Electron. Agric. **171**, 105300 (2020)
8. Zhang, H., Tian, Y., Wang, K., Zhang, W., Wang, F.-Y.: Mask SSD: an effective single-stage approach to object instance segmentation. IEEE Trans. Image Process. **29**, 2078–2093 (2019)
9. Yu-Huan, W., Liu, Y., Zhang, L., Gao, W., Cheng, M.-M.: Regularized densely-connected pyramid network for salient instance segmentation. IEEE Trans. Image Process. **30**, 3897–3907 (2021)
10. Liu, S., Qi, L., Qin, H., Shi, J., Jia, J.: Path aggregation network for instance segmentation. In: Proceedings of the IEEE Conference on Computer Vision and Pattern Recognition, pp. 8759–8768 (2018)
11. Ronneberger, O., Fischer, P., Brox, T.: U-Net: convolutional networks for biomedical image segmentation. In: Navab, N., Hornegger, J., Wells, W.M., Frangi, A.F. (eds.) MICCAI 2015. LNCS, vol. 9351, pp. 234–241. Springer, Cham (2015). https://doi.org/10.1007/978-3-319-24574-4_28

12. Yao, K., Huang, K., Sun, J., Hussain, A.: PointNu-Net: keypoint-assisted convolutional neural network for simultaneous multi-tissue histology nuclei segmentation and classification. IEEE Trans. Emerg. Top. Comput. Intell. **8**(1), 802–813 (2023)
13. Lou, W., et al.: Structure embedded nucleus classification for histopathology images. IEEE Trans. Med. Imaging **43**, 3149–3160 (2024)
14. Stevens, M., Nanou, A., Terstappen, L.W., Driemel, C., Stoecklein, N.H., Coumans, F.A.: Stardist image segmentation improves circulating tumor cell detection. Cancers **14**(12), 2916 (2022)
15. Chen, S., Ding, C., Liu, M., Cheng, J., Tao, D.: CPP-Net: context-aware polygon proposal network for nucleus segmentation. IEEE Trans. Image Process. **32**, 980–994 (2023)
16. Zhao, B., et al.: Triple U-net: hematoxylin-aware nuclei segmentation with progressive dense feature aggregation. Med. Image Anal. **65**, 101786 (2020)
17. He, H., et al.: CDNET: Centripetal direction network for nuclear instance segmentation. In: Proceedings of the IEEE/CVF International Conference on Computer Vision, pp. 4026–4035 (2021)
18. Graham, S., et al.: Hover-net: Simultaneous segmentation and classification of nuclei in multi-tissue histology images. Med. Image Anal. **58**, 101563 (2019)
19. He, K., Zhang, X., Ren, S., Sun, J.: Deep residual learning for image recognition. In: Proceedings of the IEEE Conference on Computer Vision and Pattern Recognition, pp. 770–778 (2016)
20. Simonyan, K., Zisserman, A.: Very deep convolutional networks for large-scale image recognition. arXiv preprint arXiv:1409.1556 (2014)
21. Law, H., Deng, J.: CornerNet: detecting objects as paired keypoints. In: Proceedings of the European Conference on Computer Vision (ECCV), pp. 734–750 (2018)
22. Lin, T.-Y., Goyal, P., Girshick, R., He, K., Dollár, P.: Focal loss for dense object detection. In: Proceedings of the IEEE International Conference on Computer Vision, pp. 2980–2988 (2017)
23. Kumar, N., Verma, R., Sharma, S., Bhargava, S., Vahadane, A., Sethi, A.: A dataset and a technique for generalized nuclear segmentation for computational pathology. IEEE Trans. Med. Imaging **36**(7), 1550–1560 (2017)
24. Raza, SEA.: et al. Micro-Net: a unified model for segmentation of various objects in microscopy images. Med. Image Anal. **52**, 160–173 (2019)
25. Chen, J., Huang, Q., Chen, Y., Qian, L., Yu, C.: Enhancing nucleus segmentation with HARU-Net: a hybrid attention based residual u-blocks network. arXiv preprint arXiv:2308.03382 (2023)

Skin-Adapter: Fine-Grained Skin-Color Preservation for Text-to-Image Generation

Zhuowei Chen, Mengqi Huang, Nan Chen, and Zhendong Mao[✉]

University of Science and Technology of China, Hefei, China
zdmao@ustc.edu.cn

Abstract. With the recent advancements in diffusion-based text-to-image (T2I) models, generating high-quality, human-centric images has become increasingly easy. However, T2I models cannot preserve the fine-grained skin color information from reference images, raising potential AI ethics concerns. In this work, we propose **Skin-Adapter**, the first work on preserving fine-grained skin color information in text-to-image generation. To achieve this goal, we first devise the frequency-based adaptive color histogram to accurately represent the user's skin color information. Additionally, we introduce the color distribution matching reward to explicitly enhance skin color consistency between input and output images. Experimental results show that our Skin-Adapter can maintain the fine-grained skin color information of input images in the generated images. Furthermore, we validate the superiority of our approach through quantitative and qualitative comparisons against possible alternatives.

Keywords: Text-to-Image Generation · Controllable Image Generation · AI Ethics

1 Introduction

Recently, diffusion-based text-to-image models [3, 6, 18, 19, 21] have demonstrated remarkable capabilities in generating images from textual descriptions. These models allow for the effortless creation of vivid and expressive human-centric images. However, the controllability of these models is often insufficient, leading to the failure to preserve key information from reference images when relying solely on text-to-image models in practical applications.

One of the most important pieces of information that users wish to retain is their skin color. As illustrated in Fig. 1, simply using text to describe a user's skin color often fails to achieve consistency between the input and output images due to the inherent ambiguity in textual descriptions. Such discrepancies could negatively impact user experience and pose ethical risks in AI applications. Thus, it is crucial to develop methods that enable text-to-image models to preserve the original skin color of users. However, achieving fine-grained skin color consistency between input and output images is challenging.

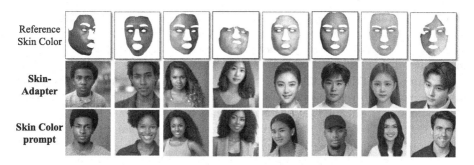

Fig. 1. Given the reference skin color images (the first row), Skin-Adapter enables text-to-image models to generate human-centric images with consistent skin color (the second row). However, naively converting the reference skin color into a text description(e.g., "a person with dark/brown/yellow/white skin...") results in a substantial color mismatch (the last row). Best viewed in color.

Firstly, **current mainstream color representation methods are inadequate for accurately representing skin color information**. To achieve color consistency between input and output images, it is essential to accurately represent the reference skin color. There are two primary methods currently in use: the color averaging method and the fixed color histogram method. The color averaging method [14] treats skin color as uniform distribution. Hence, the average color is used to represent the skin color. However, skin regions are influenced by various factors, including lighting, shadows, and reflections, making them far from uniform distribution. The fixed color histogram method [1,11] quantizes each pixel color to the nearest of N predefined colors and then counts the occurrences of each predefined color. However, the choice of N faces a dilemma: when N is too large, it leads to increased training instability and higher computational costs; on the other hand, if N is too small, the model fails to accurately represent the input color, resulting in significant color representation errors. In practice, this dilemma often forces the number of predefined colors to be limited to a few hundred [11], requiring a balance between color accuracy and training stability. We experimentally find that the representation errors of these two methods are particularly noticeable for skin color, as users are more sensitive to skin color variations than to general color distributions in natural images.

Secondly, **learning to preserve skin color is inherently difficult for text-to-image diffusion model**. Since skin color information is a local(skin area) color statistic within the image, the control signal is relatively weak. Moreover, under the current diffusion model training paradigm, the color consistency is learned implicitly through pixel-wise loss computed in the noisy latent space [9,19]. This loss does not explicitly constrain color consistency between the input and output images. As a result, skin color consistency is often compromised.

In this work, we propose the first skin color preservation model, **Skin-Adapter**, for text-to-image generation. As shown in Fig. 1, Skin-Adapter can enable text-to-image models to generate human images with consistent skin color based on the reference image. To overcome the aforementioned challenges, Skin-adapter is equipped with a novel adaptive frequency-based color histogram for skin color representation and trained using the effective color distribution matching reward. The motivation behind our adaptive frequency-based quantized color histogram is that skin regions typically consist of a few dominant colors, and the other pixels are mostly close to the dominant colors. Thus, we can use these dominant colors as pre-defined colors for quantization and only take the pixels close to the dominant colors into account. This design can not only lower the quantization error compared to the fixed color histogram but also exclude negative impacts from the skin color unrelated pixels, such as freckles or moles. Furthermore, we design a consistency reward strategy to learn the color consistency between input and output images explicitly. During training, we simulate the inference process by generating an image from random noise and reference skin color; then, we compute the color distribution distance between the generated image and the input image. The learnable parameters are optimized to narrow such distance. This explicit reward for color distribution consistency significantly enhances the skin color consistency between generated and input images. Our contributions can be summarized in three parts:

- We propose **the first skin color preservation model, Skin-Adapter, for text-to-image generation**. The generated images from this model can retain fine-grained skin color information from input images.
- We introduce the adaptive frequency-based quantized color histogram to accurately represent users' skin color information and consistency reward for improving the color distribution consistency between input and output images.
- We validate the effectiveness of our approach through extensive ablation studies and demonstrate the generalization of our design across different T2I models and color preservation tasks.

2 Related Work

2.1 Controllable Text-to-Image Generation

In recent years, various methods are proposed to enable the text-to-image model to be controlled by specific signals. Pioneering works have explored controlling spatial structures [15,28], preserving user identity information [4,8,13,25], subject information [12,20,27], and reference style [17,24]. The most closely related work is on controlling image-level color histograms in the text-to-image model [11]. Achieving fine-grained skin color control to ensure consistency between input and output images remains an unexplored problem. This gap is significant for AI fairness, as inappropriate skin color representation in generated images can result in user dissatisfaction and even ethical concerns. Our work is the first to achieve skin color preservation control in text-to-image generation.

2.2 Skin Color Representation

To ensure consistency in skin color between input and output images, accurately representing the skin color information in the input images is crucial. Color averaging methods [14] treat skin color as uniform distribution, while fixed color histogram approaches [1,11] quantize colors into one of a limited number of predefined categories. However, these methods introduce exhibit substantial errors in representing skin colors. To address these limitations, we propose an adaptive frequency-based color histogram representation method, which reduces quantization errors and mitigates the influence of outlier colors caused by inaccurate skin segmentation.

2.3 Reward for Text-to-Image Diffusion Model

Skin color is a relatively weak control signal because it represents a local statistic within the image. Current diffusion model training primarily relies on pixel-wise losses computed in the noisy latent space [9], which lacks explicit constraints to ensure consistency between the final output and the input image. Inspired by recent advancements in diffusion rewards [5,26], we design a color consistency reward to explicitly enforce color consistency between input and output images.

3 Methods

Given the reference human image, our goal is to enable the pre-trained text-to-image (T2I) model to generate human-centric images with consistent skin color under different prompts.

The overall framework is depicted in Fig. 2. Given an input image, we first extract the face area and apply the face parsing model to obtain skin-related pixels. These pixels are represented as a statistic by our proposed adaptive frequency-based (AF) color histogram, which is a more accurate and robust representation of skin color. Then, this feature is mapped by a projector with perceiver resampler architecture [2]. The visual information is injected into the text-to-image model via additional visual cross-attention layers. To explicitly learn the skin color consistency, we simulate the inference process by generating an image from random noise via T steps and calculate color distribution matching distance as the reward.

In the following sections, we first briefly describe the diffusion-based text-to-image model used in our work and how to inject visual information into the T2I model. Next, we introduce the possible ways to represent skin colors and the proposed adaptive frequency-based color histogram. Finally, we present the color distribution matching reward and how to train and infer the model.

3.1 Preliminaries

Diffusion-Based Text-to-Image Models. In this work, we adopt the open-source Stable Diffusion 1.5 (SD1.5) as our text-to-image generation model, which has been trained on billions of images and has demonstrated excellent image generation quality and prompt understanding capabilities.

SD1.5 is a Latent Diffusion Model (LDM) with UNet backbone. LDM first represents the input image x in a low-resolution latent space z through a Variational Autoencoder (VAE). Then, a text-conditional diffusion model is trained to generate the latent code of the target image from the text input c. The loss function of this diffusion model can be expressed as:

$$\mathcal{L}_{diffusion} = \mathbb{E}_{\epsilon,z,c,t}[\|\epsilon - \epsilon_\theta(z_t, c, t)\|_2^2], \quad (1)$$

where ϵ_θ is the noise predicted by the model with learnable parameters θ, ϵ is the noise sampled from the standard normal distribution, t is the timestep, and z_t is the noisy latent variable at timestep t. During inference, the latent code is generated by the diffusion model. Subsequently, the vae decoder maps the latent code to the image space.

In SD1.5, text conditions are injected into the network via cross attention:

$$CrossAttention(Q, K, V) = Softmax\left(\frac{QK^T}{\sqrt{d}}\right) V, \quad (2)$$

where Q represents features from UNet, K, V is projected text feature at each CrossAttention layer and d is the channel dimension of Q.

Introduce Vision Control via an Additional Cross-Attention Layer. To inject skin color information into the text-to-image generation model, we follow the design of IP-adapter [27], introducing another cross-attention layer for visual information on top of the text cross-attention. Specifically:

$$CrossAttention(Q, K, V, K_{img}, V_{img}) = Softmax\left(\frac{QK^T}{\sqrt{d}}\right) V$$
$$+ Softmax\left(\frac{QK_{img}^T}{\sqrt{d}}\right) V_{img}, \quad (3)$$

where $K_{img} = W_k^{img} \cdot f_{img}$, $V_{img} = W_v^{img} \cdot f_{img}$. During training, we keep the original model parameters unchanged and only learn the visual-related weights W_k^{img}, W_v^{img} and projector.

3.2 Adaptive Frequency-Based Color Histograms

In this section, we begin by introducing two mainstream methods for representing skin color, color averaging and fixed color histogram representation, and discuss their limitations. We then present our proposed adaptive frequency-based color histograms, which offers a more accurate representation of skin color.

For an input image, we first extract the <R, G, B> information H of all pixels in the facial skin area, resulting in an $N \times 3$ matrix, where N is the number of pixels in the facial skin area and 3 corresponds to the three color channels: <R, G, B>.

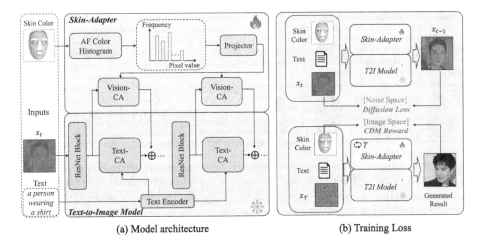

(a) Model architecture (b) Training Loss

Fig. 2. Overview of the proposed *Skin-Adapter*: (a) The model architecture of Skin-Adapter. Skin-Adapter is a plug-in module designed to equip text-to-image models with skin-color preservation capability. The reference skin color image is first represented by the proposed *Adaptive Frequency-based (AF) Color Histogram*. After projection, the visual color information is injected into the T2I model through the additional vision cross-attention (Vision-CA) layer, and the result of Vision-CA is added to the result of text cross-attention ((Text-CA)). (b) The loss function for training Skin-Adapter. In addition to the normal *Diffusion Loss* (upper) calculating the distance in noise space, We introduce a *Color Distribution Matching (CDM) Reward* to explicitly enforce the color consistency between the input image and the final image generated through T sampling steps.

Color Averaging Representation. This method is simple and straightforward, widely applied in various controllable generation tasks [7]. It directly calculates the average value for each channel in the skin area.

$$C_{mean}(i) = \frac{1}{N} \sum_n H_{n,i}, \qquad (4)$$

The limitation of this approach is that it assumes the skin color distribution to be uniform, when in reality, it is highly diverse. This oversimplification results in significant errors in representing true skin color information.

Fixed Color Histogram Representation. This method first defines a set of fixed predefined colors, then iterates over all pixels, approximating each pixel to the nearest predefined color. Finally, it counts and normalizes the occurrences of each predefined color. Specifically:

1. Initialization. The color histogram is initialized by equally quantizing each <R, G, B> channel into K values, creating a three-dimensional array C_{FQ} with dimensions $K \times K \times K$. This array represents the occurrences of each predefined color. We initialize this matrix with zeros, $C_{FQ}[R_q, G_q, B_q] = 0$

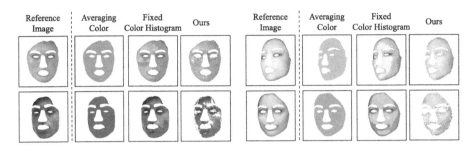

Fig. 3. The reconstruction results of different color representations. The white pixels in the above figure will not be taken into account for color representation. Our method can achieve better reconstruction results while eliminating the influence of the skin-unrelated pixels by not counting outliner pixels.

2. Counting Occurrences. For each pixel $< R_i, G_i, B_i >$ in H, the nearest predefined color $< R_q, G_q, B_q >$ is identified, and the corresponding frequency is incremented: $C_{FQ}[R_q, G_q, B_q] += 1$.
3. Normalization. Each value in the histogram C_{FQ} is then normalized by dividing it by the total number of pixels N to obtain the normalized histogram: $C_{FQ}[i, j, k] = \frac{C_{FQ}[i,j,k]}{N}$.

This method can better capture complex color distributions compared to the color averaging method. However, its accuracy is limited by the number of quantized anchor colors. If the number of quantized colors is very large, such as in an extreme case with no quantization and approximately 16 million ($255 \times 255 \times 255$) predefined colors, the high-dimensional input poses a significant challenge for training. Therefore, current practices use relatively limited quantized colors, which may be sufficient for natural images but introduce noticeable errors for sensitive information like skin color, as shown in Fig. 3.

Adaptive Frequency-Based Color Histogram Representation. To address the inaccuracies of color representation in previous color representations, we propose an adaptive frequency-based color histogram representation method. The motivation is that the skin color region comprises a few primary colors, with most colors distributed around these primary colors, and the other pixels can be viewed as skin-unrelated outlier colors. By using these primary colors as the predefined colors, most pixels can find a predefined color with a much smaller error.

We provide a detailed description in Algorithm 1. We select the most frequent unrecorded color as the quantization anchor. After counting all pixels that are close to this color, we proceed to the next most frequent unrecorded color. To exclude outlier colors, we use the pixel coverage ratio R to limit the number of colors considered.

Algorithm 1. Adaptive Frequency-Based Color Histogram Representation.

Input: Pixel coverage ratio R, the number of colors for quantization N_q, and pixel error range δ, color distance function Dis
Output: The adaptive frequency-based color histogram $C_{af} \in \mathbf{R}^{N_q \times 4}$.
1: Initialize the color histogram matrix $C_{af} \in \mathbf{R}^{N_q \times 4}$, $C_{af}[:] = 0$, current recorded pixel ratio $r_c = 0$, recorded color number as $i = 0$
2: Count the occurrence probability of each color and sort them in descending order according to occurrence, resulting in an ordered list F with pixel value and its frequency.
3: // iterate over all the unrecorded colors, record the occurrence of colors that are close to the current most frequent color $C_{af}[i]$
4: **while** $r_c < R$ **and** $i < N_q$ **and** $\text{len}(F) > 0$ **do**
5:     $D = \text{len}(F)$
6:     Extract the most frequent unrecorded color, set $C_{af}[i,:3] = F[0][:3]$
7:     $r_c \leftarrow r_c + F[0][3]$.
8:     pop $F[0]$
9:     $j=0$
10:     **while** $j < len(F)$ **do**
11:       **if** $Dis(F[j][:3], C_{af}[i,:3]) < \delta$ **then**
12:         update C_{af} with the occurrence of current unrecorded color: $C_{af}[i,3] = C_{af}[i,3] + F[j][3]$.
13:         $r_c \leftarrow r_c + F[j][3]$.
14:         pop $F[j]$
15:       **else**
16:         $j = j + 1$
17:       **end if**
18:     **end while**
19:     $i = i + 1$
20: **end while**
21: Normalize the frequency dimension of matrix C_{af}, $C_{af}[:,3] = \frac{C_{af}[:,3]}{\sum C_{af}[:,3]}$.
22: **return** adaptive frequency-Based color histogram C_{af}

Using this method, we achieve precise skin color representation and eliminate the interference from unrelated skin color regions. As shown in Fig. 3, our representation method accurately represents the input skin color while excluding noise and unrelated information caused by inaccurate skin segmentation.

3.3 Color Distribution Matching Reward

Although the adaptive frequency-based color histogram accurately represents input skin color information, generating images that reflect this in current diffusion-based text-to-image models remains challenging due to the current training paradigm. We propose a Color Distribution Matching (CDM) Reward to explicitly enforce the color consistency between generated and input images. Specifically, the final image is generated through multi-step inference, denoted as:

$$x_{pred} = T2I_{(x_T, c_{txt}, C_{af}(x_{input}))}, \tag{5}$$

where $T2I$ represents the multi-step inference process, x_T denotes the initial noise, c_{txt} is the input text and $C_{af}(x_{input})$ is the reference AF color histogram. Then, the color distribution gap between the input and output images is then minimized as a reward, formalized by the following loss function::

$$\mathcal{L}_{color} = D(C_{af}(x_{input}), C_{af}(x_{pred})), \tag{6}$$

where D denotes the l2 distance of two color histograms. Due to non-differentiable color histograms, we follow the differentiable formulation in [1] to obtain a differentiable color histogram with respect to the reference colors. We select the predefined colors from the fixed input and the AF color histogram to form the final color anchors $[p_0, p_1, ..., p_i...]$. The differentiable AF color histogram is represented as :

$$C_{af} \approx [h(p_0), h(p_1), ..., h(p_i), ...], \tag{7}$$

$$h(p_i) = \frac{1}{Z} \sum_x k(r(x), g(x), b(x), r(p_i), g(p_i), b(p_i)), \tag{8}$$

where $r(x), g(x), b(x)$ is the <r,g,b> channel of the color x. Z is the normalization term. The function k computes the distance between $r(x), g(x), b(x)$ and $r(p_i), g(p_i), b(p_i)$ using a kernel function. In this work, we adopt the inverse-quadratic kernel:

$$k(r(x), g(x), b(x), r(p_i), g(p_i), b(p_i)) = \prod_{f \in \{r,g,b\}} (1 + (\frac{|f(x) - f(p_i)|}{\sigma})^2)^{-1}, \tag{9}$$

where σ is the distance smoothing coefficient, set to 0.1.

3.4 Training and Inference

Training. We adopt a two-stage approach to train our Skin-Adapter. In the first stage, we train the model with the standard diffusion loss. In the second stage, we introduce the Color Distribution Matching Reward. The formulation is as follows:

$$\mathcal{L}_{total} = \mathcal{L}_{diffusion} + \lambda \mathcal{L}_{color}, \tag{10}$$

where λ is the reward weight. Experimentally, we found that the model converges quickly within 20,000 iterations in the second stage. Following the method in [5], We only use the gradient from the last diffusion timestep to update parameters when applying the CDM Reward.

Inference. We use Classifier-Free Guidance (CFG) [10] to generate high-quality images with consistent skin color. The CFG formula is given by:

$$\epsilon_{prd} = \epsilon_{uc} + \beta_{cfg}(\epsilon_c - \epsilon_{uc}), \tag{11}$$

where ϵ_{prd} represents the final model output, ϵ_{uc} denotes the prediction from the unconditional image and null text input, and ϵ_c represent the result of conditional image and text input. β_{cfg} is the CFG coefficient.

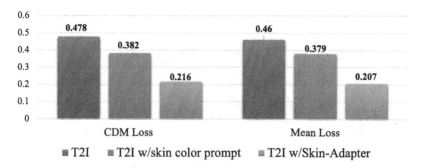

Fig. 4. Quantitative comparisons with the prompt-based method. CDM Loss and mean loss indicate the color discrepancy between the input and output images; lower values correspond to better skin color consistency. With the help of Skin-Adapter, the T2I model achieves superior skin color consistency on both metrics.

4 Experiments

4.1 Experiment Settings

Dataset. The training dataset is a subset of LAION-5B [22]. We retained high-aesthetic single-person images and filtered out images with small faces, resulting in a training dataset of approximately 500,000 images. The test image set covers diverse skin color images, with 40 images for each race (yellow/dark/white/brown), totaling 160 images. We design 10 prompts covering various life scenarios as the testing prompts and generated 4 images per prompt for each input image.

Metrics. We evaluate skin color consistency using two skin color distribution distances. The primary metric (CDM Loss) is the adaptive frequency-based color histogram distance, accurately representing the color distribution across different images. The secondary metric (Mean Loss) is the color-averaging distance, used as an auxiliary reference.

Implementation Details. We utilize the widely-used Stable Diffusion 1.5 as our text-to-image backbone. The first training stage consists of 200k iterations, followed by the second stage incorporating CDM reward, trained for approximately 20k iterations with a batch size of 8. Experiments are conducted on an 8-GPU A100 server. We use the DDIM sampler [23] with 35 steps for inference. The guidance scale is set to 7.5.

4.2 Main Results

We quantitatively compared our method with the current T2I models without skin color-related prompts/with skin color-related prompts in Fig. 4. From the quantitative experimental results, we can see that introducing the Skin-Adapter significantly improves skin color preservation. Figure 1 also verifies this conclusion. Furthermore, we can also observe that Skin-adapter can preserve the user's skin color across various input prompts in Fig. 5.

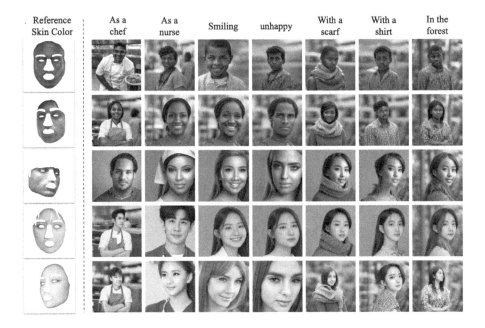

Fig. 5. Skin-adapter can generate images with consistent skin color across a wide range of input prompts and skin colors.

4.3 Ablation Study

The Effectiveness of Skin Color Representation. We compared the effects of different color encoding methods. The quantitative and qualitative results show that our approach improves the color distribution loss by 11.2% compared to the second-best color histogram method (Table 1). As shown in Fig. 6a, methods based on mean color or fixed quantized histograms exhibit significant discrepancies between the generated and original skin colors, whereas our method accurately reflects the skin color information from the original image.

The Effectiveness of Color Distribution Matching Reward. As shown in Fig. 6b, the skin-adapter trained with diffusion loss alone can lead to color deviations in the generated images. Introducing the Color Distribution Matching Reward significantly enhances skin color preservation. CDM loss is improved by 20.3% and mean loss is improved by 29.1% (Table 2).

Table 1. Ablation study on the different color representation methods.

Methods	CDM Loss ↓	Mean Loss ↓
RGB mean	0.332	0.345
fixed color	0.293	0.322
Ours	0.260	0.305

Table 2. The effectiveness of Color Distribution Matching Reward

Methods	CDM Loss ↓	Mean Loss ↓
baseline	0.260	0.305
w/reward	0.207	0.216

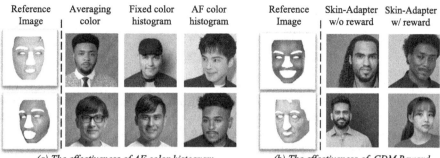

(a) The effectiveness of AF color histogram *(b) The effectiveness of CDM Reward*

Fig. 6. Qualitative results of ablation studies. (a) With the AF color histogram, the generated images show better skin color consistency. (b) Introducing CDM reward can further improve the result.

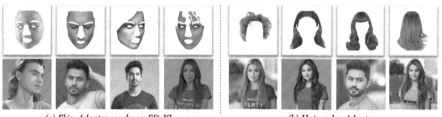

(a) Skin-Adapter works on SD-XL *(b) Hair-color Adapter*

Fig. 7. Generalization of the proposed Skin-Adapter: (a) Skin-Adapter is compatible with various T2I model architectures. We demonstrate the result of Skin-Adapter-xl trained on the SD-XL model. (b) The color preservation task can be extended from the skin domain to the hair domain, enabling the training of the Hair-color Adapter to preserve reference hair color.

The Generalization of Skin-Adapter. Fig. 7 illustrates the generalization capability of Skin-Adapter. Firstly, Skin-Adapter should work well for various T2I models. Figure 7(a) shows the results of Skin-Adapter-xl trained for SD-XL [16] arthitecture. Secondly, the generalization of Skin-Adapter's color representation and loss formulation allows for seamless extension to other color preservation tasks. For example, we trained a Hair-color Adapter to preserve the user's hair color, and the results, as shown in Fig. 7(b), demonstrate satisfactory performance.

5 Conclusion

In this study, we propose Skin-Adapter, the first approach in the text-to-image domain designed to preserve user skin tone information. We achieve precise skin tone representation through adaptive frequency-based color histograms and enhance the consistency of skin tone statistics between input and output images with a color distribution matching reward. Extensive qualitative and quantitative experiments validate the effectiveness of our method.

References

1. Afifi, M., Brubaker, M.A., Brown, M.S.: HistoGAN: controlling colors of GAN-generated and real images via color histograms. In: Proceedings of the IEEE/CVF Conference on Computer Vision and Pattern Recognition, pp. 7941–7950 (2021)
2. Alayrac, J.B., et al.: Flamingo: a visual language model for few-shot learning. Adv. Neural. Inf. Process. Syst. **35**, 23716–23736 (2022)
3. Chen, J., et al.: Pixart–α: Fast training of diffusion transformer for photorealistic text-to-image synthesis. arXiv preprint arXiv:2310.00426 (2023)
4. Chen, Z., Fang, S., Liu, W., He, Q., Huang, M., Mao, Z.: DreamIdentity: enhanced editability for efficient face-identity preserved image generation. In: Proceedings of the AAAI Conference on Artificial Intelligence, vol. 38, pp. 1281–1289 (2024)
5. Clark, K., Vicol, P., Swersky, K., Fleet, D.J.: Directly fine-tuning diffusion models on differentiable rewards. arXiv preprint arXiv:2309.17400 (2023)
6. Esser, P., et al.: Scaling rectified flow transformers for high-resolution image synthesis. In: Forty-first International Conference on Machine Learning (2024)
7. Ge, S., Park, T., Zhu, J.Y., Huang, J.B.: Expressive text-to-image generation with rich text. In: Proceedings of the IEEE/CVF International Conference on Computer Vision, pp. 7545–7556 (2023)
8. Guo, Z., Wu, Y., Chen, Z., Chen, L., He, Q.: PuLID: pure and lightning id customization via contrastive alignment. arXiv preprint arXiv:2404.16022 (2024)
9. Ho, J., Jain, A., Abbeel, P.: Denoising diffusion probabilistic models. Adv. Neural. Inf. Process. Syst. **33**, 6840–6851 (2020)
10. Ho, J., Salimans, T.: Classifier-free diffusion guidance. arXiv preprint arXiv:2207.12598 (2022)
11. Huang, L., Chen, D., Liu, Y., Shen, Y., Zhao, D., Zhou, J.: Composer: creative and controllable image synthesis with composable conditions. In: International Conference on Machine Learning, pp. 13753–13773. PMLR (2023)
12. Huang, M., Mao, Z., Liu, M., He, Q., Zhang, Y.: RealCustom: narrowing real text word for real-time open-domain text-to-image customization. In: Proceedings of the IEEE/CVF Conference on Computer Vision and Pattern Recognition, pp. 7476–7485 (2024)
13. Li, Z., Cao, M., Wang, X., Qi, Z., Cheng, M.M., Shan, Y.: PhotoMaker: customizing realistic human photos via stacked id embedding. In: Proceedings of the IEEE/CVF Conference on Computer Vision and Pattern Recognition, pp. 8640–8650 (2024)
14. Liu, L., Fu, Q., Hou, F., He, Y.: Flexible portrait image editing with fine-grained control. arXiv preprint arXiv:2204.01318 (2022)
15. Mou, C., et al.: T2I-Adapter: learning adapters to dig out more controllable ability for text-to-image diffusion models. In: Proceedings of the AAAI Conference on Artificial Intelligence, vol. 38, pp. 4296–4304 (2024)
16. Podell, D., et al.: SDXL: improving latent diffusion models for high-resolution image synthesis. arXiv preprint arXiv:2307.01952 (2023)
17. Qi, T., et al.: DEADiff: an efficient stylization diffusion model with disentangled representations. In: Proceedings of the IEEE/CVF Conference on Computer Vision and Pattern Recognition, pp. 8693–8702 (2024)

18. Ramesh, A., Dhariwal, P., Nichol, A., Chu, C., Chen, M.: Hierarchical text-conditional image generation with clip latents. arXiv preprint arXiv:2204.06125 (2022)
19. Rombach, R., Blattmann, A., Lorenz, D., Esser, P., Ommer, B.: High-resolution image synthesis with latent diffusion models. In: CVPR, pp. 10684–10695 (2022)
20. Ruiz, N., Li, Y., Jampani, V., Pritch, Y., Rubinstein, M., Aberman, K.: DreamBooth: fine tuning text-to-image diffusion models for subject-driven generation. In: Proceedings of the IEEE/CVF Conference on Computer Vision and Pattern Recognition, pp. 22500–22510 (2023)
21. Saharia, C., et al.: Photorealistic text-to-image diffusion models with deep language understanding. Adv. Neural. Inf. Process. Syst. **35**, 36479–36494 (2022)
22. Schuhmann, C., et al.: LAION-5B: an open large-scale dataset for training next generation image-text models. Adv. Neural. Inf. Process. Syst. **35**, 25278–25294 (2022)
23. Song, J., Meng, C., Ermon, S.: Denoising diffusion implicit models. arXiv preprint arXiv:2010.02502 (2020)
24. Wang, H., Wang, Q., Bai, X., Qin, Z., Chen, A.: InstantStyle: free lunch towards style-preserving in text-to-image generation. arXiv preprint arXiv:2404.02733 (2024)
25. Wang, Q., Bai, X., Wang, H., Qin, Z., Chen, A.: InstantID: zero-shot identity-preserving generation in seconds. arXiv preprint arXiv:2401.07519 (2024)
26. Xu, J., et al.: ImagereWard: learning and evaluating human preferences for text-to-image generation. In: Advances in Neural Information Processing Systems, vol. 36 (2024)
27. Ye, H., Zhang, J., Liu, S., Han, X., Yang, W.: IP-adapter: text compatible image prompt adapter for text-to-image diffusion models. arXiv preprint arXiv:2308.06721 (2023)
28. Zhang, L., Rao, A., Agrawala, M.: Adding conditional control to text-to-image diffusion models. In: Proceedings of the IEEE/CVF International Conference on Computer Vision, pp. 3836–3847 (2023)

Small Tunes Transformer: Exploring Macro and Micro-level Hierarchies for Skeleton-Conditioned Melody Generation

Yishan Lv, Jing Luo, Boyuan Ju, and Xinyu Yang[✉]

School of Computer Science and Technology, Xi'an Jiaotong University, Xi'an, China
{yisan,luojingl,juboyuan}@stu.xjtu.edu.cn, yxyphd@mail.xjtu.edu.cn

Abstract. Recently, symbolic music generation has become a focus of numerous deep learning research. Structure as an important part of music, contributes to improving the quality of music, and an increasing number of works start to study the hierarchical structure. In this study, we delve into the multi-level structures within music from macro-level and micro-level hierarchies. At the macro-level hierarchy, we conduct phrase segmentation algorithm to explore how phrases influence the overall development of music, and at the micro-level hierarchy, we design skeleton notes extraction strategy to explore how skeleton notes within each phrase guide the melody generation. Furthermore, we propose a novel Phrase-level Cross-Attention mechanism to capture the intrinsic relationship between macro-level hierarchy and micro-level hierarchy. Moreover, in response to the current lack of research on Chinese-style music, we construct our Small Tunes Dataset: a substantial collection of MIDI files comprising 10088 Small Tunes, a category of traditional Chinese Folk Songs. This dataset serves as the focus of our study. We generate Small Tunes songs utilizing the extracted skeleton notes as conditions, and experiment results indicate that our proposed model, Small Tunes Transformer, outperforms other state-of-the-art models. Besides, we design three novel objective evaluation metrics to evaluate music from both rhythm and melody dimensions.

Keywords: Symbolic Music Generation · Hierarchical Structures · Cross Attention · Chinese Folk Songs

1 Introduction

Music stands as a treasure within human civilization. In recent years, music generation has become a focus of deep learning research. Many sequence models has been employed to generate symbolic music [3,10,14,18]. Following the introduction of Music Transformer [7], which utilizes a transformer-based architecture for music generation, several Transformer-based models have made a progress in generating complete melodies [4,22,28].

Structure is of great significance to music, recently, plenty of works start to study the hierarchical structural features within the music. [25] studies the phrase-level hierarchy of music, [21,28] study the bar-level hierarchy of music, [16] studies the phrase & bar-level hierarchies of music. While beneath the bar-level hierarchy, there exists a micro-level hierarchy in which skeleton notes play an important role. In this study, we delve into the hierarchical structure, exploring the intrinsic relationship among macro-level hierarchy and micro-level hierarchy.

Figure 1 illustrates the difference between the phrase & bar-level hierarchies and our macro & micro-level hierarchies. Specifically, a melody comprises several phrases, with each phrase comprising several bars. In the phrase & bar-level hierarchies, the bar serves as the fundamental structure unit. Within a bar there are several notes, among which some play a crucial role in guiding the melody generation. These significant notes, known as skeleton notes, are extracted to establish the micro-level hierarchy in our macro & micro-level hierarchies. For Chinese Folk Songs, most phrases are relatively short and the distinction between phrase-level and bar-level hierarchies is not that obvious, so we examine phrases instead of bars as the macro-level hierarchy.

Accordingly, we conduct phrase segmentation on the melody at the macro-level hierarchy, and design a skeleton notes extraction strategy within each phrase at the micro-level hierarchy. Especially, we define a new type of skeleton note for Chinese Folk Songs. Building upon this, we propose a novel Phrase-level Cross-Attention mechanism, which enables the model a deep understanding of musical features from both macro-level and micro-level hierarchical structures.

We construct our own dataset: Small Tunes[1] Dataset (STD), and utilize it to train our model: Small Tunes Transformer (STT). Utilizing the extracted skeleton notes as conditions, STT is capable of generating Small Tunes songs with clear structure and captivating melody. We design 3 novel metrics to evaluate the quality of music from pitch and rhythm dimensions. The experiment results indicate that STT outperforms other state-of-the-art models on all 5 subjective evaluation metrics and 5 out of 6 objective evaluation metrics. Besides, we add 6 ablative groups to explore the hierarchical structural features within music.

Our main contributes can be summarized as follows:

- We propose STT, a Transformer-based model, incorporating the novel Phrase-level Cross-Attention mechanism, to explores the hierarchical structures within music from both macro-level and micro-level hierarchies.
- We design three objective evaluation metrics: Theme Pitch Corresponding (TPC) and Theme Rhythm Corresponding (TRC) evaluate the coherence corresponding to the theme from pitch and rhythm dimensions, and Pentatonic Scale Consistency (PSC) evaluates consistency in a Chinese-style scale dimension.
- We construct our own dataset: STD, a large-scale dataset containing 10088 MIDI files, covering almost all recorded Small Tunes songs in China.

---

[1] *Small Tunes*, as known as *XiaoDiao* in Chinese phonetics, is a category of Chinese Folk Songs. For details, see https://en.chinaculture.org/library/2008-01/11/content_71371_3.htm.

Fig. 1. Two multi-level hierarchies: phrase & bar-level hierarchies (left) and our macro & micro-level hierarchies (right). The dashed boxes indicate levels that are not considered in the respective hierarchies.

2 Related Work

Music Transformer [7] is the first work to utilize the Transformer-based architecture to generate music with coherent structure. Drawlody [12], a music generation system, composes music by converting a user-input melody curve into melody. MusicVAE [18] utilizes a hierarchical decoder to generate music with long-term structure. WuYun [24] leverages music theory to prioritize the generation of structurally important notes as the skeleton, gradually filling in ornamental notes to complete the melody. While WuYun effectively generates coherent melodies, it lacks consideration for structural features within music. In this paper, we build upon the principles of WuYun to explore the intrinsic relationship between macro-level and micro-level hierarchies in music.

In recent years, an increasing number of works have focused on the structural features of music. These studies can be categorized based on their exploration of intrinsic structural hierarchies into four types: phrase-level, bar-level, phrase & bar-level, and others. 1) phrase-level: MusicFrameworks [1], a Transformer-LSTM architecture, processes music sequences by incorporating chord, melody, and rhythm features. [2] generates music by imitating the structure, melody, and style of a given seed song. [25] explores the form, harmony, and texture features to enhance the structure within music. Theme Transformer [19] centers on theme-based conditioning, generating music using thematic material as the condition. 2) bar-level: Melons [28], a Transformer-based music generation model, represents music sequences as graphs based on eight custom-defined structural types. Popmnet [21] generates pop music with a well-organized structure by establishing relationships of repetition and sequence between all bars. 3) phrase & bar-level: Hyperbolic Music Transformer [8] enhances the structure of music by leveraging hyperbolic theory. [9] utilizes a data-driven approach to analyze the structure of symbolic music. [16] proposes the Phrase and Bar Countdown events to study the phrase & bar-level hierarchies within music. 4) Others: [6] explores repetitive patterns at the motif-level. [13] progressively expands a music fragment into a complete melody across the motif, phrase, and section levels. [20] explores structural elements at the note, chord, and section levels in music to enhance its quality.

Most of the aforementioned works concentrate on music generation within Western music genres such as Western pop music, while research on Chinese-style music, especially Chinese Folk Songs, remains relatively limited. Although some researchers have employed sequence-to-sequence models for Chinese-style music generation, such as MG-VAE [15] for regional-style Chinese Folk Songs composition, [27] generates melody and arrangement for Chinese pop-style songs.

3 Method

3.1 Phrase Segmentation

The structure of Chinese Small Tunes is unique, often presenting orderly structural patterns. The distinctive hierarchical structure in Chinese Small Tunes reflects traditional style of Chinese Folk Songs. Most of the phrases within Small Tunes are relatively short, and thus we examine the phrases as macro-level hierarchy.

We dedicate to produce the accurate phrase segmentation of Small Tunes, which is significant to explore the intrinsic structural features within music. We apply a deep learning method to get phrase segmentation. The model architecture we select is a convolutional neural network with conditional random field [26], and 1168 labeled Chinese Folk Songs in public data set Essen Folksong Database are used to train the model. Then the phrase segmentation of each song in our dataset can be produced using the trained model. The phrase segmentation of a song is defined as $S = \{s_1, s_2, \ldots, s_n\}$, where n is the length of sequence, and for instance, $s_i = k$ indicates that the ith note belongs to the kth segment.

3.2 Skeleton Extraction

A melody consists of structural notes and ornamental notes, these structural notes, called skeleton [17], is the underlying framework of a full melody. Based on the melodic skeleton, a full-fledged melody can be composed by filling into ornamental notes. The skeleton notes, which tend to be more prominent in auditory perception, are selected as the micro-level hierarchy for our study.

Skeleton notes can be divided into pitch and rhythm dimensions. One skeleton note extracted from the pitch dimension contributes to the stability and harmony, while one from the rhythm dimension is of importance of the rhythm of melody development.

For the pitch dimension, we define a Small Tunes Trembling Tote, which often occurs in the Chinese Small Tunes, featuring traditional Chinese style. The Small Tunes Trembling Note starts and ends with the note which has the same pitch, among them there exists some other ornamental notes with shorter duration. Figure 2b shows one piece of a famous Chinese Folk Song *Molihua* (or *Jasmine Flower*) as example.

For the rhythm dimension, we select three types of skeleton notes according to [24], which are metrical accent, syncopation, and long note. After conducting

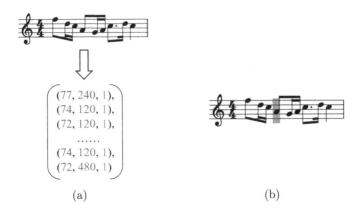

Fig. 2. (a) An example of music representation: For instance, the first note will be represented as (77, 240, 1). (b) A piece of *Molihua*, a famous Chinese Folk Song. The blue-colored $A4$ note, followed by a passing note and returning to $A4$, will be selected as a Small Tunes Trembling Note.

phrase segmentation on a single song, we extract the skeleton notes from each phrase, thereby obtaining the skeleton note sequence. Figure 3 illustrates an example of skeleton extraction result.

3.3 Music Representation

REMI [5] is a a widely used method for symbolic music representation. However, we utilize a triplet format of $\{pitch, duration, segment\}$ instead of REMI to represent symbolic music sequences for the following reasons: 1) The REMI representation results in an excessively long input sequence, complicating melody modeling. 2) Tokens such as *bar* and *position* in REMI appear irregularly at the beginning or middle of sequences, disrupting the alignment between skeleton notes and full notes sequences during Phrase-level Cross-Attention (as discussed later). Conversely, the triplet format, which includes only *pitch*, *duration*, and *segment* attributes, represents each note as a single token after concatenation. This ensures a one-to-one correspondence between the skeleton notes sequence and the full notes sequence during Phrase-level Cross-Attention, thereby enhancing modeling efficiency.

The *pitch* and *duration* values are obtained directly, while the *segment* value is derived from the outcome of phrase segmentation. After being converted into the digital format, the symbolic music token sequence can be fed into the model as input. Figure 2a illustrates the music representation.

For pitch sequence $P : \{p_1, p_2, \ldots, p_n\}$, duration sequence $D : \{d_1, d_2, \ldots, d_n\}$, segment sequence $S : \{s_1, s_2, \ldots, s_n\}$. $P, D, S \in R^{n \times 1}$, we embed them as $P_{emb}, D_{emb}, S_{emb} \in R^{n \times d_{model}}$ where d_{model} represents the embedding dimension. Then we utilize a fusion layer to merge the pitch, duration and segment information, resulting in what we denote as Music Fusion (MF) in Eq. 1, where W_{MF} represents a trainable linear, and $\oplus$ is a vector concatenation operation.

Fig. 3. An example of skeleton extraction. The skeleton notes consist of Small Tunes Trembling Note, Metrical Accent, Syncopation and Long Note.

$$MF = W_{MF} \cdot (P_{emb} \oplus D_{emb} \oplus S_{emb}) \quad (1)$$

The positional encoding is illustrated in Eq. 2, PE_i is the original positional encoding of transformer where $I = \{0, 1, \ldots, n-1\}$ represents the index of the music sequence, besides, we propose an additional positional encoding PE_s to embed the phrase segment $S : \{s_1, s_2, \ldots, s_n\}$.

$$PE = PE_i + PE_s \quad (2)$$

Now, the input of encoder and decoder block is as follows:

$$input = MF + PE \quad (3)$$

3.4 Model Architecture

We model a song from macro-level and micro-level hierarchies. At the macro-level hierarchy, a Small Tunes song consists of multiple phrases, which intricately interweave and connect with each other. At the micro-level hierarchy, skeleton notes within each phrase play a pivotal role in guiding the melody generation. In order to better study the intrinsic features among these hierarchical structures, we propose a novel Phrase-level Cross-Attention.

The skeleton notes sequence input of encoder block and the full notes sequence input of decoder block are denoted as G_{input} and H_{input} respectively. After being processed by the encoder block, G_{input} serves as the key and value inputs for the Phrase-level Cross-Attention in decoder block, denoted as $G' : \{g_1, g_2, \ldots, g_m\} \in R^{m \times d_{model}}$, and after being processed by the Masked Relative Self-Attention [7] and Add & Norm layer, H_{input} serves as the query input for the Phrase-level Cross-Attention, denoted as $H' : \{h_1, h_2, \ldots, h_n\} \in R^{n \times d_{model}}$. Where m and n are the length of skeleton notes sequence and the length of full notes sequence respectively, and d_{model} is the embedding dimension. The query (Q), key (K), and value (V) are shown in Eq. 4, where W_Q, W_K, W_V are three trainable linear layers.

$$Q, K, V = W_Q \cdot H', W_K \cdot G', W_V \cdot G' \quad (4)$$

We design a Phrase-level Mask Matrix to ensure that the melody generation of one phrase only attends the skeleton notes within the same phrase, thereby the skeleton notes can guide the melody generation of the corresponding phrase. For explanation purposes, we provide an example as follows. Given

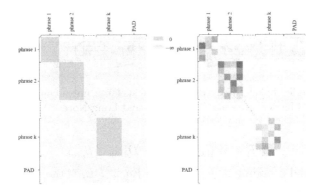

Fig. 4. Phrase-level Mask Matrix (left) and attention weights (right)

the $k^{th}(k \in 1, 2, \ldots)$ phrase, after performing phrase segmentation operations as mentioned earlier, we obtain the phrase segmentation labels: $S^g : \{s_1^g, s_2^g, \ldots, s_m^g\}$ for the skeleton notes sequence and $S^h : \{s_1^h, s_2^h, \ldots, s_n^h\}$ for the full notes sequence. Based on this result, we can extract the skeleton notes subsequence $G_k' : \{g_i, g_{i+1}, \ldots, g_j\}$ and the full notes subsequence $H_k' : \{h_p, h_{p+1}, \ldots, h_q\}$ within the kth phrase, according to Eq. 5.

$$s_i^g = s_{i+1}^g = \cdots = s_j^g = k = s_p^h = s_{p+1}^h = \cdots = s_q^h \tag{5}$$

Where i, p are the index of the first note in one phrase and j, q are the index of the last note. Furthermore, after obtaining the index i, j, p and q, the kth block matrix can be represented as Eq. 6, where r stands for row, c stands for column, 0 represents no masking required while $-\infty$ indicates masking (Fig. 4).

$$M^k = \begin{cases} 0, & p \leq r \leq q \text{ and } i \leq c \leq j \\ -\infty, & \text{others} \end{cases} \tag{6}$$

Performing the same operation on each phrase yields a total of n_p block matrices, where n_p is the number of phrases. Combining these matrices yields the Phrase-level Mask Matrix M.

Finally, the output of Phrase-level Cross-Attention can be obtained as Eq. 7. Figure 5 illustrates the architecture of our model.

$$Att(Q, K, V, M) = softmax(\frac{Q \cdot K^T}{\sqrt{d_{model}}} + M) \cdot V \tag{7}$$

4 Experiment

4.1 Experiment Settings

Dataset. There has been abundant research on Western music genres like classical and pop music, while studies on Chinese-style songs remain relatively limited.

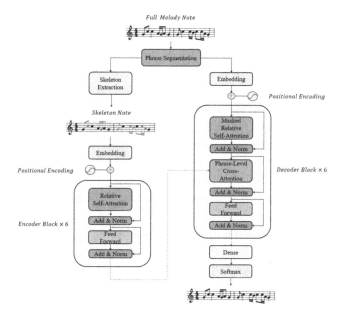

Fig. 5. Architecture of Small Tunes Transformer

Chinese Folk Songs, a unique music genre of Chinese-style songs, with strong regional characteristics [11,23], captivating melody and richest numbers, can be traced back to the *Classic of Poetry* (or *Shijing*) over 3,000 years ago. Small Tunes[1], a category of Chinese Folk Songs, is popular among towns or countries and is characterized by fixed melody and lyrics, orderly structure, and subtle, melodious tunes. Small Tunes serve as the focus of our study.

We construct our dataset, named the Small Tunes Dataset[2] (STD), a large-scale collection of 10088 Small Tunes songs. STD encompasses almost all recorded Small Tunes songs from 31 provinces in China, each meticulously transcribed into MIDI format by us. For model training, we select songs with a time signature denominator of 4.

Baseline Models. In order to explore the advantages of the model architecture, we select three models as our baseline models:

- **Music Transformer** (MT), which is the first Transformer-based model to generate symbolic music [7].
- **WuYun**, which uses the skeleton notes as a condition but lack of any segment information [24].
- **Music Transformer with Phrase and Bar Countdown events** (MT+Ph &BC), which introduces Phrase and Bar Countdown events to enhance structural coherence [16].

---

[2] https://chinglohsiu.github.io/files/MGD.html.

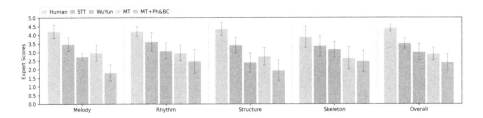

Fig. 6. Results of the subjective evaluation. Human, STT, WuYun, MT, MT+Ph&BC stand for human composition, our proposed model, WuYun architecture, Music Transformer and Music Transformer utilizing Phrase&Bar Countdown events, respectively.

Experiment Configurations. We utilize 7280 songs from our STD after data preprocessing, with 90% selected as training set to train the model and the remaining 10% as test set to evaluate the performance of the model. The number of layers for both encoder and decoder is 6. The embedding dimension d_{model} is 256, learning rate is 0.001, batch size is 16, and the optimizer we select is Adam with $\epsilon = 10^{-8}$, $\beta_1 = 0.9$, $\beta_2 = 0.999$.

4.2 Subjective Evaluation

To assess the quality of the generated music, we conduct a subjective evaluation. Specifically, we invite 10 music experts with professional music training and instrument-playing experience to rate 10 songs generated by STT, three state-of-the-art models and human composers(Ground Truth) on five aspects:

- **Melody**: Whether the melody is clear and captivating.
- **Rhythm**: Whether the rhythm features consistency.
- **Structure**: Whether the melody features a hierarchical structure in its phrases.
- **Skeleton**: Whether there are any notes that audibly stand out, playing a role of the musical skeleton.
- **Overall**: The overall auditory perception of the entire song.

Figure 6 shows the results of the subjective evaluation. The results indicate that our model, STT, outperforms other state-of-the-art models across all subjective evaluation metrics. This suggests that STT is capable of generating melodies that are more captivating, structures that are clearer, and themes that are more consistently coherent compared to other models. Specially, experts note that STT exhibits a more prominent hierarchical structure in the generated melody compared to other baseline models. However, compared to human compositions, the music generated by STT still exhibits some flaws, indicating room for improvement.

4.3 Objective Evaluation

To ensure a comprehensive assessment of the generated music, we also perform an objective evaluation using six metrics. Specifically, we propose three objective evaluation mechanisms as follows:

Theme Rhythm Correspondence (TRC)

$$TRC = \min_i D_{\text{hamming}}\left(R_{\text{theme}}, R_i\right) \tag{8}$$

We propose Theme Rhythm Correspondence to evaluate the rhythm of the generated melody in relation to the theme piece. For this study, the first two bars, as prompt during the generation phase, are selected as the theme sequence. R_{theme} is the binary onset vector of the theme piece (1 represents an onset, otherwise 0), similarly, R_i is the binary onset vector of the ith melody with the same length as the theme piece, and $D_{hamming}(R_{theme}, R_i)$ is the hamming distance to compute the difference of the two melodies R_{theme}, R_i. The smaller the TRC value, the more rhythmically similar the generated melody is to the theme, reflecting better rhythmic coherence.

Theme Pitch Correspondence (TPC)

$$TPC = \min_i D_{\text{hamming}}\left(P_{\text{theme}}, P_i\right) \tag{9}$$

Similarly, we propose Theme Pitch Correspondence to evaluate the generated melody, with P_{theme} and P_i denote the pitch sequence of theme and the ith piece, respectively.

Pentatonic Scale Consistency (PSC)

$$S_i = \begin{cases} 10, & p_i \in \{C, D, E, G, A\} \\ 6, & p_i \in \{F, F^\#, B, B^b,\} \\ -10, & \text{others} \end{cases} \tag{10}$$

$$\text{PSC} = \frac{1}{n}\left(\sum_{i=1}^{n} s_i\right) \tag{11}$$

We propose Pentatonic Scale Consistency to evaluate the consistency of generated melody in the pitch scale dimension. Traditional Chinese songs are mostly composed using the Chinese Pentatonic Scale, a distinctive system in Chinese music. This scale consists of five tones: C, D, E, G, and A, which satisfy the perfect fifth intervals. Additionally, four tones (F, $F^\#$, B and B^b) can be added to play ornamental roles. We rate each note in the melody: assign 10 points if it belongs to $\{C, D, E, G, A\}$, 6 points if it belongs to $\{F, F^\#, B, B^b\}$, and deduct 10 points if it does not adhere to the rules of the pentatonic scale. Finally, compute the average score across all notes to obtain the PSC. PSC evaluates whether a melody follows the pattern of the Chinese pentatonic scale.

Table 1. Objective evaluation results of comparative experiments. For all metrics, models with values closer to the ground truth demonstrate better performance.

Model	TPC	TRC	RC	PSC	PCE	PE
STT(ours)	**4.61**	**2.67**	85.2%	**9.75**	**2.29**	**2.66**
WuYun	6.26	3.59	85.1%	9.78	2.28	2.62
MT	5.39	3.2	86.2%	9.82	2.25	2.58
MT+Ph&BC	7.70	4.75	**86.5%**	9.90	2.23	2.53
Ground Truth	3.78	1.91	87.3%	9.73	2.31	2.65

Moreover, we utilize Rhythm Consistency (RC), Pitch Entropy (PE) and Pitch Class Entropy (PCE) from MusPy[3] to evaluate the pitch consistency of melody.

Comparison Result. To evaluate the performance, we compare our model, STT, against three baseline models and human compositions (Ground Truth). Table 1. shows the result of comparative experiment. STT outperforms other baseline models in all metrics except RC. The closest TPC and TRC values among all baseline models indicate that our model generates more coherent melodies in both pitch and rhythm dimensions. This suggests that Phrase-level Cross-Attention mechanism effectively learns structural features of Small Tunes songs at both macro level and micro level hierarchies. The PSC, PCE and PE value of our model closely match those of ground truth, indicating its capability to generate Small Tunes songs with more consistent melodies. Although STT slightly lags behind the MT model by 1.3% in the RC metric, its close proximity to the ground truth indicates that both models perform well in generating melodies with consistent rhythm.

Ablation Result. To explore the underlying features of hierarchical structure in the Chinese Small Tunes, we design 6 ablative groups focusing on two key aspects: phrase segmentation and skeleton notes extraction. In addition to the phrase segmentation utilized in our method, we also employ three phrase segmentation methods:

- No use of phrase segmentation, treating the music sequence as a single segment (abbreviated as No Segment).
- Selection of 2 bars as the phrase unit, a rule-based approach to phrase segmentation (abbreviated as 2 Bars).
- Expansion of the phrase boundaries from our phrase segmentation result, combining two phrases into a larger unit (abbreviated as Expansion).

Based on these phrase segmentation methods, we additionally design a skeleton notes extraction method, which reduces the number of extracted skeleton notes by randomly removing 50% skeleton notes within each phrase.

---

[3] https://salu133445.github.io/muspy/metrics.html

Table 2. Objective evaluation results of ablation experiments. *Phrase* and *Skeleton* are abbreviations for the ablation methods of phrase segmentation and skeleton notes extraction, respectively. For all metrics, models with values closer to the ground truth demonstrate better performance.

Group	Phrase	Skeleton	TPC	TRC	RC	PSC	PCE	PE
1(STT)	-	-	**4.61**	**2.67**	85.2%	**9.75**	2.29	**2.66**
2(WuYun)	No Segment	-	6.26	3.59	85.1%	9.78	2.28	2.62
3	2 Bars	-	6.58	3.76	84.5%	9.76	**2.30**	2.67
4	Expansion	-	5.11	3.17	85.0%	9.80	2.26	2.60
5	-	Remove 50%	4.94	2.92	85.5%	9.80	2.25	2.57
6	No Segment	Remove 50%	5.25	3.18	85.1%	9.81	2.24	2.55
7	2 Bars	Remove 50%	7.21	4.02	84.9%	9.77	2.29	2.62
8	Expansion	Remove 50%	5.23	3.12	**85.8%**	9.82	2.24	2.56
Ground Truth			3.78	1.91	87.3%	9.73	2.31	2.65

Table 2 shows the result of ablation experiment. We construct 6 ablation groups (Group 2–8) by adjusting the phrase segmentation and skeleton notes extraction strategies. Group 1–4 and 5–8 each employ the same skeleton notes extraction strategy within their groups but utilize different phrase segmentation strategies. Group 1 and Group 5 respectively achieve the best performance in TPC and TRC, indicating that our phrase segmentation strategy contributes to generating coherent melodies. Furthermore, Group 1 outperforms Group 2–4 in almost all metrics except PCE, suggesting that inappropriate segment boundaries are detrimental to capturing the structural features within Small Tunes songs. Moreover, Group 1 outperforms Group 5, indicating that an appropriate number of skeleton notes contribute to guiding the melody generation and constructing the hierarchical structure.

5 Conclusion

In order to study the hierarchical structural features within music, we delve into multi-level hierarchies: at the macro-level hierarchy, we conduct phrase segmentation algorithm to study the impact of phrase on the overall structural organization, and at the micro-level hierarchy, we design a skeleton notes extraction strategy to explore how skeleton notes within phrases influence the melody generation. Building upon this, we propose a novel Phrase-level Cross-Attention to capture the intrinsic relationship among multi-level hierarchies. Moreover, we train our proposed model: Small Tunes Transformer on our own established dataset: Small Tunes Dataset, providing a new perspective for the composition of Chinese-style music. We design three novel metrics to evaluate music from rhythm and melody dimensions. The experiment results indicate that our model outperforms other state-of-the-art models on both subjective and objective evaluations. Additionally, we add several ablative groups to deeply explore the intrinsic features within hierarchical structures. In future work, we aim to extend our study of macro and micro-level hierarchies within music, particularly focusing on polyphonic compositions.

References

1. Dai, S., Jin, Z., Gomes, C., Dannenberg, R.B.: Controllable deep melody generation via hierarchical music structure representation. In: Proceedings of the 22nd International Society for Music Information Retrieval Conference, pp. 143–150 (2021)
2. Dai, S., Ma, X., Wang, Y., Dannenberg, R.B.: Personalised popular music generation using imitation and structure. J. New Music Res. **51**(1), 69–85 (2022)
3. Dong, H.W., Hsiao, W.Y., Yang, L.C., Yang, Y.H.: MuseGAN: multi-track sequential generative adversarial networks for symbolic music generation and accompaniment. In: Proceedings of the AAAI Conference on Artificial Intelligence, vol. 32, pp. 34–41 (2018)
4. Guo, Z., Kang, J., Herremans, D.: A domain-knowledge-inspired music embedding space and a novel attention mechanism for symbolic music modeling. In: Proceedings of the AAAI Conference on Artificial Intelligence, vol. 37, pp. 5070–5077 (2023)
5. Hsiao, W.Y., Liu, J.Y., Yeh, Y.C., Yang, Y.H.: Compound word transformer: learning to compose full-song music over dynamic directed hypergraphs. In: Proceedings of the AAAI Conference on Artificial Intelligence, vol. 35, pp. 178–186 (2021)
6. Hu, Z., Ma, X., Liu, Y., Chen, G., Liu, Y., Dannenberg, R.B.: The beauty of repetition: an algorithmic composition model with motif-level repetition generator and outline-to-music generator in symbolic music generation. IEEE Trans. Multim. **26**, 4320–4333 (2024)
7. Huang, C.Z.A., et al.: Music transformer: generating music with long-term structure. In: International Conference on Learning Representations (2018)
8. Huang, W., Yu, Y., Xu, H., Su, Z., Wu, Y.: Hyperbolic music transformer for structured music generation. IEEE Access **11**, 26893–26905 (2023)
9. Jiang, J., Chin, D., Zhang, Y., Xia, G.: Learning hierarchical metrical structure beyond measures. In: Proceedings of the 23rd International Society for Music Information Retrieval Conference (2022)
10. Johnson, D.D., Keller, R.M., Weintraut, N.: Learning to create jazz melodies using a product of experts. In: ICCC, pp. 151–158 (2017)
11. Li, J., Luo, J., Ding, J., Zhao, X., Yang, X.: Regional classification of Chinese folk songs based on CRF model. Multimedia Tools Appl. **78**, 11563–11584 (2019)
12. Liang, Q., Wang, Y.: Drawlody: sketch-based melody creation with enhanced usability and interpretability. IEEE Trans. Multimedia **26**, 7074–7088 (2024)
13. Lu, P., Tan, X., Yu, B., Qin, T., Zhao, S., Liu, T.Y.: MeloForm: generating melody with musical form based on expert systems and neural networks. In: Proceedings of the 23rd International Society for Music Information Retrieval Conference, pp. 567–574 (2022)
14. Luo, J., Yang, X., Herremans, D.: BandControlNet: parallel transformers-based steerable popular music generation with fine-grained spatiotemporal features. arXiv preprint arXiv:2407.10462 (2024)
15. Luo, J., Yang, X., Ji, S., Li, J.: MG-VAE: deep Chinese folk songs generation with specific regional styles. In: Proceedings of the 7th Conference on Sound and Music Technology (CSMT) Revised Selected Papers, pp. 93–106 (2020)

16. Naruse, D., Takahata, T., Mukuta, Y., Harada, T.: Pop music generation with controllable phrase lengths. In: Proceedings of the 23rd International Society for Music Information Retrieval Conference, pp. 125–131 (2022)
17. Povel, D.J., et al.: Melody generator: a device for algorithmic music construction. J. Softw. Eng. Appl. **3**(07), 683 (2010)
18. Roberts, A., Engel, J., Raffel, C., Hawthorne, C., Eck, D.: A hierarchical latent vector model for learning long-term structure in music. In: International Conference on Machine Learning, pp. 4364–4373 (2018)
19. Shih, Y.J., Wu, S.L., Zalkow, F., Muller, M., Yang, Y.H.: Theme transformer: symbolic music generation with theme-conditioned transformer. IEEE Trans. Multimedia **25**, 3495–3508 (2022)
20. Wu, G., Liu, S., Fan, X.: The power of fragmentation: a hierarchical transformer model for structural segmentation in symbolic music generation. IEEE/ACM Trans. Audio Speech Lang. Process. **31**, 1409–1420 (2023)
21. Wu, J., Liu, X., Hu, X., Zhu, J.: PopMNet: generating structured pop music melodies using neural networks. Artif. Intell. **286**, 103303 (2020)
22. Wu, S.L., Yang, Y.H.: The jazz transformer on the front line: exploring the shortcomings of AI-composed music through quantitative measures. In: Proceedings of the 21st International Society for Music Information Retrieval Conference, pp. 142–149 (2020)
23. Yang, X., Luo, J., Wang, Y., Zhao, X., Li, J.: Combining auditory perception and visual features for regional recognition of Chinese folk songs. In: Proceedings of the 2018 10th International Conference on Computer and Automation Engineering, pp. 75–81 (2018)
24. Zhang, K., et al.: WuYun: exploring hierarchical skeleton-guided melody generation using knowledge-enhanced deep learning. arXiv preprint arXiv:2301.04488 (2023)
25. Zhang, X., Zhang, J., Qiu, Y., Wang, L., Zhou, J.: Structure-enhanced pop music generation via harmony-aware learning. In: Proceedings of the 30th ACM International Conference on Multimedia, pp. 1204–1213 (2022)
26. Zhang, Y., Xia, G.: Symbolic melody phrase segmentation using neural network with conditional random field. In: Shao, X., Qian, K., Zhou, L., Wang, X., Zhao, Z. (eds.) CSMT 2020. LNEE, vol. 761, pp. 55–65. Springer, Singapore (2021). https://doi.org/10.1007/978-981-16-1649-5_5
27. Zhu, H., et al.: Xiaoice band: a melody and arrangement generation framework for pop music. In: Proceedings of the 24th ACM SIGKDD International Conference on Knowledge Discovery & Data Mining, pp. 2837–2846 (2018)
28. Zou, Y., Zou, P., Zhao, Y., Zhang, K., Zhang, R., Wang, X.: MELONS: generating melody with long-term structure using transformers and structure graph. In: ICASSP 2022-2022 IEEE International Conference on Acoustics, Speech and Signal Processing (ICASSP), pp. 191–195 (2022)

SMG-Diff: Adversarial Attack Method Based on Semantic Mask-Guided Diffusion

Yongliang Zhang and Jing Liu[✉]

College of Computer Science, Inner Mongolia University, Hohhot, China
32209048@mail.imu.edu.cn, liujing@imu.edu.cn

Abstract. Deep Neural Networks (DNNs) exhibit vulnerability to adversarial examples, which exposes their fragility under semantic adversarial attacks. Semantic adversarial attacks can effectively manipulate the decisions of DNNs by introducing subtle modifications to critical semantic regions in images. These modifications present substantial challenges to existing defense mechanisms. However, current approaches to semantic adversarial attacks face difficulties in simultaneously achieving both stealthiness and a high success rate. To address this limitation, this paper proposes a novel semantic adversarial attack method, termed the Semantic Mask-Guided Diffusion Model (SMG-Diff). The proposed method first extracts critical semantic information from the target image and subsequently generates a semantic mask using the Salient Semantic Extraction Module (SSEM). During the reverse phase of the diffusion process, a Dynamic Mask Fusion Module (DMFM) is employed to adaptively adjust the mask, ensuring that the feature fusion process concentrates on the most critical regions of the image. By iteratively optimizing the mask and the feature fusion process, the method ultimately generates adversarial examples with enhanced stealthiness. Experimental results validate that SMG-Diff significantly enhances the stealthiness of adversarial examples while maintaining a high attack success rate.

Keywords: Deep Neural Networks · Semantic adversarial attacks · Diffusion model

1 Introduction

In recent years, Deep Neural Networks (DNNs) have exhibited exceptional performance across various domains, including autonomous driving [8], medical image analysis [31], and remote sensing [6]. However, with the increasing adoption of these technologies, the susceptibility of DNNs to adversarial examples has raised critical security concerns [4]. Adversarial examples are crafted by introducing subtle perturbations to mislead model decisions. This poses significant risks in scenarios such as illegal identity verification, where facial recognition systems could be manipulated to falsely identify individuals. These risks could also compromise the safety of autonomous vehicles. These pressing concerns have driven researchers to investigate and develop more robust models and effective defense mechanisms.

Currently, most adversarial attack methods, such as the Fast Gradient Sign Method (FGSM) [10] and Projected Gradient Descent (PGD) [19], focus primarily on pixel-level perturbations of input images. Although the images generated by these methods are almost imperceptible to the human eye, they are often easily identified and neutralized by existing defense mechanisms. In contrast, semantic-level adversarial attacks adopt a more sophisticated and difficult-to-detect strategy [14,24]. These attacks go beyond simple pixel modifications and instead alter the overall semantic content of the image. For example, they may blur or modify key features, thereby disrupting the model's accurate interpretation of the image's true intent. In practical applications, this approach not only effectively evades traditional defense measures but also exhibits significant stealthiness and powerful attack capabilities.

Although existing semantic attack methods have proven feasible, the images they generate are often easily noticeable to the human eye. To be truly effective, semantic attacks must not only successfully deceive the classifier but also ensure that the attacked images appear natural and credible. Current semantic attack strategies struggle to balance stealthiness and success rates; modifications can be either too subtle, rendering them ineffective, or too obvious, making them easily detectable by existing defense technologies. Consequently, there is an urgent need to develop more innovative techniques to overcome these challenges. As an emerging generative model, the diffusion model [12] shows unique potential in addressing these challenges. Unlike traditional methods that add perturbations at the pixel level, the diffusion model, by precisely manipulating the latent space, can generate adversarial examples that are highly realistic both visually and semantically, effectively avoiding detection and defense.

To balance attack stealthiness and attack success rates, this paper proposes an innovative method: SMG-Diff. Based on the diffusion model, this method achieves key feature fusion by accurately manipulating the latent space, guided by the semantic mask of the target image. SMG-Diff integrates the Salient Semantic Extraction Module (SSEM) and the Dynamic Mask Fusion Module (DMFM) to accurately identify and fuse important semantic regions of the target image. DMFM plays a crucial role by adaptively adjusting the semantic mask during the reverse diffusion process, ensuring that the feature fusion focuses on the most critical regions of the image. The adversarial examples generated by this strategy are highly stealthy in visual terms and difficult to detect, while maintaining the naturalness and semantic consistency of the image, effectively enhancing both the stealthiness and efficacy of the attack. Contributions of this paper are summarized as follows:

1. This paper introduces a novel semantic adversarial attack method, SMG-Diff, which addresses the challenge of balancing stealthiness and effectiveness in adversarial attacks. By precisely manipulating key regions of the image while preserving semantic consistency, SMG-Diff achieves a substantial improve-

ment in both stealthiness and the success rate of adversarial examples, as shown in Fig. 1.
2. The Salient Semantic Extraction Module (SSEM) and Dynamic Mask Fusion Module (DMFM) are introduced. SSEM identifies critical regions in the target image, while DMFM refines the masks during the reverse diffusion process to focus feature fusion on key semantic areas. These innovations enhance SMG-Diff's stealthiness and attack success rate.
3. Experiments on the CelebA-HQ [16] dataset show that SMG-Diff significantly outperforms existing methods across key metrics such as stealthiness and success rate, confirming its superior feasibility and effectiveness in practical applications (Fig. 1).

Fig. 1. The SMG-Diff attack method designed for facial identity recognition models is illustrated. The top row presents the target images, the middle row displays the source images, and the bottom row shows the generated adversarial examples.

2 Related Work

2.1 Semantic Adversarial Attacks

The current mainstream methods for semantic adversarial attacks primarily rely on altering the color or texture of images [1], or by manipulating the latent space within generative models [21], such as Generative Adversarial Networks (GANs) [9]. These approaches aim to deceive models into making incorrect judgments while maintaining the visual realism of the images. Unlike traditional pixel-level perturbation attacks that focus on minor distortions, semantic adversarial attacks target high-level features, altering the underlying meaning or perception of the image. For instance, the color shift attack proposed by Hosseini et al. [14] demonstrated that even models proficient in natural image classification are vulnerable to slight color changes, underscoring a significant weakness in handling semantic adversarial examples. Similarly, the Colorfool attack by Shamsabadi et al. [22] illustrates how subtle alterations in an image's internal colors can mislead deep learning models. These modifications, while appearing benign to human perception, pose significant challenges for automated systems, complicating the classifier's ability to defend against such seemingly innocuous changes.

2.2 Semantic Adversarial Attacks Based on Generative Models

Recent studies [28] and [32] have demonstrated methods to achieve adversarial attacks by introducing perturbations that are difficult for the model to perceive but highly effective in their impact. These methods include leveraging techniques based on perceptual similarity or finely manipulating the feature space of deep neural networks. Other studies [1] and [15] have proposed creating semantically rich adversarial examples through feature space interpolation and parametric transformations. These examples successfully mislead classification models while maintaining a natural appearance. A study [20] explored generating targeted, unrestricted adversarial attacks in the latent space of adversarial generative models (GANs) using decision-based attack algorithms. This approach maintains attack stealthiness while also challenging the robustness of the models.

2.3 Semantic Adversarial Attacks Based on Diffusion Models

Most previous work on semantic adversarial attacks has utilized generative models to alter attributes, relying on attribute annotations [15], color, or texture information [1,22]. However, diffusion models have recently gained significant attention in this field because of their ability to generate high-quality, intricate, and hard-to-detect adversarial images. A recent study [17] proposed a language-guided semantic adversarial attack method that leverages latent diffusion models for targeted manipulation. The I2A method adversarially guides the reverse diffusion process by searching for adversarial latent codes conditioned on the input image and text instructions. DiffAttack [5] generates imperceptible perturbations containing semantic cues by introducing them into the latent space of the diffusion model, rather than directly operating in the pixel space. This method preserves the content structure while enhancing the transferability and attack efficiency of adversarial examples by manipulating the diffusion model's attention mechanisms.

3 Method

We introduce SMG-Diff, a novel semantic adversarial attack method specifically designed to optimize both stealthiness and attack success rates. As depicted in Fig. 2, SMG-Diff consists of two core modules: the Salient Semantic Extraction Module (SSEM) and the Dynamic Mask Fusion Module (DMFM). The SSEM employs Layer-wise Relevance Propagation (LRP) [2] in combination with attribution techniques to generate a semantic mask that identifies critical regions influencing the model's decisions. This mask is subsequently refined by the DMFM during the reverse diffusion process, ensuring that feature fusion remains concentrated on these key areas. Through iterative optimization, SMG-Diff produces adversarial examples that maintain high stealthiness while effectively misleading the target model.

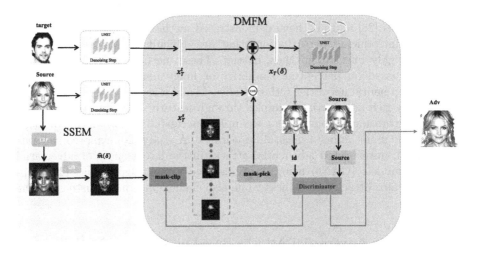

Fig. 2. The architecture of the proposed SMG-Diff framework, consisting of the Salient Semantic Extraction Module (SSEM) and the Dynamic Mask Fusion Module (DMFM). The SSEM identifies and generates the semantic mask, while the DMFM dynamically refines the mask during the adversarial example generation process, ensuring effective feature fusion and high attack stealthiness.

3.1 Salient Semantic Extraction Module

The Salient Semantic Extraction Module (SSEM) is a pivotal component of our adversarial example generation framework. Its primary objective is to accurately identify and extract key semantic regions within an image, which are then used to guide the feature fusion process in the latent space. This process is crucial for enhancing both the stealthiness and effectiveness of the generated adversarial examples.

Initial Saliency Region Selection. In SSEM, we initially employ the Layer-wise Relevance Propagation (LRP) technique to compute the saliency scores for the input image. LRP quantifies the contribution of each pixel to the model's final decision through a layer-by-layer backpropagation process. Given an input image $\mathbf{x}$ and target label y, the saliency score $\mathbf{A}$ is computed as follows:

$$\mathbf{A} = \text{LRP}(\mathbf{x}, y, \mathbf{W}) \tag{1}$$

where $\mathbf{W}$ denotes the weight matrix of a specific layer. This allows LRP to identify the regions that have the greatest impact on the model's decision-making process.

Region Refinement. After obtaining the initial saliency regions using LRP, we apply the Guided Backpropagation (GB) technique to enhance the accuracy of these regions. GB achieves this by preserving gradients that positively influence the model's output while suppressing those associated with irrelevant or noisy areas. This refinement allows for more precise localization of the pixel regions that are critical to the model's predictions. The refined saliency score using GB is expressed as:

$$\mathbf{G} = \text{GB}(\mathbf{x}, y, \mathbf{A}) \tag{2}$$

Here, $\mathbf{G}$ represents the saliency score refined by GB, which further sharpens the focus on key regions relevant to the target task.

Generation of the Saliency Mask. To generate the saliency mask, we combine the results from LRP and GB by taking the absolute values of the saliency score matrix and averaging across the channel dimension. This process yields a single-channel saliency map, which is defined as:

$$\tilde{\mathbf{G}} = \frac{1}{C} \sum_{c=1}^{C} \left| \mathbf{G}^{(c)} \right| \tag{3}$$

where $\mathbf{G}^{(c)}$ represents the saliency score for the c-th channel, and C denotes the total number of channels. The resulting saliency mask effectively narrows down the regions of the image that require further processing, thereby reducing the perturbation area. This approach not only preserves the effectiveness of the adversarial attack but also enhances its imperceptibility by limiting changes to the most critical regions of the image.

In conclusion, the generation of the saliency mask is a pivotal step in facilitating the precise and efficient application of adversarial perturbations. By concentrating on key semantic regions initially identified by LRP and further refined by GB, this two-stage optimization process generates a saliency mask that accurately focuses on the most influential areas for the model's decision-making. This targeted approach reduces unnecessary alterations to irrelevant regions, thereby significantly improving both the imperceptibility and the success rate of adversarial attacks.

3.2 Dynamic Mask Fusion Module

The Dynamic Mask Fusion Module (DMFM) is essential in the reverse diffusion process, where it plays a pivotal role in generating adversarial examples. The primary function of the DMFM is to achieve precise feature fusion between the source and target images. This is accomplished by dynamically adjusting the semantic mask throughout the diffusion process. By doing so, the DMFM ensures that the generated adversarial examples maintain a high level of stealthiness while also being highly effective in misleading the target model.

As shown in Fig. 2, the process begins by transforming the source image x_0^s and the target image x_0^t into their respective latent space representations, denoted as x_T^s and x_T^t. This transformation is achieved through the forward diffusion process $q(\cdot, T)$, which represents the diffusion process from the original image to the latent representation. Specifically, $x_T^s = q(x_0^s, T)$ and $x_T^t = q(x_0^t, T)$, where the forward diffusion process maps the original images to high-dimensional latent vectors, encoding the necessary semantic information for subsequent feature fusion.

To maintain both stealthiness and attack success rates during the feature fusion process, a saliency mask $\hat{m}(\delta)$ is generated, defined as $\hat{m}(\delta) = \text{TopK}(|m^s|, \delta)$, where the "TopK" function selects the most salient features based on the threshold δ. The "mask-clip" operation adjusts the mask to focus on the most relevant semantic regions, as illustrated in the figure. Throughout the reverse diffusion process, the mask threshold δ is dynamically increased, causing the mask region to expand progressively, as shown by the "mask-pick" process in the figure.

At each reverse diffusion step, the current mask $\hat{m}(\delta)$ is applied to perform feature fusion between the latent representations of the source image x_T^s and the target image x_T^t, forming a new latent representation $\hat{x}_T(\delta)$:

$$\hat{x}_T(\delta) = (1 - \hat{m}(\delta)) \cdot x_T^s + \hat{m}(\delta) \cdot x_T^t \tag{4}$$

This iterative process continues until the desired adversarial example $\hat{x}_0(\delta)$ is achieved, with the Discriminator validating the quality of the generated adversarial example by comparing it with the source image.

Finally, during the reverse diffusion step, the final adversarial example is produced using the optimal mask $\hat{m}(\delta^*)$:

$$\hat{x}_0(\delta^*) = p_\theta(\hat{x}_T(\delta^*), T) \tag{5}$$

Here, the "Discriminator" ensures that the generated adversarial example maintains both the required stealthiness and attack efficacy.

Through the Dynamic Mask Fusion Module (DMFM), we ensure that the adversarial examples effectively balance stealthiness and attack success rate. By dynamically adjusting the semantic mask throughout the reverse diffusion process, DMFM minimizes the visibility of perturbations while maximizing the adversarial impact. As illustrated in Fig. 3, this approach allows the generated adversarial examples to remain inconspicuous to human observers while successfully misleading the target model's predictions.

4 Experiments

Datasets. Our experiments are based on the CelebA-HQ [16] dataset, a high-quality face dataset constructed by Karras et al. (2018) on top of CelebA. The dataset contains approximately 30,000 images with a resolution of 1024×1024, representing 6,217 unique identities. For the purpose of facial identity recognition tasks, we filtered the CelebA-HQ dataset, selecting identities with at least 15 images. This resulted in a subset containing 307 identities, including 4,263 images for training and 1,215 images for testing [27].

Classification Model. To evaluate the effectiveness of our SMG-Diff method, we fine-tuned several classification models on the CelebA-HQ dataset, including DenseNet121, ResNet18, and ResNet101, which were pre-trained on the ILSVRC 2012 dataset. The images were uniformly resized to 256×256 during preprocessing. The validation accuracies of these models are summarized in Table 1. We found that the ResNet18 model demonstrated strong generalization performance; therefore, we report the primary experimental results using the fine-tuned ResNet18 model.

Experimental Details. In our experiments, we used a diffusion model pre-trained on the CelebA-HQ dataset. For the attribution method, we selected Guided Backpropagation and set 30 mask refinement steps. The PGD and MI-FGSM methods were implemented using the Torchattacks library, with the maximum perturbation set to 8/255 and 10 iterations. LatentHSJA, a query-based

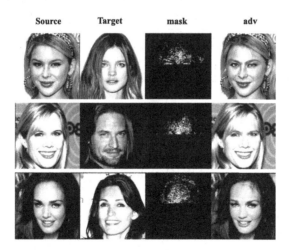

Fig. 3. Illustration of the SMG-Diff process. The "Source" column displays the original source images, the "Target" column shows the target images used to guide the generation, the "Mask" column presents the final semantic masks generated and refined by the DMFM module, and the "Adv" column demonstrates the final adversarial examples produced by the SMG-Diff framework.

Table 1. Validation Accuracy on the CelebA-HQ Dataset

Architectures	CelebA-HQ Dataset
DenseNet121	85.42%
ResNet101	86.82%
ResNet18	87.98%

black-box attack method, had its query count adjusted to 1,000 to balance quality and efficiency. The Latent Mask (LM) method also employed 30 mask refinement steps. It is important to note that the ASR and FID values reported in Table 2 were measured in our experiments. Although we closely followed the experimental settings outlined in the original papers, slight variations in conditions (e.g., hardware configurations or random initialization) may have led to differences in the reported values. To ensure fairness and reproducibility, we used the same dataset and evaluation metrics across all experiments, adhering closely to the configuration files. All experiments were conducted on a single NVIDIA V100 GPU.

4.1 Comparative Experiments

In the comparative experiments, we evaluated the attack performance of two classic adversarial attack methods (PGD, MI-FGSM) and two semantic adversarial attack methods (LatentHSJA, LM) on the CelebA-HQ dataset. To verify the imperceptibility of our proposed method in terms of image quality, we conducted a comprehensive image quality assessment on the generated adversarial examples and compared the results with those of existing adversarial attack algorithms, as shown in Fig. 4. We employed several commonly used image quality metrics, including FID [11], KID [3], PSNR [29], and SSIM [13], which effectively reflect the quality of the adversarial examples and their imperceptibility to the human eye-the higher the image quality, the more difficult it is to detect the attack perturbations. Through a comprehensive analysis of these metrics, we confirmed the superiority of our proposed method in maintaining high-quality images and stealthiness.

Table 2 summarizes the comparative performance of various adversarial attack methods. Our proposed SMG-Diff method achieved the highest ASR at 98.8%, highlighting its effectiveness in generating successful adversarial examples. Additionally, SMG-Diff excelled in preserving image quality, as evidenced by the lowest FID (17.66), KID (0.008), and the highest PSNR (25.01) and SSIM (0.9142). These results indicate that SMG-Diff not only excels in generating robust adversarial examples but also maintains the integrity of the original image content, resulting in minimal perceptual distortion. In contrast, traditional methods such as PGD and MI-FGSM exhibited higher FID and KID values, indicating greater deviations from the original images, and lower PSNR and SSIM values, reflecting a decline in image quality. Although the LM and

Table 2. Performance Comparison of Different Adversarial Attack Methods on CelebA-HQ.

Metric	ASR(%) ↑	FID ↓	KID ↓	PSNR ↑	SSIM ↑
PGD [19]	97.8	48.31	0.032	24.73	0.6087
MI-FGSM [7]	96.4	55.42	0.039	22.96	0.5578
LM [27]	97.6	25.8	0.015	21.8	0.8915
LatentHSJA [20]	98.1	52.95	0.046	23.48	0.7935
SMG-Diff	**98.8**	**17.66**	**0.008**	**25.01**	**0.9142**

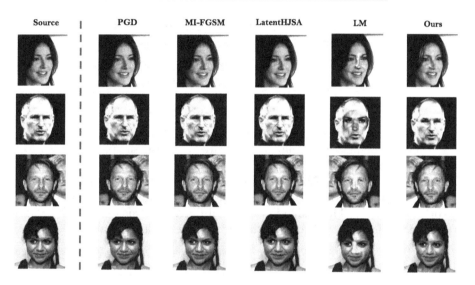

Fig. 4. Visual comparison of adversarial examples generated by different methods, including PGD, MI-FGSM, LatentHSJA, LM, and our proposed SMG-Diff. The first column shows the original source images, while the subsequent columns display the adversarial examples produced by each method. Our method (SMG-Diff) preserves the visual characteristics and quality of the original images more effectively while still causing the target model to misclassify, demonstrating both high stealthiness and attack effectiveness.

LatentHSJA methods performed better in preserving image quality compared to PGD and MI-FGSM, they still did not reach the performance level of SMG-Diff. This analysis emphasizes SMG-Diff's ability to balance high attack success rates with the preservation of image fidelity.

Efficiency Comparison. In real-world applications such as facial recognition or medical image analysis, where high-quality and semantically consistent adversarial examples are essential, the generation efficiency of semantic-level methods becomes a critical factor. To assess this efficiency, we compared SMG-Diff with other semantic-level methods LM and LatentHSJA using two key metrics: average number of queries and average generation time. As shown in Table 3, SMG-Diff not only achieved the highest attack success rate (ASR) at 98.8%, but also demonstrated superior efficiency with the lowest average query count (14.73) and the shortest average generation time (16.89 s). In contrast, methods like LatentHSJA required significantly more queries (1000) and longer generation times (45.87 s), making SMG-Diff a more practical choice for scenarios where both efficiency and high image fidelity are paramount.

Table 3. Comparison of Generation Efficiency among Semantic-Level Methods.

Method	ASR (%) ↑	Average Query ↓	Average Time (s) ↓
LM	97.6	20.17	27.03
LatentHSJA	98.1	1000	45.87
SMG-Diff	**98.8**	**14.73**	**16.89**

The experimental results further demonstrate that the SMG-Diff method not only achieves competitive attack success rates but also significantly improves the visual quality and imperceptibility of adversarial examples. By optimizing the generation process, SMG-Diff maintains a high attack success rate while effectively minimizing the visual discrepancies between adversarial examples and the original images. This reduction in visual differences makes the adversarial attacks more challenging to detect, thereby enhancing their stealthiness.

4.2 Ablation Study

To evaluate the impact of different attribution methods on the attack success rate (ASR) and the quality of adversarial examples, we conducted an ablation study by varying the perturbation percentage and employing different attribution methods to identify important regions. As shown in Table 4, we assessed the performance of several methods, including Guided Backpropagation, GradientShap, Deconvolution, Integrated Gradients, and InputXGradient across various perturbation levels.

The results show that ASR generally increases with higher perturbation percentages, with significant differences across methods. Notably, Guided Backpropagation consistently achieved higher ASR compared to other methods, especially at lower perturbation levels (e.g., 2% to 10%), where its ASR remained between 96.50% and 100%. This indicates that Guided Backpropagation is particularly effective in identifying critical regions with minimal perturbation, leading to more efficient and less perceptible adversarial attacks.

Based on these findings, we selected Guided Backpropagation as the optimal attribution method for subsequent experiments, ensuring that the generated adversarial examples not only achieve a high ASR but also preserve image quality and minimize perceptibility.

Table 4. Success Rate of Different Attribution Methods with Increasing Perturbation Percentage

Attributions	2%	5%	10%	20%	30%
GradientShap [18]	57.20	84.60	94.80	98.80	99.60
Deconvolution [30]	54.80	79.20	91.50	97.80	99.30
Integrated Gradients [26]	57.50	85.90	95.80	98.90	99.70
InputXGradient [23]	54.30	81.50	92.80	98.20	99.60
GuidedBackprop [25]	**52.90**	**86.70**	**96.50**	**99.20**	**100**

SMG-Diff focuses perturbations on the salient semantic regions of the image through precise semantic mask localization. Compared to traditional methods that apply perturbations indiscriminately, this targeted perturbation not only enhances the adversarial effectiveness of the samples but also reduces the impact on the non-essential parts of the image, thereby improving the overall quality and imperceptibility of the image without increasing the number of queries. By concentrating adjustments on the salient semantic regions of the image, SMG-Diff effectively preserves the naturalness of the image while ensuring the effectiveness of the adversarial examples.

5 Conclusion

In this paper, we propose SMG-Diff, a novel semantic adversarial attack method based on a semantic mask-guided diffusion model. The method first employs the Salient Semantic Extraction Module (SSEM) to identify and generate semantic masks, combining Layer-wise Relevance Propagation (LRP) with advanced attribution techniques to accurately pinpoint the critical regions in an image that influence the model's decision-making. During the generation of adversarial examples, the Dynamic Mask Fusion Module (DMFM) dynamically refines the semantic masks, ensuring that the feature fusion process is focused on the most crucial areas of the image. Experimental results show that SMG-Diff achieves high attack success rates while significantly enhancing stealthiness by maintaining image fidelity. Future work will focus on optimizing this approach and exploring its applicability and robustness in more complex scenarios.

Acknowledgement. This work was supported in part by the Natural Science Foundation of Inner Mongolia of China (No.2023ZD18), the Natural Science Foundation of China (No.62462047), the Engineering Research Center of Ecological Big Data, Ministry of Education, the fund of Supporting the Reform and Development of Local Universities (Disciplinary Construction) and the special research project of First-class Discipline of Inner Mongolia A. R. of China under Grant YLXKZX-ND-036.

References

1. Bhattad, A., Chong, M.J., Liang, K., Li, B., Forsyth, D.A.: Unrestricted adversarial examples via semantic manipulation. arXiv preprint arXiv:1904.06347 (2019)
2. Binder, A., Montavon, G., Lapuschkin, S., Müller, K.-R., Samek, W.: Layer-wise relevance propagation for neural networks with local renormalization layers. In: Villa, A.E.P., Masulli, P., Pons Rivero, A.J. (eds.) ICANN 2016. LNCS, vol. 9887, pp. 63–71. Springer, Cham (2016). https://doi.org/10.1007/978-3-319-44781-0_8
3. Bińkowski, M., Sutherland, D.J., Arbel, M., Gretton, A.: Demystifying mmd GANs. arXiv preprint arXiv:1801.01401 (2018)
4. Carlini, N., Wagner, D.: Towards evaluating the robustness of neural networks. In: 2017 IEEE symposium on security and privacy (SP), pp. 39–57. IEEE (2017)
5. Chen, J., Chen, H., Chen, K., Zhang, Y., Zou, Z., Shi, Z.: Diffusion models for imperceptible and transferable adversarial attack. arXiv preprint arXiv:2305.08192 (2023)
6. Chen, J., Chen, K., Chen, H., Li, W., Zou, Z., Shi, Z.: Contrastive learning for fine-grained ship classification in remote sensing images. IEEE Trans. Geosci. Remote Sens. **60**, 1–16 (2022)
7. Dong, Y., et al.: Boosting adversarial attacks with momentum. In: Proceedings of the IEEE Conference on Computer Vision and Pattern Recognition, pp. 9185–9193 (2018)
8. Feng, S., et al.: Dense reinforcement learning for safety validation of autonomous vehicles. Nature **615**(7953), 620–627 (2023)
9. Goodfellow, I., et al.: Generative adversarial networks. Commun. ACM **63**(11), 139–144 (2020)
10. Goodfellow, I.J., Shlens, J., Szegedy, C.: Explaining and harnessing adversarial examples. arXiv preprint arXiv:1412.6572 (2014)
11. Heusel, M., Ramsauer, H., Unterthiner, T., Nessler, B., Hochreiter, S.: GANs trained by a two time-scale update rule converge to a local NASH equilibrium. In: Advances in Neural Information Processing Systems, vol. 30 (2017)
12. Ho, J., Jain, A., Abbeel, P.: Denoising diffusion probabilistic models. Adv. Neural. Inf. Process. Syst. **33**, 6840–6851 (2020)
13. Hore, A., Ziou, D.: Image quality metrics: PSNR vs. SSIM. In: 2010 20th International Conference on Pattern Recognition, pp. 2366–2369. IEEE (2010)
14. Hosseini, H., Poovendran, R.: Semantic adversarial examples. In: Proceedings of the IEEE Conference on Computer Vision and Pattern Recognition Workshops, pp. 1614–1619 (2018)
15. Joshi, A., Mukherjee, A., Sarkar, S., Hegde, C.: Semantic adversarial attacks: parametric transformations that fool deep classifiers. In: Proceedings of the IEEE/CVF International Conference on Computer Vision, pp. 4773–4783 (2019)
16. Karras, T., Aila, T., Laine, S., Lehtinen, J.: Progressive growing of GANs for improved quality, stability, and variation. arXiv preprint arXiv:1710.10196 (2017)
17. Liu, J., et al.: Instruct2Attack: language-guided semantic adversarial attacks. arXiv preprint arXiv:2311.15551 (2023)
18. Lundberg, S.M., Lee, S.I.: A unified approach to interpreting model predictions. In: Advances in Neural Information Processing Systems, vol. 30 (2017)
19. Madry, A., Makelov, A., Schmidt, L., Tsipras, D., Vladu, A.: Towards deep learning models resistant to adversarial attacks. arXiv preprint arXiv:1706.06083 (2017)

20. Na, D., Ji, S., Kim, J.: Unrestricted black-box adversarial attack using GAN with limited queries. In: Karlinsky, L., Michaeli, T., Nishino, K. (eds.) European Conference on Computer Vision, vol. 13801, pp. 467–482. Springer, Cham (2022). https://doi.org/10.1007/978-3-031-25056-9_30
21. Qiu, H., Xiao, C., Yang, L., Yan, X., Lee, H., Li, B.: SemanticAdv: generating adversarial examples via attribute-conditioned image editing. In: Vedaldi, A., Bischof, H., Brox, T., Frahm, J.-M. (eds.) ECCV 2020. LNCS, vol. 12359, pp. 19–37. Springer, Cham (2020). https://doi.org/10.1007/978-3-030-58568-6_2
22. Shamsabadi, A.S., Sanchez-Matilla, R., Cavallaro, A.: ColorFool: semantic adversarial colorization. In: Proceedings of the IEEE/CVF Conference on Computer Vision and Pattern Recognition, pp. 1151–1160 (2020)
23. Shrikumar, A., Greenside, P., Kundaje, A.: Learning important features through propagating activation differences. In: International Conference on Machine Learning, pp. 3145–3153. PMlR (2017)
24. Song, Y., Shu, R., Kushman, N., Ermon, S.: Constructing unrestricted adversarial examples with generative models. In: Advances in Neural Information Processing Systems, vol. 31 (2018)
25. Springenberg, J.T., Dosovitskiy, A., Brox, T., Riedmiller, M.: Striving for simplicity: the all convolutional net. arXiv preprint arXiv:1412.6806 (2014)
26. Sundararajan, M., Taly, A., Yan, Q.: Axiomatic attribution for deep networks. In: International conference on machine learning, pp. 3319–3328. PMLR (2017)
27. Wang, C., Duan, J., Xiao, C., Kim, E., Stamm, M., Xu, K.: Semantic adversarial attacks via diffusion models. arXiv preprint arXiv:2309.07398 (2023)
28. Wang, Y., Wu, S., Jiang, W., Hao, S., Tan, Y.a., Zhang, Q.: Demiguise attack: crafting invisible semantic adversarial perturbations with perceptual similarity. arXiv preprint arXiv:2107.01396 (2021)
29. Wang, Z., Bovik, A.C., Sheikh, H.R., Simoncelli, E.P.: Image quality assessment: from error visibility to structural similarity. IEEE Trans. Image Process. **13**(4), 600–612 (2004)
30. Zeiler, M.D., Fergus, R.: Visualizing and understanding convolutional networks. In: Fleet, D., Pajdla, T., Schiele, B., Tuytelaars, T. (eds.) ECCV 2014. LNCS, vol. 8689, pp. 818–833. Springer, Cham (2014). https://doi.org/10.1007/978-3-319-10590-1_53
31. Zhang, Y., Xie, F., Song, X., Zheng, Y., Liu, J., Wang, J.: Dermoscopic image retrieval based on rotation-invariance deep hashing. Med. Image Anal. **77**, 102301 (2022)
32. Zhao, Z., Liu, Z., Larson, M.: Towards large yet imperceptible adversarial image perturbations with perceptual color distance. In: Proceedings of the IEEE/CVF Conference on Computer Vision and Pattern Recognition, pp. 1039–1048 (2020)

SPLGAN-TTS: Learning Semantic and Prosody to Enhance the Text-to-Speech Quality of Lightweight GAN Models

Ding-Chi Chang[1], Shiou-Chi Li[2], and Jen-Wei Huang[1(✉)]

[1] Department of Electrical Engineering, National Cheng Kung University, Tainan, Taiwan
jwhuang@mail.ncku.edu.tw

[2] Institute of Computer and Communication Engineering, Department of Electrical Engineering, National Cheng Kung University, Tainan, Taiwan

Abstract. Autoregressive-based models have proven effective in speech synthesis; however, numerous parameters and slow inference limit their applicability. Though non-autoregressive models can resolve these issues, speech synthesis quality is unsatisfactory. This study employed a tree-based structure to enhance the learning of semantic and prosody information using a lightweight model. A Variational Encoder (VAE) is used for the generator architecture, and a novel normalizing-flow module is used to enhance the complexity of the VAE-generated distribution. We also developed a speech discriminator with a multi-length architecture to reduce computational overhead as well as multiple auxiliary losses to assist in model training. The proposed model is smaller than existing state-of-the-art models, and synthesis performance is faster, particularly when applied to longer texts. Despite the fact that the proposed model is roughly 30% smaller than FastSpeech2 [1], its mean opinion score surpasses FastSpeech2 as well as other models.

Keywords: speech synthesis · non-autoregressive · tree-based architecture · generative adversarial networks

1 Introduction

Speech synthesis refers to the process of converting text data into corresponding speech data. Early research in this field relied on traditional serial models [2], in which the speech synthesis task is divided into multiple modules, including a speech unit library, speech selection module, and splicing module, each of which is responsible for a different task. Unfortunately, this method requires a large number of databases, and the results often lack smoothness due to the need for splicing the outputs of multiple modules.

This prompted the development of methods based on statistical parametric speech synthesis such as [3,4], in which the speech signal is decomposed into multiple statistical parameters (e.g., spectrum, pitch, and duration) that are then converted into statistical models, such as HMM modeling. This approach makes it possible to generate various styles of speech using a variety of voices or different speakers; however, the results are still easily differentiated from natural speech. Using this method to synthesize natural-sound speech patterns would require an enormous amount of data for training.

Researchers have recently developed deep learning-based Text-to-speech (TTS) models such as [1,5] capable of generating highly realistic speech results. Deep learning-based methods are generally categorized as autoregressive or non-autoregressive models. Autoregressive models, such as [6–9] can generate high-quality speech; however, the synthesis speed is slow. Although non-autoregressive models provide such as [1,5,10–13] rapid synthesis capability, the speech quality generated by non-autoregressive models is generally inferior. They also require additional modules to assist with model learning.

The high-speed parallel processing capabilities of non-autoregressive models have opened the door to real-time applications; however, these methods generate speech results from sequence to sequence, which makes it difficult to capture the dependencies and continuity between different parts in the sequence. Existing state-of-the-art (SOTA) models [1,5] now include additional modules to enhance speech quality. In many cases, these systems are able to achieve speech performance that is close to or even surpasses that of autoregressive models. Nonetheless, these models are too large for deployment on devices with limited memory capacity. They also tend to produce less natural and less realistic prosody when dealing with longer text inputs.

This paper presents a lightweight model architecture for non-autoregressive speech synthesis with superior performance. The proposed text-to-speech synthesis architecture is based on a generative adversarial network, and adds dependency tree and prosodic tree to enhance the model's semantic and prosodic information extraction, as well as a generator and normalized flow module [14] with VAE architecture [15] to reduce Computational overhead. We developed a speech discriminator with a multi-length architecture for the processing of speech information of various lengths to more finely distinguish real/fake speech outputs. We also developed several auxiliary losses to assist in model training.

In experiments, the proposed model surpassed all other models in terms of Mean Opinion Score (MOS) as well as inference time. Despite the fact that the proposed model is roughly 30% smaller than FastSpeech2, its MOS results were roughly the same. Moreover, mel spectrograms revealed that the SPLGAN-TTS results provide richer prosodic information when dealing with long texts. The contributions of this work are summarized as follows:

1. We developed a GAN-based architecture that uses dependency trees as well as prosodic trees to facilitate the learning of semantic and prosodic information. This allows our model to generate highly natural and realistic speech outputs (close to ground truth). Even when dealing with long text inputs, it can also be closer to the prosody of ground truth.
2. The proposed generator architecture employs a VAE to reduce the computational load associated with text-to-speech synthesis, making the model more lightweight. We also developed a Normalizing-Flow module to optimize the

modeling of high-dimensional data distributions to generate speech outputs of higher quality outputs.
3. The proposed multi-length discriminator allows the process of speech information of various lengths, thereby making it possible to focus on local details and global structure as well as enhancing the synthesis capabilities of the generator.
4. We introduced multiple auxiliary losses to improve training on speech prosody, naturalness, and clarity, leading to speech outputs that are more natural and realistic.
5. The proposed model outperformed FastSpeech2 and GlowTTS, despite being roughly 30% smaller.

2 Related Works

2.1 Statistical Parametric Speech Synthesis

Statistical Parametric speech Synthesis (SPSS) is a speech generation technique that uses statistical models, utilizing methods such as Hidden Markov Models (HMM) and Gaussian Mixture Models (GMM) as decribed in [3,4,16]. These approach decomposes the speech signal into spectral parameters, fundamental frequency (F0), and duration features, which are then used for speech synthesis. Compared to neural network-based models, SPSS may still fall short in terms of naturalness and audio quality. Moreover, poor performance by any module in the synthesis pipeline can have a detrimental effect on the overall results of speech synthesis.

2.2 Autoregressive Model

Autoregressive models, such as **WaveNet** [17], **Tacotron 2** [6], and **Deep Voice 3** [7], are characterized by their use of the results from the previous time step to generate the outcomes for the next time step. This step-by-step approach to data generation makes it easier for the model to more effectively capture the temporal correlations in the data and maintain dependencies over long distances. When applied to speech generation, these models are able to produce speech of high quality, while preserving textual details. Unfortunately, the need to generate results sequentially precludes the processing of data in parallel, which slows training and speech generation while introducing latency issues in real-time applications.

2.3 Non-autoregressive Model

Non-autoregressive models, such as **FastSpeech2** [1], **ParaNet** [18], **Glow-TTS** [5], and **DiffSpeech** [19], are characterized by their ability to generate the current output without depending on the output from a previous step. Unlike autoregressive models, non-autoregressive models are able to generate entire sequences simultaneously, resulting in rapid speech synthesis.

FastSpeech2 improves on FastSpeech by including duration, energy, and pitch information to make the outputs sound smoother and more natural. **ParaNet** employs a fully convolutional architecture to enable the stacking of multiple layers as well as knowledge distillation methods to train attention blocks aimed at ensuring precise alignment between text and speech.

Glow-TTS, based on the Flow-based model, uses a reversible flow structure and variational inference technology to capture the diversity of speech features to enrich synthesized speech patterns. **DiffSpeech** employs a diffusion probability model and combines diffusion and progressive noise reduction techniques during the training process to enhance the smoothness of synthesized speech. That approach also accelerates the synthesis process while reducing computational overhead.

3   SPLGAN-TTS: Learning Semantic and Prosody to Enhance the Text-to-Speech Quality of Lightweight GAN Models

In this section, we introduce the architecture of the proposed model, including the Text Encoder, Semantic and Prosody Extractor, Speech Generator, pretrained Vocoder, and Training Loss components (see Fig. 1).

A text Encoder first aligns the text data with speech data. Extract semantic and prosodic information by converting text into dependency tree and prosodic tree, and inputting it into semantic encoder and prosodic encoder, respectively. The resulting embeddings are summed and input into the GAN model to generate speech. A HiFi-GAN vocoder converts the mel spectrogram into the final waveform. The overall model architecture is shown in Fig. 1.

3.1   Text Encoder

After converting text into phoneme form, the Montreal Forced Aligner (MFA) [20] is used to align speech and text data. We then obtain the phoneme sequence corresponding to each word and the duration of each phoneme in order to calculate the time length of each word, referred to as the word boundary. The phoneme sequence is inserted into the Phoneme Encoder to obtain the phoneme embedding E_P, which is then input with the word boundary into the Duration Predictor to obtain the duration of each phoneme at the word level. After applying word pooling to the phoneme embedding and word boundary, the phoneme embeddings within each word are averaged to obtain a word-level word representation. Then, the Word Encoder encodes the word representation into a word embedding E_W (Fig. 2).

There is an enormous difference between speech and text representations in terms of dimensions. The length of the text must be adjusted to match the length of the speech, after which the Length Regulator adjusts the length of the word embedding in accordance with the word-level duration time to match the length in the mel spectrogram E_W.

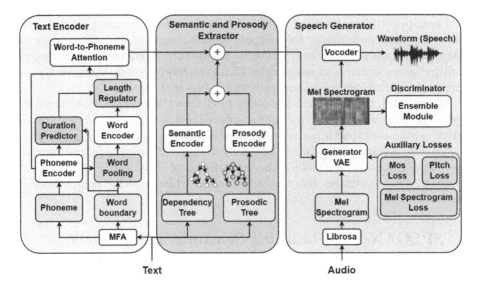

Fig. 1. Model Architecture of SPLGAN-TTS

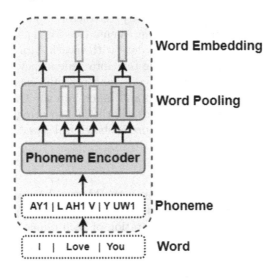

Fig. 2. The process of Word Pooling.

The fine-grained details between words can be enhanced by adding a Word-to-Phoneme Attention mechanism, using E_W as the query and E_P as the key and value. Speech is essentially a time-ordered sequence; therefore, it is crucial that the spectrogram representing the time-varying speech signal align with the time sequence of phonemes in the text. Due to the monotonic nature of text-to-spectrogram alignment, word-level relative positional encoding embeddings are added to both E_W and E_P before being input into the attention module.

For K and V, the positional encoding is $(i/L_W)E_{kv}$, where i indicates the position of the corresponding phoneme in the word w; L_w is the number of phonemes in word w; E_{kv} is a learnable embedding; and $i \in \{0, 1, \ldots, L_w - 1\}$.

For Q, the positional encoding becomes $(j/T_w)E_q$, where j indicates the position of the corresponding frame in the word w; T_w is the number of frames in word w; E_q is another learnable embedding; and $j \in \{0, 1, \ldots, T_w - 1\}$.

3.2 Semantic and Prosody Extractor

Although text encoder provides rich word-level information in single-word representations, semantic and prosodic information extraction are still insufficient to achieve natural and realistic speech synthesis. This can be attributed to the fact that the prosody of speech usually relies on the entire sentence, whereas the rhyme and rhythm of each word varies depending on the context. Many studies [21,22] on text-related problems have shown that the rich structural features of dependency trees can enhance information extraction. And adding prosodic trees can enhance the naturalness of synthesized speech. Therefore, we include dependency tree and prosodic tree in the model.

For semantic and prosodic features, we convert the text data into the forms of a dependency tree and a prosodic tree, which are then input into the Semantic Encoder and Prosody Encoder to obtain the semantic embeddings as E_{Sem} Eq. 1 and prosody embedding E_{Pro} Eq. 2.

To help the model learn relevant information more quickly, we pre-train the Semantic Encoder and Prosody Encoder using node classification and edge classification tasks. After pre-training, we perform end-to-end training of the entire model, including the Semantic Encoder and Prosody Encoder.

$$\mathbf{E}_{Sem} = SemanticEncoder(DependencyTree) \qquad (1)$$

$$\mathbf{E}_{Pro} = ProsodyEncoder(ProsodicTree) \qquad (2)$$

Finally, we sum the results of E_{Sem} and E_{Pro}, and then add this combined result to the final output of the Text Encoder. This final text embedding is then input into the generative model.

$$TextEmbedding = E_{Sem} + E_{Pro} + TextEncoderOutput \qquad (3)$$

3.3 Speech Generator

The obtained text embedding is input into the speech generator for speech synthesis; however, generating speech directly from a text embedding imposes a number of challenges. Thus, we input both the text embedding and the corresponding mel spectrogram into the model during the training phase, to assist in learning the relationship between text and speech more effectively.

To create a lightweight GAN capable of high-quality output, we adopted a VAE architecture for the generator. Note however that the Gaussian distribution typically used by conventional VAEs limits the posterior distribution and suppresses the diversity of the generated outputs. To enhance the complexity of the prior distribution and thereby improve the quality of speech generated by the VAE, we employed normalizing flow in which the simple distribution is transformed into a complex distribution through a series of invertible mappings. The increased complexity of the posterior distribution of the VAE makes it easier for the generative model to learn data distribution and generate data of higher quality. As shown in Fig. 3, the generator outputs a mel spectrogram that corresponds closely to the text. The normalizing flow is trained using the KL-divergence loss, which can be derived as follows:

$$\mathcal{L}_{\mathrm{KL}} = \mathrm{KL}(q_\phi(z \mid x) \parallel p(z)) = \mathbb{E}_{q_\phi(z \mid x)}[\log q_\phi(z \mid x) - \log p(z)] \quad (4)$$

And the generator loss is given by

$$\mathcal{L}_G = -\mathbb{E}_{z \sim p_z(z)}[\log D(G(z))] \quad (5)$$

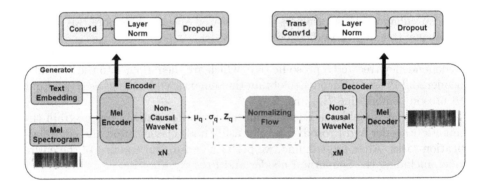

Fig. 3. The architecture of generator.

As shown in Fig. 4, we kept the model as lightweight as possible by employing a multi-length discriminator, which splits the complex discriminator into multiple lightweight sub-discriminators, each of which can focus on a mel spectrogram of a different size. In this manner, the multi-length architecture not only reduces the model size but also enables the capture of multi-scale features.

The losses for all discriminators can be obtained as follows:

$$\mathcal{L}_D = \frac{1}{N} \sum_{i=1}^{N} \mathcal{L}_{D_i} \quad (6)$$

The loss of each discriminator is derived as follows:

$$\mathcal{L}_{D_i} = -\left(\mathbb{E}_{x \sim p_{\mathrm{data}_i}(x)}[\log D_i(x)] + \mathbb{E}_{z \sim p_z(z)}[\log(1 - D_i(G(z)))]\right), \quad i = 1, 2, \ldots, N \quad (7)$$

where N indicates the number of discriminators.

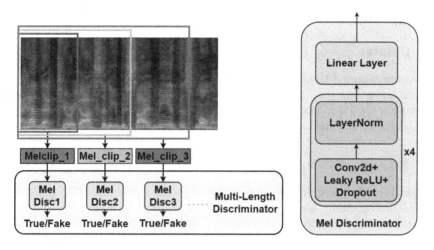

Fig. 4. The architecture of multi-length discriminator.

3.4 Auxiliary Losses Help Training Generator

The proposed model was implemented using multiple auxiliary training losses to enhance generator performance in terms of duration loss, pitch loss, mel spectrogram loss, and MOS loss. The loss formulas are detailed in Eq. 8 ~ 10.

$$\mathcal{L}_{\text{duration}} = \frac{1}{N} \sum_{i=1}^{N} (\log d_{\text{pred},i} - \log d_{\text{target},i})^2 \tag{8}$$

$$\mathcal{L}_{f_0} = \frac{1}{N} \sum_{i=1}^{N} \left| \log f_{0i} - \log \hat{f}_{0i} \right|^2 \tag{9}$$

$$\mathcal{L}_{\text{mel}} = \frac{1}{N} \sum_{i=1}^{N} |\mathbf{m}_{\text{pred},i} - \mathbf{m}_{\text{target},i}| \tag{10}$$

Previous works have shown that generated speech can achieve high naturalness and clarity, but there is still a notable gap in subjective evaluation scores. To address this issue, we augmented our model with MOS loss to facilitate training (see Eq. 11). This involved using a pre-trained MOS score prediction model (UTMOS [23]) to predict the MOS scores of the synthesized speech, which is then fed back to the generator so that it can learn the details necessary to generate realistic speech.

$$\mathcal{L}_{\text{mos}} = \sum_{i=1}^{N} \frac{1}{1 + e^{-(\mathbf{m}_i - 3)}} \tag{11}$$

The final generator training object is shown as following equation, Eq. 12.

$$\mathcal{L}_{\text{Gfinal}} = \mathcal{L}_G + \mathcal{L}_{\text{KL}} + \mathcal{L}_{\text{duration}} + \mathcal{L}_{f_0} + \mathcal{L}_{\text{mel}} + \mathcal{L}_{\text{mos}} \tag{12}$$

4 Experimental Results

4.1 Dataset

This study adopted the LJSpeech dataset, which is commonly used in research on speech synthesis and speech recognition. This dataset contains 13,100 audio files, with a total duration of approximately 24 h. The mean clip duration is 6.57 sec, and the mean words per clip is 17.23.

4.2 Evaluate Metrics

Subjective Metrics. In the field of text-to-speech synthesis, MOS is a subjective metric commonly used to evaluate the naturalness, intelligibility, and listening quality of synthesized speech. Ratings range from 1 to 5, where 1 indicates very poor quality and 5 indicates excellent quality. The MOS variant MOS-Q evaluates speech sound quality, whereas MOS-P focuses on the prosody of speech.

The Comparative Mean Opinion Score (CMOS) is a metric used to compare the differences between two different speech samples. The scores range from -3 to +3, where -3 indicates that the comparative speech is significantly inferior to the main speech, whereas +3 indicates that the comparative speech is significantly better than the main speech. The CMOS variant CMOS-Q compares two speech samples in terms of speech sound quality, whereas CMOS-P compares them in terms of prosody aspects.

Objective Metrics. Mel-Cepstral Distortion (MCD) is an objective metric commonly used in the field of speech synthesis. It measures the Euclidean distance between the Mel-Cepstral Coefficients (MCCs) of two speech samples using Eq. 13. Log-Likelihood Ratio (LLR) is another objective metric commonly used to measure the difference between synthesized speech and a reference sample using Eq. 14.

$$\text{MCD} = \frac{10}{\ln 10}\sqrt{2} \sqrt{\sum_{t=1}^{T}\sum_{m=1}^{L} (c_t[m] - \hat{c}_t[m])^2} \tag{13}$$

$$\text{LLR} = \log \frac{\mathbf{a}^T R \mathbf{a}}{\hat{\mathbf{a}}^T R \hat{\mathbf{a}}} \tag{14}$$

4.3 Experiment Setup

In this experiment, we set the audio sampling rate of the audio to 22050 Hz. Before audio data were input into the model, we sought to maintain stability and predictability during audio processing and prevent data overflow by setting the max wave value to 32768. For training, we used the Adam optimizer with beta1 set to 0.9, beta2 set to 0.98, and a batch size of 64. Dynamic learning rate adjustment was applied with an initial learning rate of 0.05 and an annealing step set to 0.3. All experiments are conducted using a single NVIDIA 3090 GPU.

4.4 Experimental Results of MOS

Performance. It is essential that the quality of speech samples undergoes evaluations using both subjective as well as objective metrics. Table 1 lists the experiment results, where 'Params' indicates the number of parameters and 'RTF' indicates the time required to synthesize 1 s of speech.

The proposed model was compared with both Autoregressive models and Non-Autoregressive models. In terms of MOS, the proposed model significantly outperformed the Autoregressive models, despite a smaller model size and lower RTF. This demonstrates that the proposed architecture makes it possible to learn rich textual information and synthesize high-quality speech using a lightweight model. The proposed model is 30–40% smaller than Fast-Speech2 and GlowTTS; however, it outperformed those models in terms of MOS, MCD, and LLR scores. This demonstrates that the proposed text encoder and semantic and prosody extractor are able to learn textual information, while the auxiliary losses made it possible to fine-tune the speech quality and prosody by fine-tuning the details during speech synthesis.

Table 1. Performance comparison of various TTS models. Note: AR = Autoregressive; NAR = Non-Autoregressive.

Method	MOS	MCD	LLR	#Params.	RTF	Inference Speedup 1	Inference Speedup 2
GT	4.50	0.00	0.00	/	/	/	/
AR Model							
Tacotron2	3.73	23.38	5.42	28.2M	0.115	/	8.3x
TransformerTTS	3.89	20.92	4.81	24.2M	0.955	0.12x	/
NAR Model							
FastSpeech2	4.25	15.53	3.92	27.0M	0.0200	5.75x	47.75x
Glow-TTS	4.14	15.34	3.80	28.6M	**0.0196**	**5.87x**	**48.72x**
DiffSpeech	4.05	17.43	3.31	27.7M	0.171	0.67x	5.58x
SPLGAN-TTS	**4.29**	**14.83**	3.60	**19.2M**	0.0212	5.42x	45.05x

4.5 Experimental Results: MOS-P and MOS-Q

In this section, we focus on evaluating the speech quality and prosody. The results are shown in Table 2. The speech samples synthesized in this study achieve the second best performance compare to other models in MOS-Q, that very close to FastSpeech2. However, from the MOS-P results, we can see that our model outperforms all other models and is even close to the Ground Truth. This demonstrates the benefits of the prosodic tree in learning prosodic information. They also show that incorporating pitch loss for auxiliary training can significantly improve prosodic performance.

Table 2. MOS-Q & MOS-P Scores for Different Methods

Method	MOS-Q	MOS-P
GT	4.54	4.29
AR Model		
Tacotron2	3.47	3.89
TransformerTTS	3.72	3.93
NAR Model		
FastSpeech2	**4.28**	4.21
Glow-TTS	4.18	4.10
DiffSpeech	4.12	3.98
SPLGAN-TTS	4.23	**4.28**

4.6 Ablation Studies

Ablation studies were conducted to assess the effects of the proposed semantic and prosody extractor and auxiliary losses on the quality of synthesized speech, the results of which are listed in Table 3. Note that CMOS was used as the metric in the ablation studies.

Removing the auxiliary losses had a negative effect on the results, particularly in terms of CMOS-Q. This confirms that the auxiliary losses help to improve the naturalness and clarity of the speech samples, thanks to the inclusion of MOS loss and mel loss.

Removing the tree-based architecture had an even more pronounced negative effect on speech results, particularly in terms of prosody performance. This confirms that the addition of the dependency tree enhanced semantic information and that the prosodic tree played an important role in capturing the contextual prosody relationships of each word in the text sentences.

Table 3. CMOS Scores under various Model Settings

Model Setting	CMOS	CMOS-Q	CMOS-P
SPLGAN-TTS	0.00	0.00	0.00
w/o Auxiliary Losses	-0.18	-0.54	-0.22
w/o Semantic and Prosody Extractor	-1.26	-0.31	-0.83

5 Conclusions

This study adopts a semantic and prosodic extractor to enhance the learning of semantic and prosodic information and uses a lightweight generative model. VAE is used for the generator architecture, and a novel normalized flow module is used to enhance the complexity of the distribution generated by the VAE. We also develop a speech discriminator with a multi-length architecture to reduce computational overhead as well as multiple auxiliary losses to assist model training. The proposed model is smaller and more synthetically efficient than existing state-of-the-art models, especially when applied to longer text samples.

References

1. Ren, Y., Hu, C., Tan, X., Qin, T., Zhao, S., Zhao, Z., Liu, T.Y.: FastSpeech 2: fast and high-quality end-to-end text to speech (2022). https://arxiv.org/abs/2006.04558
2. Hallahan, W.I.: DECtalk software: text-to-speech technology and implementation. Digit. Tech. J. **7** 1995. https://api.semanticscholar.org/CorpusID:7643037
3. Yoshimura, T., Tokuda, K., Masuko, T., Kobayashi, T., Kitamura, T.: Simultaneous modeling of spectrum, pitch and duration in hmm-based speech synthesis. In: Sixth European Conference on Speech Communication and Technology (1999)
4. Zen, H., Tokuda, K., Black, A.W.: Statistical parametric speech synthesis. Speech Commun. **51**(11), 1039–1064 (2009)
5. Kim, J., Kim, S., Kong, J., Yoon, S.: Glow-TTS: a generative flow for text-to-speech via monotonic alignment search (2020). https://arxiv.org/abs/2005.11129
6. Shen, J., et al.: Natural TTS synthesis by conditioning Wavenet on MEL spectrogram predictions (2018). https://arxiv.org/abs/1712.05884
7. Ping, W., et al.: Deep voice 3: 2000-speaker neural text-to-speech. In: International Conference on Learning Representations (2018). https://openreview.net/forum?id=HJtEm4p6Z
8. Li, N., Liu, S., Liu, Y., Zhao, S., Liu, M.: Neural speech synthesis with transformer network (2019). https://arxiv.org/abs/1809.08895
9. Wang, Y., et al.: Tacotron: towards end-to-end speech synthesis (2017). https://arxiv.org/abs/1703.10135
10. Łańcucki, A.: Fastpitch: parallel text-to-speech with pitch prediction (2021). https://arxiv.org/abs/2006.06873
11. Lim, D., Jang, W., Park, H., Kim, B., Yoon, J.: JDI-T: jointly trained duration informed transformer for text-to-speech without explicit alignment (2020). https://arxiv.org/abs/2005.07799
12. Miao, C., Liang, S., Chen, M., Ma, J., Wang, S. and Xiao, J.: Flow-TTS: a non-autoregressive network for text to speech based on flow. In: ICASSP 2020 - 2020 IEEE International Conference on Acoustics, Speech and Signal Processing (ICASSP), pp. 7209–7213 (2020). https://doi.org/10.1109/ICASSP40776.2020.9054484
13. Ren, Y.: et al.: Fastspeech: Fast, robust and controllable text to speech. In: Advances in Neural Information Processing Systems, vol. 32 (2019)
14. Setiawan, H., Sperber, M., Nallasamy, U., Paulik, M.: Variational neural machine translation with normalizing flows. arXiv preprint arXiv:2005.13978 (2020)
15. Kingma, D.P., Welling, M.: Auto-encoding variational bayes. arXiv preprint arXiv:1312.6114, 2013
16. Yu, J., Zhang, M., Tao, J., Wang, X.: A novel hmm-based TTS system using both continuous HMMS and discrete HMMS. In: 2007 IEEE International Conference on Acoustics, Speech and Signal Processing-ICASSP 2007, vol. 4, pp. IV–709. IEEE (2007)

17. van den Oord, A., et al.:. WaveNet: a generative model for raw audio (2016). https://arxiv.org/abs/1609.03499
18. Peng, K., Ping, W., Song, Z., Zhao, K.: Non-autoregressive neural text-to-speech (2020).https://arxiv.org/abs/1905.08459
19. Liu, J., Li, C., Ren, Y., Chen, F., Zhao, Z.: DiffSinger: singing voice synthesis via shallow diffusion mechanism (2022). https://arxiv.org/abs/2105.02446
20. McAuliffe, M., Socolof, M., Mihuc, S., Wagner, M., Sonderegger, M.: Montreal forced aligner: trainable text-speech alignment using kaldi. In: Proceedings of the Interspeech 2017, pp. 498–502, (2017). https://doi.org/10.21437/Interspeech.2017-1386
21. Yu, B., Mengge, X., Zhang, Z., Liu, T., Yubin, W., Wang, B.: Learning to prune dependency trees with rethinking for neural relation extraction. In: Proceedings of the 28th International Conference on Computational Linguistics, pp. 3842–3852 (2020)
22. Reichartz, F., Korte, H., Paass, G.: Dependency tree kernels for relation extraction from natural language text. In: Buntine, W., Grobelnik, M., Mladenić, D., Shawe-Taylor, J. (eds.) ECML PKDD 2009. LNCS (LNAI), vol. 5782, pp. 270–285. Springer, Heidelberg (2009). https://doi.org/10.1007/978-3-642-04174-7_18
23. Saeki, T., Xin, D., Nakata, W., Koriyama, T., Takamichi, S., Saruwatari, H.: UTMOS: UTokyo-Sarulab system for Voicemos challenge 2022. arXiv preprint arXiv:2204.02152 (2022)

SSCDUF: Spatial-Spectral Correlation Transformer Based on Deep Unfolding Framework for Hyperspectral Image Reconstruction

Hui Zhao, Na Qi, Qing Zhu$^{(\boxtimes)}$, and Xiumin Lin

College of Computer Science (School of Software Engineering, A National Pilot Software College), Beijing University of Technology, Beijing 100124, China
zhaohui212@emails.bjut.edu.cn, {qina,ccgszq}@bjut.edu.cn

Abstract. Reconstructing Hyperspectral Images (HSIs) from Coded Aperture Snapshot Spectral Imaging (CASSI) is an important yet challenging task. The core issue lies in recovering reliable and detailed 3D HSI cube from 2D measurement. Deep unfolding framework which alternates between solving data subproblems and prior subproblems has made satisfactory progress in HSIs reconstruction task. However, current methods do not fully utilize the spatial spectral prior of HSIs. To solve this problem and further enhance the spectral-spatial representation capabilities in the prior subproblems, we propose a **S**patial-**S**pectral **C**orrelation Transformer Based on **D**eep **U**nfolding **F**ramework (SSCDUF). Specifically, we introduce a multi-scale **S**patial-**S**pectral **C**orrelation Fusion **T**ransformer (SSCT) module that simultaneously utilize the similarity and correlation of spectral features as well as local and non-local spatial features, jointly using spatial and spectral prior to enhance feature representation. Moreover, we further propose an **A**daptive **A**ggregation **S**kip **C**onnection (AASC) module to adaptively aggregate spatial and spectral features in multiple scales. Extensive experimental results on both simulated and real scenes demonstrate that SSCDUF outperforms the state-of-the-art methods in terms of quantitative metrics while maintaining low parameter costs and runtime.

Keywords: Hyperspectral Image Reconstruction · Deep Unfolding Framework · Spatial-Spectral Transformer · Adaptive Aggregation Skip Connection

1 Introduction

Hyperspectral Images (HSIs) capture a wider range of spectral bands compared to ordinary RGB images, enabling them to hold more detailed information about the scene being imaged. Consequently, HSIs are widely used in various fields such as image recognition [26], target detection [10,21], medical image processing [19], and remote sensing [17]. The traditional method of capturing HSI is scanning scene, which is time-consuming and not suitable for moving scene. To overcome this limitation, CASSI [16] modulates 3D HSI into compressed 2D measurement using a coded aperture and disperser, significantly enhancing the convenience of acquiring hyperspectral images. Based on the CASSI system, numerous methods have been developed to reconstruct 3D HSI from 2D measurement, including traditional model-based approaches [6,13,24,28], end-to-end neural network methods [2,8,16–18], and deep unfolding methods [4,5,9,15,20].

Traditional model-based methods exploit various priors to solve the ill-posed inverse problem of HSI reconstruction. Commonly used priors include non-local similarity [6], total variation [24], low-rank properties [13], and sparsity [28]. However, handcrafted priors have limited generalization capabilities, which may lead to a mismatch between the prior assumptions and the problem, addressing different characteristics in various scenarios requires more time and effort.

To address the generalization capabilities problem of model-based method, many end-to-end neural network-based methods [2,8,16–18] have been successfully applied to HSI reconstruction by learning a mapping function, which can reduce the reliance on manually designed a prior knowledge and capture spatial spectral information more efficiently, with better generalization ability, higher reconstruction accuracy and faster processing speed. However, these methods ignore the working principles of CASSI system, so they lack the proven performance, interpretability and flexibility in theory.

Recently, deep unfolding methods [4,5,9,15,20] have achieved significant improvements in HSI reconstruction, which employ a multi-stage network to map measurement to HSI cube, with each stage typically consisting of a data module for linear projection and a prior module for learning underlying denoising priors. The combination of data-driven and prior-driven allows the deep unfolding framework to leverage theoretical proofs with traditional manual prior, as well as the power of deep learning to solve optimization problems.

However, previous deep unfolding methods have three problems: 1. Mainly using CNNs to capture features, which have limitations in capturing remote dependencies; 2. The straightforward utilization of local and global transformers leads to limited sensory fields and high computational costs; 3. Only one of the spatial or spectral dimensions is considered during the reconstruction process, which leads to the loss of spatial or structural correlations with texture or spectral information.

To address the above problems, we propose a multi-scale **S**patial-**S**pectral **C**orrelation Fusion **T**ransformer (SSCT) module and an **A**daptive **A**ggregation **S**kip **C**onnection (AASC) module to effectively extract and fuse features. We integrate these modules into a deep unfolding network and propose a **S**patial-**S**pectral **C**orrelation Transformer Based on **D**eep **U**nfolding **F**ramework (SSC-DUF) for HSI reconstruction. The main contributions of our work are as follows,

-We propose a multi-scale spatial-spectral correlation fusion transformer module, which enhances feature representation for HSI reconstruction in the prior module of the deep unfolding network through spatial-spectral and local-global joint attention.

-We propose an adaptive aggregation skip connection module, which adaptively aggregates spatial-spectral features from upsampling and downsampling processes, recovering low-level information lost during downsampling operations and better enhancing aggregated features.

-We conduct extensive experiments on both simulated and real-world scenarios to demonstrate the effectiveness of our method in HSI reconstruction.

2 Related Work

2.1 Model-Based Methods

Traditional model-based methods require the use of a handcrafted prior to solve the reconstruction problem, usually by separating the data fidelity term from the regularization term to optimize the objective function. Twist [1] proposed a two-step iterative shrinkage/thresholding algorithm to deal with a class of convex unconstrained optimization problems arising from image restoration and other linear inverse problems. Non-local similarity and low-rank regularization are used in [6,7], which takes into account the consideration of the HSI strong remote dependencies between pixels for more accurate results. The sparse-based method [12] relies on the sparse representation of HSIs to make assumptions. In [24], a generalized alternating projection (GAP) method for HSIs compressed perception is proposed using full variational minimization. Although these methods produce satisfactory results in appropriate cases, but their computational efficiency and generalization ability are still worth considering.

2.2 End-to-End Neural Network-Based Methods

Inspired by the excellent performance of deep neural networks, many end-to-end neural networks [2,8,16–18] have achieved satisfactory results in the field of HSI reconstruction. CNN-based approaches have shown advantages in modeling local similarity, for example in [18], λ-net reconstructs HSI through a two-step process. But due to induction bias, CNN-based approaches have some limitations in identifying non-local similarity. Considering the excellent performance of transformer in capturing long-range dependencies, some transformer-based methods have been proposed, such as MST and CST which utilize the multi-head self-attention mechanism to capture the internal similarity of HSI. TSA-Net [16] introduces spatial-spectral self-attention for sequential reconstruction of HSI. Hu et al. [8] introduces a high-resolution dual-domain learning network (HDNet) to solve the spectral compression imaging task. Despite significant progress, these brute force methods are difficult to utilize physical degradation features.

2.3 Interpretable Deep Unfolding Network Based Methods

Deep unfolding methods [4,5,9,15,20,23] have recently achieved impressive results in HSI reconstruction tasks. The core idea of deep unfolding networks is to implement model-driven iterative optimization algorithms by stacking recursive deep neural network (DNN) blocks, this makes the method of deep unfolding methods not only make good use of the traditional manual prior as a theoretical support, but also can make use of the powerful ability of depth learning to solve optimization problems, improve speed and reconstruction quality. Initially, this design was applied to deep plug-and-play methods [25,27], which utilizes pretrained denoisers to implicitly express a prior subproblems as denoising problems. For example, GAP-Net [15] unfolds a generalized alternating projection algorithm and employs a trained convolutional neural network to enhance the reconstruction quality. DGSMP [9] introduces an unfolding model estimation framework that utilizes learned Gaussian scale mixing to improve the performance of the model. DAUHST [4] introduces a novel semi-shuffled Transformer structure that is incorporated into the unfolding framework to further enhance the HSI reconstruction. On the other hand, RDLUF [5] guides cross-stage feature fusion by introducing a pixel adaptive recovery data module and utilizing frequency information. These methods usually take the sensing matrix as the degradation matrix and utilize neural networks to learn the degradation matrix in the data subproblem.

3 Proposed Method

3.1 Problem Formulation

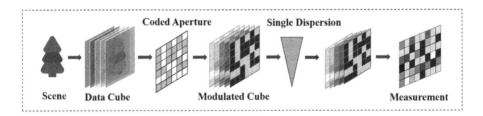

Fig. 1. Coded Aperture Snapshot Spectral Imaging (CASSI) system [16].

Figure 1 shows the coding process of the CASSI system, which is based on the theory of HSI compression to capture compressed measurements containing information covering all bands. In CASSI system, we denote the coded measurement as $\mathbf{y} \in \mathbb{R}^{H \times (W+d(N_\lambda-1))}$, where H, W and N_λ denote the height, width and wavelengths of HSI, respectively. d denote the shifting step in dispersion. According to the principle of CASSI system, the HSI signal $\mathbf{x} \in \mathbb{R}^{H \times (W+d(N_\lambda-1)) \times N_\lambda}$. For the connivance, we can describe the degradation model of CASSI system as,

$$\mathbf{y} = \Phi \mathbf{x} + \mathbf{n}, \qquad (1)$$

where **n** represents the additive noise on the measurement and $\boldsymbol{\Phi}$ is the sensing matrix which is determined by the physical mask. So the aims of HSI reconstruction is to recovering high-quality images **x** from degraded measurements **y**, which is usually an ill-posed problem.

3.2 Deep Unfolding Framework

According to the MAP, HSI reconstruction can estimate the original signal by minimizing the following energy function,

$$\hat{\mathbf{x}} = \arg\min_{\mathbf{x}} \frac{1}{2}\|\mathbf{y} - \boldsymbol{\Phi}\mathbf{x}\|_2^2 + \omega Q(\mathbf{x}), \tag{2}$$

where $\frac{1}{2}\|\mathbf{y} - \boldsymbol{\Phi}\mathbf{x}\|_2^2$ is the data fidelity term and $Q(\mathbf{x})$ is the regularizer term with parameter ω. To obtain the unfolding inference, Eq. (2) uses a version of the quadratic splitting (HQS) algorithm with simple algorithms and fast convergence rate by Introducing auxiliary variable **z**, the constrained optimization problem is reformulated as a subproblem,

$$(\hat{\mathbf{x}}, \hat{\mathbf{z}}) = \arg\min_{\mathbf{x},\mathbf{z}} \frac{1}{2}\|\mathbf{y} - \boldsymbol{\Phi}\mathbf{x}\|_2^2 + \omega Q(\mathbf{z}) + \frac{\mu}{2}\|\mathbf{z} - \mathbf{x}\|_2^2, \tag{3}$$

where μ is the penalty parameter, then **x** and **z** in Eq. (3) can be approximated and optimized by the HQS algorithm by decoupling them into the following two iterative convergence subproblems as follows,

$$\begin{aligned}\mathbf{x}^{k+1} &= \arg\min_{\mathbf{x}} \|\mathbf{y} - \boldsymbol{\Phi}\mathbf{x}\|_2^2 + \mu\|\mathbf{z}^k + \mathbf{x}^k\|_2^2, \\ \mathbf{z}^{k+1} &= \arg\min_{\mathbf{z}} \frac{\mu}{2}\|\mathbf{z}^k - \mathbf{x}^{k+1}\|_2^2 + \omega Q(\mathbf{z}^k),\end{aligned} \tag{4}$$

where k=0,1,...,K-1 index iterations, the data fidelity term **x** has a closed form solution as,

$$\mathbf{x}^{k+1} = (\boldsymbol{\Phi}^T\boldsymbol{\Phi} + \mu\mathbf{I})^{-1}(\boldsymbol{\Phi}^T\mathbf{y} + \mu\mathbf{z}^k), \tag{5}$$

where **I** is the unit matrix, and $\boldsymbol{\Phi}$ is a fat matrix.

We denote μ in k iterations as μ^k. Returning to Eq. (4), $\mathbf{z}^{k+1}$ can be reformulated as,

$$\mathbf{z}^{k+1} = \arg\min_{\mathbf{z}} \frac{1}{2(\sqrt{\omega^{k+1}/\mu^{k+1}})^2}\|\mathbf{z} - \mathbf{x}^{k+1}\|_2^2 + Q(\mathbf{z}). \tag{6}$$

According to the Bayesian probability, Eq. (6) can be equated to denoising the image $\mathbf{x}^{k+1}$ with Gaussian noise of rank $\sqrt{\omega^{k+1}/\mu^{k+1}}$. To facilitate the solution of Eq. 6, we set $\frac{1}{(\sqrt{\omega^{k+1}/\mu^{k+1}})^2}$ as the parameter estimated from CASSI. Let $\alpha^k \stackrel{\text{def}}{=} \mu^k$, $\alpha \stackrel{\text{def}}{=} [\alpha^1,...,\alpha^k]$, $\beta^k \stackrel{\text{def}}{=} \mu^k/\omega^k$, and $\beta \stackrel{\text{def}}{=} [\beta^1,...,\beta^k]$. This can then be expressed as an iterative scheme,

$$(\alpha, \beta) = \theta(\mathbf{y}, \boldsymbol{\Phi}), \mathbf{x}^{k+1} = \mathcal{P}(\mathbf{y}, \mathbf{z}^k, \alpha^{k+1}, \boldsymbol{\Phi}), \mathbf{z}^{k+1} = \mathcal{D}(\mathbf{x}^{k+1}, \beta^{k+1}), \tag{7}$$

where θ denotes the parameter estimator with the compression measurements $\mathbf{y}$ and the sensing matrix $\mathbf{\Phi}$ of the CASSI system as inputs. $\mathcal{P}$ denotes the linear projection and $\mathcal{D}$ denotes the Gaussian denoiser, respectively. $\mathbf{z}^0$ is initialized by connecting the shifted $\mathbf{y}$ with $\mathbf{\Phi}$ through a $conv1 \times 1$ (1×1 kernel convolution).

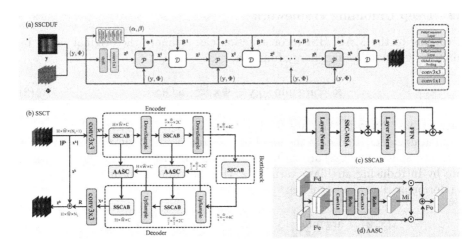

Fig. 2. Schematic diagram of SSCDUF. (a) Architecture of our K stages SSCDUF. α and β are estimated from the $\mathbf{y}$ and the $\mathbf{\Phi}$, used in subsequent iterative learning. $\mathcal{P}$ and $\mathcal{D}$ denote the linear projection and denoising network. (b) Schematic diagram of SSCT. (c) Components of SSCAB. (d) Schematic diagram of AASC.

3.3 Framework

Based on the above, we design a new transformer with spatial-spectral correlation fusion for HSI reconstruction based on deep unfolding framework. Figure 2 illustrates the general framework of our proposed approach, which consists of K stages, each containing a linear projection module $\mathcal{P}$ and a denoising network $\mathcal{D}$. The measurement $\mathbf{y}$ and the sensing matri $\mathbf{\Phi}$ are input into the framework, and they are connected by a $conv1 \times 1$ to get the initial $\mathbf{z}^0$. Then, $\mathbf{z}^0$ passes through the $\mathcal{P}$ to produce x, which is input into the $\mathcal{D}$. After K iterations between the $\mathcal{P}$ and $\mathcal{D}$, the final reconstructed HSI is obtained. $\mathbf{\Phi}$ and $\mathbf{y}$ are estimated by the θ to obtain α and β for subsequent guidance of iterative learning and to provide noise level information.

In order to solve the problems of previous methods, we propose a **S**patial-**S**pectral **C**orrelation Fusion **T**ransformer (SSCT) to act as a noise canceller, whose network architecture is shown in Fig. 2 (b).

The SSCT built by the basic unit **S**patial-**S**pectral **C**orrelation **A**ttention **B**lock (SSCAB). First of all, SSCT joins the $\mathbf{x}^k$ and β^k by $conv3 \times 3$ and maps to the feature $\mathbf{X}^0 \in \mathbb{R}^{H \times \hat{W} \times C}$, where $\hat{W} = W + d(N_\lambda - 1)$. Then, $\mathbf{X}^0$ is embedded into depth feature $\mathbf{X}^d \in \mathbb{R}^{H \times \hat{W} \times C}$ through encoder, bottleneck and decoder. Each layer of the encoder and decoder contains SSCABs and up or down sampling modules. As Fig. 2 (c) shown, SSCAB consists of SSC-MSA, feed-forward network and two layers of normalization. Finally, the residual image $\mathbf{R} \in \mathbb{R}^{H \times \hat{W} \times N_\lambda}$ is generated by $conv 3 \times 3$ operation of $\mathbf{X}^d$. The output de-noising image $\mathbf{z}^k$ is obtained by the sum of $\mathbf{x}^k$ and $\mathbf{R}$.

Spatial-Spectral Correlation Multi-head Self-attention(SSC-MSA). The most important element of SSCAB is the SSC-MSA, which is depicted in Fig. 3. The input label is $\mathbf{X}_{in} \in \mathbb{R}^{H \times \hat{W} \times C}$, and features are respectively extracted by spatial and spectral branches.

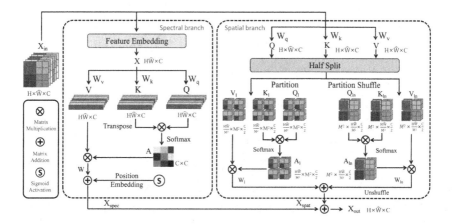

Fig. 3. Spatial-Spectral Correlation Multi-head Self-Attention (SSC-MSA).

In spatial branch, the $\mathbf{X}_{in}$ is linearly projected to the query Q, key K, value $V \in \mathbb{R}^{H \times \hat{W} \times C}$ through the learnable parameters $W_Q, W_K, W_V \in \mathbb{R}^{C \times C}$. Then Q, K, V are divided into two equal parts along the channel dimension, as $Q = [Q_l, Q_{nl}], K = [K_l, K_{nl}], V = [V_l, V_{nl}]$. After that, $Q_l, K_l, V_l \in \mathbb{R}^{H \times \hat{W} \times \frac{C}{2}}$ and $Q_{nl}, K_{nl}, V_{nl} \in \mathbb{R}^{H \times \hat{W} \times \frac{C}{2}}$ are sent into the local branch and non-local branch to capture the local content and models the non-local dependencies. In both branches, Q, K, V are divided into non-overlapping windows of M × M, and then reshaped into $\mathbb{R}^{\frac{H\hat{W}}{M^2} \times M^2 \times \frac{C}{2}}$, it is worth noting that for the non-local branch, the shape of Q_{nl}, K_{nl}, V_{nl} is changed from $\mathbb{R}^{\frac{H\hat{W}}{M^2} \times M^2 \times \frac{C}{2}}$ to $\mathbb{R}^{M^2 \times \frac{H\hat{W}}{M^2} \times \frac{C}{2}}$ to disrupt the position of the tokens and create dependencies between windows. Then, along the channel, Q, K, V is split into N heads with the dimensions $d_h = \frac{C}{2N}$. Figure 3 depicts the case of N = 1 and some details are omitted for simplification.

Self-attention A_l^i and A_{nl}^i is calculated inside each head as,

$$A_l^i = \text{Softmax}(\frac{Q_l^i K_l^{i^T}}{\sqrt{d_h}} + P_l^i)V_l^i, i = 1, ..., N,$$
$$A_{nl}^i = \text{Softmax}(\frac{Q_{nl}^i K_{nl}^{i^T}}{\sqrt{d_h}} + P_{nl}^i)V_{nl}^i, i = 1, ..., N, \quad (8)$$

where $P_j^i \in \mathbb{R}^{M^2 \times M^2}$ are learnable parameters embedding the position information. Subsequently, $A_{nl}^i \in \mathbb{R}^{M^2 \times \frac{H\hat{W}}{M^2} \times d_h}$ is unshuffled by being transposed to shape $A_{nl}^i \in \mathbb{R}^{\frac{H\hat{W}}{M^2} \times M^2 \times d_h}$. Then the output of the two branches are aggregated by linear projection as,

$$\text{SSC-MSA}_{\text{spatial}}(\mathbf{X}_{\text{in}}) = \sum_{i=1}^{N} A_l^i W_l^i + \sum_{i=1}^{N} A_{nl}^i W_{nl}^i, \quad (9)$$

where $W_l^i, W_{nl}^i \in \mathbb{R}^{d_h \times C}$ refer to learnable parameters. Reshape the result of Eq. (9), we can obtain the output of spatial branch $\mathbf{X}_{\text{spat}} \in \mathbb{R}^{H \times \hat{W} \times C}$.

In spectral branch, the input $X_{\text{in}} \in \mathbb{R}^{H \times W \times C}$ is reshaped to $X \in \mathbb{R}^{H\hat{W} \times C}$, and then X is linearly projected to query Q, key K, value $V \in \mathbb{R}^{H\hat{W} \times C}$ through the learnable parameters $W_Q, W_K, W_V \in \mathbb{R}^{C \times C}$, then we split Q, K, V into N heads along the spectral dimension, so that $Q = [Q_1, ..., Q_N]$, $K = [K_1, ..., K_N]$ and $V = [V_1, ..., V_N]$. The dimension of each head is $d_h = \frac{C}{N}$. Figure 3 depicts the case of N = 1 and some details are omitted for simplification. Unlike the MSA for the spatial branch, the spectral branch treats each spectral representation as a token and computes the self-note for each head$_j$,

$$A_j = \text{Softmax}(\tau_j K_j^T Q_j), \text{head}_j = V_j A_j, j = 1, ..., N, \quad (10)$$

where K_j^T represents the transposed matrix of K_j. τ_j is a learnable parameter which is used to reweight the matrix multiplication K_j^T. Then, N head outputs are connected along the spectrum, and positions are added for embedding,

$$\text{SSC-MSA}_{\text{spectral}}(\mathbf{X}_{\text{in}}) = (\underset{j=1}{\overset{N}{\text{Concat}}}(\text{head}_j))W + f_p(V), \quad (11)$$

where $W \in \mathbb{R}^{C \times C}$ is a learnable parameter and $f_p(\cdot)$ is a function of generating position embedding. Finally, we reshaped Eq. 11, the output feature map $\mathbf{X}_{\text{spec}} \in \mathbb{R}^{H \times W \times C}$ is obtained.

Finally, we add the output map $\mathbf{X}_{\text{spec}}$ and $\mathbf{X}_{\text{spat}}$ to obtain the final output $\mathbf{X}_{\text{out}}$.

Adaptive Aggregation Skip Connection(AASC). Generally, in U-net, the fusion of the features of the same depth encoder and decoder is realized by the additive skip connection, and the shallow features of the encoder and the high-level features of the decoder have the same weight. But this simple addition operation does not take into account that the information density of different spectral bands and characteristic channels is different, in fact, we may need to preserve more low-level features for less noisy spectral bands to preserve detail, whereas for noise spectral bands we may want to preserve more high-level features to properly remove all noise. To solve this problem, we propose the AASC that explicitly weights features of different sources using attention weights calculated from two convolution layers, which provides an efficient way to recover low-level information lost during an encoder-decoder subsampling operation. Supposing features of encoder and decoder of SSCT at the ith depth level are $F^i e$ and $F^i d$, respectively, denoting the weight of ith depth level as M_i, the AASC can be described as the following formula,

$$F = \text{LeakyReLU}(\text{Conv1} \times 1([F^i d, F^i e])), \quad (12)$$

$$Mi = \sigma(\text{Conc3} \times 3(F)), \quad (13)$$

$$F^i o = (1 - M_i) \odot F^i d + Mi \odot F^i e, \quad (14)$$

where σ represents the sigmoid function. The structure diagram of our AASC is shown in Fig. 2 (d), where the AASC adaptively enhances more important spatial location and spectral band features through element gating, resulting in better reconstruction quality.

4 Experiments

4.1 Experiment Setup

In accordance with previous works [3,8,9,15,16], we conduct both simulation and real experiments by adopting 28 wavelengths ranging from 450 nm to 650 nm for HSIs. In the implementation, we set NUM_BLOCK to 1 to get the base version of SSCDUF, and change NUM_BLOCK to 2 to get SSCDUF++ , which has more SSC-MSA modules than SSCDUF and achieves better performance. NUM_BLOCK is not set higher due to hardware device limitations.

Datasets. For the simulated experiment, we employ two commonly used HSI datasets, CAVE and KAIST, for both training and testing. The KAIST dataset includes 30 HSIs with a spatial resolution of 2704×3376 and a spectral dimension of 31. The CAVE dataset contains 32 HSIs, each with dimensions of 512×512×31. Following the settings in TSA-Net [16], we use the CAVE dataset for training and select 10 scenes from the KAIST dataset for testing. The size of each HSI patch is set to 256×256×28. For the real data experiment, five real HSIs collected by the CASSI system developed in [16] are used for testing. Each testing sample has dimensions of 660×660×28.

Evaluation Metrics. The effectiveness of HSI restoration methods will be evaluated using PSNR and SSIM [22].

Table 1. Results of our method and other SOTA methods on 10 simulated scenes, with the best results shown in bold.

Methods	Params	GFLOPs	s1	s2	s3	s4	s5	s6	s7	s8	s9	s10	Avg
GAP-TV [24]	-	-	26.82	22.89	26.31	30.65	23.64	21.85	23.76	21.98	22.63	23.10	24.36
			0.754	0.610	0.802	0.852	0.703	0.663	0.688	0.655	0.682	0.584	0.669
λ-Net [18]	62.64M	117.98G	30.10	28.49	27.73	37.01	26.19	28.64	26.47	26.09	27.50	27.13	28.53
			0.849	0.805	0.87	0.934	0.817	0.853	0.806	0.831	0.826	0.816	0.841
TSA-Net [16]	44.25M	110.06G	32.03	31.00	32.25	39.19	29.39	31.44	30.32	29.35	30.01	29.59	31.46
			0.892	0.858	0.915	0.953	0.884	0.908	0.878	0.888	0.890	0.874	0.894
HDNet [8]	2.37M	154.76G	35.14	35.67	36.03	42.3	32.69	34.46	33.67	32.48	34.89	32.38	34.97
			0.935	0.940	0.943	0.969	0.946	0.952	0.926	0.941	0.942	0.937	0.943
MST-L [3]	2.03M	28.15G	35.4	35.87	36.51	42.27	32.77	34.8	33.66	32.67	35.39	32.5	35.18
			0.941	0.944	0.953	0.973	0.947	0.955	0.925	0.948	0.949	0.941	0.948
CST-L [2]	3.00M	40.01G	35.96	36.84	38.16	42.44	33.25	35.72	34.86	34.34	36.51	33.09	36.12
			0.949	0.955	0.962	0.975	0.955	0.963	0.944	0.961	0.957	0.945	0.957
DAUHST-2stg [4]	1.40M	18.44G	35.93	36.7	37.96	44.38	34.13	35.43	34.78	33.65	37.42	33.07	36.34
			0.943	0.946	0.959	0.977	0.954	0.957	0.940	0.950	0.955	0.940	0.952
D^{2}PL-Net [23]	7.65M	84.71G	35.83	36.97	37.72	43.93	33.44	35.59	34.54	33.44	36.43	33.04	36.09
			0.949	0.956	0.961	0.983	0.957	0.965	0.938	0.958	0.954	0.950	0.957
RDLUF-3stg [5]	1.90M	231.09G	36.67	38.48	40.63	46.04	34.63	36.18	35.85	34.37	38.98	33.73	37.56
			0.953	0.965	0.971	0.986	0.963	0.966	0.951	0.963	0.966	0.950	0.963
DADF-Net [23]	19.35M	76.62G	36.95	38.90	39.47	45.39	35.22	36.56	36.17	34.41	39.29	34.21	37.66
			0.959	0.971	0.97	0.987	0.969	0.97	0.952	0.968	0.972	0.958	0.967
Ours-2stg	1.54M	20.14G	36.20	36.93	38.95	44.28	34.23	35.55	35.28	33.87	37.44	33.13	36.59
			0.945	0.950	0.963	0.976	0.956	0.959	0.945	0.952	0.956	0.942	0.959
Ours-3stg	2.28M	30.14G	36.68	37.93	39.45	44.08	34.3	36.03	35.92	34.5	38.44	33.8	37.13
			0.955	0.962	0.967	0.983	0.963	0.967	0.951	0.962	0.965	0.952	0.962
Ours++-2stg	2.66M	38.2G	36.67	37.76	39.62	45.38	34.64	36.21	36.06	34.24	38.12	33.77	37.25
			0.957	0.963	0.972	0.986	0.964	0.969	0.955	0.964	0.965	0.952	0.964
Ours++-3stg	3.97M	55.7G	**37.26**	**39.07**	**41.26**	**46.69**	**35.71**	**36.95**	**37.27**	**35.26**	**39.38**	**34.46**	**38.35**
			0.963	**0.973**	**0.977**	**0.991**	**0.973**	**0.974**	**0.963**	**0.969**	**0.971**	**0.959**	**0.971**

Implementation. Our SSDUT is implemented by Pytorch and trained with Adam optimizer [11] ($\beta_1 = 0.9$ and $\beta_2 = 0.999$) using the Cosine Annealing scheme [14] for 300 epochs on an RTX 3090 GPU. During training, the learning rate is 4×10^{-4} and the batch size is 5. In different stages, the weight of $\mathcal{D}$ are not shared. Enhancement of data through random rotations and flips. The training goal is to minimize RMSE between the reconstructed image and ground-truth as much as possible.

4.2 Quantitative Results

Table 1 compares the results of SSCDUF and 10 SOTA methods including the model-based method GAP-TV [24], five end to-end methods (λ-Net [18], TSA-Net [16], HDNet [8], MST [3], CST [2]), and four deep unfolding methods (DAUHST [4], RDLUF [5], D^{2}PL-Net [11], and DADF-Net [23]) on 10 simulation scenes. To ensure fair comparisons, all algorithms are tested with the same settings as [3,9]. From the quantitative results, we can see that our SSCDUF++ achieves the best results in terms of PSNR and SSIM in all scenes with lower computational complexity and parameters. When compared with the recent SOTA

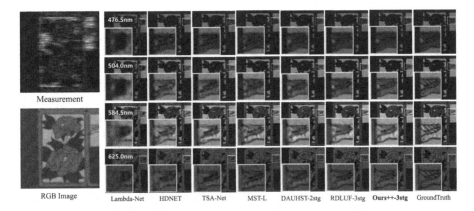

Fig. 4. Visual comparisons of simulation HSI reconstruction of our method and other SOTA methods on the KAIST dataset of scene 7 with 4 spectral channels.

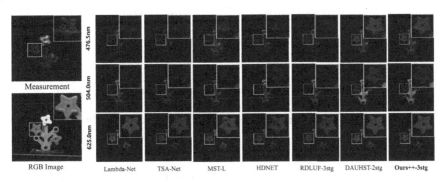

Fig. 5. Visual comparisons of real HSI reconstruction of our method and other SOTA methods on scene 1 with 3 spectral channels.

method RDLUF-3stg, our 3-stg SSCDUF++ outperforms RDLUF-3stg by 0.79 dB but only costs 24.1% (55.7/231.09) GFLOPs.

4.3 Qualitative Results

Simulation HSI Reconstruction. Figure 4 shows the subjective results of the simulation HSI reconstruction of our 3-stage SSCDUF++ compared with other SOTA methods. We choose 4 spectral channels of scene 7, and zoom in the results on a portion for better viewing and analysis. It can be seen that our 3-stage SSCDUF++ achieves excellent visual effects with clearer texture and fewer artifacts, while maintaining spatial smoothness compared with other methods. In contrast, previous approaches can produce overly smooth results that affect fine-grained structures, or bring color artifacts and textures that are not present in ground truth.

Real HSI Reconstruction. For a fair comparison, our 3-stage SSCDUF++ follow the same setting as [4], retrain on the CAVE and KAIST datasets jointly and test on real measurements. The visual result of the real-world reconstruction of the HSI is shown in Fig. 5. Most methods restore blurred images and generate incomplete responses, but our method can successfully restore clear details.

4.4 Ablation Study

To verify the effectiveness of our proposed structure, we conduct an ablation study using the 2-stage SSCDUF method. All evaluations are performed on simulated datasets. Experimental on ablation study are conducted on four sets, including model **A** with only the spectral branch, model **B** with only the spatial branch, model **C** with the spatial branch and the spectral branch, and model **D** with all the modules, including AASC, to investigate the impact of each component on higher performance. As shown in Table 2, when we combine spatial branch and spectral branch (model **C**), the result is improved by 0.67 dB and 0.33 dB compared to spatial-only (model **B**) and spectral-only (model **A**). When we add the AASC module (model **D**), both the SSIM and PSNR are improved, these results demonstrate the effectiveness of our SSCAB and AASC.

Table 2. Ablation study of different modules.

Model	Spectral branch	Spatial branch	AASC	Params	GFLOPs	PSNR	SSIM
A	✓	-	-	0.93	19.08	35.80	0.953
B	-	✓	-	1.4	18.44	36.14	0.952
C	✓	✓	-	1.63	18.76	36.47	0.956
D	✓	✓	✓	1.54	20.14	36.59	0.959

5 Conclusion

In this paper, we propose a spatial-spectral correlation fusion transformer based on deep unfolding framework for HSI reconstruction. Our method aims to solve the existing problems in the previous deep deep unfolding works, that only one of spatial or spectral dimension is considered. So we propose a multi-scale spatial-spectral correlation fusion transformer to extract spatial and spectral information in a local, non-local manner, jointly using spatial and spectral prior to enhance feature representation. Moreover, the skip connection in spatial-spectral correlation fusion transformer is improved by adaptive aggregation skip connection, which can aggregates the features from encoder and decoder with different importance weight. Comprehensive experiments show that our method surpasses the performance of other SOTA HSI reconstruction methods.

Acknowledgments. This work is supported by the Beijing Natural Science Foundation (Project No. 4232017), the National Natural Science Foundation of China under Grant 61906009, the Scientific Research Common Program of Beijing Municipal Commission of Education KM202010005018, and the International Research Cooperation Seed Fund of Beijing University of Technology (Project No. 2021B06).

References

1. Bioucas-Dias, J., Figueiredo, M.: A new TwIST: two-step iterative shrinkage/thresholding algorithms for image restoration. IEEE Trans. Image Process. **16**, 2992-3004 (Dec 2007)
2. Cai, Y., et al.: Coarse-to-fine sparse transformer for hyperspectral image reconstruction. In: Avidan, S., Brostow, G., Cissé, M., Farinella, G.M., Hassner, T. (eds.) Computer Vision – ECCV 2022, pp. 686–704. Springer, Cham (2022). https://doi.org/10.1007/978-3-031-19790-1_41
3. Cai, Y., et al.: Mask-guided spectral-wise transformer for efficient hyperspectral image reconstruction. In: 2022 IEEE/CVF Conference on Computer Vision and Pattern Recognition (CVPR), pp. 17481–17490 (2022)
4. Cai, Y., et al.: Degradation-aware unfolding half-shuffle transformer for spectral compressive imaging. In: Advances in Neural Information Processing Systems, vol. 35, pp. 37749–37761 (2022)
5. Dong, Y., Gao, D., Qiu, T., Li, Y., Yang, M., Shi, G.: Residual degradation learning unfolding framework with mixing priors across spectral and spatial for compressive spectral imaging. In: 2023 IEEE/CVF Conference on Computer Vision and Pattern Recognition (CVPR), pp. 22262–22271 (2022)
6. Fu, Y., Zheng, Y., Sato, I., Sato, Y.: Exploiting spectral-spatial correlation for coded hyperspectral image restoration. In: 2016 IEEE Conference on Computer Vision and Pattern Recognition (CVPR), pp. 3727–3736 (2016)
7. He, W., et al.: Non-local meets global: an iterative paradigm for hyperspectral image restoration. IEEE Trans. Pattern Anal. Mach. Intell. **4**, 2089–2107 (2022)
8. Hu, X., et al.: HDNET: high-resolution dual-domain learning for spectral compressive imaging. 2022 IEEE/CVF Conference on Computer Vision and Pattern Recognition (CVPR), pp. 17521–17530 (2022)
9. Huang, T., Dong, W., Yuan, X., Wu, J., Shi, G.: Deep gaussian scale mixture prior for spectral compressive imaging. In: 2021 IEEE/CVF Conference on Computer Vision and Pattern Recognition (CVPR), pp. 16211–16220 (2021)
10. Kim, M.H., et al.: 3D imaging spectroscopy for measuring hyperspectral patterns on solid objects. ACM Trans. Graph. **31**, 1–11 (2012)
11. Kingma, D., Ba, J.: Adam: a method for stochastic optimization. arXiv: Learning,arXiv: Learning (2014)
12. Lin, X., Liu, Y., Wu, J., Dai, Q.: Spatial-spectral encoded compressive hyperspectral imaging. ACM Trans. Graph. **33**, 1–11 (2014)
13. Liu, Y., Yuan, X., Suo, J., Brady, D.J., Dai, Q.: Rank minimization for snapshot compressive imaging. IEEE Trans. Pattern Anal. Mach. Intell. **41**, 2990–3006 (2019)
14. Loshchilov, I., Hutter, F.: SGDR: stochastic gradient descent with warm restarts. arXiv: Learning,arXiv: Learning (2016)
15. Meng, Z., Jalali, S., Yuan, X.: Gap-net for snapshot compressive imaging. Cornell University - arXiv, Cornell University - arXiv (2020)

16. Meng, Z., Ma, J., Yuan, X.: End-to-End low cost compressive spectral imaging with spatial-spectral self-attention. In: Vedaldi, A., Bischof, H., Brox, T., Frahm, J.-M. (eds.) ECCV 2020. LNCS, vol. 12368, pp. 187–204. Springer, Cham (2020). https://doi.org/10.1007/978-3-030-58592-1_12
17. Meng, Z., Qiao, M., Ma, J., Yu, Z., Xu, K., Yuan, X.: Snapshot multispectral endomicroscopy. Opt. Lett. **45**, 3897–3900 (2020)
18. Miao, X., Yuan, X., Pu, Y., Athitsos, V.: lambda-Net: reconstruct hyperspectral images from a snapshot measurement. In: 2019 IEEE/CVF International Conference on Computer Vision (ICCV), pp. 4058–4068 (Oct 2019)
19. Uzkent, B., Rangnekar, A., Hoffman, M.J.: Aerial vehicle tracking by adaptive fusion of hyperspectral likelihood maps. In: 2017 IEEE Conference on Computer Vision and Pattern Recognition Workshops (CVPRW), pp. 233–242 (2017)
20. Wang, L., Sun, C., Fu, Y., Kim, M.H., Huang, H.: Hyperspectral image reconstruction using a deep spatial-spectral prior. In: 2019 IEEE/CVF Conference on Computer Vision and Pattern Recognition (CVPR), pp. 8024–8033 (2019)
21. Wang, X., Zhao, L., Wu, W., Jin, X.: MCANet: multiscale cross-modality attention network for multispectral pedestrian detection. In: MultiMedia Modeling, pp. 41–53 (2023). https://doi.org/10.1007/978-3-031-27077-2_4
22. Wang, Z., Bovik, A., Sheikh, H., Simoncelli, E.: Image quality assessment: from error visibility to structural similarity. IEEE Trans. Image Process. **13**, 600–612 (2004)
23. Xu, P., Liu, L., Zheng, H., Yuan, X., Xu, C., Xue, L.: Degradation-aware dynamic fourier-based network for spectral compressive imaging. IEEE Trans. Multimedia **26**, 2838–2850 (2024)
24. Yuan, X.: Generalized alternating projection based total variation minimization for compressive sensing. In: 2016 IEEE International Conference on Image Processing (ICIP), pp. 2539–2543 (2016)
25. Yuan, X., Liu, Y., Suo, J., Dai, Q.: Plug-and-play algorithms for large-scale snapshot compressive imaging. In: 2020 IEEE/CVF Conference on Computer Vision and Pattern Recognition (CVPR), pp. 1444–1454 (2020)
26. Zhang, F., Du, B., Zhang, L.: Scene classification via a gradient boosting random convolutional network framework. IEEE Trans. Geosci. Remote Sens. **54**, 1793–1802 (2016)
27. Zhang, K., Zuo, W., Zhang, L.: Deep plug-and-play super-resolution for arbitrary blur kernels. In: 2019 IEEE/CVF Conference on Computer Vision and Pattern Recognition (CVPR), pp. 1671–1681 (2019)
28. Zhang, S., Dong, Y., Fu, H., Huang, S.L., Zhang, L.: A spectral reconstruction algorithm of miniature spectrometer based on sparse optimization and dictionary learning. Sensors **18**, 644 (2018)

SSDL: Sensor-to-Skeleton Diffusion Model with Lipschitz Regularization for Human Activity Recognition

Nikhil Sharma[1(✉)], Changchang Sun[2], Zhenghao Zhao[2], Anne Hee Hiong Ngu[3], Hugo Latapie[4], and Yan Yan[2]

[1] Department of Computer Science, Illinois Institute of Technology, Chicago, USA
nsharma20@hawk.iit.edu
[2] Department of Computer Science, University of Illinois Chicago, Chicago, USA
[3] Department of Computer Science, Texas State University, San Marcos, USA
[4] Cisco Research, San Jose, USA

Abstract. Human Action Recognition (HAR) has recently achieved significant success through the analysis of human behavior using non-visual data (e.g., sensor data) and visual data (e.g., skeleton data). However, sensor-based methods face challenges due to the inherent limitations of sensor data, including the absence of 3D body pose information, high volatility, and vulnerability to noise. Meanwhile, skeleton-based methods, while effective due to their rich spatial and temporal information, are constrained by the stringent requirements for data acquisition and problems like occlusion, limiting their feasibility in real-world outdoor scenarios. Therefore, to solve these challenges, we resort to the cross-modal generation strategy and aim to generate hard-to-collect but information-rich skeleton data conditioned on easy-to-monitor sensor data. In our work, we propose a novel Sensor-to-Skeleton Diffusion Model with Lipschitz Regularization, named SSDL. Specifically, we first design an Angular Variation module and extract angular variation information of joint movements with time information. Subsequently, noise is added to the skeleton key points and angular variation during the forward diffusion process. To address noisy sensor data and improve training stability, we incorporate Lipschitz regularization with the diffusion model's loss to prevent overfitting. We verify the generalizability and effectiveness of our methods on two benchmark multimodal human action datasets: UTD-MHAD, Berkeley-MHAD, and SmartFall-MHAD dataset. Extensive results demonstrate the superiority of leveraging generated skeleton information conditioned on the sensor data for accurate human activity recognition with limited computational demands. Code will be available at https://github.com/nikhiliit/SSDL.

Keywords: Human Activity Recognition · Diffusion Models · Angular Variations · Lipschitz Continuity

1 Introduction

Recently, **H**uman **A**ction **R**ecognition (HAR) task has gained increasing attention due to its vast potential in different domains like human-robot interaction,

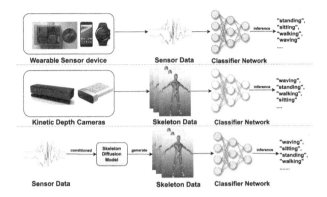

Fig. 1. Comparison of various human activity recognition (HAR) systems. Form top to bottom are systems that (1) Wearable sensor-based devices are used for accelerometer/gyroscopic readings to be fed for identifying the patterns in the velocity regarding each activity, (2) the skeleton body pose of an individual is collected using Kinetic depth camera, and (3) Our proposed diffusion method where sensor data collected from wearable devices are used to generate the skeleton body pose with help of a Skeleton Diffusion Model.

healthcare, and sports [3]. According to the type of data input as shown in Fig. 1, HAR systems can be roughly classified into two lines: non-visual modality-based system and visual modality-based one. Specifically, non-visual modalities [7] encompass sensor data such as accelerometer reading from wearable devices, gyroscopes, and Wi-Fi signals. In contrast, visual modality-based methods [7,38] utilize data such as RGB video, depth information, and skeletal representations to capture and analyze human actions using detailed visual cues.

Benefiting from its small footprint and widespread availability on numerous inexpensive sensors, accelerometer data collected from wearable sensor devices have propelled notable progress in the development of efficient, low-power, and cost-effective human activity recognition systems [4], where potential threats and abnormal activities can be monitored in real-time and preventive measures can be taken in advance to ensure the safety and well-being of individuals. Accordingly, it gives rise to the emerging exploration of deep learning techniques to handle such time-series data, including Convolution Neural Networks (CNNs) [8,27], Recurrent Neural Networks (RNNs) [26], Long Short-Term Memory (LSTM) [37], and Ensemble methods [15]. In contrast, vision-based HAR systems have made significant achievements due to their remarkable advantages of rich contextual information, such as body poses and joint movements captured from camera devices. Notably, skeleton modality can eliminate the privacy concerns that occurred in the pure video-based methods [18] and encode the trajectories of human body joints. This allows for effective capture of temporal dynamics and spatial relationships among joints, especially with devices like the Kinect camera [36], enabling continuous 3D movement characterization. Additionally, the skeleton modality's resistance to background variations has attracted significant research interest in developing robust systems.

Although existing sensor-based and skeleton-based efforts have achieved compelling success, their performance is still limited by the inherent characteristics of both the sensor and skeleton data. On the one hand, despite the sensor data being easily accessible using wearable devices, they only provide limited information and always lack intrusive detailed spatial information for recognizing complex real-world activities [1]. Therefore, the activity recognition performance can be decreased, and overfitting is likely to occur during the training process. Additionally, acquiring skeleton data is challenging and impractical for outdoor setups due to the need for controlled lighting and occlusion, making it unsuitable for real-time or mobile HAR applications [7] Meanwhile, multiple frames need to be processed to gain comprehensive spatial and temporal insights from skeleton data, leading to increasing computational requirements and potential latency.

How to leverage the advantages of both sensor data and skeleton data to solve the wearable HAR problem? Our approach leverages the cross-modal generation capabilities of generative models to produce skeleton data rich in spatial and joint information from easily accessible sensor data. This eliminates the need for complex hardware setups to capture detailed joint movements. The diffusion model, originally introduced by Ho et al. [11], serves as the foundation for this method. In fact, they have proven effective in generating diverse human actions based on text prompts [39]. Above all, in our work, we propose a novel *Sensor-to-Skeleton Diffusion Model with Lipschitz Regularization*, named SSDL. Specifically, in the forward diffusion process, we first extract the angular information of joint movements with time information from the Angular Variation module, and then consistent noise is added to the skeleton key points as well as the angular variation step by step. Hence, our skeleton diffusion model can focus on learning joint-specific movements rather than other less relevant key points Additionally, to tackle potential noise collected from wearable devices, we introduce Lipschitz regularization [20,21] as an underlying function of sensor data. The variability of the diffusion model outputs can be controlled in response to the conditioned sensor data fluctuations and the generation process can also be stabilized by compensating for noise present in the conditioning sensor data. Our main contributions can be summarized in fold:

- To the best of our knowledge, we are the first to recognize human activity based on the generated skeleton key points conditioned on the sensor data collected from the wearable devices. We used sensor data from wearable devices as a condition for the diffusion model to generate synthetic skeleton data for HAR.
- An effective loss function based on angular variations and Lipschitz regularization is added to the noise estimation loss of diffusion model to train a simple U-Net architecture, where joint-specific movements can be well learned and the generation progress can be effectively stabilized.
- Our proposed framework outperforms state-of-the-art HAR models and achieves significant performance improvement on the three MHAD datasets. As a byproduct, we will release the datasets, codes, and involved parameters to benefit other researchers and the source codes are available at here now.

2 Related Work

2.1 Human Activity Recognition

Wearable Device Based HAR. Sensor devices have emerged to be one of the most convenient choices for HAR tasks. Owing to their small size, independence from environments, and consistent class-specific patterns, regardless of size [1]. Therefore, several studies focused on optimizing human-activity recognition with the help of wearable devices [2]. Over the last decade, various approaches, including CNN-based methods [8,34], RNN-based methods [24,28,34,35] has been proposed. However, non-visual sensor HAR systems lag behind visual-based modalities such as RGBD [38] and Skeleton. This is because the sensor information only provides limited data. In this study, we aim to use important activity-related information using diffusion models to obtain skeleton data concerning the original sensor data of human activity.

Skeleton Based HAR. Compared to raw video data [30,31], which contains rich but often redundant pixel-level information, skeleton data provides a more compact and abstract representation of movement by focusing on key human joint coordinates. The use of skeleton data, which contains spatiotemporal information about human motion, has gained popularity in recent years. For such studies, focused on extracting the handcrafted features and classifiers to improve HAR [9] and involving more complex feature extractors such as RNNs [8,25,37], attention network [10] has been proposed to improve the HAR performances. Although skeleton key points can enhance action recognition models, the complexity and cost of these methods may limit their practicality compared to more user-friendly options like wrist-worn devices.

2.2 Diffusion Models

Recently, diffusion models [11,23,29] have gained widespread adoption across a range of domains, including image, text, and time-series generation, and have demonstrated success in generating conditional data based on prompts. Notably, extensive research has also been conducted to explore the potential for generating 3D human motion from text, achieving high fidelity and diversity in the generated motions, while also exhibiting effective zero-shot learning capabilities [39]. Our work utilizes the principles of conditional generation to produce 3D skeletal keypoints from wearable sensor data, effectively using accelerometer and gyroscope readings as prompts. This approach enables the model to infer activities based on body dynamics derived from data collected by wearable smartwatches on the wrist and hip, eliminating the need for a traditional setup to obtain skeleton keypoints.

3 Methods

3.1 Model Architecture

In this paper, we address the HAR task using generated skeleton data from a conditional diffusion model (DM), conditioned on sensor data from wearable devices.

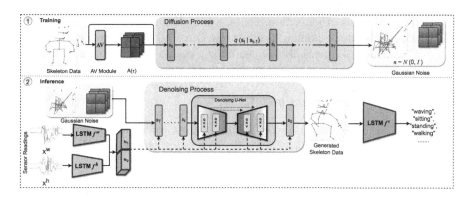

Fig. 2. Overview of the SSDL model training approach: In the first stage, skeleton data is input into the AV module to obtain angular variation $A(\tau)$, which is then processed with skeleton key points in the diffusion process up to time step T. In the second stage, sensor readings from a wearable device are used in the denoising process to generate class-specific skeleton data, which is then used for HAR via an LSTM encoder (f^s).

The participant wears a smartwatch on the left wrist and a Nexus smartphone on the right hip, secured in a harness. Given sensor data $\{(X^w, X^h)\}$, representing wrist and hip positions, we employ deep neural networks for effective sensor representation learning.

Specifically, we used two LSTM-based encoders $f^w(\Theta^w)$ and $f^h(\Theta^h)$ with parameters Θ^w and Θ^h to extract temporal features $\mathbf{A^w}$ and $\mathbf{A^h}$. Then, a simple concatenation operation is used to obtain combined features $\mathbf{Z}_0$ from $\mathbf{A^w}$ and $\mathbf{A^h}$. Furthermore, we also calculate the changes in the joint angles A using an Angular Variation (AV) module. These are then fed into the forward diffusion process along with original skeleton data (S) obtained from the Kinect camera. To improve the generation quality in terms of the skeleton structure, in the training phase, we utilize the noisy version of the original angular joints $\hat{A}_t$ at each denoising step t to calculate the angular variation loss $\mathcal{L}_c(A_t, \hat{A}_t)$. Meanwhile, we approximate the predicted noise to the added Gaussian noise with $\mathcal{L}_{\text{diff}}$ loss using a 1D-UNet architecture during the reverse diffusion process. Moreover, to increase the robustness of the denoising model, we attribute a small amount of Gaussian noise δ to the conditioned embedding $\mathbf{Z}_0$ to calculate the LR loss $\mathcal{L}_{\text{reg}}$. As for the inference phase, conditioned on a pair of multi-sensor data $\{(X^w, X^h)\}$, we generate its embedding ($\mathbf{Z}_0$) from the LSTM module, which is then taken into account as conditioning for the DM to generate a specific skeleton structure, which is then fed into the classifier to recognize the specific human activity.

3.2 Angular Variations

Suppose the skeleton data consists of a set of P key points, where P depends on the device used for capturing the data. To effectively extract the spatial and temporal features of joint movements, we implement the Angular Variation (AV) module. For three points, as illustrated in Fig. 3, the three-dimensional coordinates of these sequentially connected key points are denoted as follows:

$$P_1(x_1, y_1, z_1), \quad P_2(x_2, y_2, z_2), \quad P_3(x_3, y_3, z_3). \tag{1}$$

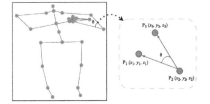

Fig. 3. The group of 3-key points selection from skeleton structure for calculating the angle between joints.

Then, we can define vectors $\boldsymbol{V_{2,1}}$ and $\boldsymbol{V_{2,3}}$ based on the (P_1, P_2, P_3) as follows,

$$\boldsymbol{V_{2,1}} = (x_1 - x_2, y_1 - y_2, z_1 - z_2), \tag{2}$$

$$\boldsymbol{V_{2,3}} = (x_3 - x_2, y_3 - y_2, z_3 - z_2). \tag{3}$$

Thereafter, we can obtain the angle value θ between two vectors with the help of dot product and we have,

$$\theta = \arccos\left(\frac{\boldsymbol{V_{2,1}} \cdot \boldsymbol{V_{2,3}}}{\|\boldsymbol{V_{2,1}}\|\|\boldsymbol{V_{2,3}}\|}\right), \tag{4}$$

where arccos stands for inverse cosine function. Here, θ is expressed in radians and we can convert it into degrees to obtain θ_{degree} as follows,

$$\theta_{\text{degree}} = \theta \times \left(\frac{180}{\pi}\right). \tag{5}$$

In our work, we use the angle θ_{degree} to capture angular variations between key points such as hands, elbows, shoulders, hips, ankles, and knees, providing crucial insights into pose configuration during the diffusion process. To avoid introducing non-informative data, we focus on a selective subset of angles. Specifically, the angles between the head and neck for all frames are almost constant over time and may not provide helpful information for the discrimination of activities, which aligns with the general consensus in video-based research [32,33]. Therefore, in our experiments, we consider 12 key points related to upper and lower body structure, i.e., "shoulder-¿elbow-¿wrist" and "hip-¿ankle-¿knee", and both the left and right side of the skeleton are considered. Hence, we can well capture the activity-specific features from the perspective of natural human skeleton movements. For example, the variations between the hip, knee, and foot can help to distinguish between sitting and standing actions.

3.3 Guidance Through Angular Variations

After accessing the angular variations from each skeleton segment, we focused on adapting our denoising learning process to generate improved relationships between different body parts. For such, as shown in Fig. 2, this angular variation matrix $A(\tau)$ is used in the forward diffusion process to obtain the noisy version $\hat{A}(\tau)$ where τ represents the number of total frames. Then $\hat{A}(\tau)$ is used to evaluate the generation quality with the help of a contrastive loss strategy [22] between the real and generated Angular Variations as shown below,

$$\mathcal{L}_c = \frac{1}{2}\left\|A(\tau) - \hat{A}(\tau)\right\|_F^2, \tag{6}$$

where $\|\cdot\|_F$ represents the Frobenius norm to measure the similarity between the generated and original angular variations across τ frames. By minimizing $\mathcal{L}_c$ loss, we encourage the model to minimize the discrepancy between the original and generated changes in the angular. In summary, we use the angular variation to guide our reverse diffusion process to generate samples following changes in key points and angular variations across time.

3.4 Lipschitzc Regularized Robust Generation

To address the issue constant noise in sensor data, we introduce a Lipschitz Regularization (LR) approach [16], hypothesizing that the generative model's response varies with noise in input data. For convenience, we represent the 1D-Unet as Θ^s and analyze how small variations in $\mathbf{Z_0}$ affect generation quality by perturbing data embeddings with random Gaussian noise δ. Mathematically, let $\mathbf{S}$ be the target time-series skeleton data and $\mathbf{A^w}$, $\mathbf{A^h}$ the conditioning time-series sensor data. We incorporate a Lipschitz regularization term into the conditional diffusion objective to regularize learning. The diffusion loss ensures the model accurately predicts the added noise, while the LR term enhances robustness to latent space perturbations. Thus, our loss function comprises: a) Noise Prediction Loss: The noise prediction loss is standard diffusion loss for reverse diffusion model to estimate the amount of noise in the current sampling step. For our case, this conditioned loss function can be represented as follows,

$$\mathcal{L}_{\text{diff}} = \mathbb{E}_{s_0,t,\epsilon}\left[\|\epsilon - \epsilon_\theta(s_t, \mathbf{Z_0}, t)\|^2\right], \tag{7}$$

where t is a timestep in the diffusion process, indicating the level of noise added to s_0 to obtain s_t, $\mathbf{Z_{0t}}$ is embedding of the sensor data to guide the reverse diffusion process from wrist and hip and ϵ is the actual noise added to the skeleton data t-th step of the forward diffusion process. b) Lipschitzc Regularised Loss: To ensure that generated skeleton data closely match the expected output given by $\mathbf{Z_0}$, we opt to minimize the MSE loss in between the generated skeleton key points from $\mathbf{Z_0}$ and $(\mathbf{Z_0} + \delta)$ as follows,

$$\mathcal{L} = \mathbb{E}_{t,s_t,\delta}\left[\|f^s\left(\hat{s}(\mathbf{Z_0}, t)\right) - f^s\left(\hat{s}(\mathbf{Z_0} + \delta, t)\right)\|^2\right], \tag{8}$$

where $\hat{s}_{\text{gen}}(\mathbf{Z_0}, t)$ represents the skeleton sequence generated by the model conditioned on Z_0, and $\hat{s}_{\text{gen}}(\mathbf{Z_0} + \delta, t)$ represents the skeleton sequence generated by the model conditioned on the perturbed embedding $Z_0 + \delta$. These skeleton data when passed as an input to the LSTM encoder f^s provide the output label for a specific activity. Through such, we explicitly encourage the model to maintain consistent HAR predictions for slightly varied conditioning embedding, enhancing robustness and stability in the generated outputs relative to perturbations in the sensor data embeddings.

3.5 Overall Training Objective

From the above motivation, the overall training objective of our model is a weighted combination of the loss of noise prediction for the standard diffusion model and the Lipschitz regularization term and we have,

$$\mathcal{L}(\Phi, \Theta) = \mathcal{L}_{\text{diff}} + \lambda \cdot \mathcal{L}_{\text{reg}} + \mathcal{L}_{\text{c}}, \tag{9}$$

where λ is a hyper-parameter that controls the weight of the $\mathcal{L}_{\text{reg}}$, $\mathcal{L}_{\text{c}}$ is the angular loss as explained earlier. One more advantage of our strategy is that the gradients of the model(s) can be easily computed with the help of traditional optimizers such as SGD and Adam, as shown in the equation below,

$$\Phi \leftarrow \Phi - \alpha_1 \cdot (\nabla_\Phi \mathcal{L}_{\text{diff}} + \nabla_\Phi \lambda \mathcal{L}_{\text{reg}} + \nabla_\Phi \mathcal{L}_{\text{c}}), \tag{10}$$

$$\Theta \leftarrow \Theta - \alpha_2 \cdot \nabla_\Theta \mathcal{L}_{\text{reg}}, \tag{11}$$

where α_1 and α_2 are the learning rates for specific DM and LSTM encoder branches. Moreover, during the experiments, we notice that such a joint training strategy may result in "modal collapse". Therefore, to address such challenges, we train both branches separately. Specifically, we first train the LSTM module using sensor data and then use these pre-trained embedding as conditions for DM to learn effective generation, and we have,

$$\nabla_\Theta \mathcal{L}\text{reg} \approx 0. \tag{12}$$

Hence, there will be no update in the sensor model's weights,

$$\Theta \leftarrow \Theta - \alpha_2 \cdot \nabla_\Theta \mathcal{L}_{\text{reg}}$$
$$\approx \Theta. \tag{13}$$

Thereafter, we allow our diffusion model to converge quickly without collapse of the loss function.

4 Experiments

4.1 Dataset

In this research, we conducted experiments on the following datasets.

Berkeley MHAD [19] comprises 11 action categories performed by 12 individuals. (7 male and 5 female) in the age range between 23 to 30 years and one aged subject. All the subjects performed five repetitions of each action, generating 660 action sequences which made up 82 min of capturing duration.

UTD MHAD [6] contains 27 actions performed by eight subjects, with each subject repeating each Action 4 times, resulting in a total of 861 sequences. The dataset was collected using a Microsoft Kinect and wearable inertial sensors indoors. It provides multiple modalities, including RGB videos, depth maps, skeleton data, and inertial sensor data, allowing researchers to compare the performance of different methods that utilize various data modalities for human action recognition.

SmartFall MHAD is a non-public dataset collected by SmartFall Research Group at Texas State University. Data are collected from 27 participants with age greater than 60 and 12 young adults with age between 20–30. The dataset used for this paper was compiled only from 12 young adult participants (7 male and 5 female). The young adult participants performed five types of falls (front, back, left, right, and rotational) and nine prescribed ADLs on five repetitions each. The collection process includes the use of four types of sensors, which are three Azure Kinect cameras, a Huawei smartwatch, a Nexus smartphone, and two Meta Sensors developed by MBIENTLAB.

4.2 Training Setup

All experiments were conducted on four Nvidia GeForce RTX A6000 GPUs using PyTorch. A sliding window technique with a three-second duration and 50% overlap was applied for each modality. For the SmartFall dataset, we collected 1,510 instances, categorized into 14 classes, and split them into training and validation sets with an 80:20 ratio, using stratified sampling to maintain class distribution. For the UTD-MHAD dataset, after removing corrupt frames, we used 861 sequences with an 80:20 split for training and testing. For the Berkeley dataset, data from the first seven participants was used for training, while the remaining participants were used for testing.

4.3 Comparison on the HAR Task

In this section, we compare our method with state-of-the-art HAR and RGB-based approaches. As shown in Table 1, our method demonstrates improved performance over most existing sensor-based models, as reported by [17,24]. This improvement is due to our model's efficient cross-modal generation of detailed skeleton key points from sensor inputs, which enhances HAR performance. Notably, our method performs competitively on the UTD-MHAD dataset, matching the results of Ni et al.'s approach [18], which uses cross-modality knowledge distillation for HAR. Additionally, on the Berkeley MHAD dataset, where human motions are captured using a single sensor, our method outperforms Ni et al.'s approach, highlighting that their method is more sensitive to practical settings, while ours remains stable across different scenarios. We also compared our method with skeleton-based HAR methods [5,8,13,27,37], further showcasing its robustness and effectiveness. In conclusion, our comparative findings with these SOTA methods not only validate the advantage of our approach but also potential solutions to address the persistent challenges within HAR systems (Table 2).

Table 1. Comparison of various models on different MHAD datasets. "-" indicates the source codes are unavailable, preventing evaluation on specific datasets. Meanwhile, "N/A" indicates that the method is not applicable to the dataset due to the unavailability of the testing modality.

Method	Required Modality	UTD Accuracy (%)	Berkeley Accuracy (%)	SmartFall Accuracy (%)
Hussein et al. [12]	RGB	85.5	-	N/A
Lin et al. [14]	RGB	-	96.16	N/A
Avigyan et al. [8]	Skeleton (Velocity CNN)	87.5	92.6	88.77
Avigyan et al. [8]	Skeleton (Angular CNN)	96.2	96.6	92.3
Wang et al. [27]	Skeleton	85.81	-	-
Chuankun et al. [13]	Skeleton	88.1	-	-
Zhao et al. [37]	Skeleton	92.1	-	-
Chao et al. [5]	Depth+Skeleton	93.26	-	N/A
Singh et al. [24]	Accelerometer + Gyroscope	91.4	-	89.03
Ni et al. [17]	Accelerometer	95.19	94.76	94.41
Ni et al. [18]	Accelerometer	**96.97**	90.18	-
Ours (SSDL)	Accelerometer	96.8	**96.9**	**94.66**

Table 2. Ablation study of Angular Variations. The performance of all methods exhibited a consistent improvement with the inclusion of angular variations.

Steps	Dataset	No Angular Variations		Angular Variations	
		Accuracy ↑	FID ↓	Accuracy ↑	FID ↓
1000	SmartFall MHAD	62.75	14.95	72.71	7.95
	Berkeley MHAD	63.24	11.2	74.91	7.28
	UTD-MHAD	62.73	15.04	77.3	6.65
4000	SmartFall MHAD	68.78	11.86	76.17	6.7
	Berkeley MHAD	68.25	9.37	78.66	5.32
	UTD-MHAD	67.84	9.11	79.6	4.67
10000	SmartFall MHAD	69.2	7.3	79.07	4.88
	Berkeley MHAD	70	9.1	84.9	4.25
	UTD-MHAD	72.64	8.1	81.11	4.60

4.4 Ablation Study

Effect of Angular Variation. This ablation study demonstrates the impact of incorporating Angular Variations (AV) into the diffusion model. Using FID metrics to evaluate generation quality, where lower scores indicate closer resemblance to real samples, we found that the baseline model consistently exhibited poor performance with high FID scores, regardless of diffusion steps (Table 1). However, incorporating AV significantly improved model performance, reducing FID scores and increasing accuracy. For instance, on the UTD-MHAD dataset, the FID score dropped from 8.1 to 4.60 over 10,000 training steps, leading to enhanced model accuracy. Therefore, by capturing joint movement dynamics through Angular Variations, the model effectively differentiates subtle movement nuances, resulting in a substantial improvement in performance.

Table 3. Comparative Analysis of Model Performance with and without Lipschitz Regularization under Joint Training and Pre-trained Embedding Scenarios.

Steps	Dataset	No Lipschitz Regularisation				With Lipschitz Regularisation			
		Joint Training		Pre-trained Embedding		Joint Training		Pre-trained Embedding	
		Accuracy ↑	FID ↓	Accuracy ↑	FID ↓	Accuracy ↑	FID ↓	Accuracy ↑	FID ↓
1000	SmartFall MHAD	77.5	4.19	83.4	2.40	76.41	4.16	84.1	1.81
	Berkeley MHAD	79.3	3.85	86.2	1.88	77.91	4.06	88.04	0.94
	UTD-MHAD	75.8	4.9	81.2	2.9	78.9	11.02	89.66	0.82
4000	SmartFall MHAD	76.4	4.65	83.16	2.35	77.23	6.45	90.44	0.85
	Berkeley MHAD	77.6	4.07	85.36	2.11	77.9	6.07	91.5	0.71
	UTD-MHAD	84.1	2.02	86.77	2.06	80.2	2.07	93.1	0.53
10000	SmartFall MHAD	80.5	2.27	85.01	2.26	77.72	4.91	94.66	0.6
	Berkeley MHAD	82	2.1	84.9	2.16	78.77	4.72	96.9	0.50
	UTD-MHAD	82.28	2.02	86.92	2.02	81.33	2.49	96.8	0.59

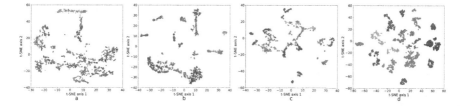

Fig. 4. T-SNE analysis of the learned features on dataset SmartFall-MHAD with different training setups. a) Diffusion model and LSTM encoder are jointly learned. b) Effect of LR into the joint training setup. c) According to Eq. 13, using a pre-trained model for sensor data guide the diffusion model. d) Use of the LR method clearly showed a better distinctive ability of the model to differentiate between different activities.

Effect of Lipschitz Regularisation. To validate the effectiveness of our Lipschitz approach against noisy issues in sensor data, we monitor the change in FID score during the training. As shown in Table 3, the performance of the diffusion model step increased as shown by higher accuracy and lower FID values. Moreover, as shown in the t-SNE plot (Fig. 4(c) and (d)) with the help of our LR method, the model learned distinctive features in the form of clear separation of clusters for each activity. Therefore, results show that our method of LR allows the model to capture meaningful patterns in the data, mitigating the impact of noise and ultimately leading to improved performance and representations of human actions.

Effect of Diffusion Steps. Additionally, we observed that increasing the number of steps in the diffusion model often improved its capacity to generate more realistic features. Although including angular variations hinted towards improvement in the performance of the generation quality related to each activity, it proved insufficient for the diffusion model to consistently converge at better solutions compared to other state-of-the-art HAR methods as shown in Table 1. However, increasing the number of diffusion steps is computationally expensive, especially for resource-constrained devices like smartwatches.

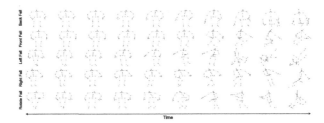

Fig. 5. Synthetic set of actions generated by our SSDL diffusion-based model trained on Smartfall MHAD dataset.

Table 4. Experiments on SOTA Generative models for SmartFall-HAR.

Method	Model parameters	FID ↓	Accuracy ↑ (%)	Train (hrs)	Inference (msec/sample)
UNet-VAE	161M	29.3	48.2	7	11
CycleGAN	295M	25.2	49.4	13	22
Conditional GAN	275M	14.4	68.1	15	53
Unconditional DM	126M	98.1	42.5	9	11
Conditional DM	133M	14.9	62.7	9	12
Ours (**SSDL**)	133M	4.9	94.6	9	12

Using Pre-trained LSTM Embedding. The results in Table 3 and Fig. 4 reveal that the jointly trained model underperforms, suggesting potential issue modal collapse. To address this, we first trained the LSTM encoder separately and used the pre-trained embeddings for reverse diffusion conditioning. This approach improved model performance, as shown in Table 3, likely due to reduced variation in the conditional signal during generation. For consistency, we used the same backbone architecture in all evaluations, varying only the sampling time steps. The results affirm that pre-trained embeddings significantly enhance skeleton data generation, closely mimicking authentic datasets, while increasing training steps further boosts the model's learning capabilities.

Qualitative Results for the Generated Samples. The figure shown in Fig. 5 depicts various synthetic action sequences that were generated using our proposed SSDL approach for different activities.

This visualization helps evaluate the model's stability in producing coherent action sets for each activity. Our SSDL method generates more naturalistic action sequences with minimal noise. As shown in the figure, the synthetic sequences derived from sensor data also enhance the model's classification performance across different classes.

4.5 Inference Speed

In the experiments summarized in Table 4, we evaluated the inference times of various generative models on the SmartFall-HAR dataset. The proposed SSDL approach excelled with an inference time of 12 ms per sample, significantly outperforming other state-of-the-art methods. In contrast, Conditional GAN and

CycleGAN had longer inference times of 53 ms and 22 ms, respectively, with lower performance metrics. While UNet-VAE and Unconditional DM achieved similar inference times of 11 ms, their accuracy was notably lower. The superior speed and accuracy of the SSDL approach make it ideal for real-time human activity recognition.

5 Conclusion

In this study, we propose a novel approach for generating skeleton critical points based on sensor data from wearable devices to solve HAR problems. The derived angular variation guided the reverse diffusion process to generate more meaningful movements. Based on our proposed framework, we generated more diverse and aligned body part movements that inherited kinetically meaningful body structures. The proposed method obtains state-of-the-art results on the two datasets and shows competitive performance in the UTD-MHAD dataset for Human Action Recognition. Our experiments demonstrated that the Lipschitz Loss improved the diffusion model's performance while also achieving low inference times for generating skeleton information from sensor data, validating the real-world applicability of our approach for a robust HAR system.

Acknowledgments. This research is supported by NSF IIS-2309073, ECCS-2123521 and Cisco Unrestricted Gift. This article solely reflects the opinions and conclusions of authors and not funding agencies.

References

1. Al-Eidan, R.M., Al-Khalifa, H., Al-Salman, A.M.: A review of wrist-worn wearable: sensors, models, and challenges. J. Sens. **2018**(1), 5853917 (2018)
2. Alexan, A.I., Alexan, A.R., Oniga, S.: Real-time machine learning for human activities recognition based on wrist-worn wearable devices. Appl. Sci. **14**(1), 329 (2024)
3. Anagnostis, A., Benos, L., Tsaopoulos, D., Tagarakis, A., Tsolakis, N., Bochtis, D.: Human activity recognition through recurrent neural networks for human-robot interaction in agriculture. Appl. Sci. **11**(5), 2188 (2021)
4. Bayoumy, K., et al.: Smart wearable devices in cardiovascular care: where we are and how to move forward. Nat. Rev. Cardiol. **18**(8), 581–599 (2021)
5. Chao, X., Ji, G., Qii, X.: Multi-view key information representation and multimodal fusion for single-subject routine action recognition. Appl. Intell. **54**(4), 3222–3244 (2024)
6. Chen, C., Jafari, R., Kehtarnavaz, N.: UTD-MHAD: a multimodal dataset for human action recognition utilizing a depth camera and a wearable inertial sensor. In: ICIP, pp. 168–172 (2015)
7. Dang, L.M., Min, K., Wang, H., Piran, M.J., Lee, C.H., Moon, H.: Sensor-based and vision-based human activity recognition: a comprehensive survey. Pattern Recogn. **108**, 107561 (2020)
8. Das, A., Sil, P., Singh, P.K., Bhateja, V., Sarkar, R.: MMHAR-EnsemNet: a multimodal human activity recognition model. IEEE Sens. J. **21**(10), 11569–11576 (2021)

9. Devanne, M., Wannous, H., Berretti, S., Pala, P., Daoudi, M., Bimbo, A.D.: 3-D human action recognition by shape analysis of motion trajectories on Riemannian manifold. IEEE Trans. Cybern. **45**(7), 1340–1352 (2015)
10. Gao, W., Zhang, L., Teng, Q., He, J., Wu, H.: DanHAR: dual attention network for multimodal human activity recognition using wearable sensors. Appl. Soft Comput. **111**, 107728 (2021)
11. Ho, J., Jain, A., Abbeel, P.: Denoising diffusion probabilistic models. In: NeurIPS, vol. 33, pp. 6840–6851 (2020)
12. Hussein, M.E., Torki, M., Gowayyed, M.A., El-Saban, M.: Human action recognition using a temporal hierarchy of covariance descriptors on 3D joint locations. In: IJCAI, pp. 2466–2472 (2013)
13. Li, C., Hou, Y., Wang, P., Li, W.: Joint distance maps based action recognition with convolutional neural networks. IEEE Signal Process. Lett. **24**(5), 624–628 (2017)
14. Lin, J., Gan, C., Han, S.: TSM: temporal shift module for efficient video understanding. In: ICCV, pp. 7082–7092 (2019)
15. Mutegeki, R., Han, D.S.: A CNN-LSTM approach to human activity recognition. In: ICAIIC, pp. 362–366 (2020)
16. Muthukumar, R., Sulam, J.: Adversarial robustness of sparse local lipschitz predictors. SIAM J. Math. Data Sci. **5**(4), 920–948 (2023)
17. Ni, J., Ngu, A.H., Yan, Y.: Progressive cross-modal knowledge distillation for human action recognition. In: MM, pp. 10–14 (2022)
18. Ni, J., Sarbajna, R., Liu, Y., Ngu, A.H., Yan, Y.: Cross-modal knowledge distillation for vision-to-sensor action recognition. In: ICASSP, pp. 4448–4452 (2022)
19. Ofli, F., Chaudhry, R., Kurillo, G., Vidal, R., Bajcsy, R.: Berkeley MHAD: a comprehensive multimodal human action database. In: WACV, pp. 53–60 (2013)
20. Shang, Y., Duan, B., Zong, Z., Nie, L., Yan, Y.: Lipschitz continuity guided knowledge distillation. In: ICCV, pp. 10655–10664. IEEE (2021)
21. Shang, Y., Xu, D., Duan, B., Zong, Z., Nie, L., Yan, Y.: Lipschitz continuity retained binary neural network. In: Avidan, S., Brostow, G.J., Cissé, M., Farinella, G.M., Hassner, T. (eds.) ECCV, vol. 13671, pp. 603–619. Springer, Cham (2022). https://doi.org/10.1007/978-3-031-20083-0_36
22. Shang, Y., Xu, D., Zong, Z., Nie, L., Yan, Y.: Network binarization via contrastive learning. In: Avidan, S., Brostow, G.J., Cissé, M., Farinella, G.M., Hassner, T. (eds.) ECCV, vol. 13671, pp. 586–602. Springer, Cham (2022). https://doi.org/10.1007/978-3-031-20083-0_35
23. Shang, Y., Yuan, Z., Xie, B., Wu, B., Yan, Y.: Post-training quantization on diffusion models. In: CVPR, pp. 1972–1981 (2023)
24. Singh, S.P., Sharma, M.K., Lay-Ekuakille, A., Gangwar, D., Gupta, S.: Deep ConvLSTM with self-attention for human activity decoding using wearable sensors. IEEE Sens. J. **21**(6), 8575–8582 (2020)
25. Veeriah, V., Zhuang, N., Qi, G.J.: Differential recurrent neural networks for action recognition. In: ICCV, pp. 4041–4049 (2015)
26. Viswambaran, R.A., Chen, G., Xue, B., Nekooei, M.: Evolutionary design of recurrent neural network architecture for human activity recognition. In: CEC, pp. 554–561 (2019)
27. Wang, P., Li, Z., Hou, Y., Li, W.: Action recognition based on joint trajectory maps using convolutional neural networks. In: MM, pp. 102–106 (2016)
28. Wei, X., Wang, Z.: TCN-attention-HAR: human activity recognition based on attention mechanism time convolutional network. Sci. Rep. **14**, 7414 (2024)

29. Wu, J., Wang, H., Shang, Y., Shah, M., Yan, Y.: PTQ4DIT: post-training quantization for diffusion transformers. CoRR abs/2405.16005 (2024)
30. Wu, Z., Sun, C., Xuan, H., Liu, G., Yan, Y.: WaveFormer: wavelet transformer for noise-robust video inpainting. In: Wooldridge, M.J., Dy, J.G., Natarajan, S. (eds.) AAAI, pp. 6180–6188 (2024)
31. Wu, Z., Sun, C., Xuan, H., Yan, Y.: Deep stereo video inpainting. In: CVPR, pp. 5693–5702 (2023)
32. Wu, Z., Xuan, H., Sun, C., Guan, W., Zhang, K., Yan, Y.: Semi-supervised video inpainting with cycle consistency constraints. In: CVPR, pp. 22586–22595 (2023)
33. Wu, Z., Zhang, K., Sun, C., Xuan, H., Yan, Y.: Flow-guided deformable alignment network with self-supervision for video inpainting. In: ICASSP, pp. 1–5. IEEE (2023)
34. Xiaochun, Y., Zengguang, L., Deyong, L., Xiaojun, R.: A novel CNN-based bi-LSTM parallel model with attention mechanism for human activity recognition with noisy data. Sci. Rep. **12**(1) (2022). https://doi.org/10.1007/s44196-024-00689-0
35. Yang, Z., Li, Y., Yang, J., Luo, J.: Action recognition with spatio-temporal visual attention on skeleton image sequences. TCSVT **29**(8), 2405–2415 (2019)
36. Zhang, Z.: Microsoft kinect sensor and its effect. IEEE Multimedia **19**(2), 4–10 (2012)
37. Zhao, R., Wang, K., Su, H., Ji, Q.: Bayesian graph convolution LSTM for skeleton based action recognition. In: ICCV, pp. 6881–6891 (2019)
38. Zhao, Z., Tang, H., Wan, J., Yan, Y.: Monocular expressive 3D human reconstruction of multiple people. In: ICMR, pp. 423–432 (2024)
39. Zhiyuan, R., Zhihong, P., Xin, Z., Le, K.: Diffusion motion: generate text-guided 3D human motion by diffusion model. In: ICASSP, pp. 1–5 (2023)

Structural Information-Guided Fine-Grained Texture Image Inpainting

Zhiyi Fang[1], Yi Qian[1(✉)], and Xiyue Dai[2]

[1] Xi'an Jiaotong University, Xian 710049, China
yqian@mail.xjtu.edu.cn
[2] NorthWest University, Xian 710127, China

Abstract. In recent years, the wide application of deep learning has made rapid progress in image inpainting technology. However, the current image inpainting algorithms still have some shortcomings in the ability to grasp the overall structure and understand the detailed information, which will lead to texture blur and structural distortion in the generated content. For the purpose of tackling the problems mentioned above, we propose a two-stage image inpainting algorithm based on structural information guidance. In the first stage, we enhance the model's capacity to capture the global structure through the guidance of gradient maps and wireframes. In the next stage, we propose the Content Filling Network based on GAN. Specifically, we propose the Prior Addition Module to avoid the information loss during the process of convolution and normalization, and we propose the Texture Generation Module to assist the model in generating more detailed texture information. In addition, we improve the discriminator to make the model generate more realistic and vivid images. Extensive experiments are carried out to demonstrate that our approach performs noticeably better than current methods in both quantitative and qualitative evaluation.

Keywords: Image inpainting · Generative adversarial network · Prior guidance

1 Introduction

Image inpainting is an image processing technology that predicts the content of the missing regions through an appropriate algorithm based on the known content of the image (as shown in Fig. 1). Its purpose is to generate the missing region with the known information structure consistent and realistic texture. Image inpainting has found extensive applications in daily life, such as cultural relics restoration, aerospace remote sensing image processing, and medical image recovery.

In the last few years, Convolutional Neural Networks (CNNs) [33] and Generative Adversarial Networks (GANs) [7] have been widely applied to image inpainting. In 2016, Pathak et al. [2] proposed the Context Encoder (CE), marking the first application of GANs to image inpainting. However, the image inpainted by this algorithm is blurred and has poor global consistency. To tackle this challenge, Iizuka et al. [3] proposed the concept of "Globally and Locally", and introduced the double discriminator structure.

The image inpainted has improved the global and local structural consistency, but the detail and texture features are weak. Yu J et al. [4] proposed to replace the conventional convolution layer with the gated convolution layer [14] to solve the shortcoming that the traditional convolution layer considers all pixels of the image as valid pixels. However, this algorithm is only applicable to the regular mask. The images generated by these single-stage algorithms often have obvious blurring and artifacts, and the generated content is inconsistent with the known information, especially in cases of large missing areas or complex structures.

(a) Input   (b) ZITS[11]   (c) Ours   (d) GT

Fig. 1. Our results compared with ZITS and the ground truth (GT).

In order to solve the shortcomings of the single-stage inpainting algorithms, the algorithms based on prior guidance have gradually emerged. Nazeri K et al. [5] proposed the Edge-Connection model to use the edge information to assist image inpainting by designing an additional auxiliary network. However, the inpainting task cannot be correctly handled by only using the edge information as the prior. Guo X et al. [6] proposed a coupling network with mutual guidance of structure and texture to complete the entire inpainting process, and designed a bidirectional gated feature fusion module to realize the fusion of structure and texture information. This method guarantees the coherence of structure and semantics, but the inpainting image's texture details are not close to the real image. Dong et al. [11] proposed an incremental inpainting method, which uses axial attention-based [13] Transformer Structure Restoration (TSR) to repair edge maps and wireframes [12], which jointly guide the subsequent inpainting. The quality of the image generated by this method is significantly higher, but it is still challenging to get satisfying results when dealing with images with complex structures and more texture details.

In this paper, we argue that the existing inpainting methods can not make good use of prior knowledge to guide the generation of texture information, resulting in problems such as texture blur and structural distortion when inpainting images with large missing areas or complex textures. In view of the weaknesses of the above algorithms, this paper proposes a two-stage image inpainting algorithm based on prior guidance. The algorithm consists of two main parts, which are prior repair and content filling. Specifically, the gradient map and wireframe serve as prior information. Firstly, the prior is repaired using the attention-based transformer, and then the completed prior is utilized to direct the generation of the missing portion of the image.

The main contributions of this paper are as follows:

- We take the wireframe and gradient map together as the prior knowledge of the model, and propose a two-stage network for structure repair and content filling guided by prior information.
- For prior knowledge, we propose the Prior Addition Module, which applies the SPADE-based adding method to add prior information to the generator, avoiding the mixing of wireframe and gradient information, so that the two kinds of priors can guide each other and be added to the generator relatively independently.
- In the content filling stage, we propose the Texture Generation Module. Fast Fourier Convolution [17] is used to better capture the context information, and skip connections are employed to fuse high-level semantic information layer by layer, realizing the guidance of structural semantic information to the low-level visual information.
- We set new state of the arts on several benchmark datasets including Places2 and CelebA.

2 Related Work

2.1 Generative Adversarial Network

Generative Adversarial Network (GAN) is composed of two interconnected networks: a generator is in charge of generating simulated data, and a discriminator, which determines whether the input data is genuine or generated. The generator continually refine its generated data to confuse the discriminator, while the discriminator strives to enhance its accuracy in differentiating between genuine and fake data. Goodfellow et al. [7] first proposed the concept of GAN in 2014 and demonstrated its potential for generating realistic images.

Since GAN was proposed, researchers have come up with many variants to improve its performance and application range. By giving the generator and discriminator extra conditional information, Conditional Generative Adversarial Networks (CGAN) [8] generate data that is more specific. Deep Convolutional Generative Adversarial Networks

(DCGAN) [9] use convolutional neural network architectures to boost the quality of image generation. By introducing Wasserstein distance, Wasserstein GAN (WGAN) [10] enhances the stability of the training process. GAN has a wide range of applications in the field of image processing, including image generation, image editing, image super-resolution, etc.

2.2 Image Inpainting

Image inpainting techniques aim to restore damaged image regions in order to obtain a complete image. Traditional inpainting algorithms can be categorized into diffusion-based methods [31] and patch-based methods [32]. The diffusion-based methods use the idea of diffusion based on partial differential equation to smoothly propagate the information of the image's known area, but it is only applicable to the scene of small area or no texture. The patch-based methods adopt the way of comparison to perform image inpainting, copying the best matching background patch from the boundary to the hole. However, this method is not applicable when the missing region contains complex structures or non-repetitive textures.

In recent years, the rise of deep learning technology has greatly propelled the development of image inpainting. Deep learning based inpainting algorithms have more powerful feature representation and learning ability compared to traditional algorithms, and can mine deeper information of the image. Xie et al. [1] proposed the Denoising Auto-Encoder Network (DAEN), which can automatically extract and match features, so as to realize image inpainting. In order to strengthen the model's comprehension of the surrounding content in the missing region, Zeng et al. [29] proposed a pyramid context encoder network to capture contextual information of various granularities through image pyramids at multiple scales. Suvorov et al. [17] proposed an inpainting algorithm on the basis of Fast Fourier Convolution (FFC), which allows the network have a large receptive field in the shallow layer. Li et al. [30] proposed a transformer-based model for high-resolution image inpainting, which improved the inpainting effect of large-area missing images.

Despite significant progress in image inpainting techniques, there are still some challenges. For example, for large-area missing or highly complex damages, existing methods may not suffice to fully recover the content and style of the original image. Therefore, we propose a prior guided two-stage image inpainting algorithm that fully utilize the structural and textural information to guide the image generation.

3 Method

3.1 Overview

Our proposed model is shown in Fig. 2. The model is divided into two parts, which are prior repair and content filling. Specifically, in the first stage, we feed the masked image I_m, gradient map C_m, wireframe W_m and mask M into the network, and use the attention-based Transformer Structure Restoration (TSR) to repair the broken gradient map and wireframe $[\tilde{C}, \tilde{W}] = TSR(I_m, C_m, W_m, M)$. Here, $\tilde{C}$ and $\tilde{W}$ represent the repaired gradient map and wireframe respectively; subsequently, the repaired gradient map $\tilde{C}$ and wireframe $\tilde{W}$ are added to the Gated Convolution-based Feature Encoding Network (GFE) in parallel, which encodes the two priors to acquire the multi-scale features of the gradient map and wireframe respectively $\hat{C}_k = GFE(\tilde{C}, M)$, $\hat{W}_k = GFE(\tilde{W}, M)$, $\{k = 0, 1, 2\}$. The encoded prior features are sent to the model's second stage to guide the network to generate the missing content that is aligned with the known information of the image. The second stage is implemented based on GAN, which feeds the masked image I_m, mask M, gradient map features $\hat{C}_k$ and wireframe features $\hat{W}_k$ into the generator to obtain the predicted image $I_{\text{pred}} = G(I_m, M, \hat{C}_k, \hat{W}_k)$, where G stands for our generator. The final completed image is $I_{\text{comp}} = I_m + I_{\text{pred}} + M$.

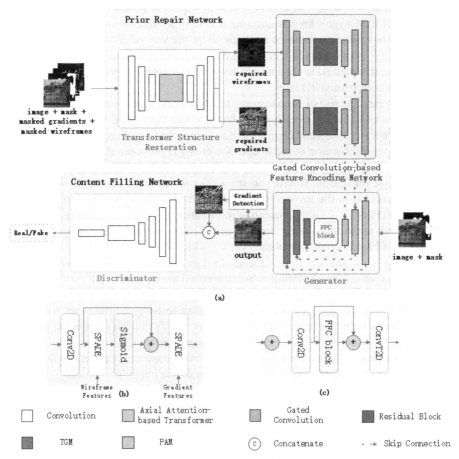

Fig. 2. (a) The overall framework of the proposed model, which consists of the Prior Repair Network and the Content Filling Network. (b) Prior Addition Module. (c) Texture Generation Module.

3.2 Prior Repair Network

Presently, although the image inpainting model can achieve semantically correct and visually pleasing results for most damaged images, the inpainting effects on images with complex backgrounds and large damaged areas are still poor. For the purpose of further enhancing the effect of image inpainting, we utilise prior knowledge to guide this process. As shown in the Fig. 3, we mainly use gradient maps and wireframes to describe the contours of the image content.

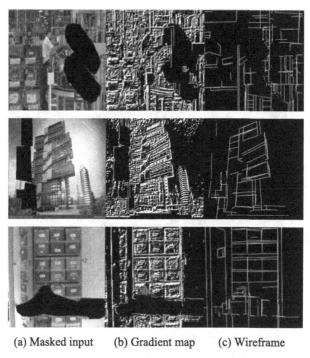

(a) Masked input   (b) Gradient map   (c) Wireframe

Fig. 3. The prior knowledge of our proposed method.

The edge map is the most commonly used prior to describe the edge structure of an image. However, compared with it, the gradient map not only conveys the necessary edge information, but also implies the texture information and high-frequency details, which is of great significance for the subsequent texture generation. As shown in the Fig. 4, common gradient extraction operators include Sobel, Scharr, Roberts and Laplacian, etc. With higher orientation sensitivity and better rotational invariance, the Scharr operator is capable of detecting gradient information more accurately and is more effective in the case of large changes in edge orientation. It can be seen that the gradient map obtained by Scharr operator has stronger contrasts and more thorough description. Therefore, we adopt the gradient map extracted by Scharr operator as one of the model's priors. The gradient map extracted by Scharr is shown in Fig. 5. This operator has two convolution kernels, which are used to calculate the gradient of the image in the horizontal and vertical directions respectively. Finally, the gradients of the two directions can be added to create the final gradient map. In addition, we also use the LSM-HAWP algorithm to extract the wireframe to further provide a more powerful structural prior for the repair network.

We apply the TSR model to inpaint the gradient map and wireframe, TSR uses the axial attention module and the standard attention module alternately. The complexity of axial attention is lower, which allows the model to handle more attention layers to promote its capability. The standard attention has significant advantages in learning global correlation. They complement each other in repairing gradient map and wireframe. Then the two priors are fed into a parallel GFE consisting of three layers of gated convolutions for downsampling, three residual modules with dilated convolutions [15], and three layers of gated convolutions for upsampling. Following the gradient map and the wireframe encoded by GFE, feature maps of sizes 32, 64, 128 and 256 are generated, which can offer significant structural guidance for the subsequent texture generation.

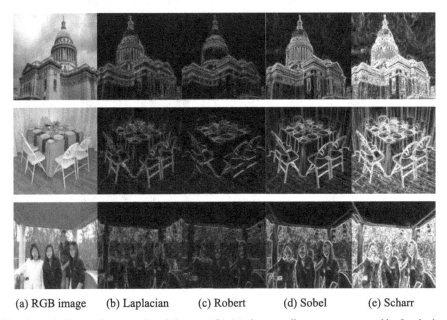

(a) RGB image　　(b) Laplacian　　(c) Robert　　(d) Sobel　　(e) Scharr

Fig. 4. (a) indicates the ground truth images, (b)-(e) show gradient maps extracted by Laplacian, Robert, Sobel and Scharr respectively.

3.3 Content Filling Network

Prior Addition Module (PAM). If the multi-scale features encoded by GFE are directly added to the generator, many guidance information will be filtered out in the subsequent convolution and normalization process. In order to avoid this information loss, we propose the Prior Addition Module (PAM). As shown in the Fig. 2, we add wireframe and gradient information separately, adding wireframe information before gradient information, wireframe can play a role in guiding gradient to a certain extent. In the meanwhile, we use Spatial-Adaptive Normalization (SPADE) [16] for normalization. Specifically, the input feature F_{in}^P is first convolved to get feature F_1, after that F_1 is sent to SPADE together with the wireframe feature $\hat{W}$ for the first normalization and then processed by Sigmoid function to acquire feature F_s. The pixel-by-pixel addition of F_s and F_1 is sent to SPADE as well as the gradient feature $\hat{C}$ for the second normalization. This process can be expressed as follows:

$$F_{out}^P = S\left(\hat{C}, Conv(F_1) \oplus \sigma\left(S\left(Conv\left(F_{in}^P\right), \hat{W}\right)\right)\right) \tag{1}$$

Here, F_{out}^P represents the output feature map of the Prior Addition Module, S stands for SPADE, $\oplus$ represents pixel-by-pixel addition, and σ is Sigmoid function.

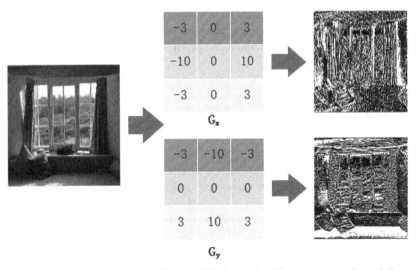

Fig. 5. Gradient map extracted from a RGB image by Scharr operators G_x and G_y.

Texture Generation Module (TGM). When there are large missing areas in the image, the current image inpainting algorithm still has issues such as obvious inpainting boundaries and artifacts. Fast Fourier Convolution has two branches: global and local. It not only has a large receptive field, but also carries out the fusion of cross-scale information right inside the convolution, which realizes the effective application of global information and local information. For the purpose of making better use of context information, we propose Texture Generation Module (TGM). In TGM, the input features F_{in}^T are skip connected with the output features F_{out}^P of the corresponding scale of the Prior Addition Module, and then they are processed by ordinary convolution, Fast Fourier Convolution (FFC), pixel-by-pixel addition and deconvolution. The process can be expressed as follows:

$$F_{out}^T = ConvT\left(Conv\left(F_{in}^T \oplus F_{out}^P\right) \oplus F_c\left(Conv\left(F_{in}^T \oplus F_{out}^P\right)\right)\right) \tag{2}$$

Here, F_{out}^T represents the output feature of the Texture Generation Module, F_c is FFC, and $\oplus$ denotes pixel-by-pixel addition.

Generator. The generation network is implemented based on the Prior Addition Module and Texture Generation Module. In the downsampling stage, the Prior Addition Module is adopted to add the gradient map and wireframe information to the generator. And in the upsampling stage, Texture Generation Module generates the fine-grained texture information, while employing skip connections to progressively integrate the high-level semantic information, realizing the guidance of structural semantic information to visual representation.

Discriminator. When the common discriminator guides the generator to repair the image, the detail information of the repaired area will be fuzzy. Therefore, we introduce the discrimination of the gradient into the discriminator. Specifically, Scharr operator is used to extract the gradient map of the generator's output image, and the gradient map as well as the generated image are spliced together and input into the discriminator. The discriminator extracts the deep semantic features of the input image through multiple downsampling layers, and then calculates the loss to reverse guide the update of network parameters.

4 Experiments

4.1 Experimental Details

Datasets: We train and evaluate our method on two benchmark datasets: Places2 [18] and CelebA [19]. We use irregular masks [20] and COCO segmentation masks [21] to process the images. All input images and masks are resized to 256 × 256 pixels.

Implementation Details: The proposed method is developed based on version 1.9.0 of Pytorch framework, and is trained and tested on the RTX 3090 GPU with 24 GB memory. In the training stage, we optimize the model using Adam optimizer. Specifically, in the first stage of prior repair, we set the learning rate to 6e−4. In the second stage of content filling, we set the learning rate of the generative network and the discriminative network to 1e−3 and 1e−4, respectively. The loss function settings are the same as [17], including L1 loss, feature match loss, adversarial loss, and high receptive field (HRF) perceptual loss. The weights of each part are $\lambda_{L1} = 10$, $\lambda_{\text{fm}} = 10$, $\lambda_{\text{adv}} = 100$, $\lambda_{\text{hrf}} = 30$, respectively.

4.2 Qualitative Comparisons

To objectively assess the inpainting performance of the proposed method, we compare it with the following advanced methods: LaMa [17], ZITS [11], CoordFill [22], SCAT [23] and HINT [24]. As shown in the Fig. 6, LaMa and SCAT cannot well understand the known information within the image, resulting in noticeable blurriness and artifacts in the generated image. Although ZITS and CoordFill can understand the context of the image to a certain degree, the generated content is still inconsistent with the surrounding information in terms of color and structure. Compared to the above methods, HINT has a significant improvement in the ability to capture global information, but for some complex structures, such as the railings and windows in the first and third rows in the Fig. 6, there are still some disadvantages. Thanks to the prior guidance and the GAN-based Content Filling Network, our model is capable of generating content that is harmonious and coherent with the known information of the image in structure and texture, whether for the restoration of natural or man-made landscapes.

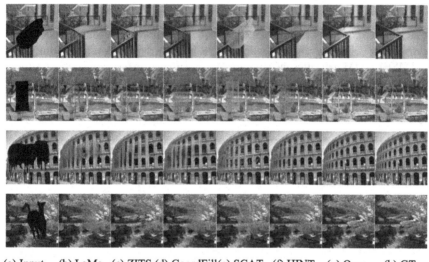

(a) Input (b) LaMa (c) ZITS (d) CoordFill (e) SCAT (f) HINT (g) Ours (h) GT

Fig. 6. Qualitative results on Places2.

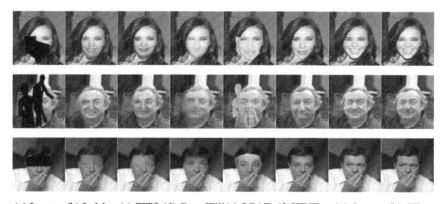

(a) Input (b) LaMa (c) ZITS (d) CoordFill (e) SCAT (f) HINT (g) Ours (h) GT

Fig. 7. Qualitative results on CelebA.

For facial inpainting, as shown in the Fig. 7, the effective content generated by CoordFill and SCAT in the missing area is very limited, with obvious blur and boundary. Although the images generated by LaMa and ZITS largely avoid artifacts, the size and shape of the generated features are obviously inconsistent. The HINT model avoids the above problems to some extent, but the faces generated still have unnatural expressions and images do not correspond to the real world. Thanks to the global receptive field and strong prior guidance, our model can not only achieve clear and natural facial inpainting, but also be closer to real life.

4.3 Quantitative Comparisons

Apart from the visual qualitative comparison, PSNR, SSIM [25], FID [26] and LPIPS [27] are selected as evaluation indicators for quantitative comparison of models. As an objective image quality evaluation index, PSNR is defined on the basis of Mean Squared Error (MSE). SSIM measures the similarity of two images in terms of contrast, structure and brightness. FID is used to assess the accuracy and variety of the generated image. LPIPS can better simulate the human visual perception of images and is sensitive to details and textures.

Table 1. Objective quantitative comparison on Places2 and CelebA.

Method	Places2				CelebA			
	PSNR	SSIM	FID	LPIPS	PSNR	SSIM	FID	LPIPS
LaMa [17]	24.108	0.874	3.537	0.119	26.213	0.912	2.576	0.082
ZITS [11]	25.801	0.885	2.641	0.111	27.049	0.925	1.776	0.071
CoordFill [22]	26.976	0.902	–	0.083	26.652	0.904	–	0.097
SCAT [23]	24.410	0.894	–	–	26.749	0.913	–	–
HINT [24]	27.523	0.914	1.758	0.103	29.682	0.946	1.135	0.058
Ours	**29.872**	**0.962**	**1.365**	**0.064**	**31.195**	**0.975**	**0.964**	**0.047**

As shown in the Table 1, our model achieves better results on the Places2 and CelebA datasets, and all indicators have been improved. This can be attributed to our model balances the structural guidance and texture generation, which not only explores a better prior type to guide the inpainting process, but also ensures that the effective information of prior knowledge will not be lost in the process of guiding the repair. Simultaneously, the model's capability to capture the image's global information is enhanced even more.

4.4 Ablation Studies

We conduct ablation experiments on Places2 and CelebA datasets to validate the effectiveness of each module proposed in this paper, and successively add Texture Generation Module (TGM), prior guidance (PG), Prior Addition Module (PAM) and discriminator on the basis of baseline. As shown in the Fig. 8, the use of FFC in Texture Generation Module (TGM) greatly enhances the model's capacity to comprehend the global information of the image, improves the coherence between the generated content of the model and the surrounding information, and reduces the artifacts in the generated area to a certain extent. The prior guidance (PG) greatly improves the model's perception capacity to the overall structure of the image, and promotes the model to generate the missing content with reasonable structure and clear texture. The Prior Addition Module (PAM) effectively protects the guidance information from being filtered out in the process of generating images, so that the model generates more coherent and coordinated content, and further improves the model's performance. The improvement of the discriminator makes the generated images more natural in color. As shown in the Table 2, each module has different degrees of improvement on the performance of the model, especially the guidance of the prior greatly improves the generation ability of our method.

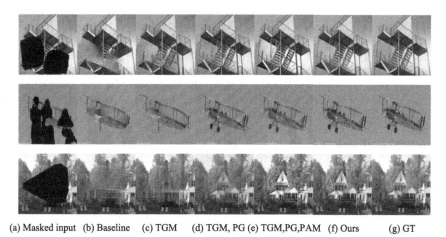

(a) Masked input (b) Baseline (c) TGM (d) TGM, PG (e) TGM,PG,PAM (f) Ours (g) GT

Fig. 8. Visualization of the effects of network architecture and individual modules on Places2. Here, TGM represents Texture Generation Module, PG denotes prior guidance, and PAM represents Prior Addition Module.

Table 2. Quantitative ablation study on Places2 and CelebA.

Model configurations	Places2				CelebA			
	PSNR	SSIM	FID	LPIPS	PSNR	SSIM	FID	LPIPS
Baseline	24.193	0.877	4.382	0.130	25.682	0.913	2.005	0.088
TGM	24.646	0.880	3.792	0.126	26.701	0.920	1.569	0.079
TGM, PG	27.097	0.934	1.981	0.085	28.952	0.956	1.124	0.058
TGM, PG,PAM	28.913	0.953	1.405	0.069	30.384	0.963	0.989	0.052
Ours	**29.872**	**0.962**	**1.365**	**0.064**	**31.195**	**0.975**	**0.964**	**0.047**

5 Conclusions

In this paper, we propose a structural information-guided fine-grained texture inpainting model, which consists of two phases, namely prior repair and content filling. In the first stage, we take the gradient map and the wireframe as the guidance information of the model, following which we use TSR to repair them. The repaired priors are encoded by GFE, and then the encoded features are sent to the second stage to guide the image inpainting. In the second stage, we propose the Prior Addition Module to minimize the loss of prior information in the inpainting process, and Texture Generation Module to enhance the model's ability to perceive context information. In addition, we add the discrimination of gradients to the discriminator to promote the model to yield more lifelike content. The experimental results demonstrate the effectiveness of the proposed method on Places2 and CelebA datasets compared with several state-of-the-art methods.

References

1. Xie, J., Xu, L., Chen, E.: Image denoising and inpainting with deep neural networks. In: Advances in Neural Information Processing Systems, vol. 25 (2012)
2. Pathak, D., Krähenbühl, P., Donahue, J., Darrell, T., Efros, A.A.: Context encoders: feature learning by inpainting. In: Proceedings of the IEEE Conference on Computer Vision and Pattern Recognition, pp. 2536–2544 (2016)
3. Iizuka, S., Simo-Serra, E., Ishikawa, H.: Globally and locally consistent image completion. ACM Trans. Graph. (ToG) **36**(4), 1–14 (2017)
4. Yu, J., Lin, Z., Yang, J., Shen, X., Lu, X., Huang, T.S.: Generative image inpainting with contextual attention. In: Proceedings of the IEEE Conference on Computer Vision and Pattern Recognition, pp. 5505–5514 (2018)
5. Nazeri, K., Ng, E., Joseph, T., Qureshi, F.Z., Ebrahimi, M.: EdgeConnect: generative image inpainting with adversarial edge learning. arXiv preprint arXiv:1901.00212 (2019)
6. Guo, X., Yang, H., Huang, D.: Image inpainting via conditional texture and structure dual generation. In: Proceedings of the IEEE/CVF International Conference on Computer Vision, pp. 14134–14143 (2021)
7. Goodfellow, I.J., et al.: Generative adversarial nets. In: Advances in Neural Information Processing Systems, vol. 27 (2014)
8. Mirza, M., Osindero, S.: Conditional generative adversarial nets. arXiv preprint arXiv:1411.1784 (2014)
9. Radford, A.: Unsupervised representation learning with deep convolutional generative adversarial networks. arXiv preprint arXiv:1511.06434 (2015)
10. Arjovsky, M., Chintala, S., Bottou, L.: Wasserstein GAN. arXiv preprint arXiv:1701.07875 (2017)
11. Dong, Q., Cao, C., Fu, Y.: Incremental transformer structure enhanced image inpainting with masking positional encoding. In: Proceedings of the IEEE/CVF Conference on Computer Vision and Pattern Recognition, pp. 11358–11368 (2022)
12. Cao, C., Fu, Y.: Learning a sketch tensor space for image inpainting of man-made scenes. In: Proceedings of the IEEE/CVF International Conference on Computer Vision, pp. 14509–14518 (2021)
13. Ho, J., Kalchbrenner, N., Weissenborn, D., Salimans, T.: Axial attention in multidimensional transformers. arXiv preprint arXiv:1912.12180 (2019)
14. Yu, J., Lin, Z., Yang, J., Shen, X., Lu, X., Huang, T.S.: Free-form image inpainting with gated convolution. In: Proceedings of the IEEE/CVF International Conference on Computer Vision, pp. 4471–4480 (2019)
15. Yu, F., Koltun, V.: Multi-scale context aggregation by dilated convolutions. arXiv preprint arXiv:1511.07122 (2015)

16. Park, T., Liu, M., Wang, T., Zhu, J.: Semantic image synthesis with spatially-adaptive normalization. In: Proceedings of the IEEE/CVF Conference on Computer Vision and Pattern Recognition, pp. 2337–2346 (2019)
17. Suvorov, R., et al.: Resolution-robust large mask inpainting with Fourier convolutions. In: Proceedings of the IEEE/CVF Winter Conference on Applications of Computer Vision, pp. 2149–2159 (2022)
18. Zhou, B., Lapedriza, À., Khosla, A., Oliva, A., Torralba, A.: Places: a 10 million image database for scene recognition. IEEE Trans. Pattern Anal. Mach. Intell. **40**(6), 1452–1464 (2017)
19. Liu, Z., Luo, P., Wang, X., Tang, X.: Deep learning face attributes in the wild. In: Proceedings of the IEEE International Conference on Computer Vision, pp. 3730–3738 (2015)
20. Liu, G., Reda, F.A., Shih, K.J., Wang, T., Tao, A., Catanzaro, B.: Image inpainting for irregular holes using partial convolutions. In: Proceedings of the European Conference on Computer Vision (ECCV), pp. 85–100 (2018)
21. Lin, T. Y., et al.: Microsoft COCO: common objects in context. In: Computer Vision–ECCV 2014: Proceedings of the 13th European Conference, Zurich, Switzerland, 6–12 September 2014, Part V 13, pp. 740–755 (2014)
22. Liu, W., Cun, X., Pun, C., Xia, M., Zhang, Y., Wang, J.: CoordFill: efficient high-resolution image inpainting via parameterized coordinate querying. arXiv preprint arXiv:2303.08524 (2023)
23. Zuo, Z., et al.: Generative image inpainting with segmentation confusion adversarial training and contrastive learning. In: Proceedings of the AAAI Conference on Artificial Intelligence, vol. 37, no. 3, pp. 3888–3896 (2023)
24. Chen, S., Atapour-Abarghouei, A., Shum, H. P.: HINT: high-quality inpainting transformer with mask-aware encoding and enhanced attention. IEEE Trans. Multimedia (2024)
25. Wang, Z., Bovik, A.C., Sheikh, H.R., Simoncelli, E.P.: Image quality assessment: from error visibility to structural similarity. IEEE Trans. Image Process. **13**(4), 600–612 (2004)
26. Heusel, M., Ramsauer, H., Unterthiner, T., Nessler, B., Hochreiter, S.: GANs trained by a two time-scale update rule converge to a local Nash equilibrium. In: Advances in Neural Information Processing Systems, vol. 30 (2017)
27. Zhang, R., Isola, P., Efros, A.A., Shechtman, E., Wang, O.: The unreasonable effectiveness of deep features as a perceptual metric. In: Proceedings of the IEEE Conference on Computer Vision and Pattern Recognition, pp. 586–595 (2018)
28. Vaswani, A.: Attention is all you need. arXiv preprint arXiv:1706.03762 (2017)
29. Zeng, Y., Fu, J., Chao, H., Guo, B.: Learning pyramid-context encoder network for high-quality image inpainting. In: Proceedings of the IEEE/CVF Conference on Computer Vision and Pattern Recognition, pp. 1486–1494 (2019)
30. Li, W., Lin, Z., Zhou, K., Qi, L., Wang, Y., Jia, J.: MAT: mask-aware transformer for large hole image inpainting. In: Proceedings of the IEEE/CVF Conference on Computer Vision and Pattern Recognition, pp. 10758–10768 (2022)
31. Ballester, C., Bertalmío, M., Caselles, V., Sapiro, G., Verdera, J.: Filling-in by joint interpolation of vector fields and gray levels. IEEE Trans. Image Process. **10**(8), 1200–1211 (2001)
32. Barnes, C., Shechtman, E., Finkelstein, A., Goldman, D.B.: PatchMatch: a randomized correspondence algorithm for structural image editing. ACM Trans. Graph. **28**(3), 24 (2009)
33. Krizhevsky, A., Sutskever, I., Hinton, G.E.: Imagenet classification with deep convolutional neural networks. In: Advances in Neural Information Processing Systems, vol. 25 (2012)

Style Separation and Content Recovery for Generalizable Sketch Re-identification and a New Benchmark

Lingyi Lu(✉), Xin Xu, and Xiao Wang

School of Computer Science and Technology, Wuhan University of Science and Technology, Wuhan 430065, China
{lly,xuxin,wangxiao2021}@wust.edu.cn

Abstract. Existing research on Sketch-Photo Person Re-identification (Sketch Re-ID) has been conducted using a publicly available dataset comprising only 200 identities, yielding satisfactory retrieval results. However, the limited size and diversity of training datasets constrain the potential for performance enhancement in current Sketch Re-ID models. When applying general deep learning models to novel datasets characterized by unknown distributions, these models often overlook the impact of domain shifts resulting from distributional discrepancies among different datasets. This oversight frequently leads to performance deterioration and diminished generalization capabilities. To address this challenge, we introduce a large-scale, multi-source, multi-view synthetic sketch re-identification dataset, designated as MSMV-SK, which comprises 3,364 synthetic sketches associated with 1,364 identities. To develop a generalizable Sketch Re-ID model utilizing this dataset, we propose a cross-modal feature learning network aimed at extracting domain-invariant features. Specifically, we implement a Style Separation and Content Recovery (SSCR) module that mitigates the effects of style variations while reconstructing identity-related components from features derived from both sketch and photo modalities. Additionally, we formulate a Bi-constrained Optimization Loss (BOL) to enhance the discriminative power of identity-related features. Extensive experimental evaluations demonstrate that our model, trained on the synthetic MSMV-SK dataset, outperforms a model trained on the Mask1k dataset when assessed on the PKU dataset. Notably, our approach achieves a 4% improvement in Rank-1 accuracy and a 3.72% increase in mean Average Precision (mAP), underscoring its superior generalization capabilities. Dataset is publicly available at: https://github.com/Lulingyi01/MSMV-SK.git.

Keywords: Sketch-photo person re-identification · Domain generalization · Instance normalization · Convolutional neural networks

1 Introduction

Person re-identification (Re-ID) [11,14,21] aims to retrieve all photos of the same identity across cameras by giving a photo of the target person. However, the application of the Re-ID technology in real-world scenarios will be extremely limited when the available query images cannot be captured due to the surveillance blind zones. Therefore, Sketch-Photo Person Re-identification (Sketch Re-ID) [12] has been proposed to match sketches of the target person with photos captured by surveillance videos. In cases where witnesses are involved to assist in the investigation, a mock sketch of the suspect drawn by a professional portraitist through communication with the witnesses will serve as a useful clue, with which the scope of the suspect can be quickly narrowed down to greatly accelerate the speed of solving the case, which has proven to be an effective criminal investigation tool.

Benefiting from the availability of real Sketch Re-ID datasets, existing state-of-the-art algorithms have achieved extraordinary human-level accuracy. Nevertheless, the generalization of these learned models to unknown target domains is still poor (see Fig. 1(a)), i.e., models trained on the source dataset perform significantly degraded on the target dataset. On the one hand, this is due to the lack of large-scale and diverse source training datasets, the existing commonly used public Sketch Re-ID dataset known as PKU-Sketch, which has only 200 person identities. Collecting sufficient photo-sketch examples is both time-consuming and labor-intensive. On the other hand, the domain shift problem caused by the distribution discrepancy between different datasets hinders the high-generalization capability, as shown in Fig. 1(b). Sketches are hand-drawn by professional artists with different drawing styles and richness of details, while photos of people taken by different cameras have considerable style differences in terms of lighting, quality, etc., and are affected by a variety of factors, such as the pose of the person, the shooting angle, and occlusion. In addition, there are obvious appearance differences between the abstract sketch modality and the concrete photo modality. It is difficult for the model to extract modality-invariant features due to the disturbance of background information, the missing of key objects and the inconsistency of detail.

To tackle the aforementioned challenges, we undertook two key initiatives: (1) We constructed a large-scale multi-source, multi-view sketch dataset, referred to as MSMV-SK, utilizing low-cost and easily accessible synthetic sketches. This dataset aims to provide a diverse range of examples, facilitating better training and evaluation for Sketch Re-ID models. (2) We developed a cross-modal identity-related generalization feature-learning network specifically designed to address the domain shifts exacerbated by the modal differences present across various data domains. This network focuses on extracting robust features that are invariant to the discrepancies between sketches and photos, ultimately enhancing the model's generalization capabilities in real-world applications.

Specifically, our focus is on developing a more practical and generalizable Sketch Re-ID framework, with the goal of designing models that can effectively generalize to datasets with unknown distributions, without requiring access to target domain data or labels. Research has indicated that models trained on synthetic datasets tend to exhibit superior generalization performance for re-identification tasks compared to those trained solely on publicly available real

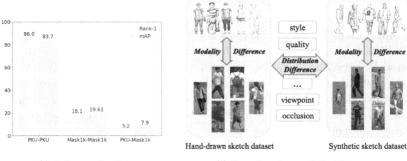

Fig. 1. Idea illustration. (a) Demonstrates the existing best performance achieved on the two Sketch Re-ID datasets PKU and Mask1k, while the performance is severely degraded when the model obtained by training on one dataset is directly tested on the other dataset. (b) Comparison of hand-drawn and synthetic sketch dataset. It shows that there are significant modal differences between the sketches and the photos. There is also a huge distribution difference between the two datasets.

datasets [7]. Our multi-view synthetic sketches not only approximate the stylistic characteristics of the sketch modality but also preserve the structural information inherent in the photo modality.

The advantages of utilizing synthetic sketches can be summarized in two key points:

- **Consistent Style.** Synthetic sketches maintain a uniform style, unlike real sketches that are hand-drawn by different artists, resulting in stylistic variations based on individual interpretations and descriptions. Such discrepancies can adversely affect matching performance, making consistent synthetic sketches more advantageous for model training.
- **Rich View Variations.** Synthetic sketches can incorporate diverse view changes, whereas hand-drawn sketches typically lack such variations. This richness in view changes helps to mitigate the challenges associated with domain transfer between sketches and view-rich real photos.

In conclusion, the use of synthetic sketches offers a more consistent style and a broader range of view variations, which can significantly enhance the model's ability to learn cross-modal identity-related generalization features, thereby improving overall performance in practical applications of Sketch Re-ID.

Based on the proposed synthetic dataset MSMV-SK, we propose a new cross-modal identity-related feature learning generalizable network. Specifically, style separation and content recovery module uses instance normalization (IN) to normalize features across spatial locations on each channel during the style separation phase, which removes the effects of style changes and reduces the impact of domain shifts on recognition performance. In order to recover useful information from the features discarded after IN, we decompose the features extracted from the two modalities, and let the model focus more on learning identity-related features. We further design a bi-constrained optimization loss to facilitate the separation of identity-related features and identity-irrelevant features.

Our main contributions are summarized as follows:

- We constructed a large-scale multi-source multi-view synthetic sketch re-identification dataset MSMV-SK by introducing synthetic sketches.
- We propose a style separation and content recovery module to learn identity-related features and a bi-constrained optimization loss for improving the generalization of the model between different modality datasets.
- We conducted extensive experimental evaluations to show that our models trained on the synthetic MSMV-SK dataset have superior generalization capability, outperforming models trained on the real Mask1k dataset.

2 Related Work

2.1 Sketch-Photo Person Re-identification

Sketch Re-ID refers to the use of professional sketches of target persons to retrieve them from photos of pedestrians taken from surveillance videos. Pang et al. [12] firstly defined the cross-domain matching problem based on professional sketches and photos as Sketch Re-ID, and they used a cross-domain adversarial learning approach to extract discriminative modality invariant in the sketches and visible image features. One way to bridge the modal gap at the image level is to convert the images of both modalities into unimodal greyscale images [4] as a preprocessing to remove the effect of color, which facilitates modality-invariant feature learning. Chen et al. [2] used cross-spectral image generation technique as a means of data augmentation to expand the training set of sketches in order to achieve recognition performance improvement. Chen et al. [1] utilized sketch synthesis methods to dynamically synthesize A-sketch-assisted modality that are more consistent with the person's representation, and then used knowledge transfer to generate stylistically consistent cross-modal representations to mitigate the effects of modal differences between sketches and photos.

The other way is to learn modal invariant features to bridge the modal gap at the feature level. To make the model pay more attention to the regions in the photo that are relevant to the sketch, Zhu et al. [25] designed a cross-domain attention mechanism and also used cross-domain center loss to capture the domain invariant features. Zhang et al. [22] proposed a cross-compatible embedding method based on the idea of non-exclusionary porting and further designed a semantic consistent feature construction scheme to enhance the feature representation via dealing with the information inconsistency at the same location of the sketch and the corresponding photo.

However, all the above methods were only evaluated on a small-scale sketch dataset, the PKU-Sketch, and have achieved exceptional human-level accuracy. Recently, Lin et al.[10] constructed a large-scale sketch dataset named Market-Sketch-1K aiming to focus on the subjectivity challenge. Based on this dataset, the author utilized attribute information as an implicit mask to align cross-domain features. Compared to the existing datasets in the field of Re-ID, the sketch dataset is significantly underpowered, which limits the development of the field. It is worth noting that none of the aforementioned datasets take into account the specificity of sketch data, and thus proposing a new benchmark applicable to the task of sketch re-identification is necessary.

2.2 Domain Generalization Person Re-identification

DG Re-ID [8,15,18] refers to deploying models trained on source domains directly to unseen target domains without prior knowledge of their distribution, while minimizing performance degradation. Current mainstream methods can be broadly categorized into two types: those based on domain-invariant feature learning, and those that rely on domain adaptation techniques.

An intuitive way to disentangle representation in domain-irrelevant feature learning is to disentangle a model into two parts, one that learns domain-specific knowledge, and the other that learns domain-irrelevant knowledge. Zhang et al. [23] developed a Structural Causal Model (SCM) to disentangle identity-specific and domain-specific factors into two separate feature spaces, and they proposed an effective backdoor adjustment approximation to achieve causal intervention. To address the Camera-to-Camera (CC) and Camera-to-Person (CP) challenges in DG Re-ID, Xu et al. [19] introduced a Bi-stream Generative Model (BGM) that learns a fine-grained representation by integrating camera-invariant global features and person-aligned local features. Jin et al. [6] proposed a generalizable person re-identification framework that separates identity-related and identity-irrelevant features through style normalization and dual causal relationship loss.

Although the above methods have yielded some achievements in the photo-to-photo generalization task, they lack exploration on larger domain transfer tasks like sketch-to-photo.

3 Multi-source Multi-view Synthetic Sketch Dataset

3.1 Dataset Description

In this paper, we collect a new challenging multi-source synthetic sketch-photo person re-identification dataset called the MSMV-SK dataset. The MSMV-SK dataset is constructed based on four public datasets VIPeR [3], Market1501 [24], SYSU-MM01 [17], and WePerson [9]. We randomly select identities from these datasets as a reference subset, and the corresponding photos constitute the photo part of the MSMV-SK dataset, with the specific information shown in Table 1.

Table 1. Comparisons between related datasets and the created MSMV-SK.

	Person Re-ID				Sketch Re-ID		
	single-modality			cross-modality	cross-modality		
attributes	VIPeR	Market1501	WePerson	SYSU-MM01	PKU-Sketch	Mask1k	Ours
ID	632	1,501	1,500	491	200	996	1364
sketch styles	–	–	–	–	5	6	1
camera views	2	6	40	6	2	6	54
rgb photos	1,264	32,668	4,000,000	287,628	400	2,000	2,874
sketches/infrared photos	–	–	–	15,792	200	4,763	3,364
virtual dataset	×	×	✓	×	×	×	✓
sketch style factor	–	–	–	–	✓	✓	✓
sketch view factor	–	–	–	–	×	×	✓

In the sketch part, we will obtain the synthetic sketches through the following three steps: (1) **Data collection.** In order to simulate the frontal sketches drawn by professional artists in real situations, we use the frontal face detector in the dlib library to filter out the frontal photos of each identity; (2) **Background erase.** Referring to the hand-drawn sketches, we firstly erase the background of the RGB image through the ZuoTang API to mitigate the effect of background noise. This API returns the foreground person image by recognising the human contours in the input image and separating them from the background. (3) **Sketch generation.** We convert photos to sketches through the InstantPhotoSketch editor, which adjusts the degree of blackness and whiteness in the final rendered sketch by moving the horizontal scroll bar. It is user-friendly and free of charge.

3.2 Evaluation Protocol

We divide the MSMV-SK dataset into a training set and a test set in a ratio of approximately 3:1. The training set contains 2,537 sketches and 2,114 photos for 1,022 identities, and the test set contains 827 sketches and 760 photos for 342 identities. In the testing phase, for each camera, we randomly select one image from each identity to form the gallery set used to evaluate the model performance. We repeat the above evaluation 10 times, randomly splitting the gallery set and reporting the average performance.

4 Method

Our goal is to design a generalizable and robust Sketch Re-ID framework to bridge the gap between different domains. Figure 2 illustrates the overall flow of our proposed cross-modal feature learning network, which consists of three main components: a) dual-stream feature extraction network; b) style separation and content recovery module; c) the bi-constrained optimization loss.

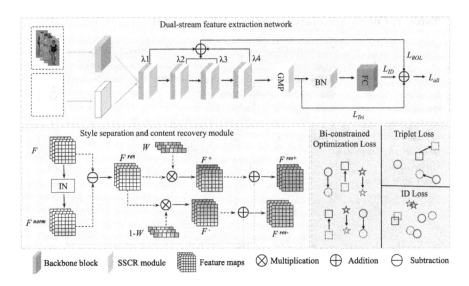

Fig. 2. Overview of our method. The backbone is a dual-stream feature extraction network with the style separation and content recovery (SSCR) module added after each block, which removes the effects of style variations and recovers the identity-related part of the features extracted from sketch and photo modality. In addition, to better disentangle identity-related and identity-irrelevant features, we design the bi-constrained optimization loss (BOL) to improve the discrimination of identity-related features.

4.1 Dual-Stream Feature Extraction Network

For a sketch-photo dataset, let $\mathcal{P} = \{p_i\}_{i=1}^{N_p}$ and $\mathcal{S} = \{s_i\}_{i=1}^{N_s}$ denote the photo and sketch modality, respectively. $N = N_s + N_p$ denotes the total number of person samples. We use a two-stream convolutional neural network [20] as the backbone for generalized sketch-photo person re-identification, extracting feature maps from the input images. First, we use two shallow convolutional blocks with individual network parameters to extract modality-specific feature representations. Then, four deep convolutional blocks with shared parameters are used to learn modality-shared feature representations. After the convolutional layer with Generalized-Mean Pooling (GMP), a shared Batch Normalization (BN) layer is added to learn the shared feature embeddings. Finally, a shared fully-connected layer is employed on top of the dual-stream feature extractor.

We denote the embedding function as $F_p(\cdot)$ for photos and $F_s(\cdot)$ for sketches. Therefore, given a photo $\{x_p^n\}$ and a sketch $\{x_s^n\}$, where n denotes the person with the identity $\{y^n\}$. The learned person features $\mathbf{f}_p$ and $\mathbf{f}_s$ in the common space can be represented as:

$$\mathbf{f}_p = F_p(x_p), \mathbf{f}_s = F_s(x_s). \tag{1}$$

Generalized-Mean Pooling. For most fine-grained image retrieval tasks, maximum pooling or average pooling is commonly used for feature extraction but neither of them can capture domain specific features. We employ GeM [13], a learnable pooling layer, formulated by

$$\mathbf{f} = [f_1 \cdots f_m \cdots f_M]^T, f_m = \left(\frac{1}{|\mathcal{X}_m|} \sum_{x_i \in \mathcal{X}_m} x_i^{p_m} \right)^{\frac{1}{p_m}}. \quad (2)$$

where f_m represents the feature map, and M is number of feature maps in the last layer. $\mathcal{X}_m$ is the set of $W \times H$ activations for feature map $m \in \{1, 2, \cdots, M\}$. p_n is a pooling hyper-parameter, which is learned in the back-propagation process [13].

4.2 Style Separation and Content Recovery

For Sketch Re-ID, there exists large modal differences (e.g., in appearance, colour, background, detail, etc.) between sketches and photos. For samples from two different datasets/domains, there are style differences between sketches hand-drawn by different artists, while photos of the person captured by different cameras show style differences (e.g., in terms of illumination, viewpoint, quality). In other words, the distribution difference between the source and target domains usually affects the generalization ability of the model.

We use Instance Normalization layer (IN) to normalize features across spatial locations on each channel during the style separation phase, which preserves spatial structure and reduces the impact of domain bias on recognition performance. However, IN can lead to serious loss of discriminative information, and even though we only add IN at a shallow level of the network. Therefore, a further content recovery phase will extract identity-related discriminative features from the IN dropped features and add them to the normalized features, which is crucial to improve the recognition capability of the model.

For an SSCR module, we denote the input by $F \in \mathbb{F}^{h \times w \times c}$ and the output by $F^{res+} \in \mathbb{F}^{h \times w \times c}$, where h, w, c denote the height, width, and number of channels, respectively. To learn domain-irrelevant features, we use IN to reduce style differences between domains. The formula is given as

$$F = \phi(\mathbf{f}_p, \mathbf{f}_s),$$
$$F^{norm} = \text{IN}(F) = \gamma \left(\frac{F - \mu(F)}{\varrho(F)} \right) + \beta, \quad (3)$$

where $\phi(\cdot)$ denote the concatenate operation, $\mu(\cdot)$ and $\varrho(\cdot)$ denote the mean and standard deviation computed across spatial dimensions independently for each channel and each instance, $\gamma, \beta \in \mathbb{F}^c$ are parameters learned from data.

For feature mapping of an image, the style information of the image is encoded in the first order statistics (mean and variance) of the convolutional feature map. Thus, by normalizing the output of the residual bottleneck, instance-specific styles can be separated. However, it inevitably discards some discriminative identity-related information, though it is used in shallow-layer networks. We extract identity-related discriminative features from the IN discarded features F^{res} and add them to the normalised features. F^{res} is defined as

$$F^{res} = F - F^{norm}, \tag{4}$$

which denotes the difference between the original input feature F and the style normalized feature F^{norm}.

In order to recover the identity-related content from the residual features F^{res}, we first extract the identity-related features using the SE-like [5] channel attention adaptive response vector W,

$$W = \omega(F^{res}) = \sigma\left(C_2 \delta\left(C_1 pool\left(F^{res}\right)\right)\right), \tag{5}$$

which consists of a global average pooling layer followed by fully connected layers that are parameterized by $C_1 \in \mathbb{F}^{(c/r) \times c}$ and $C_2 \in \mathbb{F}^{c \times (c/r)}$ which are followed by ReLU activation function $\delta(\cdot)$ and sigmoid activation function $\sigma(\cdot)$, respectively.

According to the learned channel attention vector W, we further disentangle F^{norm} into two parts: identity-related feature $F^+ \in \mathbb{F}^{h \times w \times c}$ and identity-irrelevant feature $F^- \in \mathbb{F}^{h \times w \times c}$.

$$\begin{aligned} F^+(:,:,k) &= w_k F^{res}(:,:,k), \\ F^-(:,:,k) &= (1-w_k) F^{res}(:,:,k), \end{aligned} \tag{6}$$

where $F(:,:,k) \in \mathbb{F}^{h \times w}$ denotes the kth channel of feature map F, $k = 1, 2, \ldots, c$..

By adding the distilled identity-related feature F^+ to the style normalized feature F^{norm}, we obtain the output feature F^{res^+} of the SSCR module as

$$F^{res^+} = F^{norm} + F^+. \tag{7}$$

4.3 Bi-constrained Optimization Loss

To facilitate the disentanglement of identity-related and identity-irrelevant features by comparing the discriminative ability of the features before and after restoration, we design the bi-constrained optimization loss $\mathcal{L}_{BOL}$ which consists of clarification loss $\mathcal{L}_{BOL}^+$ and destruction loss $\mathcal{L}_{BOL}^-$, i.e., $\mathcal{L}_{BOL} = \mathcal{L}_{BOL}^+ + \mathcal{L}_{BOL}^-$. It encourages higher discriminative power after recovering identity-related features and lower discriminative power after recovering identity-irrelevant features.

We define $pool(\cdot)$ as a spatial average pooling operation. Therefore, f^{res^+} can be obtained by $f^{res^+} = pool(F^{res^+})$, then it pass into a FC layer followed by softmax function $\Phi(\cdot)$. We denote an entropy function as $H(\cdot) = -p(\cdot) \log p(\cdot)$. Similarly, $f^{res^-} = pool(F^{res^-})$ can be obtained, and the style normalized feature vector is $f^{norm} = pool(F^{norm})$. $\mathcal{L}_{BOL}^+$ and $\mathcal{L}_{BOL}^-$ are defined as:

$$\begin{aligned} \mathcal{L}_{BOL}^+ &= Softplus\left(H\left(\Phi\left(f^{res^+}\right)\right) - H(\Phi(f^{norm}))\right), \\ \mathcal{L}_{BOL}^- &= Softplus\left(H(\Phi(f^{norm})) - H\left(\Phi\left(f^{res^-}\right)\right)\right), \end{aligned} \tag{8}$$

where $Softplus(\cdot) = \ln(1 + \exp(\cdot))$ is a monotonically increasing function. Intuitively, $\mathcal{L}_{BOL}$ promotes better feature disentanglement of the residual for feature recovery and makes samples with the same identity feature closer together and samples with different identities feature farther apart.

4.4 Supervision

Identity Loss. Given an image x with label y, its probability to be predicted as label y within the total n classes is p. The probability is calculated from the photo/sketch feature with a pooling layer followed by a softmax function. The identity loss is computed from the cross-entropy with label smoothing regularization, where ϵ is the label smoothing parameter.

$$\mathcal{L}_{ID} = -\sum_{i=1}^{n} y_i' \log(p_i),$$
$$y_i' = (1 - \epsilon) y_i + \frac{\epsilon}{n}, p_i = \frac{e^{x_i}}{\sum_{j=1}^{n} e^{x_j}}. \quad (9)$$

Triplet Loss. The basic idea is that the distance between the positive pair should be smaller than the negative pair by a pre-defined margin ρ.

$$\mathcal{L}_{Tri} = \frac{1}{N} \sum_{i=1}^{N} \max\left(d(x_i, x_j) - d(x_i, x_k) + \rho, 0\right), \quad (10)$$

where (x_i, x_j, x_k) represents a hard triplet within each training batch, N is sample numbers in the batch. $d(\cdot)$ measures the Euclidean distance between two samples.

Overall Loss. Overall, we combine the above objectives together as follows:

$$\mathcal{L}_{all} = \mathcal{L}_{ID} + \mathcal{L}_{Tri} + \sum_{i=1}^{4} \lambda_i \mathcal{L}_{BOL}^i. \quad (11)$$

where $\mathcal{L}_{BOL}^i$ denotes the bi-constrained optimization loss for the ith SSCR module. λ_i is a weight which controls the relative importance of the regularization at stage i. We experimentally set λ_3, λ_4 to 0.3, and λ_1, λ_2 to 0.1.

5 Experiments

5.1 Datasets and Settings

Datasets. We evaluate our proposed method on three Sketch Re-ID datasets, including PKU-Sketch [12] and Market-Sketch-1K [10], and the MSMV-SK proposed in this paper. For simplicity, PKU denotes the PKU-Sketch dataset and Mask1k denotes the Market-Sketch-1K dataset. PKU [12] is the first Sketch

Table 2. Performance comparison with the state-of-the-art methods on the DG Re-ID setting. "*" indicates that we re-implement this work based on the authors' code and FT indicates fine-tuning.

Source	Method	Target: PKU			Target: Mask1k		
		Rank-1	Rank-10	mAP	Rank-1	Rank-10	mAP
Mask1k	Subjectivity	58.00	92.00	55.61	18.10	50.75	19.61
	Subjectivity*	56.00	90.00	57.44	15.57	47.72	17.75
MSMV-SK	Ours	60.00	90.00	61.16	4.89	20.30	7.29
	Ours+FT	**76.00**	**90.00**	**69.97**	**16.16**	**49.20**	**18.84**

Re-ID dataset. It consists of 200 persons, each with two photos and one sketch. Specifically, the RGB photos are taken by two cross cameras, while the sketches are drawn by five professional artists based on eyewitness descriptions. Mask1k [10] is built based on the most popular Re-ID dataset Market-1501 [24] consisting of 996 persons with two photos and multiple sketches for each person. The dataset is divided into 6 different groups (S1 - S6) according to the sketch style.

Evaluation Metrics. We use common evaluation metrics to measure the goodness of our approach. One of them is CMC (Cumulative Matching Characteristics) [16] and the other is mAP (mean Average Precision) [24]. CMC measures the probability that a correct match appears in the top-k search results (Rank-k matching accuracy). The mAP is applied to measure the retrieval performance in multi-target detection and multi-label image classification.

Implementation Details. We use the ImageNet pre-trained Resnet-50 network [20] as the backbone. The Sketch Re-ID dataset was trained on a single Nvidia Geforce GTX 3090 GPU. In each training epoch, we randomly select 8 pedestrian identities from the entire dataset. Each identity randomly selects 4 sketches and 4 photos. All input images of the two modalities are first adjusted to 288×144. The initial learning rate as 0.05 on both datasets, and decay it by 0.1 and 0.01 at 20 and 50 epochs, respectively.

5.2 Benchmark Results

To prove the effectiveness of the proposed framework on cross-domain generalization learning, we select a dataset from Mask1k and MSMV-SK as the training set, and use PKU and Mask1k as the testing set. For simplicity, 'A→B' means the model is trained on A and tested on B. We evaluate the effectiveness of the proposed method by comparing it with the state-of-the-art methods in Table 2.

Our scheme significantly outperforms Mask1k→PKU on MSMV-SK→PKU with 4% higher Rank-1 and 3.72% higher mAP. The results demonstrate that the

Table 3. Ablation Study. Both training and testing are performed on the MSMV-SK dataset.

(a) Model ablation

IN	SSCR	BOL	mAP	Rank-1
×	×	×	53.42	52.08
✓	×	×	56.67	59.98
✓	✓	×	57.92	61.19
✓	✓	✓	**59.10**	**63.36**

(b) Loss ablation

L_{ID}	L_{Tri}	L_{BOL}	mAP	Rank-1
✓	×	×	21.59	24.91
✓	✓	×	57.38	61.06
✓	✓	✓	**59.10**	**63.36**

model obtained from training on the synthetic MSMV-SK dataset achieves better generalization results than the one trained on the real Mask1k dataset when tested directly on PKU. The direct testing on the Mask1k dataset of the model obtained by training on the MSMV-SK dataset achieved suboptimal results. We analyse that on the one hand, this is due to the fact that the two datasets are of comparable size, even if our synthetic sketch data is a bit less. On the other hand, the large intra-modal variance of the Mask1k dataset further amplifies the effect of the domain shift between the two datasets. Significantly, the fine-tuned models all achieved the SOTA generalization performance.

5.3 Ablation Study

We conduct a comprehensive ablation study utilizing the proposed MSMV-SK dataset to demonstrate the framework's capability to mitigate style differences while effectively recovering identity-related features. This dual capability significantly enhances the model's generalization and preserves its discriminative power. In Table 3(a) highlights the effectiveness of our proposed module, further underscoring its contribution to the overall performance of the framework. Additionally, in Table 3(b), we present results from sequentially removing one loss component at a time, revealing that the omission of any single loss leads to suboptimal performance.

6 Conclusions

In this paper, we explore the task of generalizable Sketch Re-identification (Sketch Re-ID). We identify a critical challenge in this domain: the scarcity of a large-scale dataset that encompasses diverse sources and rich variations in viewpoint. To address this gap, we have developed a large-scale Multi-Source Multi-View Sketch dataset (MSMV-SK), leveraging low-cost and readily accessible synthetic sketches. Furthermore, to achieve a generalizable Sketch Re-ID model utilizing this dataset, we propose an innovative cross-modal identity-related feature learning network. The network contains a Style Separation and Content Recovery (SSCR) module and introduces

Bi-constrained Optimisation Loss (BOL) to enhance the discriminative ability of the identity-related features. Comprehensive experiments validate the strong generalization performance of the model trained on the proposed dataset.

Acknowledgements. The work was supported by National Nature Science Foundation of China (62302351).

References

1. Chen, C., Ye, M., Qi, M., Du, B.: SketchTrans: disentangled prototype learning with transformer for sketch-photo recognition. IEEE Trans. Pattern Anal. Mach. Intell. (2023)
2. Chen, Q., Quan, Z., Zhao, K., Zheng, Y., Liu, Z., Li, Y.: A cross-modality sketch person re-identification model based on cross-spectrum image generation. In: International Forum on Digital TV and Wireless Multimedia Communications, pp. 312–324. Springer (2021)
3. Gray, D., Brennan, S., Tao, H.: Evaluating appearance models for recognition, reacquisition, and tracking. In: Proceedings of IEEE International Workshop on Performance Evaluation for Tracking and Surveillance (PETS). vol. 3, pp. 1–7 (2007)
4. Gui, S., Zhu, Y., Qin, X., Ling, X.: Learning multi-level domain invariant features for sketch re-identification. Neurocomputing **403**, 294–303 (2020)
5. Hu, J., Shen, L., Sun, G.: Squeeze-and-excitation networks. In: Proceedings of the IEEE Conference on Computer Vision and Pattern Recognition, pp. 7132–7141 (2018)
6. Jin, X., Lan, C., Zeng, W., Chen, Z.: Style normalization and restitution for domain generalization and adaptation. IEEE Trans. Multimedia **24**, 3636–3651 (2021)
7. Kang, C.: Is synthetic dataset reliable for benchmarking generalizable person re-identification? In: 2022 IEEE International Joint Conference on Biometrics (IJCB), pp. 1–8. IEEE (2022)
8. Kumar, D., Siva, P., Marchwica, P., Wong, A.: Fairest of them all: Establishing a strong baseline for cross-domain person ReID. arXiv preprint arXiv:1907.12016 (2019)
9. Li, H., Ye, M., Du, B.: WePerson: learning a generalized re-identification model from all-weather virtual data. In: Proceedings of the 29th ACM International Conference on Multimedia, pp. 3115–3123 (2021)
10. Lin, K., Wang, Z., Wang, Z., Zheng, Y., Satoh, S.: Beyond domain gap: exploiting subjectivity in sketch-based person retrieval. In: Proceedings of the 31st ACM International Conference on Multimedia, pp. 2078–2089 (2023)
11. Ming, Z., et al.: Deep learning-based person re-identification methods: a survey and outlook of recent works. Image Vis. Comput. **119**, 104394 (2022)
12. Pang, L., Wang, Y., Song, Y.Z., Huang, T., Tian, Y.: Cross-domain adversarial feature learning for sketch re-identification. In: Proceedings of the 26th ACM International Conference on Multimedia, pp. 609–617 (2018)
13. Radenović, F., Tolias, G., Chum, O.: Fine-tuning CNN image retrieval with no human annotation. IEEE Trans. Pattern Anal. Mach. Intell. **41**(7), 1655–1668 (2018)
14. Sarker, P.K., Zhao, Q., Uddin, M.K.: Transformer-based person re-identification: a comprehensive review. IEEE Trans. Intell. Veh. (2024)

15. Song, J., Yang, Y., Song, Y.Z., Xiang, T., Hospedales, T.M.: Generalizable person re-identification by domain-invariant mapping network. In: Proceedings of the IEEE/CVF Conference on Computer Vision and Pattern Recognition, pp. 719–728 (2019)
16. Wang, X., Doretto, G., Sebastian, T., Rittscher, J., Tu, P.: Shape and appearance context modeling. In: 2007 IEEE 11th International Conference on Computer Vision, pp. 1–8. IEEE (2007)
17. Wu, A., Zheng, W.S., Yu, H.X., Gong, S., Lai, J.: RGB-infrared cross-modality person re-identification. In: Proceedings of the IEEE International Conference on Computer Vision, pp. 5380–5389 (2017)
18. Wu, J.J., Chang, K.H., Lin, I.C.: Generalizable person re-identification with part-based multi-scale network. Multimedia Tools Appl. **82**(25), 38639–38666 (2023)
19. Xu, X., Liu, W., Wang, Z., Hu, R., Tian, Q.: Towards generalizable person re-identification with a bi-stream generative model. Pattern Recogn. **132**, 108954 (2022)
20. Ye, M., Shen, J., Lin, G., Xiang, T., Shao, L., Hoi, S.C.: Deep learning for person re-identification: a survey and outlook. IEEE Trans. Pattern Anal. Mach. Intell. **44**(6), 2872–2893 (2021)
21. Zahra, A., Perwaiz, N., Shahzad, M., Fraz, M.M.: Person re-identification: a retrospective on domain specific open challenges and future trends. Pattern Recogn. **142**, 109669 (2023)
22. Zhang, Y., Wang, Y., Li, H., Li, S.: Cross-compatible embedding and semantic consistent feature construction for sketch re-identification. In: Proceedings of the 30th ACM International Conference on Multimedia, pp. 3347–3355 (2022)
23. Zhang, Y.F., Zhang, Z., Li, D., Jia, Z., Wang, L., Tan, T.: Learning domain invariant representations for generalizable person re-identification. IEEE Trans. Image Process. **32**, 509–523 (2022)
24. Zheng, L., Shen, L., Tian, L., Wang, S., Wang, J., Tian, Q.: Scalable person re-identification: a benchmark. In: Proceedings of the IEEE International Conference on Computer Vision, pp. 1116–1124 (2015)
25. Zhu, F., Zhu, Y., Jiang, X., Ye, J.: Cross-domain attention and center loss for sketch re-identification. IEEE Trans. Inf. Forensics Secur. **17**, 3421–3432 (2022)

Synchronization and Calibration of Video Sequences Acquired Using Multiple Plenoptic 2.0 Cameras

Daniele Bonatto[1], Sarah Fachada[1], Jaime Sancho[2], Eduardo Juarez[2], Gauthier Lafruit[1(✉)], and Mehrdad Teratani[3]

[1] Université Libre de Bruxelles, Brussels, Belgium
{daniele.bonatto,sarah.fernandes.pinto.fachada,gauthier.lafruit}@ulb.be
[2] Universidad Politecnica de Madrid, Madrid, Spain
{jaime.sancho,eduardo.juarez}@upm.es
[3] Aichi University of Technology, Gamagori, Japan
teratani-mehrdad@aut.ac.jp

Abstract. Plenoptic cameras capture 3D content by utilizing micro-parallax through a micro-lens array placed in front of their sensor. While this design eliminates the need for stereo cameras or separate depth sensors to extract 3D information, certain scenarios -such as handling disocclusions, capturing non-Lambertian surfaces, or imaging large scenes from multiple perspectives- require a multi-camera setup. Capturing a scene with multiple plenoptic cameras introduces new challenges due to the complex camera design and the lack of tailored image-processing methods for the plenoptic format.

In this paper, we present the first comprehensive pipeline for acquiring and processing video sequences using multiple plenoptic cameras of different types. Our approach addresses synchronization, color correction and cameras' parameters calibration. We validate our method on a new video dataset featuring both a Raytrix R8 and Raytrix R32 cameras.

Keywords: Plenoptic 2.0 cameras · Calibration · Micro-lens array · Video · Synchronization

1 Introduction and Related Work

Plenoptic cameras [1–3] represent a significant advancement in imaging technology, capturing the light field of a scene and allowing for depth estimation, subaperture view extractions [4] and refocusing after acquisition.

While a single plenoptic camera is capable of acquiring only a limited plenoptic function [5] around each point in a scene, the use of multiple plenoptic cameras is necessary to capture the whole information of the scene, as an increased coverage reduces missing information due to disocclusions and bring more view-directional information.

However, working with plenoptic cameras presents several challenges, primarily due to the lack of ground truth images. Unlike traditional imaging techniques, the only outputs from plenoptic cameras are plenoptic images, in which all the micro-lenses of the micro-lens array (MLA) are visible. Converting this format in pinhole-like views necessitates specialized software -such as Multiview (LLMV) [6], Reference Lenslet content Convertor (RLC) [7], Plenoptic Toolbox (PT) [8], and PlenoRVS [9]- to create so-called subaperture views. Moreover, plenoptic cameras present more intrinsic parameters [10,11] than conventional cameras and the calibration of the plenoptic cameras is not yet a mature technology, as it relies either on subaperture view extraction [11–15] or pattern identification [16–20].

Additionally to the inherent problems of plenoptic cameras, we encountered challenges associated with multiview setups. Synchronization, while crucial as highlighted in existing literature on standard camera setups [21–24], is one of the issues we faced. Moreover, while color calibration is a standard procedure for pinhole cameras [25–28], it becomes particularly challenging with plenoptic images due to the need to sample a plenoptic function at each point in the scene [1]. Additionally, placing a color chart within the scene would be disruptive and inconvenient.

In our work, we calibrate a video sequence captured with two plenoptic cameras while addressing the aforementioned issues. Our current system has been successfully tested with two cameras, the Raytrix R32 and the Raytrix R8, but it is designed to be extensible to accommodate additional cameras as needed.

Our contributions are the following: an aquisition software enabling the capture of dynamic scenes with multiple Raytrix cameras, a color correction workflow for images in plenoptic format and an intrinsic and extrinsic calibration method relying on adaptation of structure-from-motion to plenoptic cameras.

2 Proposed Method

Our plenoptic multiview system (see Fig. 1) consists of the following steps:

1. Camera's MLA calibration using the RxLive software. This step give the location, diameter and rotation of the micro-images for each plenoptic camera. The remaining calibration of RxLive software are not performed (metric calibration and wavefront calibration).
2. Scene acquisition with synchronized cameras, detailed in Sect. 2.2.
3. Color correction of the plenoptic images, detailed in Sect. 2.3.
4. Plenoptic camera calibration and registration. We use a new subaperture-free and pattern-free method designed for multiple plenoptic 2.0 cameras, detailed in Sect. 2.4.

2.1 Acquisition Setup

Our acquisition system included two Raytrix [29] cameras, an R8 and an R32, as displayed in Fig. 2. Table 1 summarizes the specificities of each camera and of the lenses mounted on them for our acquisition.

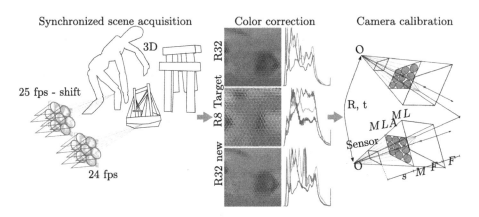

Fig. 1. Pipeline of the proposed method: 1) Synchronized acquisition of the scene. 2) Color correction. 3) Plenoptic camera calibration.

Table 1. Specification of the cameras in the setup.

	Resolution (pix)	Pixel size (mm)	Focal (mm)	Communication
R8	3740 × 2160	0.00224	25	USB
R32	6464 × 4852	0.00345	50	Ethernet

Before the acquisition, the RxLive software is used to calibrate the plenoptic cameras' MLA. During this step, a white image captured with a diffuser placed on the main lens (see Fig. 3) is analysed to 1) obtain a mask in order to debayer the raw plenoptic images, and 2) detect the micro-images' location, diameter and rotation. This step is independent to the metric calibration, hence, at this point, no intrinsic parameter of the camera is known, and no metric depth map can be obtained from the plenoptic image (only disparity map in *virtual space* [30]).

2.2 Camera Synchronization

We developed custom acquisition software specifically designed to synchronize two Raytrix cameras, addressing issues that the existing RxLive software could not handle. RxLive frequently crashed when attempting to load both cameras simultaneously, or it would only activate one camera at a time.

However, the Raytrix Light Field SDK allows to trigger the cameras with specified target framerates, shutter speeds and other camera parameters. The camera tries to capture at the specified framerate, but it fluctuates a bit. Our acquisition software implements N threads, each responsible for acquiring images from one of the N cameras. It works as following the diagram in Fig. 4.

Although the cameras are configured to capture at these target framerates, we observed that the R8 camera showed fluctuations, and captured at a slower framerate than expected, likely due to USB latency or hardware limitations. In contrast, the R32, which is monitored via an Ethernet connection, did not exhibit temporal drift.

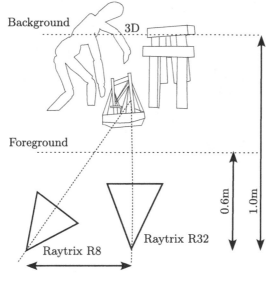

Fig. 2. Our acquisition setup: (a) Schematic of the scene, (b) Photo of the scene and cameras, (c) Camera setup: left: R32, right R8. (d-e) Moving objects of the sequence (boat and hand). (f) Static component of the scene.

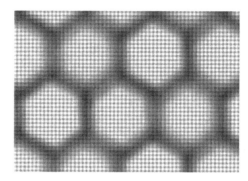

Fig. 3. Gray image used to debayer the raw image and detect the micro-images' location, rotation and diameter.

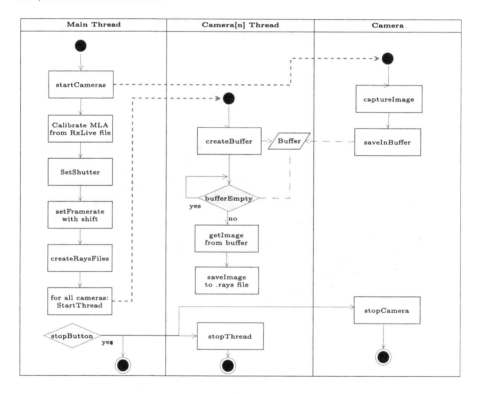

Fig. 4. Flowchart of the synchronized acquisition software.

To resolve this issue, we implemented a pre-processing step where we ran our software for 1000 frames to identify the R8's actual framerate. Using a linear regression approach, we then adjusted the R32's target framerate to match the R8's actual performance. The synchronization was verified using timestamps from both cameras, ensuring accurate alignment. During our tests, the synchronization

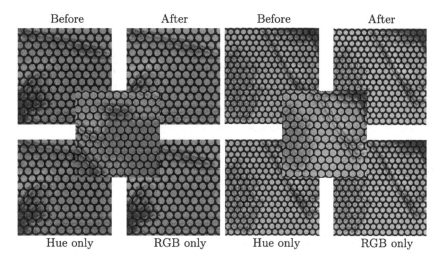

Fig. 5. Two zoomed details of the R32 color corrected image (with the area between micro-image excluded, for better visual inspection). The central image is the reference (R8).

was stable over more than 2000 frames. The flowchart in Fig. 4 illustrates the overall structure and operation of our acquisition software.

2.3 Color Correction

Plenoptic images capture the full plenoptic function from a point of view, which involves recording multiple perspectives of the same scene point through an array of micro-lenses. However, this introduces a challenge: even though these microlenses capture the same point in space, the color information they gather cannot be treated as identical due to differences in their positions and angles, especially for non-Lambertian objects. We observed color discrepancies between our two cameras, even if the prior step of Sect. 2.1 was performed in the same lighting conditions. This necessitates a color correction mechanism that can equalize these differences while preserving the integrity of the plenoptic sampling.

Instead of traditional color matching, which could compromise the plenoptic function, we opted to rectify the video frames from one of our cameras, the R32, using the R8 as the reference. The process involved computing the HSV histograms of both the R8 and R32 images. We excluded from the histogram the areas between the micro images using a mask. The mask was obtained by bluring and applying a threshold to the gray image displayed in Fig. 3. Rather than matching these histograms directly, we identified the maximum values within them and shifted the hue histogram of the R32 to align with that of the R8, similarly to [27]. This approach, relying on the cyclicity of hue values, allowed us to maintain the plenoptic nature of the captured scenes while ensuring that the main areas of the images had similar color profiles. However, discrepancies

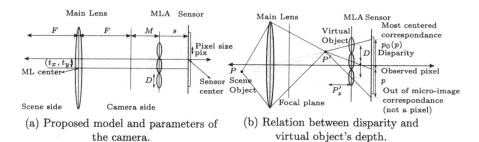

(a) Proposed model and parameters of the camera.

(b) Relation between disparity and virtual object's depth.

Fig. 6. Proposed camera model and the relation between disparity to virtual object's depth.

were still noticeable in the images. In a second step, we averaged the RGB values in both source and target images and the mean difference of each channel was added to the source image, destroying a small part of the color information available.

We noted that R32 camera has a wider field of view, which included areas of the scene that were less illuminated, particularly parts of the table supporting the scene. To enhance the quality of the data, we computed the histograms and RGB mean values on the top left portion of the R32's franes.

Figure 5 shows zoomed details on the images before color correction, after hue histogram shifting, and with RGB mean shifting only, and with our proposed method.

2.4 Camera Parameters Calibration

Now we explain our method to find the intrinsic parameters of the camera and their relative positions.

Camera Model: Our plenoptic camera model consists of five intrinsic parameters: the focal length of the main lens F, the relative position between the focal plane and the MLA (t_x, t_y, M), and the distance between the MLA and sensor s (see Fig. 6a). The main lens, MLA and sensor are assumed to be parallel. The micro-images layout is computed in a prior step using the RxLive software (see Sect. 2.1).

We model the micro-lenses as pinhole cameras seeing the virtual scene through the main lens [31]. A scene's point P is projected as follows: 1) through the main lens, modeled as a thin lens, creating an image point P', as shown in Fig. 6b. This image is then projected onto the sensor by the micro-lenses. The corresponding pixel p_m in a micro-lens m, whose center is at pixel (c_x, c_y), is given in homogeneous coordinates as:

$$p_m = K_m T_m K_M P, \tag{1}$$

where K_m is the projection matrix through a micro-lens m, T_m the coordinate transformation matrix between the main lens and the micro-lens m and K_M the projection matrix through the main lens. We have:

$$K_m = \begin{pmatrix} \frac{s}{\text{pix}} & 0 & \frac{c_x}{\text{pix}} \\ 0 & \frac{s}{\text{pix}} & \frac{c_y}{\text{pix}} \\ 0 & 0 & 1 \end{pmatrix}, T_m = \begin{pmatrix} tx - c_x \\ I_3 \ ty - c_y \\ F + M \end{pmatrix}, K_M = \begin{pmatrix} & & 0 \\ FI_3 & & 0 \\ & & 0 \\ 0 \ 0 & -1 & F \end{pmatrix}, \quad (2)$$

where pix is the pixel size, and I_3 the 3×3 identity matrix. Equation 1, when all the parameters are expressed in millimeters, allows a metric reconstruction of the scene.

Moreover, we observe on Fig. 6b that the disparity $d(p)$ between pixels corresponding to a same point only depend on the virtual object's location. It can be expressed in function of the micro-images diameter D, the virtual object's depth P'_z and the sensor-MLA distance s as:

$$d(p) = \frac{sD}{P'_z}. \quad (3)$$

Calibration: To calibrate the cameras, we propose a pattern-free and subaperture-free structure-from-motion for multiple plenoptic cameras method. It implements the steps displayed in Fig. 7.

First (B in Fig. 7), we estimate the disparity in each plenoptic camera (A) using the disparity estimator of PT [8]. As the cameras are static, we estimate the disparity on four frames for each camera to make the estimate poses more robust.

Second (C), we detect and match features between the plenoptic cameras. We detect the features only in the micro-images' center to avoid border artifacts (luminance decay or increase at the border between neighbouring micro-images). We use the SIFT descriptor [32].

Third (D), we use a pattern-free and subaperture-free calibration method [31] to estimate initial coarse camera parameters of our two cameras. This method compares the disparity of selected objects to the measured depth from the camera. Using Eq. 3, we find the distance between the main lens' focal plane and the MLA, M, and the distance between the MLA and the sensor, s.

Then (E), using the same equation, we project an initial point cloud from the first camera and its disparity map. The matches between the two cameras and Eq. 1 allow to find an initial position for the second camera, using rigid transform optimization with RANSAC.

Finally (F), these initial camera parameters M and s, as well as the relative rigid transform between the cameras are the input for our bundle adjustment process. We optimize these parameters, the cameras' principal points (t_x, t_y), and the matched 3D points' positions. The residuals are the reprojection error in the plenoptic image, and the reprojected disparity error, which we define as the difference between the point's disparity in the disparity map and its reprojected disparity obtain using Eq. 3.

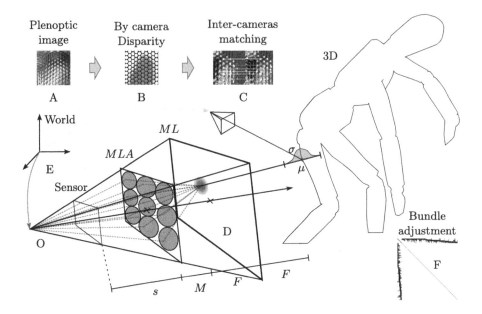

Fig. 7. Calibration. A) Captured plenoptic image, B) By camera disparity estimation, C) Inter-cameras features matching, D) Initial per-camera coarse calibration, E) Initial cameras relative position, F) Bundle adjustment. s is the distance between the MLA and the sensor, M is the distance between the MLA and the focal plane, F is the focal length of the main lens, s and M are both unknown.

In plenoptic cameras, a 3D point generally has several images, one per microlens that observes it. To compute its reprojection error, we reproject it through the micro-lens corresponding to the micro-image containing the initial pixel, even if the reprojection falls out of the micro-image. This allows to keep a continuity in the bundle adjustment procedure: otherwise the observed reprojected pixel would jump between the micro-images observing it, making a gradient difficult to compute. This reprojection error is however not sufficient to obtain convergence. We add an additional constraint over the disparity of the reprojected depth. Indeed, Eq. 3 predicts the disparity in the plenoptic image in function of the depth. Then, we target both a low reprojection error and a low disparity error.

3 Results

3.1 Description of the Dataset

This dataset presents a detailed static scene featuring an artistic representation of a boat floating on a block-colored ocean near a temple, while a giant pursues them. The boat is moved by a giant hand. The dataset is divided in two sub-datasets, captured with Raytrix R8 and R32 cameras. It also contains calibration images, including checkerboard views captured simultaneously by both

Table 2. Extrinsic camera parameters for the compared methods.

	CalLab [33]	Compote [19]	Ours (initialization)	Ours
Relative Pos. (mm)	$(-87, 215, -4)$	$(-67, 95, 2)$	$(-80, 207, 4)$	$(-86, 95, -29)$
Relative Rot. (°)	$(11.1, 2.6, 1.7)$	$(9.8, 1.5, 1.6)$	$(4.7, -1.8, 2.3)$	$(6.0, -3.5, -6.9)$

cameras. The checkerboard is made of 12 mm squares. It also contains overexposed images taken with varying apertures using a light diffuser. The dataset providing raw images, color corrected images, white images, the checkerboard images and calibrated camera parameters is available at https://zenodo.org/records/12663778.

3.2 Extrinsic Evaluation Against Checkerboard-Based Methods

To assess the effectiveness of our camera pose estimation method, we compared it with two techniques using a checkerboard: a method designed for plenoptic cameras, Compote [19,20], which optimizes the reprojection of the checkerboard's corners in the plenoptic image, and a classical method designed for pinhole cameras, CalLab [33], finding the relative poses of pinhole cameras, with the Raytrix Total focus images as input.

Our method is initialized with coarse extrinsic cameras parameters found using the rigid transform between the two matched reprojected point cloud from each camera. Our method directly estimates the camera parameters from the scene, and thus does not use the pattern images. To ensure consistency of the calibration over time, we used four frames for each plenoptic camera (8 images in total, and optimization of the intrinsic parameters for 2 cameras). As the cameras are static, after optimization, the four images corresponding to each camera should show the same extrinsics parameters.

Results of each method are reported in Table 2, showing consistency between the three methods.

3.3 Intrinsic Parameters and Reprojection Error

We run our bundle adjustment with initial intrinsic parameters found using a coarse plenoptic camera intrinsic parameter estimation [15], and report the results in Table 3. The results are consistent with the initialization calibration method but show some discrpencies with regards to Compote. This can be explained by inacurate initial focus distances and aperture given to Compote as the lens did not display it accurately.

Table 4 reports the pixel reprojection error and the disparity reprojection error after our bundle adjustment. The results show that both the reprojection and disparity error are very low. Notably, the reprojection error drastically reduced compared to the classical method showing an overall accuracy improvement. The reduction in the disparity reprojection error allows to have more

Table 3. Optimized intrinsic parameters of the camera

	M (mm)	S (mm)	principal point t_x (pix)	principal point t_y (pix)
Initial R8	0.13	0.21	0	0
Optimized R8	0.12	0.2	0.4	−1.3
Compote	0.28	0.17	3.8	−0.9
Initial R32	1.52	0.37	0	0
Optimized R32	1.55	0.39	−0.3	1.2
Compote	2.4	0.19	14.2	6.0

Table 4. Classical reprojection error, and reprojection's disparity error for the reconstructed scene after bundle adjustment.

Cameras	Images	Reproj. error (pix.)		Disp. reproj. error (pix.)	
		initial	final	initial	final
2	8	23.7	0.1	2.9	0.7

constraints on the alignment between the observed and predicted disparities and keep the scene's geometrical structure.

4 Conclusion

In this paper, we have presented the first framework for the acquisition and processing of video sequences. Our approach brought a first overview and solutions to several critical challenges of multiview plenoptic imaging, including synchronization, camera parameter calibration, and color correction. This research lays the groundwork for future exploration in multi-plenoptic camera systems.

Future work will focus on extending the system to support a greater number of cameras. Additionally, exploring real-time processing capabilities will open new avenues for plenoptic imaging applications in fields such as virtual reality, robotics, and advanced multimedia systems.

Acknowledgments. Sarah Fachada is a Postdoctoral Researcher of the Fonds de la Recherche Scientifique - FNRS, Belgium. This work was supported in part by the HoviTron project (N° 951989); in part by the FER 2021 project (N° 1060H000066-FAISAN), in part by the Emile DEFAY 2021 project (N° 4R00H000236), and in part by the FER 2023 project (N° 1060H000075). Additionally, this work has been funded by the project AIMS5.0, supported by the Chips Joint Undertaking and its members, including top-up funding by National Funding Authorities from involved countries (N° 101112089), and the European project STRATUM (N° 101137416). The robotic bench was funded by "Programa Propio UPM" in the call "convocatoria de ayudas a centros e institutos de I+D+i".

Disclosure of Interests. All authors declare that they have no conflicts of interest.

References

1. Adelson, E., Wang, J.: Single lens stereo with a plenoptic camera. IEEE Trans. Pattern Anal. Mach. Intell. **14**(2), 99–106 (1992)
2. Ng, R., Levoy, M., Brédif, M., Duval, G., Horowitz, M., Hanrahan, P.: Light Field Photography with a Hand-held Plenoptic Camera. Ph.D. dissertation, Stanford university (2005)
3. Lumsdaine, A., Georgiev, T.: The focused plenoptic camera. In: 2009 IEEE International Conference on Computational Photography (ICCP). San Francisco, CA, USA: IEEE (2009)
4. Georgiev, T., Lumsdaine, A.: Reducing Plenoptic Camera Artifacts. Comput. Graph. Forum **29**(6), 1955–1968 (2010)
5. Adelson, E.H., Bergen, J.R.: The Plenoptic Function and the Elements of Early Vision. Vision and Modeling Group, Media Laboratory, Massachusetts Institute of Technology (1991)
6. Bonatto, D., et al.: Multiview from micro-lens image of multi-focused Plenoptic camera. In: 2021 International Conference on 3D Immersion, Brussels, Belgium (2021)
7. Fujita, S., Mikawa, S., Panahpourtehrani, M., Takahashi, K., Fujii, T.: Extracting multi-view images from multi-focused plenoptic camera. In: Fujita, H., Lin, F., Kim, J.H. (eds.) International Forum on Medical Imaging in Asia 2019, Singapore, Singapore: SPIE (2019)
8. Palmieri, L., Koch, R., Veld, R.O.H.: The Plenoptic 2.0 toolbox: benchmarking of depth estimation methods for MLA-based focused plenoptic cameras. In: 2018 25th IEEE International Conference on Image Processing (ICIP). IEEE, pp. 649–653 (2018)
9. Fachada, S.: Photo-realistic depth image-based view synthesis with multi-input modality. Ph.D. dissertation, Université Libre de Bruxelles, Brussels, Belgium (2023)
10. Dansereau, D.G., Pizarro, O., Williams, S.B.: Decoding, calibration and rectification for Lenselet-based Plenoptic cameras. In: 2013 IEEE Conference on Computer Vision and Pattern Recognition, Portland, OR, USA, pp. 1027–1034. IEEE (2013)
11. Heinze, C., Spyropoulos, S., Hussmann, S., Perwaß, C.: Automated robust metric calibration algorithm for multifocus plenoptic cameras. IEEE Trans. Instrum. Measur. **65**(5), 1197–1205 (2016). publisher: IEEE
12. Strobl, K.H., Lingenauber, M.: Stepwise calibration of focused plenoptic cameras. Comput. Vis. Image Underst. **145**, 140–147 (2016)
13. Zeller, N.: Direct Plenoptic Odometry-Robust Tracking and Mapping with a Light Field Camera. PhD Thesis, Technische Universität München (2018)
14. Fachada, S., Losfeld, A., Senoh, T., Lafruit, G., Teratani, M.: A Calibration Method for Subaperture Views of Plenoptic 2.0 Camera Arrays. In: IEEE 23nd International Workshop on Multimedia Signal Processing, p. 2021. Tampere, Finland (2021)
15. Fachada, S., Bonatto, D., Losfeld, A., Lafruit, G., Teratani, M.: Pattern-free Plenoptic 2.0 camera calibration. In: EEE 24th International Workshop on Multimedia Signal Processing, p. 2022. Shanghai, China (2022)
16. Bok, Y., Jeon, H.-G., Kweon, I.S.: Geometric calibration of micro-lens-based light field cameras using line features. IEEE Trans. Pattern Anal. Mach. Intell. **39**(2), 287–300 (2017)

17. Nousias, S., Chadebecq, F., Pichat, J., Keane, P., Ourselin, S., Bergeles, C.: Corner-based geometric calibration of multi-focus plenoptic cameras. In: 2017 IEEE International Conference on Computer Vision (ICCV). Venice, pp. 957–965. IEEE (2017)
18. Noury, C.-A., Teuliere, C., Dhome, M.: Light-field camera calibration from raw images. In: 2017 International Conference on Digital Image Computing: Techniques and Applications (DICTA). Sydney, NSW, IEEE (2017)
19. Labussière, M., Teulière, C., Bernardin, F., Ait-Aider, O.: Blur Aware Calibration of Multi-Focus Plenoptic Camera. In: Proceedings of the IEEE/CVF Conference on Computer Vision and Pattern Recognition (CVPR), pp. 2545–2554 (2020)
20. Labussière, M., Teulière, C., Bernardin, F., Ait-Aider, O.: Leveraging blur information for plenoptic camera calibration. Int. J. Comput. Vision **2012**, 1–23 (2022)
21. Shrstha, P., Barbieri, M., Weda, H.: Synchronization of multi-camera video recordings based on audio. In: Proceedings of the 15th ACM International Conference on Multimedia. Augsburg Germany: ACM, pp. 545–548 (2007)
22. Shrestha, P., Weda, H., Barbieri, M., Sekulovski, D.: Synchronization of multiple video recordings based on still camera flashes. In: Proceedings of the 14th ACM international conference on Multimedia. Santa Barbara CA USA, pp. 137–140 (2006). ACM
23. Sinha, S., Pollefeys, M.: Synchronization and calibration of camera networks from silhouettes. In: Proceedings of the 17th International Conference on Pattern Recognition, 2004. ICPR 2004., vol. 1, pp. 116–119 (2004). iSSN: 1051-4651
24. Whitehead, A., Laganiere, R., Bose, P.: Temporal synchronization of video sequences in theory and in practice. In: Seventh IEEE Workshops on Applications of Computer Vision (WACV/MOTION'05) - Volume 1. Breckenridge, CO: IEEE, Jan. 2005, pp. 132–137 (2005)
25. Vrhel, M.J., Trussell, H.: Color correction using principal components. Wiley: Color Research & Application, vol. 17, no. 5, pp. 328–338 (1992)
26. Ye, S., Lu, S.-P., Munteanu, A.: Color correction for large-baseline multiview video. Sig. Process. Image Commun. **53**, 40–50 (2017)
27. Senoh, T., Tetsutani, N., Yasuda, H.: Proposal of Trimming and Color Matching of Multi-View Sequences [M47170]. ISO/IEC JTC1/SC29/WG11 (2019)
28. Dziembowski, A., Mieloch, D., Różek, S., Domański, M.: Color correction for immersive video applications. IEEE Access **9**, 75:626–75:640 (2021)
29. Raytrix. Raytrix (2019). https://raytrix.de
30. Perwass, C., Wietzke, L.: Single lens 3D-camera with extended depth-of-field. In: Proceedings, vol. 8291, p. 829108. Burlingame, California, USA, Human Vision and Electronic Imaging XVII (2012)
31. Fachada, S., Bonatto, D., Lafruit, G., Teratani, M.: Pattern and Subaperture-free Calibration of Plenoptic 2.0 cameras. In: 12th European Workshop on Visual Information Processing, p. 2024. Switzerland, Sep, Geneva (2024)
32. Lowe, D.G.: Distinctive image features from scale-invariant keypoints. Int. J. Comput. Vision **60**(2), 91–110 (2004)
33. Strobl, K.H., Hirzinger, G.: Optimal hand-eye calibration. In: IEEE/RSJ International Conference on Intelligent Robots and Systems. IEEE 2006, 4647–4653 (2006)

Target-Oriented Dynamic Denosing Curriculum Learning for Multimodel Stance Detection

Zihao Suo and Shanliang Pan(✉)

Ningbo University, Ningbo 315211, China
{2311100303,panshanliang}@nbu.edu.cn

Abstract. Multimodal Stance Detection aims to classify public opinions on specific targets in social media, incorporating both text and image data. However, prior studies have overemphasized the significance of images, neglecting the presence of irrelevant images in the dataset. Moreover, previous research has shown that employing the Chain of Thought approach with large language models can introduce noise from the generated text as well as noise from the image modality into the text input modality. These noises can degrade the performance of multimodal models. Additionally, both image and text modalities exhibit complex data patterns, resulting in significant disparities in training difficulty across the dataset. To address these issues, We proposed Target-Oriented Dynamic Denoising Curriculum Learning(TODDCL), which effectively measures different types of noise and automatically tackles the noise in both image and text modalities based on their relevance to the target, in an escalating complexity order. Experimental results on five benchmark datasets demonstrate that our proposed TODDCL method achieves state-of-the-art performance in multimodal stance detection.

Keywords: Multimodal stance detectoin · Image denosing · Curriculum learning

1 Introduction

Stance detection is a crucial task that aims to identify the authors' attitudes or positions (Pro (support), Con (oppose), Neu (neutral)) towards a specific target, such as an entity or a topic. In recent years, with the rapid development of multimodal technologies and the increasing number of platforms allowing users to share multimodal information, research efforts have shifted towards multimodal stance detection [12]. Unlike traditional stance detection, multimodal stance detection not only focuses on analyzing the textual modality of users' posts on social platforms but also incorporates information from the imaging modality to determine the stance towards a given target.

In Table 1, the post consists of both textual and image components. Extracting meaningful information solely from the textual modality is insufficient to capture the relevant opinions of the target. It is only through the integration of information from both the image and text modalities that we can accurately

Table 1. An example post of a user expressing a "Favor" stance towards "Donald Trump" using multimodal information.

Target:	Donald Trump
Stance:	Pro
Text:	This election has already ended.
Image:	

identify the stance of the given target. Therefore, leveraging multiple modalities can assist in detecting users' stances towards specific topics, enabling a better understanding of public opinions on social media platforms that include both image and text modalities. However, the presence of noise in the imaging modality, caused by factors such as the data collection process, can significantly impact the prediction results.

Previous stance detection models have explored the incorporation of external background knowledge to enhance their performance. One approach, CKENet [24] integrates target-related knowledge from ConceptNet, a knowledge graph based on common sense. Similarly, BS-GGCN [14] simplifies the ConceptNet graph to a sentence-related graph, enabling more efficient knowledge embedding for stance detection. The "chain of thought" approach [25] extracts information from large language models as input. However, the potential for large language models to generate hallucinations or make non-factual statements can introduce noise into the text modality. In addition, the presence of noise in the image modality can also negatively impact the model's results.

Inspired by Curriculum Learning, we introduce it to the task of multimodal stance detection. The denoising approach of Curriculum Learning focuses on adjusting the presentation of training data [21]. Specifically, instead of presenting training data in a completely random order, Curriculum Learning encourages more training time on less noisy data and reduces training time on noisy data, thereby addressing the noise issue [6]. The application of Curriculum Learning generally involves two steps: defining a measure to distinguish clean and noisy data and determining a suitable denoising strategy for the task. Analyzing the characteristics of multimodal stance detection, we design a customized target-oriented dynamic denoising curriculum.

The contributions of our method can be summarized as follows:

1. We introduce the approach of Curriculum Learning into the task of multimodal stance detection, enabling the model to effectively denoise image data that contains noise.
2. For the scenario where image noise is introduced into the input text modality through the COT method as well as the inherent limitations of the COT method itself, we design a noise metric calculation approach to quantitatively rank the noise in the auxiliary text from the COT method based on the relationship between the text and the target.
3. We design a dynamic multimodal denoising curriculum that dynamically selects denoising strategies based on the model's differential performance on the validation set.

2 Related Work

In this section, we provide a brief overview of common solutions employed in stance detection and multimodal stance detection tasks. Additionally, we will introduce the theoretical foundations and some relevant applications of Curriculum Learning methodologies.

2.1 Multimodal Stance Detection

Stance detection is a task that involves identifying attitudes towards a specific context or topic, typically framed as a stance classification problem for neural network-based models. This task covers a wide range of applications, including rumor stance detection [11], fake news stance detection [5], disinformation/misinformation stance detection [7], and zero-shot stance detection [1]. Early studies in stance detection employed various machine learning and deep learning techniques [7], while the emergence of large language models has further advanced the state-of-the-art performance. Some stance detection problems require domain-specific solutions that incorporate world knowledge into the systems [8,13].

Recently, multi-modal stance identification has gained attention due to the increasing amount of multi-modal stance information shared by users on social media. Detecting users' stances based solely on textual modality may not be accurate enough to meet practical needs, thus motivating the research on multi-modal stance detection. The MultiStanceCat task at IberEval 2018 [18] aimed to pioneer the evaluation of stance detection in tweets regarding the Catalan First of October Referendum from a multimodal perspective. Weinzierl and Harabagiu [22] discussed the multi-modal stance towards frames of communication, focusing on the frames of communication within multi-modal posts, which differs from conventional stance detection tasks that concentrate on predefined targets.

Wu et al. [12] proposed a framework called Targeted Multi-modal Prompt Tuning (TMPT) for multimodal stance detection tasks on Twitter. They created five new multimodal stance detection datasets and designed the TMPT model to learn stance features from both textual and visual modalities by leveraging target information. The TMPT framework utilizes target information to prompt pre-trained models for learning multimodal stance features. In addition, TMPT also utilized the chain-of-thought method from GPT-4-Version to generate auxiliary text related to stance detection based on the text, image, and target content. TMPT combines the chain of thought with the text as an overall input for the text modality. Our curriculum learning method can be directly applied to this framework.

2.2 Curriculum Learning

As deep learning progresses and requires extensive datasets for training, the collection of large-scale datasets often includes noisy data that is unrecognizable or mislabelled. In the context of Curriculum Learning (CL), this noisy data corresponds to more challenging examples in the dataset, while cleaner data represents the easier part. The CL strategy encourages more training on the easier data, leading to the intuitive hypothesis that CL learners would spend less time dealing with difficult and noisy examples, resulting in faster training. This assumption suggests that CL has a denoising effect on noisy data [2].

To gain a deeper understanding of this denoising mechanism, Gong et al. [6] proposed a hypothesis-based theory that highlights the bias between the distributions of training and testing data caused by noisy or mislabelled training data. Intuitively, the training and target/test distributions share a region with higher confidence that is denser, representing the easier examples in CL. Therefore, starting training from these easier examples through the CL strategy simulates learning from this high-confidence common region, approximating the target distribution. This guides learning in the direction of the desired target and reduces the negative impact of low-confidence noisy examples.

In the field of Natural Language Processing (NLP), CL has shown promising performance, aiding in reducing training time and improving model performance. It is employed in tasks involving noisy or heterogeneous data, such as natural language understanding [23], reading comprehension [19], named entity recognition [15]. In this work, we adopt the idea of CL to mitigate the negative effects of noisy images and text in multimodal stance detection. To the best of our knowledge, this is the first attempt to utilize CL for denoising in the multimodal stance detection task.

3 Approach

3.1 Overview

We proposed a target-oriented dynamic denoising curriculum for the multimodal stance detection task, consisting of a noise metric selection module and a denoising curriculum module. Specifically, When the relevance of the images or text to the target is lower, the images or text may contain more noise. Based on this observation, we define separate noise measurements for the image and text modalities to quantify the level of noise in each training instance.

In the denoising curriculum, we design a separate curriculum for each noise measurement. We first rank all training instances based on the magnitude of the noise measurement. Since these two noise measurements capture the noise level of images from different modalities, a natural idea is to combine them to achieve better denoising performance. To accomplish this, we extend the single denoising curriculum to a multimodal dynamic denoising curriculum.

3.2 Noise Metrics

In this section, we will introduce two modal noise metric standards in detail. The input process of the noise metric selection workflow for the text and image modalities is shown in Fig. 1.

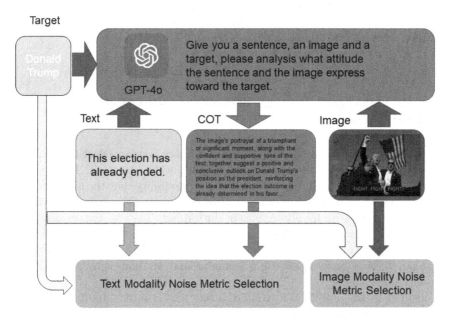

Fig. 1. The input source of the noise metric selection for different modalities.

Text Modality Noise Metric: In general stance detection tasks, it is common to use a chain-of-thought approach that generates new auxiliary sentences incorporating information from both images and text. However, due to the presence of noise in the images, the auxiliary sentences generated through the chaining process can be contaminated by the noise in the images, which in turn affects the text modality input. In addition, the chain of thought generated by large models may exhibit hallucination phenomena, which can also introduce noise into the text. It is necessary for the model to discern the noise in the text modality. Hence, we proposed a text modality noise metric.

Following this principle, we merge the sentences from the dataset and the sentences generated through COT as the input for the text modality. Specifically, for a given target-sentence pair (T_i, S_i) in the dataset, which includes N samples. We separately input the target text, the sentence, and the sentences generated by COT into the text modality encoder of the original multimodal stance detection model to obtain the corresponding text features: H_T^i, H_S^i, and H_C^i. We then use cosine similarity to measure the degree of correlation between the target text and the text modality inputs:

The cosine similarity measures the degree of correlation between the target and text modality inputs.

$$sim(H_T^i, H_S^i \oplus H_C^i) = \cos(H_T^i, H_S^i \oplus H_C^i) = \frac{H_T^i \cdot (H_S^i \oplus H_C^i)}{\|H_T^i\| \|H_S^i \oplus H_C^i\|}, \quad (1)$$

where $\cos(\cdot)$ is a cosine similarity function. After that, considering the lower the similarity between the text and the target the more likely the text contains noise, we define a text modal noise metric as follows:

$$d_t = 1.0 - \frac{sim(H_T^i, H_S^i \oplus H_C^i)}{\max_{k \in N} sim(H_T^k, H_S^k \oplus H_C^k)}, \quad (2)$$

where d_t is normalized to [0.0, 1.0]. Here, a lower score indicates that the target-sentence pair is clean data.

Image Modality Noise Metric: When the image cannot reflect the content of the text, the similarity between the two modalities will decrease. Therefore, the lower the image-text similarity, the greater the possibility that the image contains noise. Based on this principle, we measure the image noise by calculating the image-target similarity for all samples in the dataset. The conventional approach for multimodal similarity modeling involves directly inputting the image and text data into separate encoders of a pre-trained model like CLIP [17]. The image noise can then be calculated using formulas and methods similar to the text modality noise metric.

Inspired by the detection method of image text relationship in AoM [26], we utilize the multimodal stance detection model as the primary architecture for the image-text similarity modeling. We pre-train the image-text similarity model on the TRC dataset [20], which includes labels for both modality similarity and whether the image enhances the semantic meaning of the text. To better pre-train the target-image pair similarity detection model, we also use the target-image pairs from the related multimodal stance detection task to initialize the model training. For example, in the MTSE dataset, we used the two official portraits as the target-image pairs. Here is the detailed process of obtaining the image modality noise. Firstly, we extract the hidden features from the fusion layer of the original multimodal detection model.

$$h = h^T \oplus h^V \quad (3)$$

The representation h is then passed through a fully-connected layer with a softmax function specifically designed for text-image similarity detection, which yields probabilities $y \in \mathbb{R}^2$ for image-text similarity($P(s)$) and dissimilarity($P(\text{dis})$).

$$y = \text{softmax}(W^s h + b) \tag{4}$$

We train our model using the cross-entropy loss, which can be represented as:

$$\mathcal{L} = -[\hat{y}\log(P(s)) + (1-\hat{y})\log(P(dis))] \tag{5}$$

where $\hat{y}$ is the ground-truth label, indicating whether the image-text pair is similar ($\hat{y} = 1$) or dissimilar ($\hat{y} = 0$). Finally, for a target-image pair $(T_i, I_i) \in D$, we can obtain image modality noise for all N samples in the dataset.

$$d_i = \frac{1}{2}[1.0 - \frac{P_i(s)}{\max_{k \in N} P_k(s)} + \frac{P_i(dis)}{\max_{k \in N} P_k(dis)}] \tag{6}$$

where d_i is normalized to $[0.0, 1.0]$. Here, a lower score indicates that the target-image pair is clean data.

By incorporating this approach, we aim to leverage the strengths of our multimodal stance detection model to effectively capture the similarity between images and target, thus enabling accurate noise measurement and denoising in the imaging modality.

3.3 Denoising Curriculum

In this section, we initially present the training process of denoising using multiple metrics for each modality sequentially. Subsequently, we extend this approach to multimodal dynamic denoising curriculum learning.

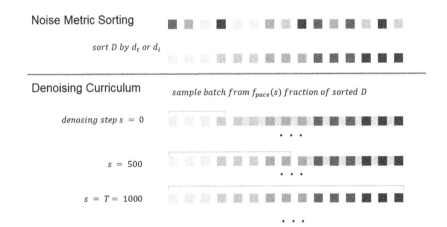

Fig. 2. Noise metric and the utilization of the denoising speed function.

Target-Oriented Single Modal Denoising Curriculum: During the training phase, we begin by assigning the noise metrics mentioned earlier, such as d_i for sentence-image noise in the imaging modality, d_t for noise in the target-image noise in the text, to each example in the dataset D. We then utilize a denoising scheme to regulate the learning pace of the model, allowing it to adapt from instances with lower noise to those with higher noise. The denoising function is defined as follows:

$$f_{pace}(s) = \min\left(1, \delta^{\left(1-\frac{s}{T}\right)}\right), \tag{7}$$

Here, δ represents a pre-defined initial value, and T represents the total number of time steps involved in the denoising curriculum learning during the training process. At each time step s, $f_{pace}(s)$ denotes the range of noise metric, allowing the model to train only on dataset instances with noise metric below N_i for the image modality, or N_t for the text modality. Once s exceeds the value of T, the model can train on the entire dataset. By employing this denoising function, the model no longer randomly samples batches from the training set, but instead emphasizes training on instances with lower noise. This progressive exposure to more complex data instances with higher noise levels demonstrates the effectiveness of curriculum learning in denoising noisy data. Figure 2 illustrates the two fundamental steps in single modality denoising learning: noise measurement and the utilization of the denoising speed function.

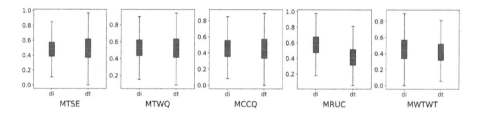

Fig. 3. Box plot of two types of modality noise distributions in different datasets.

Target-Oriented Dynamic Denoising Curriculum: In the previous single modal denoising curriculum, we only trained using a single noise indicator from one modality. A natural idea is to include multiple noise metrics from different modalities in the same training process to denoise multiple modalities affected by noise. However, due to variations in the distribution of the same noise indicator across different datasets and within the same dataset, as depicted in Fig. 3, determining which modality's noise indicator to use at each time step becomes a challenge.

One simple approach is to combine these noise indicators or select them in a predetermined order. However, this raises the issue of determining the sequence in which the noise indicator is chosen at each time step. To address this, we proposed a dynamic multimodal denoising curriculum, inspired by the concept of automatic curriculum learning [21].

In essence, we adopt a greedy-like approach where, at each time step, we compare the performance of models trained using different noise metrics from various modalities on the validation set. Based on this evaluation, we determine which modality's noise metric to select. This dynamic selection process allows us to choose the most appropriate noise indicator for each time step during the training process.

4 Experiments

In this section, we conduct experiments to evaluate the performance of our approaches on multimodal stance detection datasets. We first introduce the experimental settings and then compare other state-of-the-art methods with ours. Finally, we analyze the ablation studies.

4.1 Experimental Settings

Datasets. To verify the effectiveness of our method for different domains, We conducted experiments on five publicly available multimodal datasets: MTSE, MCCQ, MWTWT, MRUC, and MTWQ [12]. Each dataset was divided into training, development, and testing sets with a 7:1:2 ratio. Due to limitations on Twitter, we obtained complete posts containing both images and text by using the provided Post IDs. We also utilized the chain-of-thought method from GPT-4-Version to generate auxiliary text from the acquired text and images, which were used as part of the input for the text modality.

The statistical attributes of the datasets are presented in Table 2. Across all five datasets, approximately half of the stance detection information originated from the visual modality, and over half of the images did not explicitly exhibit a clear stance toward the target. Moreover, there was a substantial disparity in the distribution of image types among the different datasets, underscoring the necessity of our proposed image-denoising method using curriculum learning to address the challenges posed by complex image modalities.

Evaluation Metrics And Implementation Details. For a fair comparison we employed the same evaluation metrics used in TMPT. We use the Macro F1-score to measure the model performance. For a fair during training, we train each model for a fixed 30 epochs and then select the model with the best F1 score on the validation set. Finally, we evaluate its performance on the test set. For the MTSE, MCCQ, and MRUC, δ is set at 0.2. For the MWTWT and MTWQ, δ is set at 0.3. Our model utilizes the Adam optimizer [10] and is trained on a single RTX 3090 GPU.

Table 2. The statistics(%) of whether the image can convey stance information and the type of each image.

	Image	Person	Events	Words	Memes	Mixed
MTSE	43.7	32.7	8.2	24.0	25.0	10.1
MCCQ	44.1	12.2	22.9	26.7	16.3	21.9
MWTWT	46.2	26.4	16.8	38.4	5.9	12.5
MRUC	54.8	32.1	20.7	20.4	18.5	8.3
MTWQ	41.0	36.8	42.9	8.1	2.4	9.8

4.2 Comparison with Other Methods

Baselines. To assess the generalization capability of our proposed method across different models, we utilized the following models as base models: 1) BERT+ViT, which combines BERT [3] as the textual encoder and ViT [4] as the visual encoder. We concatenated the [CLS] vectors of the textual and visual modalities for stance detection; 2) ViLT [9], specifically the vilt-b32-mlm model; 2) CLIP [17], specifically the clip-vit-base-patch32 model; 4) We directly integrated our denoising method TODDCL into these multimodal baselines to obtain the final results.

Table 3. Test results on the five datasets for Multimodal Stance Detection (%). The best results are in bold. Models with * are from TMPT [12], while other results are reproduced under the same conditions.

MODALITY	MTSE		MCCQ	MWTWT					MRUC		MTWQ	
	DT	JB	CQ	CA	CE	AC	AH	DF	RUS	UKR	MOC	TOC
GPT4-Vision*	70.46	72.82	61.63	44.59	47.07	57.47	57.90	37.61	44.83	56.40	66.72	56.90
BERT+ViT	52.15	56.26	70.31	72.18	55.1	65.20	57.34	79.70	42.54	55.20	58.32	52.75
+TODDCL	52.17	57.25	71.35	74.91	57.98	66.12	58.12	81.7	43.94	58.14	59.27	54.65
ViLT	46.20	58.91	57.30	71.40	55.82	67.5	65.26	78.26	41.74	53.57	55.53	59.72
+TODDCL	49.28	59.43	58.01	72.25	57.98	69.62	66.44	79.65	42.96	54.52	58.91	60.26
CLIP	59.43	67.28	74.25	73.86	68.54	68.51	71.24	72.26	45.07	61.27	64.74	53.49
+TODDCL	61.82	68.76	**75.72**	75.42	69.58	69.52	**73.81**	73.91	45.72	62.31	66.09	55.98
TMPT*	66.61	68.75	71.79	74.40	69.96	68.43	63.00	82.71	45.04	60.52	68.95	59.87
+TODDCL	67.73	71.91	72.00	**76.12**	**72.41**	**70.20**	64.12	**83.06**	**45.91**	**62.51**	**71.23**	**61.78**

Results of Multimodal Stance Detection. The main experiment results are shown in Table 3. Based on these results, we can make a couple of observations: (1) Compared to the base model BERT+ViT and ViLT, BERT+ViT + TODDCL and ViLT + TODDCL achieve better results on most datasets. This outcome reveals the effective denoising effect of our denoising curriculum; (2) Compared to the base model TMPT, TMPT + TODDCL achieve better results on all datasets. The performance of TMPT+TODDCL is inferior to that of

GPT4-o on the MTSE dataset, which may be attributed to the dataset's inherent complexity and relatively lower information content regarding stance detection; These results affirm the robustness of the TODDCL, showcasing its ability to consistently generate substantial enhancements when applied to the current state-of-the-art model.

4.3 Ablation Study

We introduce a multimodal denoising curriculum learning approach where different methods are employed to obtain noise metrics for the image and text modalities. The dynamic noise curriculum we devised aids the model in achieving improved performance and faster convergence during training. To assess the impact of the noise metrics from different modalities and the effectiveness of the designed dynamic curriculum, we conducted ablation experiments.

Table 4. Ablation study of TODDCL on the five datasets for Multimodal Stance Detection (%). The best results are in bold. Models with * are from TMPT [12], while other results are reproduced under similar conditions.

MODALITY	MTSE		MCCQ	MWTWT					MRUC		MTWQ	
	DT	JB	CQ	CA	CE	AC	AH	DF	RUS	UKR	MOC	TOC
GPT4-Vision*	**70.46**	**72.82**	61.63	44.59	47.07	57.47	57.90	37.61	44.83	56.40	66.72	56.90
BERT	56.12	57.14	68.49	74.95	64.81	63.92	60.21	82.09	43.95	47.08	63.57	58.63
BERT+Text Noise	57.61	57.61	67.81	76.04	63.98	63.55	59.68	83.02	43.32	48.23	64.33	58.31
ViT	51.61	50.31	59.64	45.84	52.94	42.36	48.69	52.36	35.69	40.69	41.08	43.31
ViT+Image Noise	54.21	51.25	60.38	44.31	54.88	44.55	50.19	53.77	35.86	43.33	43.55	41.86
TMPT*	66.61	68.75	71.79	74.40	69.96	68.43	63.00	82.71	45.04	60.52	68.95	59.87
+Text Noise	66.49	71.44	71.87	73.66	71.84	69.24	64.03	82.88	45.26	60.10	70.07	61.74
+Image Noise	66.62	68.83	71.85	75.61	72.18	69.28	62.47	82.47	45.72	61.39	70.15	59.68
+TODDCL	67.73	71.91	72.00	**76.12**	**72.41**	**70.20**	64.12	**83.06**	**45.91**	**62.51**	**71.23**	**61.78**

The Effect of Different Modal Noise Metrics and Denoising Curriculum Learning. We conducted ablation experiments as presented in Table 4. Firstly, we conducted two separate ablation experiments for each modality. To validate the effectiveness of the noise indicator for the text modality, we compared the results of Bert and TMPT with BERT+Text Noise and TMPT+Text Noise. The comparison revealed that the design of the text noise indicator is appropriate, as the presence of noise in the text modality caused by the generation of auxiliary text using the COT method can be reduced using our proposed text denoising approach. Similarly, we compared ViT and TMPT with ViT+Image Noise and TMPT+Image Noise. Similar to the text modality, the presence of direct noise in images can affect model training. Therefore, incorporating image noise metrics into the denoising curriculum can mitigate the impact of irrelevant image noise and improve model performance. Furthermore, we compared TMPT with TMPT+TODDCL trained using dynamic noise metric. It was observed that the dynamic noise-denoising curriculum can effectively enhance model performance. This can be attributed to the utilization of optimal noise metrics for each training step.

The Effect of the TODDCL in Reducing Training Experiment Time. Curriculum Learning has been demonstrated to be effective in steering the learning process away from poor local optima, enabling the model to converge faster [2,16]. As a result, the training time for the denoising phase is reduced compared to regular training. Figure 4 shows the difference in training time between TMPT and TMPT+TODDCL in the MTSE dataset.

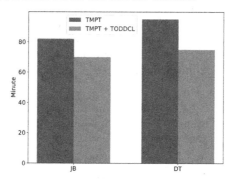

Fig. 4. Training time for TMPT and TMPT+TODDCL in the Multimodal Stance Detection task of the MTSE dataset.

5 Conclusion

In this paper, we proposed a Target-Oriented Dynamic Denoising Curriculum Learning (TODDCL) method for the Multimodal Stance Detection task. Specifically, We design two target-oriented noise metric measurement methods to quantify the noise level in the correspondence modalities of the samples. Utilizing the noise measurement, we design an unimodal denoising curriculum. To better reduce the noise present in different modalities of the samples, we design a dynamic multimodal denoising curriculum to help the model automatically learn to remove sample noise. Extensive experimental results demonstrate that our dynamic denoising method can help the multimodal stance detection model achieve better accuracy and good generalization performance. Further experimental analysis shows that the method can help the model fit the training data in a shorter time during the training phase. Supported by the Science and Technology Innovation 2025 Major Project of Ningbo City (grant number 2022Z189).

References

1. Allaway, E., McKeown, K.: Zero-shot stance detection: a dataset and model using generalized topic representations. In: Proceedings of the 2020 Conference on Empirical Methods in Natural Language Processing (EMNLP), pp. 8913–8931. Association for Computational Linguistics, Online (2020). https://doi.org/10.18653/v1/2020.emnlp-main.717

2. Bengio, Y., Louradour, J., Collobert, R., Weston, J.: Curriculum learning. In: Proceedings of the 26th Annual International Conference on Machine Learning, pp. 41–48. ACM, Montreal Quebec Canada (2009). https://doi.org/10.1145/1553374.1553380
3. Devlin, J., Chang, M.W., Lee, K., Toutanova, K.: BERT: pre-training of Deep Bidirectional Transformers for Language Understanding (2019). https://doi.org/10.48550/arXiv.1810.04805
4. Dosovitskiy, A., et al.: An image is worth 16x16 words: transformers for image recognition at scale. In: International Conference on Learning Representations (2020)
5. Gatto, J., Sharif, O., Preum, S.M.: Chain-of-Thought Embeddings for Stance Detection on Social Media (2023)
6. Gong, T., Zhao, Q., Meng, D., Xu, Z.: Why curriculum learning & self-paced learning work in big/noisy data: a theoretical perspective. Big Data Inf. Analytics **1**(1), 111–127 (2015). https://doi.org/10.3934/bdia.2016.1.111
7. Hardalov, M., Arora, A., Nakov, P., Augenstein, I.: A Survey on Stance Detection for Mis- and Disinformation Identification. In: Findings of the Association for Computational Linguistics: NAACL 2022, pp. 1259–1277. Association for Computational Linguistics, Seattle, United States (2022). https://doi.org/10.18653/v1/2022.findings-naacl.94
8. He, Z., Mokhberian, N., Lerman, K.: Infusing Knowledge from Wikipedia to Enhance Stance Detection (2022)
9. Kim, W., Son, B., Kim, I.: ViLT: vision-and-language transformer without convolution or region supervision. In: Proceedings of the 38th International Conference on Machine Learning, pp. 5583–5594. PMLR (2021)
10. Kingma, D.P., Ba, J.: Adam: A Method for Stochastic Optimization (2017)
11. Küçük, D., Can, F.: Stance detection: a survey. ACM Comput. Surv. **53**(1), 1–37 (2021). https://doi.org/10.1145/3369026
12. Liang, B., et al.: Multi-modal stance detection: new datasets and model. In: Ku, L.W., Martins, A., Srikumar, V. (eds.) Findings of the Association for Computational Linguistics ACL 2024, pp. 12373–12387. Association for Computational Linguistics, Bangkok, Thailand and virtual meeting (2024)
13. Liu, R., Lin, Z., Tan, Y., Wang, W.: Enhancing zero-shot and few-shot stance detection with commonsense knowledge graph. In: Zong, C., Xia, F., Li, W., Navigli, R. (eds.) Findings of the Association for Computational Linguistics: ACL-IJCNLP 2021, pp. 3152–3157. Association for Computational Linguistics, Online (2021). https://doi.org/10.18653/v1/2021.findings-acl.278
14. Luo, Y., Liu, Z., Shi, Y., Li, S.Z., Zhang, Y.: Exploiting Sentiment and Common Sense for Zero-shot Stance Detection (2022)
15. Pavlova, V., Makhlouf, M.: BIOptimus: pre-training an optimal biomedical language model with curriculum learning for named entity recognition. In: Demner-fushman, D., Ananiadou, S., Cohen, K. (eds.) The 22nd Workshop on Biomedical Natural Language Processing and BioNLP Shared Tasks. pp. 337–349. Association for Computational Linguistics, Toronto, Canada (2023). https://doi.org/10.18653/v1/2023.bionlp-1.31
16. Platanios, E.A., Stretcu, O., Neubig, G., Poczos, B., Mitchell, T.M.: Competence-based Curriculum Learning for Neural Machine Translation. https://arxiv.org/abs/1903.09848v2 (2019)
17. Radford, A., et al.: Learning transferable visual models from natural language supervision. In: Proceedings of the 38th International Conference on Machine Learning, pp. 8748–8763. PMLR (2021)

18. Taule, M., Rangel, F., Martı, A., Rosso, P.: Overview of the Task on Multimodal Stance Detection in Tweets on Catalan #1Oct Referendum (2018)
19. Tay, Y., et al.: Simple and effective curriculum pointer-generator networks for reading comprehension over long narratives. In: Korhonen, A., Traum, D., Màrquez, L. (eds.) Proceedings of the 57th Annual Meeting of the Association for Computational Linguistics, pp. 4922–4931. Association for Computational Linguistics, Florence, Italy (2019). https://doi.org/10.18653/v1/P19-1486
20. Vempala, A., Preoţiuc-Pietro, D.: Categorizing and inferring the relationship between the text and image of twitter posts. In: Korhonen, A., Traum, D., Màrquez, L. (eds.) Proceedings of the 57th Annual Meeting of the Association for Computational Linguistics, pp. 2830–2840. Association for Computational Linguistics, Florence, Italy (2019). https://doi.org/10.18653/v1/P19-1272
21. Wang, X., Chen, Y., Zhu, W.: A survey on curriculum learning. IEEE Trans. Pattern Anal. Mach. Intell. 1–1 (2021). https://doi.org/10.1109/TPAMI.2021.3069908
22. Weinzierl, M., Harabagiu, S.: Identification of multimodal stance towards frames of communication. In: Bouamor, H., Pino, J., Bali, K. (eds.) Proceedings of the 2023 Conference on Empirical Methods in Natural Language Processing, pp. 12597–12609. Association for Computational Linguistics, Singapore (2023). https://doi.org/10.18653/v1/2023.emnlp-main.776
23. Xu, B., Zhang, L., Mao, Z., Wang, Q., Xie, H., Zhang, Y.: Curriculum learning for natural language understanding. In: Jurafsky, D., Chai, J., Schluter, N., Tetreault, J. (eds.) Proceedings of the 58th Annual Meeting of the Association for Computational Linguistics, pp. 6095–6104. Association for Computational Linguistics, Online (2020). https://doi.org/10.18653/v1/2020.acl-main.542
24. Yan, M., Zhou, J.T., Tsang, I.W.: Collaborative Knowledge Infusion for Low-resource Stance Detection (2024). https://doi.org/10.26599/BDMA.2024.9020021
25. Zhang, B., Ding, D., Jing, L.: How would Stance Detection Techniques Evolve after the Launch of ChatGPT? (2023). https://doi.org/10.48550/arXiv.2212.14548
26. Zhou, R., Guo, W., Liu, X., Yu, S., Zhang, Y., Yuan, X.: AoM: detecting aspect-oriented information for multimodal aspect-based sentiment analysis. In: Rogers, A., Boyd-Graber, J., Okazaki, N. (eds.) Findings of the Association for Computational Linguistics: ACL 2023, pp. 8184–8196. Association for Computational Linguistics, Toronto, Canada (2023). https://doi.org/10.18653/v1/2023.findings-acl.519

TDM: Temporally-Consistent Diffusion Model for All-in-One Real-World Video Restoration

Yizhou Li[✉], Zihua Liu, Yusuke Monno, and Masatoshi Okutomi

Institute of Science Tokyo, Tokyo, Japan
{yli,zliu,ymonno}@ok.sc.e.titech.ac.jp, mxo@ctrl.titech.ac.jp

Abstract. In this paper, we propose the first diffusion-based all-in-one video restoration method that utilizes the power of a pre-trained Stable Diffusion and a fine-tuned ControlNet. Our method can restore various types of video degradation with a single unified model, overcoming the limitation of standard methods that require specific models for each restoration task. Our contributions include an efficient training strategy with Task Prompt Guidance (TPG) for diverse restoration tasks, an inference strategy that combines Denoising Diffusion Implicit Models (DDIM) inversion with a novel Sliding Window Cross-Frame Attention (SW-CFA) mechanism for enhanced content preservation and temporal consistency, and a scalable pipeline that makes our method all-in-one to adapt to different video restoration tasks. Through extensive experiments on five video restoration tasks, we demonstrate the superiority of our method in generalization capability to real-world videos and temporal consistency preservation over existing state-of-the-art methods. Our method advances the video restoration task by providing a unified solution that enhances video quality across multiple applications.

Keywords: Multi-task Video Restoration · Diffusion Models · ControlNet

1 Introduction

Video restoration is crucial as videos often lose quality due to factors like adverse weather, noise, compression, and limited sensor resolution, which can significantly hamper computer vision tasks such as object detection and video surveillance. Existing video restoration methods [4,5,9,17,18,30,34,35] have shown progress but are limited to specific degradations, requiring separate models for each restoration task. This is costly and impractical for real-world applications where multiple degradations may occur. A unified all-in-one video restoration model is highly demanded for practical use. Recent attempts to create a single model for all-in-one image restoration have been promising [16,25], but they fall short in video applications due to the challenge of maintaining temporal consistency. The main demands for video restoration are (1) developing a single model that can handle various degradations, (2) ensuring temporal consistency of restored videos, and (3) achieving robust real-world performance.

In recent years, diffusion models have markedly advanced image generation [27] and video generation [14,32,38], becoming potential keystones in vision-based AI. Extensive research has demonstrated their ability to parse and encode diverse visual

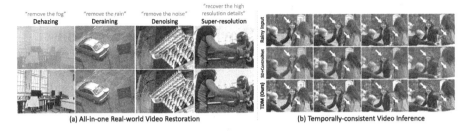

Fig. 1. Our Temporally-consistent Diffusion Model (TDM) has two main features: (a) Our model is all-in-one and can restore various real-world video degradation with a single diffusion model under the guidance of task prompts. (b) Our model can generate temporally consistent video frames with better preservation of original contents included in the input video.

representations from massive text-to-image datasets, enriching downstream applications with strong real-world generalization for tasks such as image editing [3,22] and classification [6]. However, the random nature of diffusion models often disrupts the preservation of original image contents, presenting challenges for image restoration tasks. Some works have developed a diffusion model for each specific task [26,28] by training it from scratch on a certain dataset, limiting its utility to a single task. Alternatively, approaches like Denoising Diffusion Implicit Models Inversion (DDIM Inversion) [8,22] and a content-preserving module with a task-wise plugin [21] guide the diffusion process using input images for multi-target image editing and restoration. However, when these methods are directly applied to video restoration through image-by-image inference, they show insufficient temporal consistency.

To address the above issues, we propose the first diffusion-based all-in-one video restoration method, named Temporally-consistent Diffusion Model (TDM). TDM utilizes the capabilities of the pre-trained text-to-image Stable Diffusion (SD) model [27] alongside a fine-tuned ControlNet [36]. TDM is designed to offer a simple, efficient, and easy-to-extend framework for video restoration, focusing on improving training and inference strategies for a diffusion model without introducing complex modules.

For the training phase, we fine-tune a single-image-based ControlNet [36] on multiple tasks using text prompts referring to the task name, such as "remove the noise" for video denoising and "remove the rain" for video deraining. We refer to this as Task Prompt Guidance (TPG), which directs the diffusion process toward a specific restoration task. This approach leverages the strong zero-shot classification capability of the pre-trained SD model [6]. This simple strategy enables robust all-in-one restoration on multiple tasks without additional computation time and parameters, as demonstrated in Fig. 1(a). By focusing on fine-tuning ControlNet rather than training an entire diffusion model, we harness the generative power of the pre-trained SD model. This approach improves the quality of video restoration and enhances the robustness against real-world data. Furthermore, because ControlNet is fine-tuned using single-image inputs, our strategy eliminates the extreme memory requirements needed to process multiple video frames together during training, which is common in earlier methods [2,10,31] for ensuring temporal consistency. This allows our model to be trained on a single

GPU using common single-image restoration datasets, simplifying data preparation and enhancing accessibility for future research.

In the inference phase, we aim to ensure content preservation and temporal consistency. Previous text-to-video studies [14,38] addressed temporal consistency by replacing the self-attention in U-Net with cross-frame attention during inference. However, the method [14] struggles with large motions between video frames and the method [38] requires extensive memory and computational resources. Also, both methods do not consider content preservation. To address these issues, we propose a Sliding Window Cross-Frame Attention (SW-CFA) mechanism, combined with DDIM Inversion [8,22]. This combination effectively tackles both content preservation and temporal consistency. Specifically, SW-CFA extends the concept of reference frames to a sliding window around the current frame, employing mean-based temporal smoothing in the attention calculation with minimal computational increase. This allows SW-CFA to handle large motions better than [14], achieving effective zero-shot image-to-video adaptation by adjusting the attention mechanism during inference. As discussed in [21], adding random Gaussian noise to the input latent can reduce the fidelity of diffusion-generated images. Therefore, we introduce DDIM Inversion for better content preservation. The combination of SW-CFA and DDIM Inversion maintains a deterministic and close distribution of input noises across adjacent frames, resulting in a more coherent video output. As illustrated in Fig. 1(b), compared with a straightforward combination of SD and ControlNet, our proposed TDM with SW-CFA and DDIM Inversion generates more temporally consistent video frames.

In summary, we build the first all-in-one video restoration diffusion model in this work. Our main contributions are as follows:

- Training strategies: We achieve efficient training by fine-tuning a single ControlNet using single-image inputs without requiring video inputs. The proposed TPG offers a simple way to achieve cross-task robustness.
- Inference strategies: We incorporate DDIM Inversion and introduce a novel SW-CFA mechanism for zero-shot video inference. This combination ensures accurate content preservation and robust temporal consistency in restored videos.
- Scalability: Our proposed TDM can be trained with a single GPU and is adaptable for video inference after being trained on single-image restoration datasets. This flexibility allows straightforward expansion to other video restoration tasks using only single-image datasets.
- Generalization performance: We conduct extensive experiments on five restoration tasks, demonstrating strong generalization performance of our TDM to real-world data over existing regression-based and diffusion-based methods.

2 Methodology

In this section, we first introduce the preliminary on latent diffusion models in Sect. 2.1. Then, we introduce the details of our method in Sect. 2.2 (Training) and Sect. 2.3 (Inference).

2.1 Preliminary

Latent Diffusion Model. (LDM) [27] significantly advances traditional diffusion models by operating in a latent space. LDM employs an encoding mechanism to transform an image x into a compressed latent representation $z = E(x)$, thereby facilitating the learning of the latent codes' distribution, denoted as $z_0 \sim p_{\text{data}}(z_0)$ where p_{data} denotes the latent distributions of training images, following the Denoising Diffusion Probabilistic Model (DDPM) strategy introduced by [12]. First, LDM follows a forward phase that introduces Gaussian noise progressively over time steps t to derive z_t:

$$q(z_t|z_{t-1}) = \mathcal{N}(z_t; \sqrt{1-\beta_t}z_{t-1}, \beta_t I), \tag{1}$$

with the noise scale represented by $\{\beta_t\}_{t=1}^T$, the gaussian distribution by $\mathcal{N}$, an all-one tensor with the same shape as z_t expressed by I, and the total diffusion steps denoted by T. Then, LDM follows a backward phase, where the model strives to reconstruct the preceding less noisy state z_{t-1} as follows:

$$p_\theta(z_{t-1}|z_t) = \mathcal{N}(z_{t-1}; \mu_\theta(z_t, t, \tau), \Sigma_\theta(z_t, t, \tau)), \tag{2}$$

where μ_θ and Σ_θ are mean and variance of the current state characterized by learnable parameters θ, and implemented with a noise prediction model ϵ_θ. To generate novel samples, initialization starts with a Gaussian sample $z_T \sim \mathcal{N}(0,1)$, followed by a DDIM backward process of z_{t-1} for preceding time steps:

$$z_{t-1} = \sqrt{\alpha_{t-1}}\left(\frac{z_t - \sqrt{1-\alpha_t}\epsilon_\theta(z_t, t, \tau)}{\sqrt{\alpha_t}}\right) + \sqrt{1-\alpha_{t-1}} \cdot \epsilon_\theta(z_t, t, \tau), \tag{3}$$

where the cumulative product $\alpha_t = \prod_{i=1}^t(1-\beta_i)$ is used for simplicity to indicate the transition towards z_0 at step t. The most well-known latent diffusion model is SD [27], which exemplifies text-to-image LDMs trained on a vast corpus of image-text pairs, with τ representing the text prompt. Moreover, it demonstrates its scalability across various other tasks [13,19,20].

ControlNet. [36] enhances the capabilities of SD for more precise text-to-image synthesis by incorporating additional inputs such as depth maps, body poses, and images. While retaining the U-Net architecture identical to that of SD, ControlNet is fine-tuned to accommodate specific conditional inputs, modifying the function from $\epsilon_\theta(z_t, t, \tau)$ to $\epsilon_\theta(z_t, t, c, \tau)$, where c represents these additional conditional inputs.

DDIM Inversion. [8,29] is an inverse process of the DDIM forward process, based on the assumption that the ordinary differential equation can be reversed in the limit of small steps:

$$z_{t+1} = \sqrt{\alpha_{t+1}}\left(\frac{z_t - \sqrt{1-\alpha_t}\epsilon_\theta(z_t, t, \tau)}{\sqrt{\alpha_t}}\right) + \sqrt{1-\alpha_{t+1}} \cdot \epsilon_\theta(z_t, t, \tau). \tag{4}$$

In other words, the diffusion process is performed in the reverse direction, that is z_0 to z_T instead of z_T to z_0, where z_0 in our case is set to be the encoding of the given input degraded image. Despite using ControlNet, we observe that employing noises generated through DDIM Inversion, rather than relying on random Gaussian noises, provides stable structural guidance that better preserves the contents of the input image.

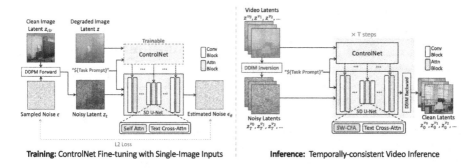

Fig. 2. Overall architecture of our proposed temporally-consistent diffusion model (TDM).

2.2 Training: Task Prompt Guided ControlNet Fine-Tuning with Single-Image Inputs

The pre-trained SD model excels at generating high-quality images without degradation, a feature we strive to maintain, particularly in the face of severe degradations such as sensor noise, low resolution, and haze. To this end, as shown in Fig. 2, we leverage ControlNet [36] which is a copy of the SD U-Net architecture. We fine-tune the ControlNet model initialized as the official tile resample model from ControlNet for SD version 1.5 because it is a solid foundation for image-to-image tasks. This ControlNet is then fine-tuned on our datasets, which include pairs of degraded images and their clean ground-truth images, focusing on image restoration.

Task Prompt Guidance (TPG). In our experiments, we consider five video restoration tasks: dehazing, deraining, denoising, MP4 compression artifact removal, and super-resolution. Leveraging the capacity of SD to generate diverse image styles through different text prompts, we recognize their potential to address a variety of tasks within a single model framework. To enhance this versatility, we introduce TPG, a method that employs specific task descriptions as input text prompts (τ) for both training and inference, as shown in Fig. 1 and Fig. 2. Unlike the approach in [21], which relies on separate plugin modules for each task and a classifier to select the appropriate plugin based on the prompt, we utilize the proven proficiency of the pre-trained SD model to interpret and classify directly from text prompts as noted in studies [6,7].

Single-Image Inputs During Training. As illustrated in Fig. 2, our proposed method requires only single-image inputs for video restoration tasks during the fine-tuning of the ControlNet. Drawing inspiration from previous works [14,38], we achieve significant temporal consistency through training-free operations at the inference stage, thereby circumventing the high memory demands of processing multiple video frames simultaneously during training, which is a common requirement in earlier approaches [2,10,31] for achieving temporal consistency. This strategy allows our model to be trained efficiently on a single GPU using single-image restoration datasets, significantly reducing data preparation complexity and making it more accessible for future research. This approach paves the way for future research to easily append novel tasks with training on corresponding single-image datasets.

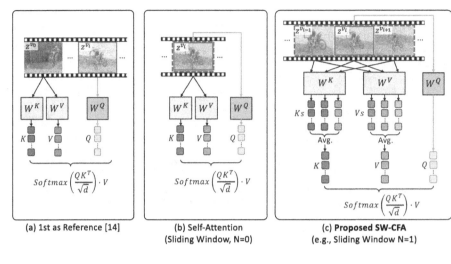

Fig. 3. Proposed SW-CFA compared with exisiting cross-frame attention.

2.3 Inference: Training-Free Content-Preserved Temporal Consistency for Larger Motion

As mentioned earlier, we only use single-image inputs in the training phase, so temporal consistency is addressed through a training-free approach during the inference. Previous studies [14,38] achieve this by replacing self-attention layers in denoising U-Nets with cross-frame attention layers. LDM models utilize a U-Net architecture with downscaling and upscaling phases, enhanced with skip connections. The architecture includes 2D convolutional residual blocks and transformer blocks, each containing a self-attention layer, a cross-attention layer, and a feed-forward network. Self-attention formulates spatial correlations within the feature map, while cross-attention maps relationships between the feature map and external conditions like text prompts. For a video frame v_i with latent representation z^{v_i}, as shown in Fig. 3(b), self-attention is defined as: $Attention(Q, K, V) = Softmax\left(\frac{QK^T}{\sqrt{d}}\right) \cdot V$, where

$$Q = W^Q z^{v_i}, K = W^K z^{v_i}, V = W^V z^{v_i}, \tag{5}$$

with W^Q, W^K, and W^V being trainable matrices for query, key, and value projections, respectively, and d is the dimension of key and query features. Previous studies [14] reorganize self-attention into cross-frame attention using a single reference latent $z_{v_{ref}}$ as both the key (K) and value (V). Typically, this reference frame is set as the first frame (v_0) as shown in Fig. 3(a):

$$Q = W^Q z^{v_i}, K = W^K z^{v_0}, V = W^V z^{v_0}, \tag{6}$$

This captures the spatial relationship between the current frame's query and the reference frame's key, maintaining the appearance, structure, and identities across frames, thus enhancing temporal consistency. However, using only v_0 as the reference limits motion range, as large motions may result in limited overlap between v_0 and subsequent frames. Additionally, the consistency between later consecutive frames is not explicitly addressed, further limiting overall temporal consistency even with small motions.

Sliding Window Cross-Frame Attention (SW-CFA). We propose a novel cross-frame attention mechanism named SW-CFA to accommodate a wider range of motions. As shown in Fig. 3(c), instead of relying on a single fixed reference frame for the key (K) and value (V), we extend the reference frames to include frames within a local window. Specifically, we average the keys and values inside the window, which includes both the preceding and succeeding N frames. The formulation is as follows:

$$Q = W^Q z^{v_i}, K = \frac{1}{2N+1} \sum_{j=i-N}^{i+N} W^K z^{v_j}, V = \frac{1}{2N+1} \sum_{j=i-N}^{i+N} W^V z^{v_j}, \quad (7)$$

which can be further simplified as

$$Q = Q_i, K = \frac{1}{2N+1} \sum_{j=i-N}^{i+N} K_j, V = \frac{1}{2N+1} \sum_{j=i-N}^{i+N} V_j, \quad (8)$$

and our SW-CFA is formulated as follows:

$$Attention(Q, K, V) = Softmax\left(Q_i \frac{1}{2N+1} \sum_{j=i-N}^{i+N} K_j^T\right) \left(\frac{1}{2N+1} \sum_{j=i-N}^{i+N} V_j\right). \quad (9)$$

By averaging K_j and V_j across a local window of frames, SW-CFA smooths over rapid variations, focusing on stable and consistent features. This is not merely filtering but a principled integration of temporal information, enhancing the model's ability to prioritize relevant information across temporal axis.

Mathematically, this averaging process reduces variance in the attention mechanism's input, effectively functioning as a temporal low-pass filter. This reduction allows the attention mechanism to produce outputs less sensitive to frame-to-frame fluctuations, increasing temporal consistency. Within the softmax function, averaging indirectly weights each nearest N frames' K and V matrices according to their temporal proximity and similarity to the current frame i, leveraging the softmax function's property of amplifying significant signals while attenuating weaker ones. Thus, SW-CFA captures both spatial dependencies within frames and temporal dependencies. This results in a robust attention mechanism that enhances temporal consistency through a simple yet effective modification of the previous cross-frame attention paradigm.

Combination with DDIM Inversion. Our proposed SW-CFA mechanism significantly enhances temporal consistency. However, random Gaussian noise added to input latents, as noted in [21], can reduce the fidelity of diffusion-generated images. To address this, we integrate DDIM Inversion [8,29] to provide stable input noise for each video frame's latents, serving as a solid structural guide, as shown in Fig. 2. DDIM Inversion ensures the noise added to each input frame's latents is deterministic, consistent, and derived from the latents themselves. Since these input latents originate from temporally coherent video frames, the noise introduced by DDIM Inversion also exhibits temporal coherence. This synergy eliminates the disruption caused by random Gaussian noise, enabling SW-CFA to more precisely capture temporal relationships. Therefore, combining DDIM Inversion not only improves content preservation but also further strengthens temporal consistency.

3 Experiments

3.1 Settings

Here, we outline our experimental settings, including the datasets, implementation details, and evaluation metrics.

Datasets. We evaluated the proposed TDM for five video restoration tasks. To train the TDM and the compared methods, we utilized representative datasets for each task, dehazing: REVIDE [37], deraining: NTURain-syn [35], denoising: DAVIS [15], MP4 compression artifacts removal: MFQEv2 [9], and super-resolution (SR): REDS [23]. Only REVIDE is a real-world dataset and the others are synthetic datasets. To avoid unbalanced training due to largely different amounts of training images for each task, we adjusted the number of images from each dataset to around 5,000-6,000 images. Then, we combined all of those images to construct the training dataset, resulting in a total of 27,843 training images. For testing, we used real-world benchmark datasets for each task, dehazing: REVIDE [37] (284 images), denoising: CRVD [34] (560 images), deraining: NTURain-real [35] (658 images), MP4 compression artifacts removal: MFQEv2 [9] (1,080 images), and SR: UDM10 [33] (320 images).

Implementation. During the training and the inference phase, we adjusted the size of each image by resizing it to ensure that the shorter side is 512 pixels. For training, we randomly cropped patches of 512×512 pixels from resized images. We utilized the AdamW optimizer with its default settings, including betas and weight decay. The training of our ControlNet was carried out with a constant learning rate of 1e-5 and a batch size of 4, on a single RTX 4090 GPU, and we incorporated gradient checkpoints for efficiency. The training process lasts for 25 epochs. In the inference stage, we set the window radius N for SW-CFA as 3 for all experiments. For the sampling technique, we initially applied DDIM Inversion [8] with 10 timesteps, followed by DDIM backward sampling [29] using 32 timesteps. Our proposed TDM model can process a 15-frame video with resolutions of 512×896 pixels in under 30 s with a single RTX 4090 GPU while requiring 10GB of GPU memory.

Metrics. We followed [21,27] to employ widely adopted non-reference perceptual metrics, FID [11] and KID [1], to evaluate our TDM, as the ground-truth images are not always available for real-world datasets. For easier view, the KID value is scaled by $100\times$. For temporal consistency evaluation, we followed [38] to estimate (i) Frame consistency (FC): the average cosine similarity between all pairs of consecutive frames, and (ii) Warping Error (WE): the average mean squared error of consecutive frames after aligning the next frame to the current one using estimated optical flow. For easier view, FC is scaled by $10\times$ and WE is scaled by $1000\times$.

3.2 Comparison with State-of-the-Art Methods

We compared our TDM with six state-of-the-art methods for image/video restoration: AirNet [16], PromptIR [25], VRT [17], RVRT [18], WeatherDiff [24], and InstructP2P [3]. As the category of each method, AirNet, PromptIR, VRT, and RVRT are regression-based methods, such as based on CNNs or Transformers, while WeatherDiff, InstructP2P, and our TDM are diffusion-based methods. As the input for the inference, AirNet, PromptIR, WeatherDiff, InstructP2P use a single-image input, while VRT, RVRT, and our TDM use a video input. We trained all methods using the same training dataset and tested them for real-world testing datasets, as explained in the previous subsection.

Quantitative Results. We provide a quantitative comparison in Table 1. It demonstrates the superiority of diffusion-based methods in producing higher-quality images, as evidenced by better FID and KID scores compared to regression-based methods. Compared with other regression- and diffusion-based methods, our proposed TDM achieves the best results on average. Although InstructP2P also generates high-quality images with low FID and KID, it tends to alter the original contents, as seen in Fig. 4. Compared to the regression-based methods including AirNet and PromptIR, which are designed for generalization across different tasks, our TDM still demonstrates robust cross-task performance with the guidance of the proposed TPG. Furthermore, while video restoration methods such as VRT and RVRT are designed to utilize video temporal information effectively during the training phase, they struggle with multi-task handling. In contrast, our TDM consistently delivers state-of-the-art performance in video restoration tasks, even though it is trained on single-image inputs.

Table 1. Quantitative comparison with state-of-the-art methods (Red: rank 1st; Blue: rank 2nd).

Method Types	Methods	Dehazing		Deraining		Denoising		MP4		SR ×4		Average	
		FID↓	KID↓	FID↓	KID↓	FID↓	KID↓	FID↓	KID↓	FID↓	KID↓	FID↓	KID↓
Single-image regression	AirNet [16]	83.34	7.77	90.86	4.24	80.63	4.41	104.48	6.56	88.09	3.07	89.47	5.21
	PromptIR [25]	75.21	5.73	88.47	5.28	82.11	4.95	102.73	6.68	89.37	3.12	87.57	5.15
Video regression	VRT [17]	79.88	7.03	88.36	4.69	81.94	4.88	107.00	6.56	88.97	2.71	89.23	5.17
	RVRT [18]	79.27	6.55	94.13	5.44	83.59	4.57	107.37	6.55	88.51	2.93	90.57	5.20
Single-image diffusion	WeatherDiff [24]	74.42	6.22	88.20	4.85	80.11	4.18	102.24	6.67	88.07	2.77	86.28	4.93
	InstructP2P [3]	74.02	6.01	81.51	3.36	79.71	3.90	101.62	7.70	88.21	2.62	85.01	4.71
Video diffusion	**TDM (Ours)**	73.68	6.42	81.27	3.79	78.63	4.09	100.91	6.59	88.04	2.55	84.50	4.68

Qualitative Results. Figure 4 illustrates the robust performance of our TDM across five challenging real-world video restoration tasks. In the denoising task (1st row), while most regression-based methods trained on synthesized Gaussian noise struggle to recognize real-world noise patterns, all diffusion-based methods including TDM successfully eliminate the real-world noise. However, TDM distinctly outperforms WeatherDiff and InstructP2P, which both leave artifacts post-denoising. In the dehazing task (2nd row), TDM excels by clearing heavy fog to reveal the sharpest and the most haze-free images. For the MP4 compression artifact removal task (3rd row), TDM outperforms regression-based methods in detail restoration. Compared with diffusion-based

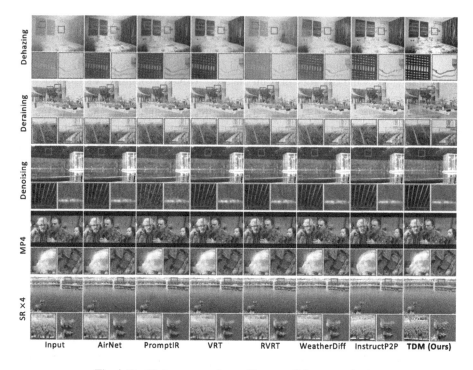

Fig. 4. Qualitative comparison with state-of-the-art methods.

InstructP2P, which inaccurately alters hairstyles, our TDM can restore sharp details and maintain consistency with the original input. In the SR task (4th row), TDM achieves the clearest detail enhancement, whereas other methods produce comparatively blurred results. Finally, in the challenging heavy rain scenario (5th row), while other models fail to detect and remove the rain, TDM significantly reduces rain artifacts, demonstrating its robustness in severe weather conditions. Overall, our TDM consistently delivers superior restoration quality with remarkable detail preservation and consistency across different video restoration challenges.

Consistency Evaluation. Although diffusion-based methods typically yield images of high visual quality, their inherent randomness often leads to poor temporal consistency in video processing. In Table 2, we compare our TDM against other diffusion-based techniques using two temporal consistency metrics: WE and FC. The results show that TDM consistently outperforms the others in maintaining temporal consistency according to both metrics. Figure 5 provides a visual comparison that supports these results. While the methods WeatherDiff and InstructP2P can remove the noise, they exhibit notable inconsistencies between the frames. WeatherDiff (2nd row) exhibits heavy changes in the appearance of building frames in the background. For InstructP2P (3rd row), despite using image-based classifier-free guidance, it still displays fluctuation across frames. In contrast, TDM, utilizing DDIM inversion to eliminate randomness and SW-CFA to bolster temporal stability, maintains consistent features across all frames, leading to stable denoising results. This robustness underscores TDM's superior temporal consistency compared to other diffusion-based methods. Also, in Table 3 and Fig. 6, we evaluate our proposed SW-CFA against the cross-frame attention from Text2Video-zero [14] (referred to as "1st as Ref") and the standard self-attention ($N=0$). While "1st as Ref" enhances consistency by using the first frame as a reference, it overlooks the consistency of consecutive frames. In contrast, our SW-CFA, which averages key-value pairs within a sliding window, achieves more substantial consistency improvements. It demonstrates a superior ability to maintain uniformity across frames, effectively handling larger motions and providing robust temporal stability.

Table 2. Image quality and consistency comparison with other diffusion-based methods.

Methods	Dehazing			Deraining			Denoising			MP4			SR ×4		
	FID↓	FC↑	WE↓	FID↓	FC↑	WE↓	FID↓	FC↑	WE↓	FID↓	FC↑	WE↓	FID↓	FC↑	WE↓
WeatherDiff [24]	74.42	9.759	8.537	88.20	9.682	3.652	80.11	9.439	2.182	102.24	9.908	1.282	88.07	**9.682**	5.751
InstructP2P [3]	74.02	9.663	6.994	81.51	9.676	4.339	79.71	**9.436**	2.628	101.62	9.877	2.126	88.21	9.628	5.929
TDM (Ours)	**73.68**	**9.849**	**5.464**	**81.27**	**9.700**	**3.415**	**78.63**	9.296	**2.008**	**100.91**	**9.921**	**1.138**	**88.04**	9.651	**5.741**

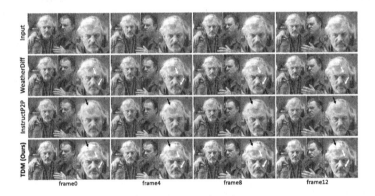

Fig. 5. Consistency comparison (MP4) with other diffusion-based methods.

Table 3. Image quality and consistency comparison with state-of-the-art zero-shot cross-frame attention (by replacing the self-attention with according cross-frame attention during inference).

Methods	Dehazing			Deraining			Denoising			MP4			SR ×4		
	FID↓	FC↑	WE↓	FID↓	FC↑	WE↓	FID↓	FC↑	WE↓	FID↓	FC↑	WE↓	FID↓	FC↑	WE↓
1st as Ref. [14]	74.37	9.712	6.432	82.39	9.693	3.526	80.12	9.229	2.219	**99.63**	9.901	1.387	88.62	9.622	6.350
Self-Attn. ($N=0$)	**73.36**	9.703	6.801	83.06	9.690	3.596	80.17	9.213	2.368	100.56	9.896	1.459	88.30	9.618	6.769
SW-CFA ($N=3$)	73.68	**9.849**	**5.464**	**81.27**	**9.700**	**3.415**	**78.63**	**9.296**	**2.008**	100.91	**9.921**	**1.138**	**88.04**	**9.651**	**5.741**

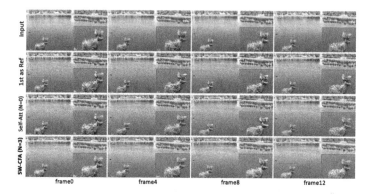

Fig. 6. Consistency comparison (SR×4) with other zero-shot cross-frame attention.

3.3 Ablation Study

In Table 4, we present an ablation study to assess the contribution of each component of our proposed TDM. We test three configurations: T+I, which omits the SW-CFA; T+S, which excludes DDIM Inversion (labeled as Inv.); and I+S, which lacks the TGP. The results show that removing either SW-CFA or DDIM Inversion diminishes temporal consistency, while their combination significantly enhances it. This indicates that DDIM Inversion, by stabilizing the input noise across frames, substantially supports SW-CFA in maintaining consistency. Furthermore, the setup without TPG achieves comparable consistency scores, but there is a noticeable degradation in image quality, as evidenced by the increased FID scores. This underscores the critical role of TPG in boosting cross-task generalization and improving overall image quality in various restoration tasks.

Table 4. Ablation study.

Methods	Proposals			Dehazing			Deraining			Denoising			MP4			SR ×4		
	TPG	Inv.	SW-CFA	FID↓	FC↑	WE↓	FID↓	FC↑	WE↓	FID↓	FC↑	WE↓	FID↓	FC↑	WE↓	FID↓	FC↑	WE↓
T+I	✓	✓		**73.36**	9.703	6.801	83.06	9.690	3.596	80.17	9.213	2.368	**100.56**	9.896	1.459	88.30	9.618	6.769
T+S	✓		✓	73.45	9.738	7.121	82.79	9.673	4.125	80.07	9.207	3.792	101.47	9.866	1.841	88.35	9.552	8.418
I+S		✓	✓	77.56	9.838	5.653	87.28	9.698	3.437	82.68	9.273	2.296	102.44	9.909	1.190	91.02	9.630	5.874
Ours	✓	✓	✓	73.68	**9.849**	**5.464**	**81.27**	**9.700**	**3.415**	**78.63**	**9.296**	**2.008**	100.91	**9.921**	**1.138**	**88.04**	**9.651**	**5.741**

4 Conclusion

In conclusion, we introduced the Temporally-consistent Diffusion Model (TDM) for all-in-one video restoration, utilizing a pre-trained Stable Diffusion model and fine-tuned ControlNet. Our method handles various video degradations with a single model, using Task Prompt Guidance (TPG) for training and combining DDIM Inversion with Sliding Window Cross-Frame Attention (SW-CFA) for enhanced temporally-consistent video inference. Experiments across five tasks exhibits proposed TDM's superior generalization over existing methods, setting a new standard for video restoration. However, our method still falls short of regression-based methods in temporal consistency. Future work will focus on addressing this gap to further improve effectiveness.

References

1. Bińkowski, M., Sutherland, D.J., Arbel, M., Gretton, A.: Demystifying mmd gans. arXiv preprint arXiv:1801.01401 (2018)
2. Blattmann, A., et al.: Stable video diffusion: Scaling latent video diffusion models to large datasets. arXiv preprint arXiv:2311.15127 (2023)
3. Brooks, T., Holynski, A., Efros, A.A.: Instructpix2pix: Learning to follow image editing instructions. In: Proceedings of the IEEE/CVF Conference on Computer Vision and Pattern Recognition, pp. 18392–18402 (2023)
4. Cao, J., et al.: Learning task-oriented flows to mutually guide feature alignment in synthesized and real video denoising. arXiv preprint arXiv:2208.11803 (2022)
5. Chan, K.C., Zhou, S., Xu, X., Loy, C.C.: Basicvsr++: improving video super-resolution with enhanced propagation and alignment. In: Proceedings of the IEEE/CVF Conference on Computer Vision and Pattern Recognition, pp. 5972–5981 (2022)
6. Chen, H., et al.: Robust classification via a single diffusion model. arXiv preprint arXiv:2305.15241 (2023)
7. Clark, K., Jaini, P.: Text-to-image diffusion models are zero shot classifiers. Adv. Neural Inform. Process. Syst. **36** (2024)
8. Dhariwal, P., Nichol, A.: Diffusion models beat gans on image synthesis. Adv. Neural. Inf. Process. Syst. **34**, 8780–8794 (2021)
9. Guan, Z., Xing, Q., Xu, M., Yang, R., Liu, T., Wang, Z.: Mfqe 2.0: a new approach for multi-frame quality enhancement on compressed video. IEEE Trans. Pattern Analy. Mach. Intell. **43**(3), 949–963 (2019)
10. Guo, Y., et al.: Animatediff: Animate your personalized text-to-image diffusion models without specific tuning. arXiv preprint arXiv:2307.04725 (2023)
11. Heusel, M., Ramsauer, H., Unterthiner, T., Nessler, B., Hochreiter, S.: Gans trained by a two time-scale update rule converge to a local nash equilibrium. Adv. Neural Inform. Process. Syst. **30** (2017)
12. Ho, J., Jain, A., Abbeel, P.: Denoising diffusion probabilistic models. Adv. Neural. Inf. Process. Syst. **33**, 6840–6851 (2020)
13. Ke, B., Obukhov, A., Huang, S., Metzger, N., Daudt, R.C., Schindler, K.: Repurposing diffusion-based image generators for monocular depth estimation. In: Proceedings of the IEEE/CVF Conference on Computer Vision and Pattern Recognition (2024)
14. Khachatryan, L., et al.: Text2video-zero: Text-to-image diffusion models are zero-shot video generators. In: Proceedings of the IEEE/CVF International Conference on Computer Vision, pp. 15954–15964 (2023)
15. Khoreva, A., Rohrbach, A., Schiele, B.: Video object segmentation with language referring expressions. In: Asian Conference on Computer Vision, pp. 123–141 (2019)
16. Li, B., Liu, X., Hu, P., Wu, Z., Lv, J., Peng, X.: All-in-one image restoration for unknown corruption. In: Proceedings of the IEEE/CVF Conference on Computer Vision and Pattern Recognition, pp. 17452–17462 (2022)
17. Liang, J., et al.: Vrt: A video restoration transformer. arXiv preprint arXiv:2201.12288 (2022)

18. Liang, J., et al.: Recurrent video restoration transformer with guided deformable attention. Adv. Neural. Inf. Process. Syst. **35**, 378–393 (2022)
19. Liu, R., Wu, R., Van Hoorick, B., Tokmakov, P., Zakharov, S., Vondrick, C.: Zero-1-to-3: zero-shot one image to 3d object. In: Proceedings of the IEEE/CVF International Conference on Computer Vision, pp. 9298–9309 (2023)
20. Liu, Y., Lin, C., Zeng, Z., Long, X., Liu, L., Komura, T., Wang, W.: Syncdreamer: Generating multiview-consistent images from a single-view image. arXiv preprint arXiv:2309.03453 (2023)
21. Liu, Y., Liu, F., Ke, Z., Zhao, N., Lau, R.W.: Diff-plugin: Revitalizing details for diffusion-based low-level tasks. arXiv preprint arXiv:2403.00644 (2024)
22. Mokady, R., Hertz, A., Aberman, K., Pritch, Y., Cohen-Or, D.: Null-text inversion for editing real images using guided diffusion models. In: Proceedings of the IEEE/CVF Conference on Computer Vision and Pattern Recognition, pp. 6038–6047 (2023)
23. Nah, S., et al.: Ntire 2019 challenge on video deblurring and super-resolution: dataset and study. In: Proceedings of the IEEE/CVF Conference on Computer Vision and Pattern Recognition Workshops (2019)
24. Özdenizci, O., Legenstein, R.: Restoring vision in adverse weather conditions with patch-based denoising diffusion models. IEEE Trans. Pattern Anal. Mach. Intell.dd (2023)
25. Potlapalli, V., Zamir, S.W., Khan, S., Khan, F.S.: Promptir: prompting for all-in-one blind image restoration. arXiv preprint arXiv:2306.13090 (2023)
26. Ren, M., Delbracio, M., Talebi, H., Gerig, G., Milanfar, P.: Multiscale structure guided diffusion for image deblurring. In: Proceedings of the IEEE/CVF International Conference on Computer Vision, pp. 10721–10733 (2023)
27. Rombach, R., Blattmann, A., Lorenz, D., Esser, P., Ommer, B.: High-resolution image synthesis with latent diffusion models. In: Proceedings of the IEEE/CVF Conference on Computer Vision and Pattern Recognition, pp. 10684–10695 (2022)
28. Saharia, C., Chan, W., Chang, H., Lee, C., Ho, J., Salimans, T., Fleet, D., Norouzi, M.: Palette: Image-to-image diffusion models. In: Proceedings of ACM SIGGRAPH Conference, pp. 1–10 (2022)
29. Song, J., Meng, C., Ermon, S.: Denoising diffusion implicit models. arXiv preprint arXiv:2010.02502 (2020)
30. Tassano, M., Delon, J., Veit, T.: Fastdvdnet: towards real-time deep video denoising without flow estimation. In: Proceedings of the IEEE/CVF Conference on Computer Vision and Pattern Recognition, pp. 1354–1363 (2020)
31. Wang, J., Yuan, H., Chen, D., Zhang, Y., Wang, X., Zhang, S.: Modelscope text-to-video technical report. arXiv preprint arXiv:2308.06571 (2023)
32. Wu, J.Z., et al.: Tune-a-video: One-shot tuning of image diffusion models for text-to-video generation. In: Proceedings of the IEEE/CVF International Conference on Computer Vision, pp. 7623–7633 (2023)
33. Yi, P., Wang, Z., Jiang, K., Jiang, J., Ma, J.: Progressive fusion video super-resolution network via exploiting non-local spatio-temporal correlations. In: Proceedings of the IEEE/CVF International Conference on Computer Vision, pp. 3106–3115 (2019)
34. Yue, H., Cao, C., Liao, L., Chu, R., Yang, J.: Supervised raw video denoising with a benchmark dataset on dynamic scenes. In: Proceedings of the IEEE/CVF Conference on Computer Vision and Pattern Recognition, pp. 2301–2310 (2020)
35. Yue, Z., Xie, J., Zhao, Q., Meng, D.: Semi-supervised video deraining with dynamical rain generator. In: Proceedings of the IEEE/CVF Conference on Computer Vision and Pattern Recognition, pp. 642–652 (2021)
36. Zhang, L., Rao, A., Agrawala, M.: Adding conditional control to text-to-image diffusion models. In: Proceedings of the IEEE/CVF International Conference on Computer Vision, pp. 3836–3847 (2023)

37. Zhang, X., et al.: Learning to restore hazy video: a new real-world dataset and a new method. In: Proceedings of the IEEE/CVF Conference on Computer Vision and Pattern Recognition, pp. 9239–9248 (2021)
38. Zhang, Y., Wei, Y., Jiang, D., Zhang, X., Zuo, W., Tian, Q.: Controlvideo: training-free controllable text-to-video generation. arXiv preprint arXiv:2305.13077 (2023)

Temporal Closeness for Enhanced Cross-Modal Retrieval of Sensor and Image Data

Shuhei Yamamoto[1](✉) and Noriko Kando[2]

[1] Institute of Library, Information and Media Science, University of Tsukuba, Tsukuba, Japan
syamamoto@slis.tsukuba.ac.jp
[2] Information and Society Research Division, National Institute of Informatics, Tokyo, Japan

Abstract. This paper presents a new approach to dense retrieval across multiple modalities, emphasizing the integration of images and sensor data. Traditional cross-modal retrieval techniques face significant challenges, particularly in processing non-linguistic modalities and creating effective training datasets. To address these issues, we propose a method that uses a shared vector space, optimized with contrastive loss, to enable efficient and accurate retrieval across diverse modalities. A key innovation of our approach is the introduction of a temporal closeness metric, which evaluates the relationship between data points based on their timestamps. This metric helps automatically extract positive and hard negative samples related to the query, improving the training process and enhancing the retrieval model's performance. We validate our approach using the Lifelog Search Challenge 2024 (LSC'24) dataset, one of the largest multi-modal datasets, including non-linguistic data such as egocentric images, heart rate, and location information. Our evaluation shows that incorporating temporal closeness into the dense retrieval process significantly improves retrieval accuracy and robustness in real-world, multi-modal scenarios. This paper's contributions include developing a novel dense retrieval framework, introducing the temporal closeness metric, and successfully applying these innovations to a comprehensive multi-modal dataset.

Keywords: Dense retrieval · Cross-modal retrieval · Temporal closeness · Contrastive loss

1 Introduction

Large language models (LLMs) have recently become a central focus of artificial intelligence research [1]. These models learn language patterns from extensive datasets and have demonstrated significant improvements in tasks such as natural language understanding and generation. Notably, models like Generative Pre-trained Transformer (GPT) [18] exhibit human-like performance across a range of language-related tasks by combining broad knowledge training with task-specific fine-tuning.

The advancements in LLMs have paved the way for cross-modal information processing, enabling interactions across various modalities, including text, images, and audio. For example, technologies such as visual question answering, which allows users to ask text-based questions about images [7], automatic video captioning [4], and context-aware information retrieval [27] are all active areas of research. These advancements in LLMs have enabled applications that were previously unattainable with traditional cross-modal processing techniques.

Despite these advancements, the methodology required for effective cross-modal information access remain unclear [24]. While large text and image corpora exist (e.g., Microsoft COCO [16] and Flickr30k [26]), they often lack corresponding data from other modalities, limiting their utility in developing robust cross-modal information access technologies. Additionally, it remains a challenge to design models capable of processing non-linguistic modalities, such as sensor data (e.g., heart rate and accelerometer ratings) or location information (e.g., latitude and longitude), which are crucial for considering user context. Consequently, there are significant research challenges surrounding the development of LLM-based cross-modal information access technologies.

This paper introduces a dense retrieval method for retrieving relevant data across multiple modalities. Dense retrieval involves learning a shared vector space between queries and documents by training an encoder on paired datasets [11]. In our approach, image and sensor data are projected into this shared vector space using modality-specific encoders. The model is optimized using contrastive loss, which penalizes the distance between hard negative samples and rewards the proximity of positive samples.

One key challenge in implementing this method is the absence of readily available positive and hard negative samples for contrastive learning [25], particularly due to lack of established datasets paired with non-linguistic modalities. To address this, we propose a metric called *temporal closeness* to evaluate the association between image and sensor data. Based on our hypothesis that data recorded at closely spaced time points are more likely to exhibit similar characteristics across different modalities, temporal closeness is calculated as the difference between the timestamps of two data points. This metric is expected to facilitate the automatic extraction of positive and hard negative samples, thereby enabling the construction of a highly accurate dense retrieval model for cross-modal information access.

We evaluate our method using the Lifelog Search Challenge 2024 (LSC'24) dataset [5], one of the largest multi-modal datasets derived from users' daily activities. This dataset includes non-linguistic modalities such as ego-centric images, heart rate, and location information (e.g., latitude/longitude), providing a robust foundation for testing our dense retrieval approach in a real-world, multi-modal context.

Our contributions are as follows:

1. We introduce a novel metric, temporal closeness, to assess the association between different modalities. This metric aids in the automatic extraction of

positive and hard negative samples, thereby improving the training process and overall performance of the retrieval model.
2. We show that in cross-modal search tasks with non-verbal modalities, temporal closeness effectively identifies hard negative samples for contrastive learning, as demonstrated by real-world data experiments.

2 Related Works

This study centers on cross-modal retrieval based on dense vector search. To establish the context of our research, we review relevant literature on cross-modal retrieval systems in Sect. 2.1. Furthermore, we highlight the novelty of our approach to extracting positive and negative samples based on temporal closeness by comparing it with existing studies on contrastive learning in Sect. 2.2.

2.1 Cross-Modal Retrieval

Transformer models [22] have significantly advanced the field of cross-modal retrieval. [17] introduced a transformer-based architecture that leverages natural language supervision to learn transferable visual models, significantly improving the alignment between text and images. [20] developed LXMERT, which leverages transformers to learn cross-modality encoder representations, resulting in substantial improvements in retrieval performance. Standardized benchmarks, such as Microsoft COCO [16] and Flickr30k [26], offer consistent evaluation frameworks for these methods, specifically focusing on text-image retrieval. The adoption of attention mechanisms, particularly cross-attention layers, has become crucial for capturing interactions between different modalities. [15] further demonstrated the effectiveness of VisualBERT, which incorporates cross-attention layers to fuse visual and textual information more effectively.

While existing cross-modal retrieval systems primarily focus on text and images, they typically do not address non-linguistic modalities. This paper introduces a novel approach that extends cross-modal retrieval to include sensor and image data, thus addressing a gap in the current research landscape.

2.2 Contrastive Learning

Contrastive learning [25] has emerged as a leading technique in self-supervised learning, particularly within the context of cross-modal retrieval. The core idea behind contrastive learning is to learn representations by contrasting positive pairs (similar samples) against negative pairs (dissimilar samples). For example, SimCLR [2] presents a simple framework for contrastive learning by augmenting images and learning invariant representations through a contrastive loss, maximizing the agreement between differently augmented views of the same image. ALIGN [9] extends vision-language representation learning by utilizing noisy text supervision from a large-scale dataset, training the model to associate images with their corresponding textual descriptions through contrastive loss.

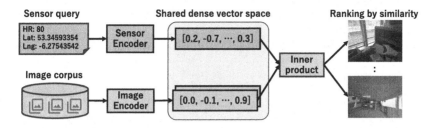

Fig. 1. Overview of our task.

The effectiveness of contrastive learning is often enhanced by improving the selection process for positive and hard negative samples. VSE++ [3] advances traditional visual-semantic embedding techniques by incorporating a hard negative mining strategy, which focuses on learning of cross-modal embeddings from the most challenging negative samples. InfoMin [21] proposes a data augmentation strategy that optimizes the mutual information between images before and after augmentation, resulting in more effective embedded representations.

While contrastive learning has been applied in various tasks, the distinguishing feature of this study is its method of extracting hard negative and positive samples based on the temporal closeness between data pairs.

3 Method

3.1 Task Definition

Given a non-linguistic modality query $\mathbf{q}$ and a corpus $\{\mathbf{d}_1, \mathbf{d}_2, \cdots, \mathbf{d}_n\}$ consisting of n data points in another modality, the goal of cross-modal retrieval is to find the k data points that are most relevant to the query $\mathbf{q}$, where $k \ll n$. The task in this paper is to efficiently generate the top-k candidates that are relevant to the query based on the similarity metirc $\text{sim}(\mathbf{q}, \mathbf{d}) \in \mathbb{R}$. In this context, the query $\mathbf{q}$ and data $\mathbf{d}$ can refer to sensor data $\{\mathbf{q}^{\text{sen}}, \mathbf{d}^{\text{sen}}\} \in \mathbb{R}^S$ or image data $\{\mathbf{q}^{\text{img}}, \mathbf{d}^{\text{img}}\} \in \mathbb{R}^{3 \times W \times H}$, where S denotes the number of dimensions of sensor data, and W and H represent the width and height of the image data. An overview of our task is shown in Fig. 1.

3.2 Cross-Modal Retrieval

To embed image and sensor data into a shared dense vector space with E dimensions, we build two types of deep neural network (DNN) encoders f and g. Specifically, for the image encoder, we utilize pre-trained models such as ResNet [6] and Vision Transformer [12], both of which are based on ImageNet [14]. Using these encoders, we obtain two dense vectors: $g\left(\mathbf{q}^{\text{sen}}\right) = \mathbf{h}_q^{\text{sen}}$ and $f\left(\mathbf{d}^{\text{img}}\right) = \mathbf{h}_d^{\text{img}}$.

The relevance of $\mathbf{q}^{\text{sen}}$ to $\mathbf{d}^{\text{img}}$ is computed as the dot product of their corresponding dense representations, $\{\mathbf{h}_q^{\text{sen}}, \mathbf{h}_d^{\text{img}}\} \in \mathbb{R}^E$ as $\text{sim}(\mathbf{q}^{\text{sen}}, \mathbf{d}^{\text{img}}) = \langle \mathbf{h}_q^{\text{sen}}, \mathbf{h}_d^{\text{img}} \rangle$. Similarly, the relevance of $\mathbf{q}^{\text{img}}$ to $\mathbf{d}^{\text{sen}}$ is calculated using the same process.

3.3 Contrastive Learning

The model is optimized end-to-end using InfoNCE loss [8]:

$$\mathcal{L}\left(\mathbf{q}^{\text{sen}}, \mathbf{d}_+^{\text{img}}, D_-^{\text{img}}\right) = \\ -\log \frac{\exp\left(\text{sim}(\mathbf{q}^{\text{sen}}, \mathbf{d}_+^{\text{img}})\right)}{\exp\left(\text{sim}(\mathbf{q}^{\text{sen}}, \mathbf{d}_+^{\text{img}})\right) + \sum_{\mathbf{d}_-^{\text{img}} \in D_-^{\text{img}}} \exp\left(\text{sim}(\mathbf{q}^{\text{sen}}, \mathbf{d}_-^{\text{img}})\right)}. \quad (1)$$

Here, $\mathbf{d}_+^{\text{img}}$ represents an image that is relevant to the sensor query $\mathbf{q}^{\text{sen}}$, typically determined by human labels. D_-^{img} denotes a set of data points that are not relevant to the query. In general, dense retrieval methodologies for documents, the set of negative includes both hard negatives, which are sampled from the top-ranking results of an existing retrieval system, and in-batch negatives, which are derived from the positive documents and hard negative documents associated with other queries in the same training batch. In practice, dense retrieval training tends to benefit from a larger set of hard negatives and in-batch negatives.

However, in cross-modal dense retrieval, positive (relevant) and hard negative (irrelevant) samples are not directly available. The standard approach of using an existing retrieval system is not applicable because no cross-modal retrieval system capable of handling non-linguistic modalities currently exits. While human-based relevance assessments are possible, they are time-consuming, costly, and prone to error. Therefore, for queries given by sensors $\mathbf{q}^{\text{sen}}$ or images $\mathbf{q}^{\text{img}}$, it is necessary to extract an appropriate set of hard negative samples $\left(D_-^{\text{img}}, D_-^{\text{sen}}\right)$ and positive sample $\left(\mathbf{d}_+^{\text{img}}, \mathbf{d}_+^{\text{sen}}\right)$.

3.4 Temporal Closeness-Based Sampling

Here, we present a method for extracting positive and hard negative samples from spatio-temporal multimodal data. Our hypothesis is that data recorded at similar time points have a high degree of similarity, and this relationship holds across different modalities. By identifying data with high temporal similarity as positive samples and those with low similarity as hard negative samples, we can construct high-quality training data to enhance contrastive learning. To achieve this, we introduce a measure of temporal closeness between two data points.

Temporal closeness evaluates not only the simple temporal distance but also the periodicity of the data. This consideration of periodicity is important because similar data may be recorded at regular intervals, such as daily or weekly. For example, in lifelog data that tracks individual behavior, it is likely that similar activities (e.g., sleeping, eating, commuting) occur at similar times each day, resulting in similar recorded data. Simple temporal distance alone may not capture the similarity of such data, but by incorporating periodicity, we can evaluate temporal closeness on a daily, weekly, or monthly basis.

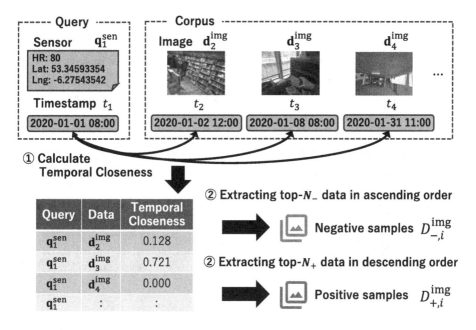

Fig. 2. Overview of our method of extracting negative and positive samples.

An overview of our method for extracting hard negative and positive samples is shown in Fig. 2. To introduce temporal closeness, the dataset is set up as follows: $\{(\mathbf{q}_1^{\text{sen}}, \mathbf{d}_1^{\text{img}}, t_1), (\mathbf{q}_2^{\text{sen}}, \mathbf{d}_2^{\text{img}}, t_2), \cdots, (\mathbf{q}_n^{\text{sen}}, \mathbf{d}_n^{\text{img}}, t_n)\}$, where t_i denotes the i-th timestamp. Let Ω be a set of periodic time expressions (e.g., $\Omega = \{\text{hour}, \text{day}, \text{week}, \text{month}\}$). For each $\omega \in \Omega$ and a timestamp of dataset t_i, we define $p_\omega(t_i)$ as the value in time expression ω of t_i. For example, if t_i is August 19, 2024, 12:30, then $p_{\text{month}}(t_i) = 8$, $p_{\text{week}}(t_i) = 1$, $p_{\text{day}}(t_i) = 19$, and $p_{\text{hour}}(t_i) = 12$. The temporal representation vector $\mathbf{r}_i$ is then calculated as:

$$\mathbf{r}_i = \left[\sin\left(\frac{2\pi p_\omega(t_i)}{\tau_\omega}\right), \cos\left(\frac{2\pi p_\omega(t_i)}{\tau_\omega}\right)\right]_{\omega \in \Omega}, \quad (2)$$

where τ_ω is a normalization term that scales $p_\omega(t_i)$ into range $[0, 1]$ (e.g., $\tau_{\text{month}} = 12$ and $\tau_{\text{hour}} = 24$). This temporal representation captures periodic components in specific time expressions using trigonometric functions. The temporal representation can also include absolute time information by combining representations for each time expression $\omega \in \Omega$. The temporal closeness $c(t_i, t_j)$ between two timestamps t_i and t_j is calculated as follows: $c(t_i, t_j) = \langle \mathbf{r}_i, \mathbf{r}_j \rangle$.

For a query $\{\mathbf{q}_i^{\text{sen}}, t_i\}$, we can compute the temporal closeness of the image set $\{(\mathbf{d}_1^{\text{img}}, t_1), (\mathbf{d}_2^{\text{img}}, t_2), \cdots, (\mathbf{d}_n^{\text{img}}, t_n)\}$, sort the image set in ascending order by temporal closeness, and extract top-N_- to obtain hard negative samples $\left(\left|D_{-,i}^{\text{img}}\right| = N_-\right)$. Conversely, positive samples $D_{+,i}^{\text{img}}$ can be obtained by sorting

Table 1. Datails of the experimental dataset.

Data split	Period	# Days	# Topics	# Data
Training	Jan. 01, 2020 ~ Feb. 29, 2020	60	–	27,907
Validation	Mar. 01, 2020 ~ Mar. 16, 2020	16	32	21,198
Test	Mar. 16, 2020 ~ Mar. 31, 2020	15	30	10,599

the image set in descending order by temporal closeness and extracting the top-N_+ images $\left(\left|D^{\text{img}}_{+,i}\right| = N_+\right)$. Our contrastive loss can be calculated as follows:

$$\mathcal{L}\left(\mathbf{q}^{\text{sen}}_i, D^{\text{img}}_{+,i}, D^{\text{img}}_{-,i}\right) = \\ -\frac{1}{\left|D^{\text{img}}_{+,i}\right|} \sum_{\mathbf{d}^{\text{img}}_+ \in D^{\text{img}}_{+,i}} \log \frac{\exp\left(\text{sim}(\mathbf{q}^{\text{sen}}_i, \mathbf{d}^{\text{img}}_+)\right)}{\exp\left(\text{sim}(\mathbf{q}^{\text{sen}}_i, \mathbf{d}^{\text{img}}_+)\right) + \sum_{\mathbf{d}^{\text{img}}_- \in D^{\text{img}}_{-,i}} \exp\left(\text{sim}(\mathbf{q}^{\text{sen}}_i, \mathbf{d}^{\text{img}}_-)\right)}.$$

(3)

4 Experimental Evaluation

4.1 Dataset

The experimental evaluation utilizes LSC'24 dataset [5], which is a multi-modal dataset capturing users' daily activities. The dataset consists of images captured by egocentric camera and sensor data by a smart tracker. We extracted heart rate and longitude/latitude as sensor data. Both image and sensor data are associated by minute ID and include timestamp information. Details of the experimental dataset are provided in Table 1.

To assess the effectiveness of our model, we created 30 retrieval tasks (topics) from the validation and test datasets. Each topic was generated on a daily basis, and sensors or images recorded at the same time as the query were extracted as relevant (correct) data. Thus, the retrieval target for each topic is the set of data recorded on each day, with only one relevant data set for each topic. Additionally, we designed two types of retrieve tasks: one that retrieves images from sensors (referred to as sen2img) and another that retrieves sensors from images (referred to as img2sen) (Fig. 3).

4.2 Parameter Settings

Each image had a resolution of 1024 × 768 in RGB format. For visual feature extraction, linearly transformed images (224 × 224 byte resolution) were processed by two Resnet50 models pretrained on the ImageNet [14] and Places365 [28] datasets. The feature vectors obtained through the pre-trained models were combined and transformed into a 512-dimensional feature vector

Topic ID	Timestamp	Query	Relevant Data
sen2img_0316	2020-03-16 16:50:00	hr: 0.015 lat: -0.121 lng: 0.055	20200316_165048.jpg
img2sen_0319	2020-03-19 09:03:00	20200319_090305.jpg	hr: 1.891 lat: -0.121 lng: 0.055

20200316_165048.jpg

20200319_090305.jpg

Fig. 3. Examples of created topics. Timestamps denote the time when the data was recorded. "hr", "lat", and "lng" represent heart rate, latitude, and longitude, respectively.

using a single multilayer perception (MLP) layer. The sensor data were normalized so that each data distribution had a mean of 0 and a standard deviation of 1. Subsequently, these data were transformed into 512-dimensional vectors through two MLP layers.

For the DNN in the proposed method, we set the number of feature vectors in each MLP layer to $E = 512$ with Dropout [19] probability of $p = 0.5$. We optimized the DNN using Adam [13], applying contrastive loss gradients calculated via the back propagation method. The mini-batch size was set to $B = 256$, with 1000 backpropagation iterations. For Resnet50, we utilized the Caffe [10] model, updating the output layer parameters through fine-tuning.

4.3 Evaluation Metric

To quantitatively evaluate our method, we employed mean reciprocal rank (MRR) as the evaluation metric [23]. MRR is the average of the reciprocal ranks of results for a sample of queries Q: $MRR = \frac{1}{|Q|} \sum_{i=1}^{|Q|} \frac{1}{\text{rank}_i}$, where rank_i refers to the rank position of the first relevant document for the i-th query.

4.4 Baseline Methods

We prepared the following three baseline methods:

Table 2. Multiple comparisons of MRR between proposed and baseline methods (p-values adjusted with Bon-ferroni correction). The table denotes significance "*" for $p < 0.05$ level, "**" for $p < 0.01$ level, and "***" for $p < 0.001$. "n.s." indicates no significant difference.

Method	sen2img			img2sen		
	None	Batch	Distant	None	Batch	Distant
Proposed(N)	***	*	**	***	*	***
Proposed(N+P)	***	n.s.	**	***	*	***

None [2]: No negative samples. This method performs contrastive learning without negative samples.

Batch [8]: In-batch negative samples. This method assigns all but the positive samples in a batch as negative samples during training. The batch data is randomly extracted from the training data, excluding the positive samples.

Distant: Temporal distance-based negative sampling. This method extracts negative samples based on their temporal distance from the query's timestamp in the training dataset. Essentially, this is a version of the proposed method that does not account for periodicity.

In addition to these baselines, we developed two variants of the proposed method. The first, **Proposed(N)**, extracts negative samples based on temporal closeness. The second, **Proposed(N+P)**, also extracts positive samples. The number of negative samples N_- in these methods was set to the batch size, excluding relevant data ($N_- = B - 1$). The number of positive samples N_+ was selected based on the validation data to maximize MRR from the set $\{1, 5, 10, 20\}$, with N_+ ultimately set to $N_+ = 5$. To calculate temporal closeness, we used four datetime expressions: hour, week, day, and month. The combination of $\Omega = \{\text{hour}, \text{week}\}$, which yielded the highest performance in the validation data, was employed.

4.5 Results

We present the retrieval performance of the proposed and baseline methods in Fig. 4. The results of significance testing between the proposed methods and the baseline methods using a t-test are shown in Table 2. **Proposed(N)** and **Proposed(N+P)** achieved the highest values among the methods compared in the sen2img and img2sen tasks, respectively. Among the baseline methods, the method with random negative samples (**Batch**) outperformed those without negatives (**None**) and with temporally distant negatives (**Distant**). This result suggests that cross-modal retrieval between sensor and image data benefits from the inclusion of negative samples, but the selection of appropriate negative samples is crucial. The results indicate that the negative sample selection in the proposed method is effective for dense retrieval between sensors and images. However, increasing the number of positive samples using temporal closeness did not significantly impact retrieval performance.

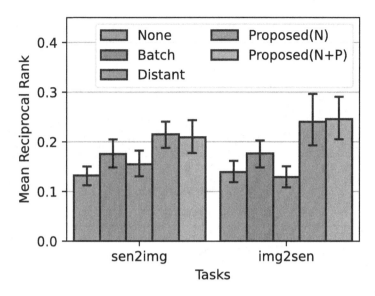

Fig. 4. Mean reciprocal rank of each method.

The breakdown of reciprocal rank by topic is shown in Fig. 5. In the sen2img and img2sen tasks, **Proposed(N)** and **Proposed(N+P)** were the methods that most frequently achieved the highest reciprocal rank. The importance of the negative samples is evident, as **None** did not achieve a maximum value for any topic. The difficulty of each topic appears consistent; for example, the reciprocal rank for topics sen2img_0326 and sen2img_0326 is approximately 0.1 across all methods. However, for the img2sen_0331 topic, **Proposed(N)** and **Proposed(N+P)** achieved nearly a fivefold improvement over **Batch**.

We evaluated the retrieval performance of the **Proposed(N)** method using different sets of datetime expressions Ω. Overall, the combination HW (hour and week) demonstrated the best performance. Among single datetime expressions, W showed the highest performance, while M showed the lowest. These results suggest that hour and week expressions adequately captured the periodicity in the experimental dataset, while the month and day expressions were less effective. One reason explanation is the relatively short duration of the experimental dataset, which spanned only two months. In a dataset with a longer time span, the periodicity captured by D and M expressions might positively impact the performance of the proposed method.

Finally, we examined the retrieval performance of the **Batch** and **Proposed(N)** methods across different numbers of hard negative samples (batch size), as shown in Fig. 7. The performance of **Proposed(N)** varied significantly

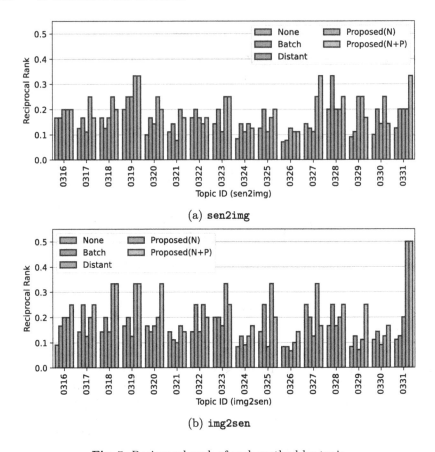

Fig. 5. Reciprocal rank of each method by topic.

with the number of negative samples, peaking at a batch size of 256. In contrast, the performance of **Batch** remained relatively stable across different batch sizes, with MRR also reaching its maximum value at 256. These findings suggest that a small number of negative samples does not provide sufficient contrastive learning, while too many negative samples increase the likelihood of including inappropriate ones. By using temporal closeness, more suitable negative samples can be prioritized. It may also be possible to automatically determine the optimal number of negative samples by setting an appropriate temporal closeness threshold, thought this remains an area for future work.

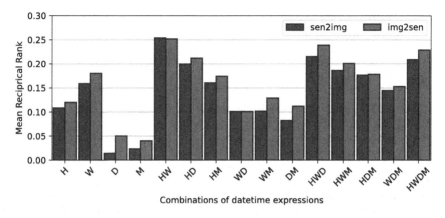

Fig. 6. Mean reciprocal rank of each set of datetime expressions in the **Proposed(N)** method. "H", "W", "D", and "M" represent hour, week, day, and month, respectively.

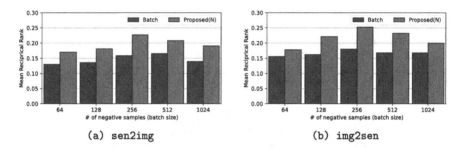

Fig. 7. Mean reciprocal rank for each number of negative samples in the **Proposed(N)** and **Batch** methods.

5 Conclusion

This paper presented a novel approach to dense retrieval across multiple modalities, addressing critical challenges in cross-modal information access. Our methodology employs a shared vector space and contrastive loss to effectively integrate and retrieve data from diverse sources, such as images and sensor data. A key innovation is the introduction of the temporal closeness metric, which facilitates the automatic extraction of positive and negative samples during training. This metric, based on the temporal closeness of data points, has been shown to significantly improve the accuracy and reliability of cross-modal retrieval.

We validated our approach using the Lifelog Search Challenge 2024 (LSC'24) dataset, one of the most comprehensive multi-modal datasets available. The evaluation results demonstrate that our proposed method not only enhances retrieval performance but also provides a robust framework for handling complex, real-world data scenarios. These findings highlight the potential of incorporating temporal information to better associate data across different modalities, contributing to the advancement of cross-modal information retrieval.

Acknowledgements. The research was partly supported by ROIS NII Open Collaborative Research 2024 (Grant Number: 24S0503).

References

1. Chang, Y., et al.: A survey on evaluation of large language models. ACM Trans. Intell. Syst. Technol. **15**(3) (2024)
2. Chen, T., Kornblith, S., Norouzi, M., Hinton, G.: A simple framework for contrastive learning of visual representations. In: ICML'20, pp. 1597–1607. PMLR (2020)
3. Faghri, F., Fleet, D.J., Kiros, J.R., Fidler, S.: VSE++: improving visual-semantic embeddings with hard negatives. In: BMVC'18 (2018)
4. Ge, Y., Zeng, X., Huffman, J.S., Lin, T.Y., Liu, M.Y., Cui, Y.: Visual fact checker: enabling high-fidelity detailed caption generation. In: CVPR'24, pp. 14033–14042 (2024)
5. Gurrin, C., et al.: Introduction to the seventh annual lifelog search challenge, lsc'24. In: ICMR'24, pp. 1334—1335 (2024)
6. He, K., Zhang, X., Ren, S., Sun, J.: Deep residual learning for image recognition. In: CVPR'16, pp. 770–778 (2016)
7. Hu, W., Xu, Y., Li, Y., Li, W., Chen, Z., Tu, Z.: BLIVA: a simple multimodal LLM for better handling of text-rich visual questions. In: AAAI'24, pp. 2256–2264 (2024)
8. Jaiswal, A., Babu, A.R., Zadeh, M.Z., Banerjee, D., Makedon, F.: A survey on contrastive self-supervised learning. Technologies **9**(1), 2 (2020)
9. Jia, C., et al.: Scaling up visual and vision-language representation learning with noisy text supervision. In: ICML'21, pp. 4904–4916. PMLR (2021)
10. Jia, Y., et al.: Caffe: convolutional architecture for fast feature embedding. In: ACM MM'14, pp. 675–678 (2014)
11. Karpukhin, V., et al.: Dense passage retrieval for open-domain question answering. In: EMNLP'20, pp. 6769–6781 (2020)
12. Khan, S., Naseer, M., Hayat, M., Zamir, S.W., Khan, F.S., Shah, M.: Transformers in vision: a survey. ACM Comput. Surv. (CSUR) **54**(10s), 1–41 (2022)
13. Kingma, D.P., Ba, J.: Adam: a method for stochastic optimization. In: ICLR'14, pp. 1–15 (2014)
14. Krizhevsky, A., Sutskever, I., Hinton, G.E.: ImageNet classification with deep convolutional neural networks. In: NIPS'12. vol. 25 (2012)
15. Li, L.H., Yatskar, M., Yin, D., Hsieh, C.J., Chang, K.W.: What does BERT with vision look at? In: ACL'20, pp. 5265–5275 (2020)
16. Lin, T.Y., et al.: Microsoft COCO: common objects in context. In: ECCV'14, pp. 740–755. Springer (2014)
17. Radford, A., et al.: Learning transferable visual models from natural language supervision. In: ICML'21, pp. 8748–8763 (2021)
18. Radford, A., Narasimhan, K., Salimans, T., Sutskever, I., et al.: Improving language understanding by generative pre-training. https://s3-us-west-2.amazonaws.com/openai-assets/research-covers/language-unsupervised/language_understanding_paper.pdf (2018)
19. Srivastava, N., Hinton, G., Krizhevsky, A., Sutskever, I., Salakhutdinov, R.: Dropout: a simple way to prevent neural networks from overfitting. J. Mach. Learn. Res. **15**(1), 1929–1958 (2014)

20. Tan, H., Bansal, M.: LXMERT: learning cross-modality encoder representations from transformers. In: EMNLP-IJCNLP'19, pp. 5100–5111 (2019)
21. Tian, Y., Sun, C., Poole, B., Krishnan, D., Schmid, C., Isola, P.: What makes for good views for contrastive learning. In: NeurIPS'20, pp. 6827–6839 (2020)
22. Vaswani, A., et al.: Attention is all you need. In: NIPS'17. vol. 30. Curran Associates, Inc. (2017)
23. Voorhees, E.M., et al.: The TREC-8 question answering track report. In: TREC. vol. 99, pp. 77–82 (1999)
24. Wang, K., Yin, Q., Wang, W., Wu, S., Wang, L.: A comprehensive survey on cross-modal retrieval. arXiv preprint arXiv:1607.06215 (2016)
25. Wu, Z., Xiong, Y., Yu, S.X., Lin, D.: Unsupervised feature learning via non-parametric instance discrimination. In: CVPR'18, pp. 3733–3742 (2018)
26. Young, P., Lai, A., Hodosh, M., Hockenmaier, J.: From image descriptions to visual denotations: new similarity metrics for semantic inference over event descriptions. Trans. Assoc. Comput. Linguist. **2**, 67–78 (2014)
27. Zhai, C.: Large language models and future of information retrieval: opportunities and challenges. In: SIGIR'24, pp. 481–490 (2024)
28. Zhou, B., Lapedriza, A., Khosla, A., Oliva, A., Torralba, A.: Places: a 10 million image database for scene recognition. IEEE Trans. Pattern Anal. Mach. Intell. **40**(6), 1452–1464 (2017)

Open Access This chapter is licensed under the terms of the Creative Commons Attribution 4.0 International License (http://creativecommons.org/licenses/by/4.0/), which permits use, sharing, adaptation, distribution and reproduction in any medium or format, as long as you give appropriate credit to the original author(s) and the source, provide a link to the Creative Commons license and indicate if changes were made.

The images or other third party material in this chapter are included in the chapter's Creative Commons license, unless indicated otherwise in a credit line to the material. If material is not included in the chapter's Creative Commons license and your intended use is not permitted by statutory regulation or exceeds the permitted use, you will need to obtain permission directly from the copyright holder.

The Right to an Explanation Under the GDPR and the AI Act

Bjørn Aslak Juliussen[✉][ID]

Department of Computer Science, UiT The Arctic University of Norway, Tromsø, Norway
bjorn.a.juliussen@uit.no

Abstract. The article provides a comprehensive overview of European regulations, the GDPR and the AI Act, focusing on the right to explanation for individual decisions inferred from high-risk AI systems and automated decision-making. It analyses the concept of the right to explanation in automated decision-making processes, emphasizing the legal obligations surrounding the provision of meaningful information pre- and post-decision. The paper examines explainable AI (XAI) methods in this context, categorizing them as intrinsic and post-hoc, with examples like decision trees, Shapley values and Local Interpretable Model-Agnostic Explanations (LIME). By analysing the legal and technical dimensions together, insights into the complex interplay between data protection, AI regulation, and the quest for transparency in the EU acquis are made.

Keywords: The Right to an Explanation · EU Law · XAI

1 Introduction

Imagine these two scenarios: You receive a warning that your current employment contract is going to be terminated due to poor work performance, or you get a notification from the tax authority that they will be conducting a manual inspection of your tax returns for the last 10 years. A natural response to either of these two notifications is to ask: Why? In both scenarios, the majority of people would most likely view it as unacceptable to receive a response that only states that an AI system recommended terminating the contract and manually reviewing the tax returns.

The following sections evaluate if the affected person has a right to a meaningful explanation in the scenarios presented. Specifically, the right to an explanation under the GDPR (sec. 2) and the EU AI Act (sec. 3) are analysed [1,2]. Due to their interrelation, both rights are examined together. The subsequent sections (sec. 4) assess various XAI methods to determine their compliance with these rights. Finally, Sect. 5 summarizes the findings.

2 The Right to an Explanation of Automated Decision-Making Under the GDPR

The right to an explanation of automated decision-making in the GDPR presupposes that the overall scope of the GDPR - the territorial and material scope - enters into effect and that the scope of Article 22 of the GDPR is activated.

Article 22 prohibits automated individual decision-making, including profiling, with several exemptions. In order for Article 22 to enter into effect, the decision must be based solely on automated processing and the decision must produce legal effects concerning a natural person or similarly significantly affect him or her.

The right to an explanation of an individual automated decision made in accordance with Article 22 of the GDPR is not part of the wording of Article 22. The right to an explanation of such a decision is part of Article 13-15 of the GDPR. The right to be informed of the reasons behind an automated decision is, thus, separated from the right not to be subject to such a decision.

Article 13 of the GDPR requires that controllers - the natural or legal person that determines the purpose of the processing and the means applied - provide specific information to data subjects when collecting their personal data directly from them.

Article 14 of the GDPR concerns the information the controller is required to inform the data subject about when the personal data has not been obtained from the data subject, e.g., from other data subjects, other controllers, where personal data is collected from sensors, or similar.

Both Articles 13 and 14 regulate the information that needs to be given to the data subject by the controller ex officio. The data subject is not required to perform any action, such as making a request, to obtain the information covered in Articles 13 and 14. Article 15, however, provides the right to access information for the data subject and regulates the information that the controller is required to provide when requested by the data subject.

Articles 13 and 14 require the controller to provide the data subject with the information at the time of collection of the personal data, while Article 15 requires the controller to provide the information at the time of the request for information, and no later than one month after the request has been submitted, under Article 12 (3).

Article 13 (1) (f), Article 14 (2) (g), and 15 (1) (h) all have the same wording regarding automated decision-making under Article 22:

> The controller shall provide the data subject with (...), **meaningful information about the logic involved**, as well as the significance and the envisaged consequences of such processing for the data subject".

The phrase 'meaningful information about the logic involved' is debated in the scholarship. Some view it as a right to explanations of automated decisions [3]. Others argue the right is minimal or nearly non-existent [4]. Some scholars suggest a contextual interpretation [5].

Article 22 – interpreted in light of recent case law from the Court of Justice of the European Union (CJEU) – is a right with a corresponding prohibition the controller needs to implement. The controller therefore needs to map out their processing operations and to have control over whether or not they have individual automated decision-making processes implemented. The complicating factor of the right to an explanation is the requirement to provide the data subject with "meaningful information about the logic involved" in automated decision-making.

What does "meaningful information about the logic involved" entail when a deployed AI system is used for automated individual decision-making? The right to meaningful information about the logic involved was not included in the predecessor of the GDPR, the data protection directive [6]. The right to meaningful information about the logic involved has not been interpreted by the CJEU, and there is only some scarce guidance on the right to meaningful information from the European Data Protection Board (EDPB) [7]. The right to receive meaningful information about the logic involved will, therefore, be interpreted in line with the existing legal sources in the following sections, mostly the wording of the GDPR - including the wording of the different official language versions - and legal literature.

A key question is whether "meaningful information" should be interpreted differently across articles 13, 14, and 15, despite the identical wording.

It is necessary to interpret the different articles not just according to their wording, but also in light of their objectives and context. Articles 13 and 14 of the GDPR regulate the information the controller needs to give to the data subject when collecting personal data, typically through the information provided to the data subject in a privacy policy. The objective of providing this information is to make the data subject aware of how his or her personal data is going to be processed. Since the information is given at the time of collection, before the actual processing by the use of automated individual decision-making has taken place, the "meaningful" criterion needs to be interpreted in this context.

The information given to the data subject about the intended automated processing at the time of collection under Articles 13 and 14 needs to be meaningful to enable the data subject to decide whether or not to consent to the processing - if the processing relies on Article 6 (1) (a). Moreover, the information provided to the data subjects should enable the data subjects to assess whether or not they should invoke specific rights when the processing has commenced.

Since the processing of personal data in the automated decision-making process has not commenced when the personal data is collected, meaningful information about the logic involved under Articles 13 and 14 would entail a general description of the overall AI system intended to be used in the automated decision-making process. Such a description could include information about how the AI model is trained, i.e. the training data and the type of AI algorithm used, typical outputs of the AI model, i.e. if the output is a prediction, classification, or generated content, and the sensitivity and the positive predictive value of the finished trained AI model, if possible. These various types of information would be examples of information given that would enable the data subject to invoke their rights at the time of collecting the personal data and throughout the processing lifecycle.

The right to receive meaningful information about automated decision-making under Article 15 is part of a reactive right after the processing of personal data has taken place. According to Recital (63) of the GDPR, the purpose of the right to access in Article 15 is to enable the data subject to "verify the lawfulness of the processing". When the data subject submits an access request, personal data has been processed and the data subject wants to receive information about the actual processing.

"Meaningful information about the logic involved" will, thus, differ across Articles 13 and 14 and Article 15. What constitutes meaningful information will differ when the information is general information about the overall processing in an AI system prior to the processing, compared to information about the actual processing of personal data that has taken place within a system pursuing an access request from the data subject.

To illustrate with an example, when a data subject has had their loan application rejected due to profiling in an automated credit scoring process and submits an access request, the data subject does not – most likely – have a general curiosity about the processing but is enquiring about the individual decision that has taken place and why it was rejected.

To better understand the term "meaningful" in Article 15 (1) (h), it can be helpful to consider other official language versions of the GDPR, as all versions carry equal authenticity and require a uniform interpretation based on the real intention of their author, as established by the CJEU [8].

The term "meaningful" in the English version of the GDPR, is "aussagekräftige informationen" in the German version, "information utiles" in the French, "nuttige informative" in the Dutch, and "meningsfulde" in the Danish language versions of the GDPR. Some nuances are present in each of these language versions. The French, Dutch and Danish wording entails that the information should be understandable, helpful, and useful for the data subject. The German version entails that the information given should be sound and reliable. The information explaining the logic involved in the automatic decision-making process, therefore, needs to be an actual and reliable representation of the automated processing. The German wording "aussagekräftige" also supports the utility of the information. The information provided under an access request should enable the data subject to assess the lawfulness of the processing [9]. The "meaningful information" condition under Article 15 therefore carries elements of understandability, usefulness, reliability, and utility.

The meaningful information should be about the "logic" involved in the automated decision-making process. The wording of Article 15 concerns the "logic" of the automated processing and not the specific technology applied. Hence, the data subject does not have the right to obtain the name of the AI system or the AI system providers' name under an access request.

According to Recital (63) of the GDPR, the right to receive meaningful information about the logic involved in automated decision-making should not "adversely affect the rights or freedoms of others, including trade secrets or intellectual property and in particular the copyright protecting the software. However, the result of those considerations should not be a refusal to provide all information to the data subject". The controller could therefore refuse to give information about trade secrets and copyright-protected material under the access request, but could not refuse an overall request for meaningful information about the logic involved by reference to Recital (63).

The information about the logic involved must be meaningful for the data subject. The explanation of the logic could be descriptive and at a high level. However, the level of abstraction in the explanation must be interpreted in line with the purpose of Article 15, to enable the data subject to assess the lawfulness of the processing, according to Recital (63). This needs to be assessed contextually and on a case-to-case basis. However, two examples can be given that do not allow data subjects to assess the lawfulness of the processing for automated decision-making. An explanation of the logic such as that the automated decision-making process "applies machine learning" or "applies AI" is too abstract for the data subject to assess lawfulness. On the other hand, descriptions such as "the automated decision-making processes uses a support vector machine to assess whether individual data points are placed on the maximum-margin hyperplane during the perceptron of optional stability". The latter description is too advanced for it to be meaningful for the data subject and the description does not make it possible for them to assess the lawfulness of the processing.

To recapitulate, both Articles 13, 14, and 15 contain a right to receive meaningful information about the logic involved when processing personal data as part of automated decision-making under Article 22 of the GDPR. However, Articles 13 and 14 of the GDPR apply to the collection of personal data, while Article 15 contains reactive rights that are dependent on requests from the data subject. Hence, Articles 13 and 14 on the one hand and Article 15, on the other hand, will provide different types of explanations where the first is more general ex-ante AI system descriptions and the latter is ex-post explanations of the output.

Article 15 requires the controller to provide "meaningful" information. This condition entails that the information must be understandable, useful, reliable, and helpful for assessing the lawfulness of the data processing for the subject. It must be evaluated contextually on a case-by-case basis.

3 The Right to an Explanation Under the AI Act

An affected person has a right to an explanation of individual decision-making under the AI Act. According to Article 86 (1) of the AI Act:

> "[a]ny affected person subject to a decision which is taken by the deployer on the basis from the high-risk AI systems listed in Annex III (...), which

produces legal effects or similarly significantly affects that person in a way that they consider to have adverse impact on their health, safety or fundamental rights shall have the right to obtain from the deployer clear and meaningful explanations of the role of the AI system in the decision-making procedure and the main elements of the decision taken".

High-risk AI systems in Annex III point 2 - critical infrastructure such as safety components in the management of critical digital infrastructure, road traffic, or in the supply of water, gas, heating and electricity - are not covered by the right to an explanation of an individual decision-making process under Article 86, according to Article 86 (1).

The right to an explanation is part of the remedies subsection in the enforcement section of the AI Act. The purpose of the right to an explanation is, thus, to act as a prerequisite to the providers and the deployers complying with the AI Act through affected persons requesting information. When the conditions under Article 86 - elaborated below - are fulfilled, the activating criterion for the scope of the right to an explanation is that the affected persons "consider" that the output from the AI system has an adverse impact on their health, safety or fundamental rights.

One relevant question is whether the deployer is only required to explain the role of the output of the AI system in the later decision process or whether the logic of the AI system needs to be explained. The wording of Article 86 (1) only requires explaining the role of the AI system in the decision-making process. However, if Article 86 (1) is interpreted in line with Recital (171) of the AI Act, it becomes evident that the explanation should be "clear and meaningful" and provide "a basis on which the affected persons are able to exercise their rights". The right to an explanation under Article 86 (1) does, thus, not only cover the simple role the output from the AI system had in the decision process but also - as far as it is feasible - the data, algorithm type, and other relevant aspects of the inferred output from the AI system.

Moreover, the right to an explanation under the AI Act "shall apply only to the extent" that the right is not covered under Union law, according to the AI AI Act Article 86 (3). If a data subject under the GDPR has the right to meaningful information about the logic involved in automated individual decision-making under Article 15 (1) (h) of the GDPR, the right to an explanation under Article 86 of the AI Act does not apply, according to Article 86 (3). The former section concluded that Article 15 (1) (h) of the GDPR contains a right to meaningful information about the logic involved in automated decisions and that this right contains a right to an ex-post explanation of individual decisions. The remainder of the legal analysis will presuppose that Article 15 (1) (h) of the GDPR provides a right similar to Article 86 (1) of the AI Act.

In which cases could affected persons rely on Article 86 (1) of the AI Act because their right to an explanation is not covered by the GDPR? The wording of Article 86 of the AI Act and the wording of Article 22 of the GDPR is not completely interchangeable. While Article 22 (1) of the GDPR covers a "decision based solely on automated processing", Article 86 (1) of the AI Act has a broader scope and covers decisions which are taken "on the basis of the output from a high-risk AI system". In the Schufa Holding AG case [10], the CJEU has included profiling conducted by another entity as a decision under Article 22 (1) where the controller draws strongly on the conducted profiling. The wording of the GDPR "decision based solely" on automated processing could, thus, not be interpreted strictly by its wording and also covers situations where it is a market standard to draw strongly on a profiled or automated decision, even though the first decision is conducted by another entity. The AI Act Article 86 only covers decisions taken on the basis of a high-risk AI system. To illustrate the relationships between the right to meaningful information about the logic involved under the GDPR and the right to an explanation under the AI Act, consider these scenarios:

Imagine that an AI system is used for recruitment to filter applications and to analyse and evaluate job candidates. Such an AI system is classified as a high-risk AI system according to Annex III Sect. 4 (a). The system is fully automated and decisions to reject applications or accept them for interviews are based solely on the AI system. In this instance, the job selection process is based solely on automated decision-making under Article 22 (1) of the GDPR. Since the data subject has a right to receive meaningful information about the logic involved under Article 15 (1) (h) of the GDPR, the right to explanation under Article 86 (1) could not enter into effect, under Article 86 (3). When automated decision-making enters the scope of Article 22 (1) of the GDPR and the AI system applied is regarded as high-risk under the AI Act, Article 15 of the GDPR prevails.

Secondly, consider that an AI system is used solely to decide prices in an online e-commerce setting. The prices are based on the personal data submitted to the e-commerce site. According to guidance from the EDPB, such price setting could be regarded as a decision with similar effect under Article 22 (1) of the GDPR if the pricing discriminates certain people and groups from buying the product [11]. If such automated decision-making enters the scope of the GDPR Article 22 (1), the data subject can request an explanation about the logic involved under Article 15 (1) (h). However, the affected person does not have a right to an explanation under Article 86 (1) of the AI Act since the AI system used for price setting is not regarded as high-risk in accordance with Annex III of the AI Act. When automated decision-making enters the scope of the GDPR, but the AI system applied is not high-risk under the AI Act, the affected person has a right to an explanation about the logic involved under the GDPR and not under the AI Act.

Suppose that an AI system is used to influence the outcome of a local referendum. Such an AI system is regarded as a high-risk, according to Annex III 8 (b) of the AI Act. If such an AI system processes personal data, it could be argued that the automated decision-making does not legally or significantly factually affect a data subject, since the decision is a referendum and not a decision directed towards the data subject. Hence, the high-risk AI system in this specific use case does not enter the scope of Article 22 (1) of the GDPR. In this example, the affected person has a right to an explanation under Article 86 (1) of the AI Act, but not under the GDPR.

Consider an AI system that is high-risk under the AI Act, but not entering the scope of the GDPR. One example could include high-risk AI systems used for law enforcement purposes outside the scope of the GDPR. Another example could include high-risk AI systems not processing personal data. One example is if an AI system is used by a judicial authority to interpret the law, under Annex III 8 (a) and does not process personal data. In such an instance, the affected person has a right to explanation under the AI Act Article 86 (1) of the AI Act and not under the GDPR.

If an AI system infers a decision while not entering GDPR Article 22 (1) and is not regarded as a high-risk AI system under Annex III of the AI Act, the affected person does not have a right to explanation under either rule set. For instance, if a tax authority has a profiling system that flags individuals who are subject to manual inspection. It is possible to argue that the manual inspection is not a legal decision or a decision similarly affecting the individual, under Article 22 (1) of the GDPR. At the same time, the Schufa Judgement is not completely transferrable since it is not certain whether the authority "draws strongly" on the flag. Annex III does not list such an AI system as high-risk, meaning the affected person may not have a right to an explanation.

To conclude, the right to an explanation under Article 86 (1) of the AI Act covers outputs from high-risk AI systems in Annex III when the output forms the basis for a decision affecting natural persons. Such a right applies only to the extent that the right to an explanation is not otherwise provided in Union law, for instance in Article 15 (1) (h) of the GDPR. The examples above have covered situations where both the AI Act and the GDPR enter into effect and what this overlap in scope signifies for the right to an explanation. The next section will address how the right to an explanation in the GDPR and AI Act relates to various methods of XAI.

4 XAI Methods and the Right to Explanation Under the GDPR and the AI Act

4.1 Explainable AI (XAI) Methods and Interpretable Methods

Based on the above legal analyses, the following sections will examine XAI methods in light of the right to an explanation under the GDPR and the AI Act.

XAI is an expanding and evolving research field [12–20]. One motivation behind the development of methods explaining the outputs of AI systems is to comply with regulations such as the GDPR and the AI Act [21,22].

In the previous sections, it is established that an explanation under Article 15 (1) (h) is required to provide useful, reliable, and understandable information about the logic involved in automated decision-making. This explanation should provide the data subject with enough information to make the data subject able to assess the lawfulness of the processing of personal data in the automated decision-making process. An explanation under Article 86 (1) of the AI Act should be a "meaningful explanation of the role of the AI system in the decision-making procedure and the main elements of the decision taken".

Which XAI methods could provide such meaningful explanations under Article 15 (1) (h) of the GDPR and Article 86 (1) of the AI Act?

XAI is a term used for methods and ML models used for making ML models and their outputs understandable for natural persons [12]. In the XAI field, the terms interpretable and explainable are sometimes used interchangeably and sometimes used to denote different notions. In the next sections, the term interpretable will be used in line with the definition from Miller as "[t]he degree to which a human can understand the cause of a decision" [13]. Explainability will be applied as a term that relates to the interpretability of individual outputs from an AI system [12].

There are various methods to achieve XAI. Generally, it is possible to divide current XAI methods into intrinsic or post-hoc XAI methods [23,24]. Intrinsic methods are interpretable on their own, due to their "simple" structure and self-explainable structure. Post-hoc methods apply such intrinsic methods on top of an uninterpretable AI method or utilise other methods to explain AI models that are not interpretable.

4.2 Instrinsic XAI Methods

Another typical manner to distinguish XAI methods is between XAI methods that make the whole trained AI model interpretable, and XAI methods that make the model output explainable. Since both Article 15 (1) (h) of the GDPR and Article 86 (1) of the AI Act revolve around the explainability of individual decisions, the next sections will focus on XAI models for model output explainability. Both intrinsic and post-hoc methods will be put under scrutiny.

An example of a potential intrinsic interpretable XAI method is to use "simple" models that are in themselves interpretable to draw inferences. One example is a type of supervised machine learning known as decision trees [12,25]. A decision tree can, e.g., be applied to predict outcomes or classify data. When decision trees are applied as a supervised machine learning method, the algorithm discovers and represents the relationships between the data in the decision tree model [25]. The tree representation of the model makes the relationships between the different data interpretable and the individual output explainable, as long as the decision tree is not too large. One algorithm typically applied in decision trees, is the CART (Classification and Regression Trees) algorithm [26].

In short, the CART algorithm "builds" the three and the internal nodes by analysing how often a data point occurs in the training data. This is done until a pre-defining stopping criterion is reached. The CART algorithm decides the cut-off values by splitting the data into clusters of similar data and deciding which splits results into the most homogeneous, "similar", subnodes [26]. The decision on maximising the similarity in each of the two subnodes is made according to an index. In this index, the split on a specific data point results in the most different data in the two subnodes, of the data in the data set [12,25].

When explaining an individual output of a decision tree, the deployer of a tree-based model starts in the inferred decision, the leaf node and goes back in the tree model through the internal nodes to the start. To provide a textual interpretation, the different intermediate subsets are connected with "and". An intrinsic interpretable XAI method, such as a decision tree, thus, makes it possible with an explanation of the specific predicted output.

There are several other examples of intrinsic XAI methods [12]. However, to interpret whether intrinsic XAI methods comply with the identified legal requirements under Article 15 (1) (h) of the GDPR and Article 86 (1) of the AI Act, the logic established with the decision tree is sufficient.

The legal analysis of Article 15 (1) (h) of the GDPR established that the purpose of the access right is for the data subject to assess the lawfulness of the processing. Moreover, in order to assess the lawfulness, the explanation of the output of an individual decision-making process needs to be understandable, reliable, and have utility for the data subject.

Intrinsic models, such as decision trees, represent the relationships between the data points in a manner that corresponds to the inference being made. In contrast to post-hoc explanations, the explanations are therefore reliable. Moreover, provided that the decision tree is not too large, it is also a pedagogical and easily understandable explanation. Intrinsic explanations, provided that they are not too advanced, would represent explanations that comply with the objective and purpose of the right to receive meaningful information about the logic involved in individual automated decision-making in Article 15 (1) (h) of the GDPR.

In relation to Article 86 (1) of the AI Act, the purpose of the explanation is to provide a remedy for the affected persons who have been affected by high-risk AI systems in a manner that they consider have had an adverse impact on their health, safety, and fundamental rights. An explanation that explain the role the AI system has in the decision being inferred by a high-risk AI system, needs to - as long as it is feasible - be an actual representation of how the data is being processed within the AI system. An intrinsic XAI method such as a decision tree would thus be a method to explain an individual decision from a high-risk AI system in compliance with Article 86 (1) of the AI Act.

4.3 Post-hoc XAI Methods

Post-hoc explanations, also referred to as model-agnostic XAI methods, separate the explanation of individual decisions from the model that is applied to draw inference [12]. The model that provides the interpretability and the explanations are put on top of the AI model that performs the tasks or inference. The next sections will address two such methods typically applied, Shapley values and local surrogate models, and examine if a post-hoc XAI method complies with Article 15 (1) (h) of the GDPR and Article 86 (1) of the AI Act. The reason behind the choice of these two methods is that they are local, meaning that they explain the individual decisions made rather than provide for the overall global model to be interpretable, and because they have a clear logic.

Shapley values is a theoretical concept from collaborative game theory used in the XAI field to provide an explanation of how much "influence" each parameter in the AI model has on the output of the model [22,27] The best manner to explain Shapley values is through the use of an example. Suppose that a natural person has applied for a loan and therefore has undergone a credit scoring process using ML. The credit score was 100 and negative and the application was rejected. An average person in the same neighbourhood, at the same age, and with similar income has an average credit score of 200 which will get the application accepted. An individual having their loan rejected would be curious about which of these features are most important for the output, which features they could improve to get their loan accepted, and - as in the example above why their application got rejected and their neighbours accepted. Shapley values is originally a method to calculate the division of a price between players that have won a game together based on how much each player contributed to winning the game [27].

In a deployed ML setting, the "price" refers to the individual inferred prediction, the "game" refers to the ML model, the gain refers to the actual predicted value minus the average predicted value, and the players are the different features in the ML model that "collaborate" to reach the specific output. The objective of the Shapley values is to explain the discrepancy between the inferred prediction, in the credit score example 100, and the average predicted credit score of 200 [12,22].

Shapley values – building on collaborative game theory - is the average value of one player in a game, calculated on the performance of the collation with and without the specific player [12,27]. In XAI, Shapley values make it possible to evaluate the value or contribution of specific features in the output of an ML model. This evaluation is done by calculating the average marginal contribution of the feature across possible coalitions with other features.

Suppose that years of education is a feature in the credit scoring. For simplicity, suppose that the credit score depends on three features: net salary, age, and years of education. We want to calculate the contribution of the *education* feature to the output of the credit scoring model. By selecting random data from the data set, *education* is replaced with random data points and the output is calculated. Then, the output of the model with and without the feature is calculated. However, the various features are interrelated and it is not just as simple as calculating the average of the education feature. The overall credit score would, e.g., be low even with a high education feature but with a low net salary. Thus, *education* needs to be interpreted in various coalitions with the other features, and the outcome of these different coalitions also needs to be averaged to calculate the Shapley value of the feature. This step is repeated and the average "value" referring to the increase or decrease in the output, the credit score, is repeated across different possible coalitions between features and values for education.

In terms of explainability, Shapley values are the average contribution of the feature across different coalitions and not the value of the feature if the feature is removed. Shapley values therefore offer explainability of individual decisions, but they are only an approximation of the feature's importance [12,28].

Shapley values are an example of an explainability method that calculated feature importance, how important one feature is to reach a specific output of an ML model. Another method to explain individual outputs from black box ML models is local surrogate models. One such surrogate model is LIME (Local Interpretable Model-agnostic Explanations) [12,29]. When it is not possible to interpret an ML model because it is a black box model, LIME make explanations possible by perturbing the input to the black box model and tests how the model performs around a specific output when the input is changed. This perturbed input data and corresponding output data from the model is used to create a model on top of the black box model that is intrinsically interpretable [12,29]. The explanation of the individual decision from the black box model can then be interpreted by the intrinsically interpretable model trained on the input and output data from the uninterpretable model.

Are model-agnostic explanation methods acceptable explanations under Article 86 (1) of the AI Act and Article 15 (1) (h) of the GDPR? It was established under the German language version of Article 15 (1) (h) of the GDPR that the meaningful information about the logic involved must be "reliable". When using a surrogate model to explain another model, there is no guarantee that the explanation "matches" the processing conducted in the underlying black box model. The models on top of the black box model, such as the LIME method and Shapley values, are just approximations of input-output data in LIME and feature importance in Shapley values. However, the wording of both the GDPR and the AI Act in relation to the right to an explanation is open-ended. In the GDPR only information about the "logic involved" is required and in the AI Act, information about the role of the AI system in the decision-making is required to be given to affected persons. As a general conclusion both intrinsic explanations and post-hoc explanation methods comply with the right to explanation of decisions under the GDPR and the AI Act.

XAI is an evolving research field and the methods are becoming more advanced. As it becomes more technically feasible to explain individual automated decisions, the corresponding rights to receive such explanations, for instance in the GDPR and the AI Act, should evolve too. Today, these rights are open-ended reflecting the difficulty in explaining outputs from AI models. However, if such explanations are becoming more and more feasible as the XAI technology improves, a natural response is to specify and strengthen the right for natural persons to receive explanations of outputs inferred from deployed AI models, both under data protection law and in the AI Act.

5 Conclusion

The "right to an explanation" in the EU, the GDPR and the AI Act, are closely interrelated. Due to the open-ended wording of the two rule sets, both intrinsic and post-hoc explanations could be applied to comply with the requirements in Article 15 (1) (h) of the GDPR and Article 86 of the AI Act.

References

1. Regulation (EU) 2016 of the European Parliament and of the Council of 27 April 2016 on the protection of natural persons with regard to the processing of personal data and on the free movement of such data, and repealing Directive 95/46/EC (General Data Protection Regulation) [2016] OJ L 199/1
2. Regulation (EU) 2024/1689 of the European Parliament and of the Council of 13 June 2024 laying down harmonised rules on artificial intelligence and amending Regulations (EC) No 300/2008, (EU) No 167/2013, (EU) No 168/2013, (EU) 2018/858, (EU) 2018/1139 and (EU) 2019/2144 and Directives 2014/90/EU, (EU) 2016/797 and (EU) 2020/1828 (Artificial Intelligence Act) OJ L 2024/1689
3. Goodman, B., Flaxman, S.: EU regulations on algorithmic decision-making and a "right to explanation". In: ICML Workshop On Human Interpretability In Machine Learning (WHI 2016), New York, NY (2016). https://arxiv.org/abs/1606.08813 V1
4. Wachter, S., Mittelstadt, B., Floridi, L.: Why a right to explanation of automated decision-making does not exist in the general data protection regulation. Int. Data Priv. Law. **7**, 76–99 (2017). https://doi.org/10.1093/idpl/ipx005
5. Selbst, A., Powles, J.: Meaningful information and the right to explanation. Int. Data Priv. Law **7**, 233–242 (2017). https://doi.org/10.1093/idpl/ipx022
6. Directive 95/46/EC of the European parliament and of the council of 24 october 1995 on the protection of individuals with regard to the processing of personal data and of the free movement of such data [1995] OJ L 281/31
7. EDPB Guidlienes 01/2022 on data subject rights-Right of access. https://www.edpb.europa.eu/our-work-tools/our-documents/guidelines/guidelines-012022-data-subject-rights-right-access_en
8. Parliament, E.: Legal aspects of EU multilingualism. https://www.europarl.europa.eu/RegData/etudes/BRIE/2017/595914/EPRS_BRI%282017%29595914_EN.pdf
9. Custers, B., Heijne, A.: The right of access in automated decision-making: the scope of article 15(1)(h) GDPR in theory and practice. Comput. Law Secur. Rev. **46**, 105727 (2022). https://www.sciencedirect.com/science/article/pii/S026736492200070X
10. Judgement of the court (first chamber) in Case C-634/21 OQ v Land Hessen and SCHUFA Holding AG ECLI:EU:C:2023:957
11. EDPB guidelines on automated individual decision-making and profiling for the purposes of regulation 2016/679 (wp251rev.01). https://ec.europa.eu/newsroom/article29/items/612053
12. Molnar, C. Interpretable machine learning. (Lulu.com, 2020) (2020)
13. Miller, T.: Explanation in artificial intelligence: insights from the social sciences (2018)
14. Gunning, D., Stefik, M., Choi, J., Miller, T., Stumpf, S., Yang, G.: XAI-Explainable artificial intelligence. Sci. Robot. **4**, eaay7120 (2019)
15. Speith, T.: A review of taxonomies of explainable artificial intelligence (XAI) methods. In: Proceedings Of The 2022 ACM Conference On Fairness, Accountability, And Transparency, pp. 2239–2250 (2022)
16. Albahri, A., et al.: A systematic review of trustworthy and explainable artificial intelligence in healthcare: assessment of quality, bias risk, and data fusion. Inf. Fusion **96**, 156–191 (2023)
17. Dwivedi, R., et al.: Explainable AI (XAI): core ideas, techniques, and solutions. ACM Comput. Surv. **55**, 1–33 (2023)

18. Ali, S., et al.: Explainable artificial intelligence (XAI): What we know and what is left to attain trustworthy artificial intelligence. Inf. Fusion. **99**, 101805 (2023)
19. Islam, M., Ahmed, M., Barua, S., Begum, S.: A systematic review of explainable artificial intelligence in terms of different application domains and tasks. Appl. Sci. **12**, 1353 (2022)
20. Linardatos, P., Papastefanopoulos, V., Kotsiantis, S.: Explainable AI: a review of machine learning interpretability methods. Entropy **23**, 18 (2020)
21. Panigutti, C., et al.: The role of explainable AI in the context of the AI act. In: Proceedings Of The 2023 ACM Conference On Fairness, Accountability, And Transparency, pp. 1139-1150 (2023)
22. Hjelkrem, L., Lange, P.: Explaining deep learning models for credit scoring with SHAP: a case study using Open Banking Data. J. Risk Finan. Manag. **16**, 221 (2023)
23. Colaner, N.: Is explainable artificial intelligence intrinsically valuable? AI Soc. **37**, 1–8 (2021). https://doi.org/10.1007/s00146-021-01184-2
24. Vale, D., El-Sharif, A., Ali, M.: Explainable artificial intelligence (XAI) post-hoc explainability methods: risks and limitations in non-discrimination law. AI And Ethics. **2**, 815–826 (2022)
25. Rokach, L., Maimom, O.: Data Mining with decision trees- theory and applications. (World Scientific, 2014) (2014)
26. Crawford, S.: Extensions to the CART algorithm. Int. J. Man Mach. Stud. **31**, 197–217 (1989). https://www.sciencedirect.com/science/article/pii/0020737389900278
27. Shapley, L.: A Value for N-Person Game. (1952). https://www.rand.org/pubs/papers/P295.html
28. Huang, X., Marques-Silva, J.: On the failings of shapley values for explainability. Int. J. Approximate Reasoning 109112 (2024). https://www.sciencedirect.com/science/article/pii/S0888613X23002438
29. Zafar, M., Khan, N.: Deterministic local interpretable model-agnostic explanations for stable explainability. Mach. Learn. Knowl. Extract. **3**, 525–541 (2021)

Toward Appearance-Based Autonomous Landing Site Identification for Multirotor Drones in Unstructured Environments

Joshua Springer[✉][iD], Gylfi Þór Guðmundsson[iD], and Marcel Kyas[iD]

Reykjavik University, Reykjavik, Iceland
{joshua19,gylfig,marcel}@ru.is

Abstract. A remaining challenge in multirotor drone flight is the autonomous identification of viable landing sites in unstructured environments. One approach to solve this problem is to create lightweight, appearance-based terrain classifiers that can segment a drone's RGB images into safe and unsafe regions. However, such classifiers require data sets of images and masks that can be prohibitively expensive to create. We propose a pipeline to automatically generate synthetic data sets to train these classifiers, leveraging modern drones' ability to survey terrain automatically and the ability to automatically calculate landing safety masks from terrain models derived from such surveys. We then train a U-Net on the synthetic data set, test it on real-world data for validation, and demonstrate it on our drone platform in real-time.

Keywords: Autonomous drone · terrain classifier · landing site · synthetic dataset · image segmentation · real-world validation

1 Introduction

Autonomous multirotor drones are widely used in many fields, primarily as remote sensor platforms. While they excel at automated data collection, surveys, and other in-flight tasks, autonomous landing remains a challenge in locations other than the initial takeoff site or specially marked locations. Furthermore, many drones are blind to obstacles and other hazardous terrain conditions, as they primarily rely on GPS positioning at takeoff and landing. It is possible to increase landing accuracy and offer some obstacle avoidance by marking a safe landing site with a visual pattern, which the drone can locate via its (typically gimbal-mounted) RGB camera. However, this requires that the landing site should be known beforehand and that additional infrastructure – possibly even requiring power – should be in place. To make landing in an unstructured, previously unknown environment viable, the drone must first analyze its environment in flight using more complex sensors, e.g., LiDAR or stereo depth cameras. Such advanced sensors are effective but also computationally expensive and thus power-hungry, reducing task-oriented mission time. Alternatively, the heavy computation can be offloaded to a ground station, but this requires added infrastructure and introduces problems with transmission overhead, connectivity, and latency.

We approach autonomous landing site identification as an image segmentation problem, with the goal of creating an appearance-based classifier that distinguishes safe and unsafe landing sites in images from our drone's gimbal-mounted RGB camera. Image segmentation often requires a labeled data set, which is potentially prohibitively expensive to create manually. We can exploit drones' strength in surveying terrain to easily build terrain models from which we can automatically generate a synthetic, labeled data set to circumvent this challenge. We prioritize keeping the computational overhead as low as possible as we run our solution onboard a drone in real time. We target the typical, gimbal-mounted RGB camera as our primary sensor instead of LiDAR, as the RGB camera is the most common drone peripheral sensor and will make the solution more generalizable and lightweight. Our contributions are as follows: (1) we present a pipeline to create synthetic image segmentation data sets for determining landing safety from terrain surveys (see Sects. 3.3 and 3.4). (2) We present a method for creating small, but effective, real-world validation sets of videos taken of known-safe and known-unsafe landing sites in the real world to determine whether our method can bridge the gap from simulation to reality (see Sect. 3.2). Finally, (3) we showcase a U-Net trained on one such data set that is able to correctly classify 15 of 18 validation cases, which run in real-time onboard our drone platform.

We evaluate our work using the drone platform described in [23], which has previously been demonstrated successfully in the context of autonomous landing with an RGB camera and visual markers [22]. We have added a Google Coral TPU accelerator to run our tiny terrain classifier (under 1 MB in size). While the small classifiers are best suited to specific environment types, they are small enough that we can store multiple classifiers on an embedded computer that can use the most relevant one. On the other hand, the method for creating the classifiers is generalizable to any environment where there is some relationship between visual appearance and landing safety. The final stage of this process is the actual autonomous landing of the drone, which is out of the scope of this paper. Our approach of identifying a viable landing site in RGB video as an image segmentation problem, where we mark each pixel as safe or unsafe, will allow our system to choose a safe pixel position representing a target landing location. With minor future adaptations, our system will then be able to control the drone and carry out landings using the method described in [22], which lands the drone at a particular site specified by a pixel position.

2 Related Work

Landing site detection is similar to the notion of *traversability*, which has been explored extensively on ground vehicles. [4] presents a survey of ground vehicle traversability methods, dividing them generally into appearance-based, geometric-based, and mixed methods. Appearance-based methods analyze terrain using visual sensors, whereas geometric-based methods require sensors such as LiDAR or RGBD that can extract a 3D representation of the environment. Some methods show success in traversability analysis, where training data is collected and labeled directly with 3D sensors, and terrain classification is performed with visual sensors only [3,26]. Some others classify 3D data from LiDAR and RGBD cameras directly [25] There is also a tendency to classify each pixel into one of three groups, e.g., traversable, not traversable, and unknown. These methods typically do not analyze reconstructed terrain models, but instead, analyze the terrain directly via some sensor and apply that analysis to label RGB images. Many traversability methods are developed and tested in simulation, but the gap from simulation to reality is often not overcome. While appearance-based methods can lack accuracy, they only require simple RGB camera sensors that are abundant, cheaper, and have a lower processing overhead than the hardware needed for geometric-based methods [4]. Many of the methods presented require specialized data sets with multiple data sources, i.e., RGB and at least one LiDAR or RGBD, to determine a label for the terrain.

Most work in autonomous landing site identification in unstructured environments is geometric-based, requiring real-time LiDAR or RGBD analysis, full-size GPUs, etc. [5,12]. However, some methods visually locate their starting location using its visual appearance only [19]. More general appearance-based methods in this context are less widespread, as they often depend on expensive, manually-labeled data sets. In this context, some methods exist to sparsely label video frames manually, and then propagate the labels from frame to frame [16]; others use previously trained neural networks to add new labels to drone videos [18]. These methods have not yet been applied to autonomous landing but have potential in identifying viable landing sites. Many methods are tested only in simulation or laboratory settings on real drone video but are not embedded onto the drone [18,24]. SafeUAV serves as an initial proof of concept for our proposed method [15]. It uses existing synthetic data sets from Google, making it possible to quickly extract training data from many locations but making the data less dynamic when the environment has recently changed. Their setup requires a camera at a fixed angle of 45° below the horizontal, and their classifiers predict the scene's 3D depth and landing safety. Crucially, they consider the problem of embedding their classifiers on a drone in the real world and therefore test them on embeddable hardware (an NVIDIA Jetson TX2) and actual drone footage, although they do not deploy their solution on a drone.

We take inspiration from many of these papers and seek to create a full-pipeline approach that allows for flexibility in data requirements, minimizes manual labeling as much as possible, and ultimately produces a viable embedded terrain classifier that can run in real-time onboard a drone. To satisfy the real-time aspect, we prefer to use an appearance-based approach such that the classifier can perform inference on RGB images, which are lightweight to analyze when compared to, e.g., LiDAR data. This also makes the method easier to generalize to drone platforms without LiDAR or RGBD cameras. For flexibility in data requirements

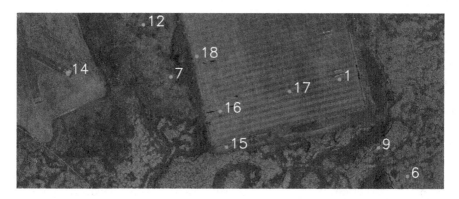

Fig. 1. One of three data collection sites. We manually and randomly picked 18 validation sites, and we number them as follows, marking safe with an S and unsafe with a U: (1U) – an archery target on a soccer field, (2U) – a bush, (3S) – a flat, dirt area, (4U) – a large, cracked rock mound, (5U) – high vegetation area, (6S) – flat, mossy area in a lava field, (7S) – flat, grassy area, (8S) – flat, mossy area in a lava field, (9U) – crack in a lava field, (10S) – dirt patch in a lava field, (11S) – road, (12U) – person, (13U) – very rough lava field, (14S) – model aircraft runway, (15U) – sloped, gravel edge of a soccer field, (16S) – green spot in a soccer field, (17S) – middle of a soccer field, (18U) – soccer goal. Map source: Loftmyndir ehf. [14] (Color figure online)

and ease in manual labeling, we generate intermediate 3D models from which we produce a synthetic data set of RGB images and masks. This gives the advantage that we can use many different data sources to generate the intermediate models, e.g., photogrammetry, LiDAR, RGBD, etc. This also means that we do not necessarily have to collect our data but can use openly available, standard-format, unlabeled data sets from terrain surveys. We also allow for the ability to quickly add manual labels to the intermediate models one time; such labels then propagate to all of the images generated. Importantly, we can vary the angle of our camera, which inherently makes the classifier more flexible than that in [15]. Finally, while we do not prescribe a particular, optimal classifier architecture, we create a successful U-Net that is relatively tiny (on the order of 1 MB instead of more than 1 GB) and can be deployed on power-efficient hardware compared to all methods described earlier. We also showcase how our classifier can be embedded onboard a drone.

3 Methods

We describe the process of collecting and transforming terrain data for our image segmentation purposes. We have automated this process except for logistical tasks such as transporting the drone to the survey location and optional, manual label refinement.

3.1 Data Acquisition – Terrain Surveys

The first step in generating data for the image segmentation problem is to collect terrain data from an environment similar to where a drone will need to land autonomously. For this purpose, we test both photogrammetry and LiDAR surveys, which have been automated in the case of many drones so that an operator needs only to select an area for surveying and then can deploy the drone to do the survey automatically. Photogrammetry is the process of combining images to create composite images or 3D models. We use ortho and oblique images in photogrammetry because this results in better 3D reconstructions of surfaces of all orientations. For example, when using primarily orthogonal images for terrain reconstruction, vertical surfaces can be severely distorted, as mentioned in [15]. LiDAR surveys provide similar terrain reconstruction capabilities as photogrammetry, but often with higher quality, higher data collection speed, and a higher price point – although this is not a hard rule. LiDAR produces point cloud data, which can be colorized by registering the points with RGB images collected simultaneously.

3.2 Data Acquisition – Validation Dataset

To generate a validation data set without frame-by-frame, manual labeling, we collect 10-second videos of particular validation sites in the field. In each video, the drone's camera holds the validation site in the center of the frame, such that it collects many different frames from many different angles as the drone moves. The camera's tilt is between 45° below the horizon and vertically down. We set the validation site to be either an obstacle or a clearing and record the classifier's predictions of the center pixels of the video for all the frames, adjusting the central region's size according to the validation site's size. We tag each video as a whole according to whether it shows a safe landing site or not, and we compare our manual classification to the classifier's predictions over the frames of the video. The prediction describes the classification of the majority of pixels in the central region of the video, and we aggregate it over time to determine a simple, binary prediction for the entire video. Some of the sites are hand-selected over a range of anticipated difficulties, considering the appearance-based classifier's lack of geometric understanding of the scene it is classifying. For example, we expect the classifier to easily determine that an open field is safe and that a large crack in the ground is unsafe. On the other hand, we expect that it should be hard to classify slanted dirt areas as unsafe, since they appear visually similar to the safe, level dirt areas.

3.3 3D Terrain Model Reconstruction

The second step in generating our dataset, after conducting a survey or downloading an openly available survey, is to generate an RGB mesh that is a colorized, 3D depiction of the terrain. We create this mesh from photogrammetry data using WebODM [27], or from LiDAR data by performing a Poisson reconstruction [13] in Cloudcompare [1].

The third step is to create a "label mesh" with the same topography as the RGB mesh that is marked according to which regions are safe and unsafe according to our geometric specifications. We sample the RGB mesh to create a point cloud with uniform density to calculate geometric features. This is necessary to remove "fuzziness" in point clouds derived from photogrammetry, and to remove scanning overlap in point clouds collected via LiDAR, since such variations can influence geometric features. We then calculate the normal vectors and geometric features of *verticality* – how slanted a region is – and *surface variation* – how rough a region is [10]. We compute a binary safety metric over the mesh, marking as unsafe all areas with verticality > 0.01, surface variation > 0.002 (experimentally determined). We further eliminate unsafe areas that have been classified as safe geometrically, because of their low verticality and surface variation, e.g., lakes and rivers, by simply selecting them in CloudCompare and adding a manual classification to all the points representing the problematic surface. This process is quick, requiring only about 2 min to manually label the river of several hundred meters in Fig. 2. We then apply Gaussian smoothing to the safety metric as the new grayscale texture for the label mesh. Finally, we slice the meshes into smaller chunks that are manageable on our hardware.

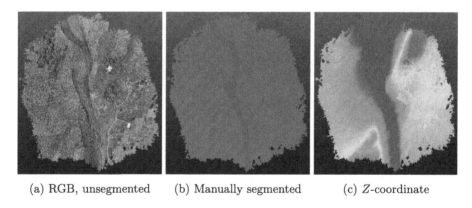

(a) RGB, unsegmented   (b) Manually segmented   (c) Z-coordinate

Fig. 2. Example of manual segmentation of a river in the summer house dataset by isolating the river and adding a manual classification to it in CloudCompare. It is not feasible to isolate the river by simply filtering on the altitude above sea level (ASL) since the terrain has a significant slope.

3.4 Synthetic Segmentation Dataset Generation

The fourth and last step in generating our data set is to create synthetic aerial images representing a drone's view of the terrain and corresponding masks that specify which pixels in those images represent safe landing sites. Using NVIDIA

Isaac Sim [7], we create a scene with the RGB and label meshes at the same position and orientation, with only one visible at a time. We position a virtual camera randomly in the scene, and aim it at a random location on the meshes, ensuring that the camera is between a minimum and maximum height above the terrain and between a minimum and maximum angular deflection from vertical down. We set the RGB mesh as visible and take a picture, effectively creating a typical aerial picture of a given terrain area. Then, we set the label mesh as visible and take another picture, creating a safety mask for the RGB image. We repeat this process for each slice to generate the labeled data set as many times as necessary. Figure 3 visualizes this process and Fig. 4 shows examples from a dataset, where the terrain is shown on top, and masks are shown on the bottom. White and black areas indicate safe and unsafe areas for landing, respectively.

The following is a non-exhaustive list of changeable parameters which we have set experimentally and which may require special attention depending on the specific scenario: data collection altitude, image/pointcloud overlap, method and radius for calculating point cloud normal vectors, Poisson reconstruction parameters, radii for calculating verticality and planarity (as well as the corresponding thresholds), slice size, maximum height and deflection of the virtual camera. These will significantly affect the quality of the datasets generated, and they may vary case by case.

3.5 Terrain Classifiers

We use deep learning methods to generate classifiers for our image segmentation dataset described in Sect. 3.4. Although this is a typical segmentation problem, and there are many deep learning frameworks we could use, we are targeting a specific inference platform: the Google Coral TPU [9], which imposes a particular toolchain revolving around Tensorflow [2,8], as well as a particular set of available operations, input sizes, etc.

Table 1. Datasets used to train the terrain classifiers. We use both LiDAR and photogrammetry (p-gram.) data.

	Location	Type	Sensor	Data points	Source
1	RC airfield	p-gram. (oblique)	DJI H20T	750	own
2	RC airfield	LiDAR	DJI L2	750	own
3	Sheffield Cross	p-gram. (ortho)	unknown	241	WebODM
4	Soccer field	LiDAR	DJI L2	750	own
5	Soccer field	p-gram. (oblique)	DJI H20T	750	own
6	Summer house	p-gram. (oblique)	DJI H20T	750	own

We test the U-Net architecture for this task based on its well-known strong performance in segmentation problems. However, this is a non-exhaustive list,

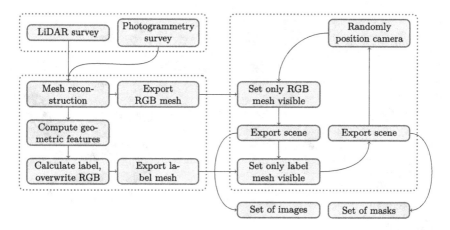

Fig. 3. Pipeline for creating labeled image datasets from terrain surveys.

and other architectures could be applicable, e.g., auto-encoders, fully convolutional, and others. Additionally, initial experimentation revealed a few key notions. First, the little available memory on the Google Coral imposed size limitations of about 8 MB on our models and the input image resolution. Second, longer chains of operations, e.g., more sections in the U-Net architecture, significantly affected the framerate at which the Google Coral could perform inference, even when all of the overhead operations, such as playing/resizing video and passing it to the TPU, were performed on a desktop machine and not on the more limited Raspberry Pi 5 to be used in the end product. Third, using RGB images as input to these small networks quickly resulted in them overfitting to both color and brightness. Therefore, we limited the image input size to 512×512 pixels, process grayscale images instead of RGB, and limited the number of U-Net sections to 3–4. We evaluate our models in terms of categorical accuracy and cross-entropy.

4 Results

4.1 Evaluation of Synthetic Segmentation Dataset

We have collected terrain data in Iceland at 3 locations via photogrammetry, and at 2 locations via LiDAR. We also use two open-source photogrammetry datasets available through WebODM. Table 1 lists the data sets used, and Fig. 4 shows example pairs of images and masks from the soccer field dataset. Notably, the data sets from WebODM tend to contain only orthogonal photography, such that the models they produce appear realistic from a top-down view but less realistic from an angle, especially on vertical surfaces. We note additionally that the LiDAR reconstructions tend to capture smaller obstacles and deeper cracks much more accurately and make the surveys and subsequent data processing much quicker, on the scale of minutes to hours.

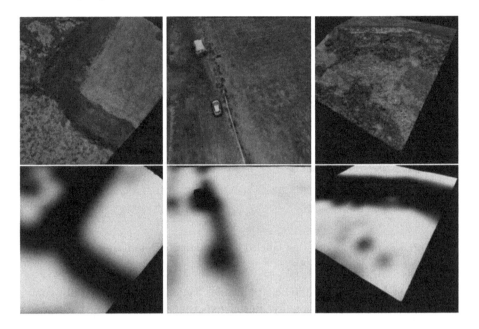

Fig. 4. Example images and masks from the synthetic data set.

4.2 Synthetic Evaluation of Terrain Classifiers

We train a 4-stage U-Net with 512×512 resolution synthetic training and testing images for a maximum of 100 epochs with early stopping (monitoring the testing loss) with a patience of 15 epochs and with a learning rate of 5×10^{-5} with the Adam optimizer. We use categorical accuracy and categorical cross-entropy as our accuracy and loss metrics, respectively. The U-Net has four encoder steps, a bridge, and four decoder steps. The encoder block pipeline has a 2D convolution, batch normalization, leaky ReLU, 2D convolution, batch normalization, leaky ReLU, and maxpool. The decoder block pipeline has an upsampling layer, concatenation, 2D convolution, batch normalization, leaky ReLU, 2D convolution, batch normalization, and leaky ReLU. To conform with the available operations for the Google Coral, we remove batch normalization layers, replace upsampling with transpose convolution, and replace each leaky ReLU layer with a PReLU layer with an untrainable alpha parameter. We train in Keras [6] and quantize classifiers destined for the Google Coral to a TFLite model. This requires full quantization to unsigned 8-bit integers instead of the standard 32-bit float and also requires a representative dataset to tune the network during the conversion. The last step is to use the edge TPU compiler to convert the TFLite model into one compatible with the Google Coral. For each experiment, we select a subset of the datasets in Table 1 for training and testing. We run each experiment 10 times, taking the model with the lowest testing loss. This is feasible because the model is small, and the training time is 10–20 min. The result is presented in Table 2.

Table 2. Best classifier accuracy, loss, and real-world validation accuracy. The safety and danger thresholds for V2 are 0.6 and 0.4 respectively.

Classifier	Accuracy		Loss		Real-World	
	Training	Testing	Training	Testing	V1	V2
Best U-Net	0.667	0.815	0.613	0.373	0.778	0.833

4.3 Post-processing

Figure 5a shows an example prediction from the network at the 10-second mark (the last frame) of a video from the validation set. The network outputs two masks: one for safety and one for danger. For clarity, we only show the danger masks and other parts can be assumed safe. Setting thresholds for converting these to a binary mask is yet another parameter; we conduct a coarse parameter sweep, with the safety threshold θ_s being $[0.1, 0.2, 0.3, 0.4, 0.5, 0.6, 0.7, 0.8, 0.9]$ and the danger threshold being $1 - \theta_s$. The central square region represents a candidate landing site that we want to evaluate – in this case, an archery target (unsafe for landing) – and the evaluation is shown in the top left of the image. As explained in Sect. 3.2, this evaluation is determined by whether the majority of pixels in the square are considered safe or unsafe. The square is white if the area is considered safe and black if it is considered unsafe. Although many of the pixels in the box in Fig. 5a are red, they are too sparse to reject the site – this can be seen in the safety prediction of 1.00, indicating that is has been deemed safe throughout the entire video. In the spirit of erring on the side of caution, we would like to reject unsafe landing sites despite the typical sparsity of unsafe classifications. We thus propose two post-processing enhancements: E1 is a box blur with the job of patching holes between rejected regions to create contiguous rejected regions, and E2 is temporal smoothing, with the job of keeping track of temporally sparse unsafe classifications. These are shown to reduce the safety prediction in Fig. 5b and Fig. 5c respectively. We experimented with values ranging from 1–10 frames for the temporal history and 7–19 for the box blur kernel size, ultimately choosing a box blur kernel size of 15 for E1 and a temporal history of 5 frames for E2. While E1 and E2 alone both reduce the safety prediction, only the combination of both completely reject the unsafe landing site with a prediction of 0.00, as shown in Fig. 5d.

4.4 Real-World Evaluation of Terrain Classifiers

Using the landing site safety prediction method defined in Sect. 4.3, we conduct a secondary validation phase for the most promising classifiers. These results are presented in the last two columns of Table 2 – V1 represents the validation performance with safety and danger thresholds of 0.5, and V2 represents the validation performance with thresholds tuned through a coarse parameter search. We compare the network's aggregated prediction to our knowledge of the area, obtained by going to the area in person, and determine the network's successful prediction rate over the validation locations in Fig. 1. Overall, the best U-net

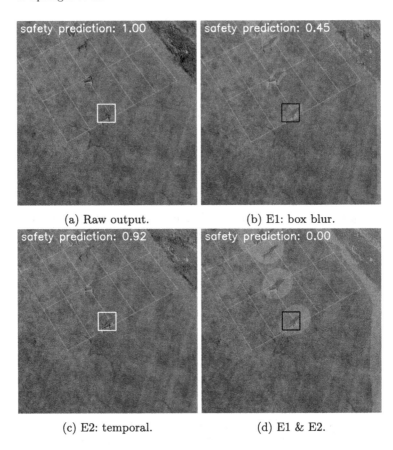

Fig. 5. Example predictions with and without post-processing

correctly classifies 15 of 18 of the 10-second validation videos, i.e., it achieves an accuracy of 0.83, which we consider promising. The classifiers can run at a rate of about 3.4 Hz on the drone payload, which is fast enough for our task given that we can add filtering methods to make the approach smooth.

There were some important trends in the real-world evaluation. First, the classifiers almost universally classified a safe runway as unsafe and the unsafe, slanted gravel edge of a soccer field as safe. This seems to be a limitation of the fact that the classifier is appearance-based; the straight, monochromatic white runway markings seem to appear as tall structures, and the gravel does not change appearance significantly when slanted compared to when it is level (e.g., in a parking lot). Further, orientation and altitude are important factors, and performance is unsurprisingly much better if these are within the ranges of the synthetic training data, i.e., between 45 °C below the horizon and vertically down, and at a distance of between 5 and 20 m from the target. For example, many classifiers correctly classified tall vegetation as unsafe when adequately close, and as safe when farther away. Finally, the small networks occasionally experience a "burn-in" where they produce persistently safe or unsafe classifications in particular locations of the image, regardless of the content of the image. This is most likely attributable to (1) disadvantageous initial conditions in the training process, and (2) biases in the data set, e.g., the fact that edges of the images are often unsafe as a result of being outside the generated terrain area, as shown in Fig. 4.

In addition to lab tests, we conducted flights at multiple validation locations in the testing area, with the method running onboard the drone. The best way to showcase these experiments is with video: https://vimeo.com/j0shua/mmm2025-demo. The code for this project is available on Github [11].

5 Conclusion and Future Work

We presented a pipeline for generating terrain classifiers to find landing sites for a drone autonomously by analyzing video from the drone's RGB camera. We automatically generated a synthetic dataset for image segmentation by reconstructing the terrain surveyed by the drone, labeling the reconstructed terrain geometrically for landing safety, and generating images and masks in simulation. We used both photogrammetry and LiDAR datasets and used one openly-available photogrammetry dataset from WebODM to show the method's flexibility. We trained a U-Net on the synthetic dataset and quantified its performance in the real world by performing inference on 10-second drone videos of 18 known-safe and known-unsafe validation sites. The U-Nets correctly classified a maximum of 15 validation sites. Finally, we ran the method in real-time onboard a drone equipped with a Raspberry Pi 5 and Google Coral TPU, motivating our creation of a less than 1 MB network.

Future work should include collecting data from more real-world environments, and generating more synthetic data sets for training. More classifiers should be tested, e.g., auto-encoders, fully convolutional networks, and other segmentation methods such as support vector machines – with consideration of the constraints of embedded hardware so they can be embedded onboard a drone. It is also a point of interest to determine whether there is a difference in performance to be gained by either photogrammetry or LiDAR over the other. Finally, we will adapt this method so that it can generate commands to control the drone and execute the landings on its own, similar to the method in [22].

References

1. CloudCompare (version 2.12) [GPL software]. http://www.cloudcompare.org/ (2023). Retrieved from https://github.com/CloudCompare/CloudCompare
2. Authors, T.: TensorFlow Lite: Lightweight solution for mobile and embedded devices. https://www.tensorflow.org/lite (2017). https://www.tensorflow.org/lite, version 2.11.0
3. Barnes, D., Maddern, W., Posner, I.: Find your own way: weakly-supervised segmentation of path proposals for urban autonomy. In: 2017 IEEE International Conference on Robotics and Automation (ICRA), pp. 203–210 (2017). https://doi.org/10.1109/ICRA.2017.7989025

4. Beycimen, S., Ignatyev, D., Zolotas, A.: A comprehensive survey of unmanned ground vehicle terrain traversability for unstructured environments and sensor technology insights. Eng. Sci. Technol. Int. J. **47**, 101457 (2023). https://doi.org/10.1016/j.jestch.2023.101457
5. Chen, L., Xiao, Y., Yuan, X., Zhang, Y., Zhu, J.: Robust autonomous landing of UAVs in non-cooperative environments based on comprehensive terrain understanding. Sci. China Inf. Sci. **65**(11), 212202 (2022). https://doi.org/10.1007/s11432-021-3429-1
6. Chollet, F., et al.: Keras. https://keras.io (2015)
7. Corporation, N.: Isaac Sim (2024). https://developer.nvidia.com/isaac-sim
8. Developers, T.: Tensorflow (2024). https://doi.org/10.5281/zenodo.12726004
9. Google: Coral Edge TPU (2023). https://coral.ai/. Accessed 02 Aug 2023
10. Hackel, T., Wegner, J.D., Schindler, K.: Contour detection in unstructured 3D point clouds. In: 2016 IEEE Conference on Computer Vision and Pattern Recognition (CVPR), pp. 1610–1618 (2016). https://doi.org/10.1109/CVPR.2016.178
11. Joshua Springer: Unstructured Landing Site Identification Repository (2024). https://github.com/uzgit/unstructured_landing_site_identification_mmm2025. Accessed 17 Oct 2024
12. Kakaletsis, E., et al.: Computer vision for autonomous UAV flight safety: an overview and a vision-based safe landing pipeline example. ACM Comput. Surv. **54**(9) (2021). https://doi.org/10.1145/3472288
13. Kazhdan, M., Bolitho, M., Hoppe, H.: Poisson surface reconstruction. In: Proceedings of the Fourth Eurographics Symposium on Geometry Processing, pp. 61–70. SGP '06, Eurographics Association, Goslar, DEU (2006)
14. Loftmyndir ehf.: (2024). https://map.is/. Accessed 17 Oct 2024
15. Marcu, A., Costea, D., Licăreţ, V., Pîrvu, M., Sluşanschi, E., Leordeanu, M.: SafeUAV: learning to estimate depth and safe landing areas for UAVs from synthetic data. In: Leal-Taixé, L., Roth, S. (eds.) Computer Vision - ECCV 2018 Workshops, pp. 43–58. Springer International Publishing, Cham (2019)
16. Marcu, A., Licaret, V., Costea, D., Leordeanu, M.: Semantics Through Time: Semi-supervised Segmentation of Aerial Videos with Iterative Label Propagation, pp. 537–552 (2021). https://doi.org/10.1007/978-3-030-69525-5_32
17. Maturana, D., Scherer, S.: 3D convolutional neural networks for landing zone detection from LiDAR. Proc. IEEE Int. Conf. Robot. Autom. **2015**, 3471–3478 (2015). https://doi.org/10.1109/ICRA.2015.7139679
18. Mitroudas, T., Tsintotas, K.A., Santavas, N., Psomoulis, A., Gasteratos, A.: Towards 3D printed modular unmanned aerial vehicle development: the landing safety paradigm. In: 2022 IEEE International Conference on Imaging Systems and Techniques (IST), pp. 1–6 (2022). https://doi.org/10.1109/IST55454.2022.9827665
19. Pluckter, K., Scherer, S.: Precision UAV landing in unstructured environments. In: Xiao, J., Kröger, T., Khatib, O. (eds.) Proceedings of the 2018 International Symposium on Experimental Robotics. Springer Proceedings in Advanced Robotics, vol. 11, pp. 177—187. Springer, Cham (2018). https://doi.org/10.1007/978-3-030-33950-0_16
20. Ronneberger, O., Fischer, P., Brox, T.: U-Net: convolutional networks for biomedical image segmentation. In: Navab, N., Hornegger, J., Wells, W.M., Frangi, A.F. (eds.) Medical Image Computing and Computer-Assisted Intervention – MICCAI 2015, pp. 234–241. Springer International Publishing, Cham (2015). https://doi.org/10.1007/978-3-319-24574-4_28

21. Springer, J., Þór Guðmundsson, G., Kyas, M.: A Precision Drone Landing System using Visual and IR Fiducial Markers and a Multi-Payload Camera (2024). https://arxiv.org/abs/2403.03806
22. Springer, J., Þór Guðmundsson, G., Kyas, M.: A Precision Drone Landing System using Visual and IR Fiducial Markers and a Multi-Payload Camera (2024). https://arxiv.org/abs/2403.03806
23. Springer, J., Þór Guðmundsson, G., Kyas, M.: Lowering Barriers to Entry for Fully-Integrated Custom Payloads on a DJI Matrice (2024). https://arxiv.org/abs/2405.06176
24. Symeonidis, C., Kakaletsis, E., Mademlis, I., Nikolaidis, N., Tefas, A., Pitas, I.: Vision-based UAV safe landing exploiting lightweight deep neural networks. In: Proceedings of the 2021 4th International Conference on Image and Graphics Processing, pp. 13–19. ICIGP '21, Association for Computing Machinery, New York, NY, USA (2021). https://doi.org/10.1145/3447587.3447590
25. Sánchez, M., Martínez, J.L., Morales, J., Robles, A., Morán, M.: Automatic generation of labeled 3D point clouds of natural environments with gazebo. In: 2019 IEEE International Conference on Mechatronics (ICM). vol. 1, pp. 161–166 (2019). https://doi.org/10.1109/ICMECH.2019.8722866
26. Tang, L., Ding, X., Yin, H., Wang, Y., Xiong, R.: From one to many: unsupervised traversable area segmentation in off-road environment. In: 2017 IEEE International Conference on Robotics and Biomimetics (ROBIO), pp. 787–792. https://doi.org/10.1109/ROBIO.2017.8324513
27. Toffanin, P., et al.: OpenDroneMap/WebODM: 2.5.4 (2024). https://doi.org/10.5281/zenodo.12775235

Towards Inclusive Education: Multimodal Classification of Textbook Images for Accessibility

Saumya Yadav[1,6](✉), Élise Lincker[2], Caroline Huron[3], Stéphanie Martin[4], Camille Guinaudeau[5,6], Shin'ichi Satoh[6], and Jainendra Shukla[1]

[1] HMI Lab, IIIT, Delhi, India
saumya@iiitd.ac.in
[2] Cedric, CNAM, Paris, France
[3] SEED, Inserm, Université Paris Cité/Learning Planet Institute, Paris, France
[4] Le Cartable Fantastique, Paris, France
[5] Japanese French Laboratory for Informatics, CNRS, Tokyo, Japan
guinaudeau@limsi.fr
[6] National Institute of Informatics, Tokyo, Japan

Abstract. To foster inclusive education, accessible educational materials must be tailored to meet the diverse needs of students, especially students with disabilities. Images in educational materials play an important role in understanding textbook exercises but can also distract such students and make them more challenging. Therefore, we need an adaptive system that identifies non-essential images to reduce cognitive load and distractions for students with disabilities. Accordingly, this work proposes a computational framework to categorize textbook exercise images, facilitating their inclusion in accessible textbooks and enhancing learning for visually impaired and neurodevelopmental disorders students. Using three French textbook exercise datasets of 652 (text, image) pairs, we compared monomodal (text-only) and multimodal (text and image) classification approaches. We found that text-based models, particularly CamemBERT, excel in classifying images, achieving an accuracy rate of 85.25% on French text data. We also use Local Interpretable Model-agnostic Explanation for model interpretability and conduct qualitative analyses to deepen insights into model performance. This work is only a very first step towards an automatic translation of inclusive textbooks. Moreover, this paper revealed that this first step is already very challenging. We hope to draw more researchers' attention to this problem.

Keywords: Educational Technology · Textbook Image Classification · Inclusive Education · Adaptive Learning · Multimodal Classification

1 Introduction

As per the United Nations Convention on the Rights of Persons with Disabilities, inclusive education asserts that it is the right of every student, not merely a privilege [13,21], underscoring the importance of ensuring that educational opportunities are accessible to all individuals, irrespective of their abilities or disabilities [19]. The right to education is universal, transcending limitations imposed by physical or cognitive disabilities. However, conventional educational resources, mainly textbooks, are not inherently designed to cater to the diverse needs of all learners, especially those with disabilities [19,33]. Inclusive education addresses this disparity by ensuring that educational materials are accessible and accommodating to all students, regardless of their disabilities.

Students benefiting from inclusive education include those with visual impairments and Neuro-Developmental Disorders (NDDs). Visual impairments range from difficulties with images, blurry vision, and partial sight to complete blindness, while NDDs affect brain function, impacting physical, social, academic, and occupational functioning [22]. Common NDDs in childhood include Attention Deficit and Hyperactivity Disorder (ADHD), Autism Spectrum Disorders (ASD), and Developmental Coordination Disorder (DCD) [5,6]. Children with visual impairment and NDDs often face academic challenges that do not reflect their true abilities due to visual and fine motor skill difficulties [2,18]. Similarly, visually impaired students encounter obstacles in accessing irrelevant visual materials and materials without suitable adaptations. Therefore, classrooms should consider these students' visual, reading, and gaze coordination difficulties when utilizing textbooks to promote inclusive education.

Recognizing their needs lays the foundation for building a computational system that automatically adapts textbooks, addressing reading, motor coordination, and attention difficulties to foster inclusivity in classrooms by tailoring educational materials for children with disabilities. Some non-profit organizations, such as *Le Cartable Fantastique* [16], have started creating adapted digital textbooks for children with DCD. However, manually transforming materials poses challenges due to diverse textbooks, frequent updates, and resource-intensive processes. The optimal solution would be for publishers to create inherently tailored textbooks for pupils with disabilities, which is currently unattainable. Therefore, automating textbook adaptation using Artificial Intelligence (AI) and Machine Learning (ML) is essential for improving accessibility.

A significant challenge in creating tailored textbooks is the manual annotation of images for diverse students. Professional educators and annotators currently spend considerable time and effort to classify images in textbooks, a process that is both labour-intensive and prone to errors [16]. This highlights the importance of automating image classification to ensure consistency and efficiency. Textbook exercise image classification is crucial to address this challenge, especially for the visually impaired and NDD students. By accurately identifying and categorizing images into different classes, we can remove those that are not required to understand the text. This approach ensures that essential visual information is preserved while managing non-essential images to reduce cognitive load and distractions. To address this, we propose a novel image classification framework for textbook exercise images. This framework categorizes images into *Essential, Informative*, and *Decorative* classes with expert help. It aids textbook adaptation by determining image inclusion in the user interface and allowing for customization based on user needs and preferences. Images can remain unaltered, be placed at the end of exercises, or be substituted with alternative text. Excluding *Decorative* images ensures document clarity and accessibility [28], particularly for students with distraction disabilities like NDD.

The main contributions of this work are: (1) the development of a novel computational framework for classifying textbook illustrations into distinct categories; (2) an empirical evaluation of monomodal and multimodal approaches for classifying (text, image) pairs; and (3) a thorough interpretability analysis using Local Interpretable Model-agnostic Explanations (LIME) [27].

2 Related Work

The work presented in this article is related to different fields: Natural Language Processing (NLP) applied to textbooks and text-image similarity and interaction. Limited research exists on NLP applied to textbooks, with some studies focusing on question generation and interactive content integration [1,3,7]. Interest in adaptive textbooks is increasing, yet manual concept indexing remains challenging. Recent advancements introduce ML methods like FACE [4] for automatic concept extraction, enhancing adaptive textbook technologies. The proliferation of MOOCs presents challenges with unstructured data, but ML frameworks for concept extraction show promise in addressing these issues [11]. Moreover, the research explores ML methods to automate the labour-intensive process of labelling educational data, particularly in Japanese schools [32]. While educational image classification primarily focuses on image type identification, recent work like [20] introduces datasets for illustration classification and [31] summarizes progress in chart classification. Additionally, [8] proposes a novel semi-supervised image classification method using curriculum learning, MMCL, outperforming five state-of-the-art methods on eight image datasets.

Similar to our goal of adapting textbooks for children with disabilities, recent studies have focused on modelling and extracting content from textbooks [16] or classifying exercises based on their adaptation for children with DCD [15]. However, these works focus on layout and textual content, not the textbook's images. To our knowledge, there is a gap in research regarding image classification for textbook adaptation to promote inclusivity. Two recent papers explore the relationship between text and image in image-text retrieval and classification [23,25]. In [23], authors present a classification framework analyzing semantic relationships between images and textual descriptions, defining eight classes based on cross-modal mutual information, semantic correlation, and status metrics. While they utilize publicly available datasets, they mainly comprise single-sentence captions or labels, differing from the context-rich text in our exercises. Several efforts, such as the Web Form Accessibility Framework for the Visually Impaired (WAFI) [9], have been made to improve website accessibility for visually impaired individuals. *W3C Web Accessibility Initiative*[1] provides guidelines for replacing images with alternative text on web pages. Although not designed

---

[1] https://www.w3.org/WAI/tutorials/images/.

for educational contexts, these guidelines can be adapted to ensure essential visual information is retained, enhancing comprehension for all.

No prior research explores the relationship between images and text in educational materials. While strides have been made in adapting textbooks for children with disabilities, a gap remains in understanding how images contribute to inclusive learning. Existing studies focus on layout, textual content, and semantic relationships between images and text but often overlook the nuanced interaction between images and more extensive textual content in educational exercises. To address these challenges, we propose a computational framework for multimodal classification of textbook illustrations for adaptive learning environments.

3 Problem and Data Challenges

Images in textbooks serve diverse roles, making the complex task of annotating them essential for automated adaptation to meet diverse needs. This process involves classifying images into categories such as *Essential*, *Informative*, or *Decorative*, based on their educational roles. Differentiating between *Essential* and *Informative* images involves understanding their educational value and context. An *Informative* image is not crucial for completing an activity but serves an informative purpose, such as providing clues for solving an exercise or depicting a concept unfamiliar to students. The *Essential* and *Informative* categories aim to provide subtle adaptations for diverse visual impairments, ensuring optimal accessibility to educational materials.

The challenge also lies in the variability of content and layouts across textbooks, which makes standardization difficult. Data challenges include dealing with inconsistent image quality and varying levels of detail, which complicate automated classification. Accurate manual annotation requires significant expertise for consistent classification but is time-consuming and resource-intensive, especially for large datasets. This process presents scalability issues and cannot efficiently handle high volumes of images. Variability in image types and content further complicates manual annotation, necessitating adaptable methods. Ensuring consistent quality across datasets is challenging due to potential errors and inconsistencies. Automated systems can address these issues by providing initial classifications that experts can review, thereby enhancing efficiency and consistency. Automation is crucial for managing large datasets and improving accessibility, as ML algorithms can classify images based on visual and contextual features, reducing manual effort and supporting the creation of inclusive educational materials.

4 Dataset

Our dataset consists of exercises with images from three elementary-grade language-learning French textbooks in PDF format. For the extraction process, each PDF is parsed to an XML file using pdfalto[2] and MuPDF[3] tools. The

---
[2] https://github.com/kermitt2/pdfalto
[3] https://github.com/ArtifexSoftware/mupdf

Table 1. Categorization of Images and Text in the Adaptation Process

Class	Essential	Informative	Decorative
Images			
Text	Write the sound common to the three words represented by the drawings.	Text: During prehistoric times, people painted cave paintings on the walls of their caves. Q: Find the verb. What tense is it conjugated to?	Copy the sentences if you recognize the verb "to go". (a) I leave at the same time every morning. (b) Saturday I weeded the garden path.

extracted words are then grouped into text segments, which are grouped into activity blocks based on layout, font style, and spacing features. Images are associated with blocks according to their position on the page. Then, two experts from *Le Cartable Fantastique* defined three classes and performed the manual annotation for each image associated with its respective text. The classes are as follows:

- **Essential Images**: These are compulsory for understanding or resolving an activity. They will be incorporated into the adaptation.
- **Informative Images**: They contribute to understanding the text and provide supplementary information, such as clues for solving an exercise or imparting knowledge. Although not essential, these images can be placed after the exercise.
- **Decorative Images**: They are unrelated to the overall exercise and irrelevant to the text. They may be removed when adapting the activity to streamline the interface.

Some contend that streamlining presentation by omitting decorative images could be beneficial [28]; while others claim that all users should be offered the same experience, including the option to receive descriptions of decorative elements. For visually impaired students, varying detail in image descriptions based on their relevance is crucial. Excluding *Decorative* images ensures only relevant content is retained, boosting accessibility alongside materials like braille and screen readers. Classifying illustrations helps NDD students by allowing educators to include images that support educational objectives while excluding those categorized as *Decorative*. An example of the categorization of images with their respective text is shown in Table 1. In the *Essential* class, the image is mandatory to solve the exercise, while the purpose of the image in the *Informative* class is to give additional information to the student, who may not know exactly what is a cave painting. Finally, for the *Decorative* class, the image associated with the text is not required to solve the exercise and only has a decorative purpose.

Our study used three elementary-grade French language textbooks, two from the same editor for training and validation with an 80:20 split, and one from

Table 2. Distribution of unique labels of train+validation set and test set

	Essential	Informative	Decorative
Test Set	131	75	38
Train+Validation Set	257	93	58

Table 3. Most frequent words in each class

Test			Train+Validation		
Essential	Informative	Decorative	Essential	Informative	Decorative
name: 29	word: 77	text: 14	write: 160	word: 93	verb: 21
drawing: 27	text: 46	verb: 11	drawing: 103	text: 47	text: 20
write: 24	verb: 41	sentence: 10	word: 101	verb: 42	sentence: 16
find: 23	observe: 34	recopy: 8	name: 89	observe: 24	word: 15
give: 17	sentence: 25	combine: 7	use: 75	write: 23	write: 15
associate: 17	red: 20	complete: 6	find: 66	name: 21	complete: 13
word: 16	name: 20	name: 6	represent: 66	remove: 17	find: 13
describe: 16	letter: 18	word: 6	sound: 64	other: 14	recopy: 9
complete: 15	read: 15	write: 5	letter: 46	sound: 14	name: 8
sentence: 14	c: 15	personal: 5	sentence: 41	pink: 14	remove: 8

a different editor as the test set to ensure unbiased evaluation. The test set includes data from unseen textbooks to assess model performance on diverse publishers. After removing blank entries from the PDFs, the processed data's final size is shown in Table 2. Although the distribution appears nearly uniform, we acknowledge the impact of topics and image characteristics on model effectiveness and ensured consistency by evaluating with a separate textbook. Table 3 highlights the most frequent words, revealing distinctive patterns across *Essential*, *Informative*, and *Decorative* categories. Notably, 'drawing' and 'write', are most prevalent in *Essential*, 'word', and 'text' in *Informative*, and 'text', and 'verb' in *Decorative*, offering insights into textual elements characterizing each class.

5 Approaches

For the textbook illustration classification, we utilized various modalities, including both multimodal approaches, which integrate images and text, and monomodal approaches that consider text or images independently. The workflow for the final approaches used in this work is shown in Fig. 1.

5.1 Multimodal Approaches

For the multimodal approach, we used the CLIP model [26], which excels in bridging the semantic gap between images and textual descriptions through zero-shot transfer, natural language supervision, and multimodal learning. Our text

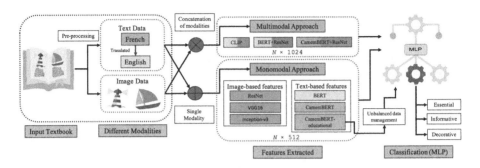

Fig. 1. Workflow for Multimodal Textbook Illustration Classification

data is extracted from French textbooks, so we translated the text into English using the open-source offline translation library Argos Translate[4] with OpenNMT [14] to align with CLIP's requirement for English text input. We chose the RN101 variant in CLIP, which performed better on our dataset. This variant has a 512-dimensional embedding, processes 224 × 224 images with a backbone of (3, 4, 23, 3) layers, and includes a text transformer with 12 heads, 512 width, and eight layers, enhancing multimodal comprehension.

We utilized the CLIP model to compute the cosine similarity between images and texts, establishing their relationship. For example, the cosine similarity between text and image features in Table 1 is 0.42, 0.48, and 0.41 for the *Essential*, *Informative*, and *Decorative* classes. Since some of our textual data exceeded CLIP's default token length of 77, we used truncation and segmentation for text data. Truncation shortens texts to the default length, while segmentation divides texts into segments of the default length. We calculated average, maximum, and truncated cosine similarity scores for text-image pairs. Descriptive statistics reveal that the *Essential* class has the highest mean average similarity score (0.42), slightly above *Decorative* (0.419), with *Informative* showing lower central tendencies. For maximum similarity scores, *Informative* leads with a mean of 0.438, followed by *Decorative* (0.431) and *Essential* (0.422). The *Informative* class also has the highest truncated similarity score (0.433). Despite *Informative* showing higher scores, the close mean values across classes indicate that CLIP-based similarity alone may not fully capture distinctions. To improve classification, we used the RN101 variant of CLIP to extract 512-size feature vectors and applied a Multi-Layer Perceptron (MLP) as shown in Fig. 1.

5.2 Monomodal Approaches

Text-Based Features Encoding: We strategically used domain-specific language models for text feature extraction: BERT for English text [12] and CamemBERT for its French counterpart [17]. These state-of-the-art language models provided comprehensive text understanding and nuanced insights into linguistic complexity. We followed the same procedure for classification as we did for our

---

[4] https://github.com/argosopentech/argos-translate.

multimodal approach. We extracted features using *bert-base-uncased* for English and *camembert-base* for French. Larger models diminished performance, likely due to overfitting and increased computational demands. We then tokenized English text using *bert-base-uncased* and French text using *CamembertTokenizer*, processed tokens through all layers of BERT and CamemBERT, and extracted features from the last hidden layer. Adaptive average pooling was used across the sequence length dimension for fixed-size representation. The textual embeddings obtained through these models are then fed to a MLP for classification.

To further enhance our exercise representation, we follow the work of [15] and fine-tune CamemBERT's language model on educational data: lessons and activities from four textbooks (two textbooks from the collection used for training, excluding the exercises used to build our dataset, and two other unseen textbooks), 1293 Fantastiques Exercices provided by the organization *Le Cartable Fantastique*, and the 79 original reading texts from *Alector*.

Image-Based Features Encoding: For Image feature extraction, we used CNN-based models, specifically ResNet [10], VGG16 [29], and Inception-v3 [30], renowned for their proficiency in image recognition tasks. We used pre-trained ResNet-50, VGG16, and Inception-v3 models, adjusting their final layers to produce 512-dimensional feature vectors that matched CLIP's size. All layers, except the modified final one, are frozen to preserve the knowledge encoded in earlier layers. The images were pre-processed by resizing to ensure compatibility with the ResNet model's input expectations.

We extracted the features using these models, generating the feature vectors of dimensions $N \times 512$, with N representing the total number of data instances. Following the same method we used for the CLIP model, we extracted data using these models, ensuring uniformity in our approach. Subsequently, we used MLP to train the extracted data, enhancing the analytical capabilities of our research.

5.3 Unbalanced Data Management

The initially processed data is highly imbalanced, as shown in Table 2, which can detrimentally affect model performance by favouring the majority class. To mitigate this issue, we used two strategies, either jointly or separately:

- `Class_weight` **strategy**: We applied the *class_weight* strategy using scikit-learn [24] to prioritize the minority class. The 'balanced' strategy dynamically adjusts class weights based on their distribution in the training data, giving higher weights to underrepresented classes.
- `Data generation`: We augment the initial training set with 150 instances from the *Decorative* class, consisting of text-only data and random images from textbooks. We then merge this augmented data with the original set, extract features, and apply the same procedure as previous models, passing it through the MLP.

6 Result and Discussions

6.1 Experimental Setups and Ablation Study

Our MLP has an input layer of size 512, two hidden layers with 256 and 128 neurons, respectively, and an output layer for 3 classes, with ReLU activations in between. It was trained using CrossEntropyLoss and Adam optimizer, with a batch size of 32 over 30 epochs. Based on validation performance, early stopping with patience of 5 epochs was applied. As part of our ablation study, we experimented with different numbers of hidden layers in the MLP: 1, 2, 3, 4, and 6 hidden layers. The results indicated that the MLP with 2 hidden layers achieved the best performance.

We also conducted an ablation study on fusion methods to evaluate their impact on model performance, using both Early fusion and Late fusion of text and image modalities. In Early fusion, we combined modalities by computing the maximum, minimum, and concatenation of two features. In Late fusion, we applied the same MLP used previously to the extracted data of both modalities, averaging and taking the weighted average of both outputs to predict the result. The best accuracy result obtained from these methods on the test data was 57.14%, which was not as good as the performance of the Early fusion approach. Therefore, we generated all results using the Early fusion concatenation method. We chose concatenation because it preserves information from both modalities, enriching the feature space and capturing complementary data. The concatenated features, sized 1024, were then input into an MLP.

6.2 Results

The results section presents predictions on test data from a third textbook by a distinct editor. Table 4 shows that the text-based models outperform the image-based models. The image-based classification (ResNet, VGG16, and Inception-v3) gives lower results than the majority class classifier, showing that image data alone is insufficient for classifying images as *Essential*, *Informative*, or *Decorative* in the context of an exercise, indicating the need for additional data or features to enhance classification accuracy. The French language model CamemBERT achieved the highest accuracy in text-based classification without fine-tuning, likely due to the original data being in French, highlighting the importance of linguistic compatibility in model performance. While CamemBERT-educational was expected to perform better, its lower performance suggests that semantic features of text exercises may not be crucial in *image classification*.

The bottom part of Table 4 shows that early fusion of image and textual data does not enhance performance relative to text-based classification. The best multimodal accuracy is achieved with the CLIP model, slightly surpassing the concatenation of CamemBERT and ResNet features. As previously shown in Fig. 1, similarities values between text and image computed with the CLIP model have nearly similar values for all three classes, contrary to our intuition that the *Decorative* images had a lower semantic cosine similarity with the image

Table 4. Monomodal and Multimodal Classification Results with CamemBERT, Using Unbalanced Data Strategies (CW: Class Weight, DA: Data Augmentation)

Models	Modality	Accuracy	F1-Score
Majority Class	–	0.727	
BERT	text	0.816	0.818
CamemBERT	text	0.836	0.831
CamemBERT-educational	text	0.80	0.798
ResNet	image	0.525	0.431
Inception-v3	image	0.504	0.421
VGG16	image	0.50	0.421
BERT+ResNet	text+image	0.754	0.755
CamemBERT+ResNet	text+image	0.803	0.796
CLIP	text+image	0.807	0.79
CamemBERT−CW⁻−DA⁻		0.836	0.831
CamemBERT−CW⁺−DA⁻		0.816	0.83
CamemBERT−CW⁻−DA⁺		0.828	0.815
CamemBERT−CW⁺−DA⁺		**0.853**	0.849

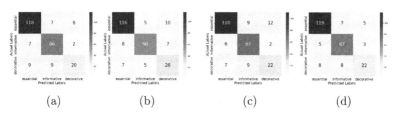

Fig. 2. Confusion matrices for the CamemBERT model only (a), with data augmentation (b), with class weight strategy (c) and both data augmentation and class weight strategies (d).

(redundancy) where the necessary images had a higher semantic cosine similarity (complementarity). Further, Table 4 also indicates that CLIP's higher accuracy stems from effectively integrating textual and image features, whereas CamemBERT+ResNet achieves a higher F1 score due to its superior precision and recall in capturing text-image relationships.

Finally, Table 4 presents the results obtained with the different strategies for dealing with our data unbalanced issues. The best results are obtained when both class weight and data augmentation strategies are used, achieving an accuracy of 85.25% when applied on our best model (CamemBERT only)[5]. From a qualitative point of view, as pictured in the confusion matrices in Fig. 2, the data augmentation tends to classify more examples in the *Decorative* class, both correctly and incorrectly (Fig. 2b), while the class weight strategy tends to improve the number of correctly classified instances of the under-represented classes (*Decorative* and *Informative*) at the expense of the *Essential* class (Fig. 2c). It's worth

---

[5] Similar tendencies are found when applied on the other models.

Table 5. Comparison of results from text-based, image-based, and multimodal models. ✓(✗) indicates correct (incorrect) labelling of the (text, image) pair.

	Exercise 1	Exercise 2	Exercise 3
Text	Decipher this puzzle.	Choose the correct adjectives to describe the princess.	What type of art is this?
Images			
Text-based	✗	✓	✓
Image-based	✓	✓	✗
Multimodal	✗	✗	✗
Cosine similarity	0.403	0.456	0.422

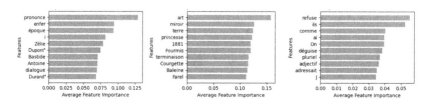

Fig. 3. LIME explanation of top features for each class (a) Essential, (b) Informative, and (c) Decorative, based on average importance. *: anonymized

noting that utilizing either the class weight or data augmentation strategy alone decreases performance compared to the initial result. Finally, combining both strategies improves the number of correctly classified instances for all classes.

Furthermore, these findings are significant for individuals with visual impairments and NDD students. By removing *Decorative* images, our approach ensures that only essential visual information is retained, enhancing accessibility and catering to the specific needs of students with NDD. This deliberate curation of visual data fosters inclusivity and improves learning outcomes for these marginalized groups.

6.3 Qualitative Analysis and Explainability

Table 5 compares results obtained with text-based, image-based, or multimodal models on three random (text, image) pairs. For the given exercises, the text-based model incorrectly labels Exercise 1, related to *image description*, while the image-based model performs well in both Exercise 1 and Exercise 2, i.e., in *image description and image reading* but fails to classify Exercise 3. Notably, images are required for all exercises. The cosine similarities for Exercises 1, 2, and 3 are 0.403, 0.456, and 0.422, respectively, reflecting the varying degrees of alignment between text and image features. The text-based model's misclassification of Exercise 1 highlights its limitation in understanding the role of visual content in specific tasks. The multimodal model struggles with all exercises, reflecting difficulties in effectively integrating text and image modalities.

We use LIME [27] to enhance model interpretability for text classification. It provides localized explanations for the predictions, improving transparency. In Table 4, CamemBERT, incorporating CW and DA, achieved the best results, prompting us to perform LIME analysis on our French text data. Figure 3 illustrates the top 10 words for each class, with their importance scores, where the Y-axis displays the top 10 contributing words and the X-axis shows their weights.

For the *Essential* class, Family names and firstnames like "Zélie", "Bastide", and "Antoine" appear in the top features. Besides, the feature "i", which is the number 1 in Roman letters, indicates that these texts are closely related to key educational elements like multiple-choice questions. Interestingly, for the *Informative* class, all top features-"art" (art), "miroir" (mirror), "terre" (earth), "princesse" (princess), "Fourmis" (ants), "terminaison" (ending), "Courgette" (zucchini), "Baleine" (whale), and "Farel" (a proper noun, likely a name) are nouns, with the exception of "1881", which is a number. Unlike the other two classes, no verbs are present in this class. Finally, it seems that the *Decorative* class contains more grammatical words like "ils" (they), "comme" (like), "ai" (have), and "on" (we) compared to the other two classes and also words related to grammar exercises ("adjectif" (adjective), "pluriel" (plural)).

Overall, the LIME analysis on CamemBERT reveals how the model differentiates between content in children's textbooks. Higher weights for the *Informative* class indicate the model's focus on contextually significant terms, while lower weights for the *Decorative* highlight its ability to identify less critical features. These insights are specific to the classification task, and domain expertise remains essential to fully understand the model's decision-making process.

In addition to the promising results, it is essential to acknowledge the limitations of our study. We recognize the dataset's limited size, which may affect the findings. However, given the novel nature of this work, we focused on assessing the feasibility of our methods. Encouraged by the positive outcomes, we plan to expand and validate our approach using larger datasets in future research. Additionally, due to the presence of private data in our dataset, we cannot share it with the community, hindering the reproducibility of our experiments. We are working on annotating publicly available data to share with the community.

7 Conclusion

Our study proposes an automatic system to classify textbook images into *Essential*, *Informative*, and *Decorative* classes, aiding inclusive education for visually impaired and NDD students. Using a French textbook dataset of 652 (text, image) pairs, we found that text-based models, particularly CamemBERT on French text data, outperformed multimodal methods. Further, the LIME analysis revealed that for *Essential* class, the family names and first names "Zélie" and "Bastide" appear in top features, as well as "i", likely denoting the numeral for multiple-choice questions. Interestingly, for the *Informative* class, we can see that only nouns are accounted as top features. Finally, it seems that the *Decorative* class contains more grammatical words and those related to grammar exercises. This approach ensures that only images with educational content are retained. In the future, we plan to expand the dataset, incorporating additional sources such as [23] and exploring alternative similarity measures. We aim to explore alternative text generation to enhance textbook adaptation for visually impaired and NDD students. Additionally, future work could introduce a new dataset of images paired with contextual descriptions for the visually impaired, which may be shared with the community.

References

1. Alpizar-Chacon, I., et al.: Transformation of PDF textbooks into intelligent educational resources. In: Proceedings of the 2nd International Workshop on Intelligent Textbooks, 21st International Conference on Artificial Intelligence in Education (2020)
2. Babij, S., et al.: Cumulative prenatal risk factors and developmental coordination disorder in young children. Matern. Child Health J. 1–7 (2023)
3. Ch, D.R., Saha, S.K.: Generation of multiple-choice questions from textbook contents of school-level subjects. IEEE Trans. Learn. Technol. (2022)
4. Chau, H., Labutov, I., Thaker, K., He, D., Brusilovsky, P.: Automatic concept extraction for domain and student modeling in adaptive textbooks. Int. J. Artif. Intell. Educ. **31**, 820–846 (2021)
5. Craig, F., Savino, R., Trabacca, A.: A systematic review of comorbidity between cerebral palsy, autism spectrum disorders and attention deficit hyperactivity disorder. Eur. J. Paediatr. Neurol. **23**(1), 31–42 (2019)
6. Delgado-Lobete, L., Santos-del Riego, S., Pértega-Díaz, S., Montes-Montes, R.: Prevalence of suspected developmental coordination disorder and associated factors in Spanish classrooms. Res. Dev. Disabil. (2019)
7. Gerald, T., Ettayeb, S., Le, H.Q., Vilnat, A., Paroubek, P., Illouz, G.: An annotated corpus for abstractive question generation and extractive answer for education. In: Conférence sur le Traitement Automatique des Langues Naturelles (2022)
8. Gong, C., Tao, D., Maybank, S.J., Liu, W., Kang, G., Yang, J.: Multi-modal curriculum learning for semi-supervised image classification. IEEE Trans. Image Process. **25**(7), 3249–3260 (2016)
9. Hakami, W.A.S., Al-Aama, A.Y.: A framework to improve web form accessibility for the visually impaired. IEEE Access (2023)
10. He, K., Zhang, X., Ren, S., Sun, J.: Deep residual learning for image recognition. In: IEEE Conference on Computer Vision and Pattern Recognition (2016)
11. Jiang, Z., Zhang, Y., Li, X.: MOOCon: a framework for semi-supervised concept extraction from MOOC content. In: Database Systems for Advanced Applications: DASFAA 2017 International Workshops: BDMS, BDQM, SeCoP, and DMMOOC, Suzhou, China, March 27-30, 2017, Proceedings 22, pp. 303–315. Springer (2017)
12. Kenton, J.D.M.W.C., Toutanova, L.K.: BERT: pre-training of deep bidirectional transformers for language understanding. In: Proceedings of NAACL-HLT, pp. 4171–4186 (2019)
13. Kielblock, S., Woodcock, S.: Who's included and who's not? an analysis of instruments that measure teachers' attitudes towards inclusive education. Teach. Teach. Educ. **122**, 103922 (2023)
14. Klein, G., Kim, Y., Deng, Y., et al.: OpenNMT: open-source toolkit for neural machine translation. In: Association for Computational Linguistics - System Demonstrations (2017)

15. Lincker, É., Guinaudeau, C., Pons, O., et al.: Noisy and unbalanced multimodal document classification: Textbook exercises as a use case. In: 20th International Conference on Content-based Multimedia Indexing (2023)
16. Lincker, E., Pons, O., Guinaudeau, C., et al.: Layout-and activity-based textbook modeling for automatic pdf textbook extraction. In: Intelligent Textbooks 2023
17. Martin, L., Muller, B., Suárez, P.J.O., et al.: CamemBERT: a tasty French language model. In: Annual Meeting of the Association for Computational Linguistics (2020)
18. Missiuna, C., Rivard, L., Pollock, N.: They're bright but can't write: developmental coordination disorder in school aged children. Teach. Except. Child. Plus **1**(1), n1 (2004)
19. Miyauchi, H.: A systematic review on inclusive education of students with visual impairment. Educ. Sci. **10**(11), 346 (2020)
20. Morris, D., Müller-Budack, E., Ewerth, R.: Slideimages: a dataset for educational image classification. In: Advances in Information Retrieval: 42nd European Conference on IR Research, ECIR 2020, Lisbon, Portugal, April 14–17, 2020, Proceedings, Part II 42, pp. 289–296. Springer (2020)
21. Mugambi, M.M.: Approaches to inclusive education and implications for curriculum theory and practice. Int. J. Humanit. Soc. Sci. Educ. **10**(4), 92–106 (2017)
22. Nowell, K.P., Bodner, K.E., Mohrland, M.D., Kanne, S.M.: Neurodevelopmental disorders (2019)
23. Otto, C., Springstein, M., Anand, A., Ewerth, R.: Characterization and classification of semantic image-text relations. Int. J. Multimedia Inf. Retrieval (2020)
24. Pedregosa, F., et al.: Scikit-learn: machine learning in python. J. Mach. Learn. Res. **12**, 2825–2830 (2011)
25. Qu, L., Liu, M., Wu, J., Gao, Z., Nie, L.: Dynamic modality interaction modeling for image-text retrieval. In: 44th International ACM SIGIR Conference on Research and Development in Information Retrieval (2021)
26. Radford, A., Kim, J.W., Hallacy, C., et al.: Learning transferable visual models from natural language supervision. In: International Conference on Machine Larning (2021)
27. Ribeiro, M.T., Singh, S., Guestrin, C.: "why should i trust you?" Explaining the predictions of any classifier. In: International Conference on Knowledge Discovery and Data Mining (2016)
28. Sanchez, C.A., Wiley, J.: An examination of the seductive details effect in terms of working memory capacity. Memory Cogn. **34**, 344–355 (2006)
29. Simonyan, K., Zisserman, A.: Very deep convolutional networks for large-scale image recognition. In: 3rd International Conference on Learning Representations (ICLR 2015). Computational and Biological Learning Society (2015)
30. Szegedy, C., Vanhoucke, V., Ioffe, S., Shlens, J., Wojna, Z.: Rethinking the inception architecture for computer vision. In: IEEE Conference on Computer Vision and Pattern Recognition (2016)
31. Thiyam, J., Singh, S.R., Bora, P.K.: Chart classification: a survey and benchmarking of different state-of-the-art methods. Int. J. Doc. Anal. Recogn. (IJDAR) **27**(1), 19–44 (2024)
32. Tian, Z., Flanagan, B., Dai, Y., Ogata, H.: Automated matching of exercises with knowledge components. In: 30th International Conference on Computers in Education Conference Proceedings, pp. 24–32 (2022)
33. Wambaria, M.W.: Accessible digital textbook for learners with disabilities: opportunities and challenges. Educ. Rev. USA **3**(11), 164–174 (2019)

Towards Visual Storytelling by Understanding Narrative Context Through Scene-Graphs

Itthisak Phueaksri[1,3](✉)[iD], Marc A. Kastner[1,2][iD], Yasutomo Kawanishi[1,3][iD], Takahiro Komamizu[1][iD], and Ichiro Ide[1][iD]

[1] Nagoya University, Nagoya, Aichi, Japan
[2] Hiroshima City University, Hiroshima, Japan
[3] RIKEN, Kyoto, Seika, Japan
phueaksrii@cs.i.nagoya-u.ac.jp

Abstract. VIsual STorytelling (VIST) is a task that transforms a sequence of images into narrative text stories. A narrative story requires an understanding of the contexts and relationships among images. Our study introduces a story generation process that emphasizes creating a coherent narrative by constructing both image and narrative contexts to control the coherence. First, the image contexts are generated from the content of individual images, using image features and scene graphs that detail the elements of the images. Second, the narrative context is generated by focusing on the overall image sequence. Ensuring that each caption fits within the overall story maintaining continuity and coherence. We also introduce a narrative concept summary, which is external knowledge represented as a knowledge graph. This summary encapsulates the narrative concept of an image sequence to enhance the understanding of its overall content. Following this, both image and narrative contexts are used to generate a coherent and engaging narrative. This framework is based on Long Short-Term Memory (LSTM) with an attention mechanism. We evaluate the proposed method using the VIST dataset, and the results highlight the importance of understanding the context of an image sequence in generating coherent and engaging stories. The study demonstrates the significance of incorporating narrative context into the generation process to ensure the coherence of the generated narrative.

1 Introduction

VIsual STorytelling (VIST) is a task to generate coherent and detailed narratives from an image sequence in the form of text. The main goal is to enable machines to present information in a natural language format, which is valuable for content generation in applications, such as photo album summarization [26,33]. This enables users to easily generate and comprehend the content of an image sequence through text. This means not only describing each image but also connecting them into a smooth story that shows the flow and main idea of the images. By achieving this, VIST helps users easily understand visual content in the form of text, making it more meaningful and easier to grasp.

The challenge of the VIST task is to generate a coherent narrative from an image sequence by understanding the context, temporal relationships, and subtleties of human emotions and interactions. To tackle this challenge, existing methods [3,12,14,18,19,27,29,30,32] are based on two main processes. First, an *Encoder* extracts and encodes image features from each image. Second, a *Decoder*, often implemented with a Recurrent Neural Network (RNN) such as Long Short-Term Memory (LSTM), processes all encoded features to generate a coherent story. This methodology has demonstrated its effectiveness in producing compelling narratives.

Despite significant achievements in recent VIST methods, the main goal of storytelling is to generate a coherent narrative based on the connections within an image sequence, ensuring the correctness of content details throughout the sequence. Traditional approaches [3,14,18,19,30] generate a story by creating a caption for each individual image, which does not emphasize the understanding of the relationships between objects. Meanwhile, some recent approaches [27,29] use scene graphs for a structured representation of the relationships within objects in an image, improving the understanding of each image context. Additionally, external knowledge is integrated to enhance understanding of the context of an image sequence. Thus, the key idea of the proposed method is to use a narrative context to ensure coherence during the generation process.

In this study, we introduce a VIST method that emphasizes understanding both the individual context of each image, referred to as image contexts, and the overall image context integrated with external knowledge, referred to as the narrative context. The proposed method consists of three main processes. First, **Image Contextualization** structures the contexts of each image from an image sequence consisting of image features and scene graphs. Second, **Narrative Contextualization** constructs narrative context by aggregating all image contexts and the knowledge-graph. Third, **Story Generation** is adapted from GLocal Attention Cascading Networks (GLACNet) [14] to generate a caption for each image in an image sequence. To improve content understanding and accurately generate captions for each image, in this work, we not only use image features but also incorporate scene-graph generation. The generated scene graphs and image features from each image are aggregated to represent the image context. Since the objective is to generate a coherent story from an image sequence, we introduce a narrative context to guide and enhance the coherence of the generated narrative. The narrative context aggregates the overall context of the image sequence by combining all image contexts, including detected image features, scene graphs, and external knowledge. In the proposed method, we introduce external knowledge represented in the form of a knowledge graph, aggregated from VIST external knowledge annotations in PR-VIST [12] and commonsense knowledge from ConceptNet [24].

The contribution of this work consists of the following:

- We propose an image contextualization process in the story generation framework by utilizing scene graphs within each image context to accurately understand and generate a caption for each image.

- We introduce a novel form of external knowledge as a knowledge graph corresponding to the VIST dataset to improve the coherency of narrative generation.
- We propose a narrative contextualization process to utilize all generated information and knowledge-graph that allows to enhance the coherency of the generation process.

2 Related Work

Because the objective of the proposed method is to utilize a scene-graph to generate a narrative of an image sequence, we summarize two research literatures. One is **VIsual STorytelling (VIST)**, which is the task of generating a story from a sequence of images. The other is **Scene-Graph Generation**, which is the task of describing an image in the form of a graph including objects and their relationships.

2.1 VIsual STorytelling (VIST)

The VIST task involves the generation of coherent and contextually relevant narratives based on a sequence of images. The target is to understand the visual content and narrative structure to generate a story from a sequence of images. Recently, the VIST dataset [13] has become one of the most significant benchmarks for this task. Most efficient VIST approaches [19,29,30] have focused on understanding both the overall and individual contexts of each image. Meanwhile, recent approaches [3,12,18,27,32] enhance the understanding of the overall context of an image sequence by incorporating external knowledge to improve the coherence of the generated narrative. GLACNet [14] was introduced to generate visual stories by combining global and local attentions with context-cascading mechanisms that incorporate two levels of encoding: overall contents (global) and image features (local). Plot and Rework modeling storyline for VIsual STorytelling (PR-VIST) [12] leverages scene-graphs to capture the relationships between objects in a sequence of images, and uses these relationships to determine an optimal storyline path for story generation. Chen et al. [2] proposed a method to enhance the diversity and informativeness of generated stories by introducing concept selection from candidate concepts provided by a commonsense knowledge-graph. Hse et al. [11] introduced a story generation method by enriching external knowledge-graphs distilled from a set of representative words derived from input prompts. In addition, it incorporates terms and object information to iteratively refine and enhance the generated story for greater coherence and relevance. Lin et al. [19] proposed an Attention Output Gate Long Short-Term Memory (AOG-LSTM) by integrating adaptive attention mechanisms and utilizing an attribute-object graph to better capture the relationships between different attributes and objects in images.

2.2 Scene-Graph Generation

Scene-graph generation [9] is a technology to describe an image using triplets of subject, predicate, and object. The process involves two stages: detecting objects and their regions with Fast R-CNN (Object Detector) [7], and estimating relationships between objects using their context (Relation Predictor). Neural Motifs [34] is a scene-graph generation method using a stacked Motif Network (MotifNet) [6] to predict relationships by estimating object features, labels, and regions as contexts. Considering the long-tail problem of the scene-graph dataset, recent work [25] aims to introduce a technique to alleviate the bias of the dataset. Relation Transformer for Scene-Graph Generation (RelTR) [4] is a one-stage end-to-end scene-graph generation technique that uses an attention mechanism and gives a fixed number of subjects, objects, and relationships to generate a scene-graph. These methods are often trained and evaluated on the Visual Genome (VG) [16] and Open Images [17] datasets.

Additionally, scene-graph generation has been widely implemented in VIST to estimate the coherent connections between the generated stories of each image, thereby helping to understand the relationships between images. Wang et al. [29] introduced a method for visual storytelling by converting images into scene-graphs, explicitly encoding objects and relationships, and using Graph Convolution Networks (GCN) and Temporal Convolution Networks (TCN) to refine these representations for coherent narrative generation. Xu et al. [32] proposed a framework for visual storytelling that constructs storylines and uses relational reasoning with external knowledge scene-graph and event-graph to generate coherent and informative narratives from image sequences.

3 Proposed Method

The proposed method for generating a story from a sequence of images involves three main processes as illustrated in Fig. 1. First, (A) **Image Contextualization** generates contextual features for each image in an image sequence by extracting image features and structuring image scene-graphs as the *Image Context*. Second, (B) **Narrative Contextualization** generates a narrative context by aggregating all image context and external knowledge-graph to improve the coherent generation of the story generation process. Third, (C) **Story Generation** involves generating a story from an image sequence. This process includes encoding all image contexts with an encoder and decoding them to produce the story text.

3.1 Image Contextualization

Image Contextualization is employed to aggregate two distinct contexts for each image in an image sequence: image features and image scene-graph. In this step, object detection and scene-graph generation are employed to extract image features and scene-graphs from each image. Image Context $\mathbf{c}_n$ of the n-th image in

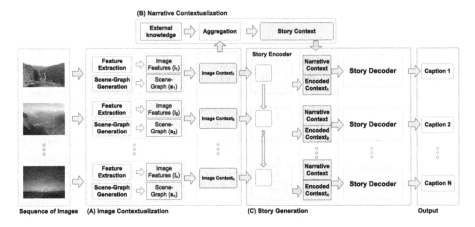

Fig. 1. Overview of the proposed method: Story generation framework that processes a sequence of images to generate a coherent narrative. (A) **Image Contextualization** extracts image features, generates scene-graph representations, and then combines them into image context. (B) **Narrative Contextualization** aggregates all image contexts and incorporate external knowledge to construct the narrative context. (C) **Story Generation** generates a story by having a story encoder to encode all image contexts combined with the narrative context and a story decoder to generate a caption for each image.

N images is generated by concatenating both image feature $\mathbf{i}_n$ and scene-graph feature $\mathbf{s}_n$ as a single feature as:

$$\mathbf{c}_n = [\mathbf{i}_n; \mathbf{s}_n], \qquad (1)$$

where $[\cdot;\cdot]$ denotes the concatenation of vectors.

Image Feature Representation represents each image by image feature $\mathbf{i}_n$ extracted using ResNeXt152-FPN [31], pre-trained on the ImageNet dataset [5]. This model captures and encodes all details of each image into a feature vector.

Image Scene-Graph Representation represents each image as a scene-graph as follows. First, *Scene-Graph Generation* is used to extract object features and their relationships from each image. Next, a *Graph Encoder* is utilized to encode each scene-graph into a feature vector $\mathbf{s}_n$.

Scene-Graph Generation is carried out using Neural Motif [34] with ResNet50-FPN [10], pre-trained on the Visual Genome dataset [16] to generate a scene-graph representation of each image by detecting object features and their relationships.

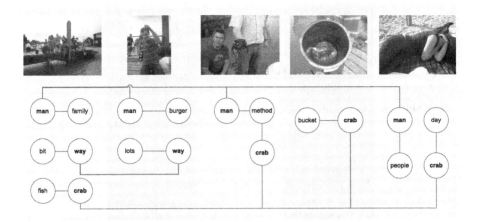

Fig. 2. Example of knowledge-graph of an image sequence, demonstrating relationships between objects among images.

Graph Encoder is constructed using a Graph Convolutional Network (GCN) architecture. This process encodes all object features and their relationships by transforming the graph representation into a feature vector.

3.2 Narrative Contextualization

The global context **g** of a given sequence is introduced to maintain the coherence of the story generation process. This is achieved by concatenating a collection of image contexts $(c_1, c_2, \ldots, c_N)$ and the image knowledge-graph representation e_{kg} as:

$$\mathbf{g} = [[\mathbf{c}_1; \mathbf{c}_2; \cdots ; \mathbf{c}_N] ; \mathbf{e}_{kg}], \tag{2}$$

where $[\cdot ; \cdot]$ denotes the concatenation of vectors.

External Knowledge Integration integrates external knowledge in the form of a graph into the proposed model by aggregating and encoding all knowledge-graphs from an image sequence. We introduce two processes to incorporate the knowledge-graph into the proposed method. *Knowledge-graph construction* involves constructing the knowledge-graph for the VIsual STorytelling (VIST), and *Knowledge-graph integration* involves encoding the knowledge-graph into a feature vector to be aggregated into the narrative contextualization process.

Knowledge-Graph Construction. In this research, we introduced external knowledge in the form of graphs by constructing them from external knowledge annotations from PR-VIST [12] and ConceptNet [24]. In the construction process, we first build a knowledge-graph from PR-VIST knowledge, where objects and their relationships are assigned to each image, including the connections between the subject and object across images. Next, we expand each node in the knowledge-graph by integrating ConceptNet, focusing on semantic relationships such as "*relatedTo*," "*similarTo*," and "*synonym*." Finally, we establish relationships between object nodes based on semantic relationships by connecting nodes that represent semantically similar objects. Each image sequence consists of an individual image knowledge-graph $P_{(x,y)}$ and the connections between objects by giving a class pair (x, y) and employing the connection between them as (V_x, V_y). Figure 2 demonstrates an example of a knowledge-graph, illustrating each image and the connections between images.

Knowledge-Graph Integration. The knowledge-graph encoder processes the knowledge-graph before incorporating it into the narrative contextualization process. In this process, each node in the knowledge-graph is encoded into a feature vector using Global Vector (GloVe) [22]. To integrate the knowledge-graph into the proposed model, we build a graph encoder using a GCN and employ the GlobalSortPool operator [35]. All knowledge-graphs of class pairs are encoded into a feature vector as:

$$\mathbf{H}^{(l+1)} = \sigma(\widehat{A}\mathbf{H}^{(l)}\mathbf{W}^{(l)}),$$
$$\mathbf{e}_{\text{kg}}^{(x,y)} = \text{GlobalSortPool}(\mathbf{N}^{(x,y)}), \quad (3)$$

where H is a matrix of node feature, $\widehat{A}$ is the adjacency matrix of the graph, W is the weight matrix, l is the l-th layer, and $\sigma(\cdot)$ is an activation function. $\mathbf{N}^{(x,y)}$ represents all embedding nodes from $P_{(x,y)}$.

3.3 Story Generation

We propose a method for story generation by adapting GLocal Attention Cascading Network (GLACNet) [14]. The architecture consists of two processes: *Story Encoder* encodes both story and image contexts, and *Story Decoder* generates a narrative.

Story Encoder processes and encodes the image and narrative contexts for the decoder. In this process, Bi-directional Long Short-Term Memory (BiLSTM) [8] is employed to encode all local features. Given a set of input features $\mathbf{c}_n$ at step n, the calculation process is as:

$$\mathbf{h}_n = \left[\overrightarrow{\text{LSTM}}^{(f)}(\mathbf{c}_n); \overleftarrow{\text{LSTM}}^{(b)}(\mathbf{c}_n) \right], \quad (4)$$

where $\overrightarrow{\text{LSTM}}^{(f)}(\cdot)$ and $\overleftarrow{\text{LSTM}}^{(b)}(\cdot)$ represent the state of the forward and the backward LSTM, respectively.

Next, each encoded feature at time step $\mathbf{e}_n$ representing each image context is constructed by concatenating the hidden state $\mathbf{h}_i$ and the narrative context $\mathbf{g}$ as:

$$\mathbf{e}_n = [\mathbf{h}_i; \mathbf{g}]. \quad (5)$$

Story Decoder is implemented using a Soft Attention Decoder inspired by GLocal Attention Cascading Networks (GLACNet) [14] via a cascading mechanism to generate a caption for each image. To generate a caption of each image, we first calculate alignment score $a_{t,i}$ to measure the relevance of the hidden state from the story encoder as:

$$a_{t,i} = \mathbf{v}^\top \tanh(\mathbf{W}\mathbf{e}_{t-1} + \mathbf{U}\mathbf{e}_n + \mathbf{b}), \tag{6}$$

where $\mathbf{v}$ is a learnable weight vector. $\mathbf{W}$ and $\mathbf{U}$ are learnable weight matrices, and $\mathbf{b}$ is a bias vector.

Next, attention weight $\alpha_{t,i}$ is calculated as:

$$\alpha_{t,i} = \frac{\exp(a_{t,i})}{\sum_{k=1}^{T_x} \exp(a_{t,k})}. \tag{7}$$

Then, the context information c_t for the input sequence is calculated to emphasize the most relevance for the output as:

$$c_t = \sum_{i=1}^{T_x} \alpha_{t,i} \bar{\mathbf{e}}_i. \tag{8}$$

Lastly, context information is used to generate a caption via Long Short-Term Memory (LSTM) as:

$$\begin{aligned} \mathbf{h}_t &= \text{LSTM}(y_{t-1}, [\mathbf{h}_{t-1}; \mathbf{c}_t]), \\ \mathbf{y}_t &= \text{softmax}(\mathbf{W}\mathbf{h}_t + \mathbf{b}), \end{aligned} \tag{9}$$

where $\mathbf{y}_t$ is a probability vector, with each value corresponding to the probability of a specific word in the vocabulary, and word$_{t'}$ with the maximum probability is selected as:

$$t' = \arg\max_t(\mathbf{y}_t). \tag{10}$$

4 Evaluation

Experiments are conducted to evaluate the proposed method in the VIsual STorytelling (VIST) task.

4.1 Experimental Conditions

Dataset. We use the VIST dataset [13] for training, validation, and evaluation. It comprises of over 200K images sourced from Flickr[1] and includes more than 50K human-annotated stories. It consists of approximately 40K stories and 164K images for training, 5K stories and 20K images for validation, and 5K stories and 20K images for testing. Each story is composed of 5 images, with each image accompanied by a corresponding sentence, resulting in a coherent narrative structure.

---
[1] https://www.flickr.com/ (Accessed Oct. 21, 2024).

Training Strategy. The proposed model is trained and validated using the VIST dataset, with distinct splits for training and validation. A learning rate of 0.001 is set. Cross-entropy loss is used as the loss function, and the Adam optimizer [15] is used during the training process.

Comparison. We chose five existing methods for story generation to evaluate the proposed method. We first adopt a common practice in generating stories. Adversarial REward Learning (AREL) [30] is a method that leverages adversarial training to generate coherent and engaging narratives from a sequence of images. Next, we consider an approach that integrates local context for story generation, while utilizing global context to ensure coherency throughout the generation process. This method is closely aligned with the proposed method. GLACNet [14] effectively combines global and local attention mechanisms to generate accurate, descriptive stories. Lastly, we explore existing approaches that incorporate external knowledge to enhance story generation. Plot and Rework: modeling storylines for VIsual STorytelling (PR-VIST) [12] is a method that generates coherent and engaging narratives from a sequence of images by employing progressive reinforcement learning to iteratively refine the storytelling process. Maximal Clique Selection Module (MCSM) [2] constructs a commonsense graph to generate a story based on two novel modules for concept selection. Knowledge-Enriched (KE) Visual Storytelling [11] is a method that utilizes external knowledge, including three stages of word distillation from input prompts, word enrichment using knowledge-graph, and story generation. Social interaction COmmonsense knowledge-based VIsual STorytelling (SCO-VIST) [27] is a multi-stage framework that utilizes social interaction knowledge to enhance commonsense reasoning in stories. However, it was evaluated on a testing set of the VIST dataset, with noise image collections filtered out.

Metrics. Five evaluation metrics are employed to evaluate the proposed method compared with the baselines. BiLingual Evaluation Understudy (BLEU) [21] and Evaluation MEtric of Translation with Explicit ORdering (METEOR) [1] are traditional metrics that measure the overlap between generated and reference texts, focusing on precision and recall, respectively. BiLingual Evaluation Understudy with Representations from Transformers (BLEURT) [23] leverages pre-trained transformer models to evaluate the fluency and relevance of generated narratives. Measure of Lexical Text Diversity (MLTD) [20] assesses the diversity of vocabulary used in storytelling to ensure varied and rich language use. Finally, Reinforcement-based VIST (RoVIST) [28] is specifically designed for VIST, evaluating the coherence and engagement of narratives derived from sequences of images by incorporating both textual and visual elements.

4.2 Quantitative Results

We first examined the quantitative results of the proposed method compared with existing methods in generating a narrative story as shown in Table 1. From

Table 1. Evaluation on the VIST dataset compared to other baselines. The bold is the best score, and the underlined is the second best. An asterisk (*) indicates scores that are evaluated on different testing set sizes. w/o KG indicates the proposed method implemented without the use of Knowledge Graphs. w/ KG indicates the proposed method, which is implemented using Knowledge Graphs.

Methods	Text Generation Eval.			Story Coherency Eval.	
	BLEU-4 ↑	METEOR ↑	BLEURT ↑	MLTD ↑	Ro-VIST ↑
AREL [30]	**14.40**	<u>35.40</u>	0.52	22.45	57.28
GLAC [14]	10.70	33.70	0.71	32.87	68.43
PR-VIST [12]	9.93	32.20	<u>1.37</u>	40.52	70.00
KE [11]	<u>13.09</u>	14.80	1.01	17.23	64.86
MCSM [2]	13.00	**36.01**	0.85	36.54	66.59
SCO-VIST [27]	—	27.50*	—	—	70.40*
Proposed (w/o KG)	7.65	22.23	1.29	<u>46.97</u>	70.40
Proposed (w/ KG)	10.09	11.01	**1.39**	**46.98**	**70.41**

the results, the proposed method with Knowledge-Graph (w/ KG) outperformed the other methods in narrative story generation in both story coherency evaluation metrics MLTD and Ro-VIST. On the other hand, PR-VIST achieved the second-best in MLTD, and SCO-VIST and the proposed method without Knowledge-Graph (w/o KG) achieved the second best in Ro-VIST. This indicates that the proposed method could generate a more coherent story. Meanwhile, for text generation evaluation metrics, the proposed method achieved the best in BLEURT, whereas PR-VIST achieved the second best. From the results, AREL achieved the best in BLEU-4, while KE achieved the second best. MCSM achieved the best in METEOR, whereas AREL achieved the second best.

4.3 Qualitative Results

Next, we demonstrate the qualitative results of generating a story for a sequence of images compared with existing methods. Figure 3 (A) shows that the proposed method effectively describes the environment of a sequence of images featuring the *city*, *people*, and *fireworks*. The proposed method was able to capture the content of *city*, including its environment, such as the *crowd*. In addition, it could describe the *firework* scene in the image sequence. Compared to the existing methods, only MCSM could accurately describe this situation, while the other baselines were unable to do so. Figure 3 (B) shows the ability of the proposed method to describe the details of a baseball game. Most existing methods, as well as the proposed method, could accurately describe the content of the *baseball game*, except for GLAC and KE which were only able to grasp the general content of the game. Also, the fourth image compares to the ground truth in describing *pitcher*, which only the proposed method and AREL could accurately describe. Figure 3 (C) demonstrates the proposed method's effectiveness in describing the

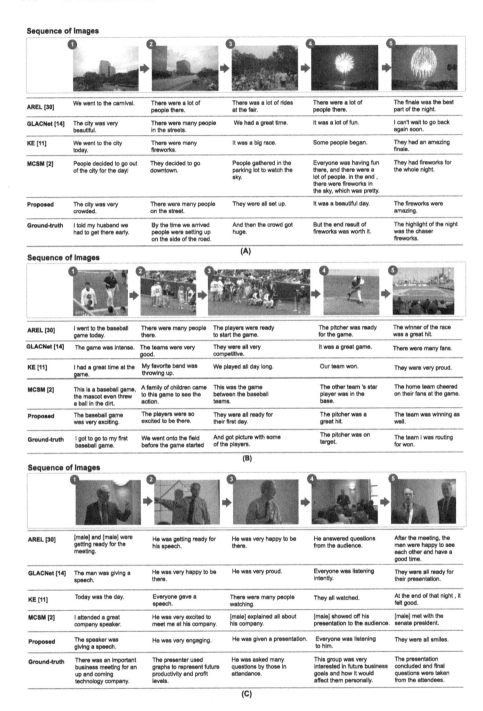

Fig. 3. Example of the output of the proposed method generating a narrative story of a sequence of images.

details of a *speaker* giving a speech and a presentation. The proposed method could describe the content of the *speaker* delivering the speech and presentation, similar to the other baseline methods. However, most existing methods and the proposed method could not generate complete information when compared to the ground truth.

5 Conclusion

We proposed a narrative generation framework by generating both image and narrative contexts. Additionally, we introduced external knowledge represented as knowledge-graphs corresponding to the VIsual STorytelling (VIST) dataset [13] to enhance the story generation process. In the proposed method, image context was generated by utilizing image features and scene-graph. A narrative context was generated by aggregating all image contexts and incorporating external knowledge. This approach allowed the generation of more accurate content for each image and also controlled the coherency of the generated story. Experimental evaluation demonstrated promising results of the proposed method on the VIST dataset. In the future, we plan to implement Large Language Models (LLM) in the story generation process by transferring their capabilities.

Acknowledgments. This work was supported in part by the Ministry of Education, Culture, Sports, Science and Technology (MEXT), Japan, through Grants-in-Aid for Scientific Research JP21H03519, JP23K16945, and JP24H00733.

References

1. Banerjee, S., Lavie, A.: METEOR: an automatic metric for MT evaluation with improved correlation with human judgments. In: Proceedings of 43rd Annual Meeting of the Association for Computational Linguistics, pp. 65–72 (2005)
2. Chen, H., Huang, Y., Takamura, H., Nakayama, H.: Commonsense knowledge aware concept selection for diverse and informative visual storytelling. In: Proceedings of 35th AAAI Conference on Artificial Intelligence, pp. 999–1008 (2021)
3. Chen, W., Li, X., Su, J., Zhu, G., Li, Y., Ji, Y., Liu, C.: TARN-VIST: Topic aware reinforcement network for VIsual STorytelling. In: Proceedings of 2024 Joint International Conference on Computational Linguistics, Language Resources and Evaluation, pp. 15617–15628 (2024)
4. Cong, Y., Yang, M.Y., Rosenhahn, B.: RelTR: relation TRansformer for scene graph generation. IEEE Trans. Pattern Anal. Mach. Intell. **45**(9), 11169–11183 (2023)
5. Deng, J., Dong, W., Socher, R., Li, L.J., Li, K., Fei-Fei, L.: ImageNet: a large-scale hierarchical image database. In: Proceedings 2009 IEEE Computer Society Conference on Computer Vision and Pattern Recognition, pp. 248–255 (2009)
6. Dunne, C., Shneiderman, B.: Motif Simplification: improving network visualization readability with fan, connector, and clique glyphs. In: Proceedings of 31st Annual SIGCHI Conference on Human Factors in Computing Systems, pp. 3247–3256 (2013)

7. Girshick, R.: Fast R-CNN. In: Proceedings of 15th IEEE International Conference on Computer Vision, pp. 1440–1448 (2015)
8. Graves, A., Schmidhuber, J.: Bidirectional LSTM networks for improved phoneme classification and recognition. In: Proceedings of 15th International Conference on Artificial Neural Networks, pp. 799–804 (2005)
9. Han, X., Yang, J., Hu, H., Zhang, L., Gao, J., Zhang, P.: Image Scene Graph Generation (SGG) benchmark. *Comput. Res. Reposit. arXiv Preprint*, arXiv:2107.12604 (Jul 2021)
10. He, K., Zhang, X., Ren, S., Sun, J.: Deep residual learning for image recognition. In: Proceedings of 2016 IEEE Conference on Computer Vision and Pattern Recognition, pp. 770–778 (2016)
11. Hsu, C.C., et al.: Knowledge-enriched visual storytelling. In: Proceedings of 34th AAAI Conference on Artificial Intelligence, pp. 7952–7960 (2020)
12. Hsu, C.Y., Chu, Y.W., Huang, T.H., Ku, L.W.: Plot and rework: modeling storylines for visual storytelling. In: Proceedings of 2021 Findings Association for Computational Linguistics, 11th International Joint Conference on Natural Language Processing, pp. 4443–4453 (2021)
13. Huang, T.K., et al.: Visual storytelling. In: Proceedings of 15th North American Chapter of the Association for Computational Linguistics: Human Language Technologies, pp. 1233–1239 (2016)
14. Kim, T., Heo, M., Son, S., Park, K., Zhang, B.: GLAC Net: GLocal attention cascading networks for multi-image cued story generation. In: Proceedings of 17th North American Chapter of the Association for Computational Linguistics (Workshop), pp. 1–6 (2018)
15. Kingma, D.P., Ba, J.: Adam: a method for stochastic optimization. In: Proceedings of 3rd International Conference on Learning Representations, pp. 1–13 (2014)
16. Krishna, R., et al.: Visual Genome: connecting language and vision using crowdsourced dense image annotations. Int. J. Comput. Vis. **123**(1), 32–73 (2017)
17. Kuznetsova, A., et al.: The Open Images dataset V4: unified image classification, object detection, and visual relationship detection at scale. Int. J. Comput. Vis. **128**(7), 1956–1981 (2020)
18. Li, T., Wang, H., He, B., Chen, C.W.: Knowledge-enriched attention network with group-wise semantic for visual storytelling. IEEE Trans. Pattern Anal. Mach. Intell. **45**(7), 8634–8645 (2022)
19. Liu, H., et al.: AOG-LSTM: an adaptive attention neural network for visual storytelling. Neurocomputing **552**(126486), 1–13 (2023)
20. McCarthy, P.M., Jarvis, S.: MTLD, VOCD-D, and HD-D: a validation study of sophisticated approaches to lexical diversity assessment. Behav. Res. Methods **42**(2), 381–392 (2010)
21. Papineni, K., Roukos, S., Ward, T., Zhu, W.J.: BLEU: a method for automatic evaluation of machine translation. In: Proceeding of 40th Annual Meeting of the Association for Computational Linguistics, pp. 311–318 (2002)
22. Pennington, J., Socher, R., Manning, C.D.: GloVe: global vectors for word representation. In: Proceedings of 2014 Conference on Empirical Methods in Natural Language Processing, pp. 1532–1543 (2014)
23. Sellam, T., Das, D., Parikh, A.P.: BLEURT: Learning robust metrics for text generation. In: Proceedings of 58th Annual Meeting of the Association for Computational Linguistics, pp. 7881–7892 (2020)
24. Speer, R., Chin, J., Havasi, C.: ConceptNet 5.5: an open multilingual graph of general knowledge. In: Proceedings of 31st AAAI Conference on Artificial Intelligence, pp. 4444–4451 (2017)

25. Tang, K., Niu, Y., Huang, J., Shi, J., Zhang, H.: Unbiased scene graph generation from biased training. In: Proceedings of 2020 IEEE Conference on Computer Vision and Pattern Recognition, pp. 3716–3725 (2020)
26. Wang, B., Ma, L., Zhang, W., Jiang, W., Zhang, F.: Hierarchical photo-scene encoder for album storytelling. In: Proceedings of 33rd AAAI Conference on Artificial Intelligence, pp. 8909–8916 (2019)
27. Wang, E., Han, C., Poon, J.: SCO-VIST: social interaction COmmonsense knowledge-based VIsual STorytelling. In: Proceedings of 18th Conference of the European Chapter of the Association for Computational Linguistics, pp. 1602–1616 (2024)
28. Wang, E., Han, S.C., Poon, J.: RoViST: learning robust metrics for visual STorytelling. In: Proceedings of 13th Conference of the North American Chapter of the Association for Computational Linguistics: Human Language Technologies, pp. 2691–2702 (2022)
29. Wang, R., Wei, Z., Li, P., Zhang, Q., Huang, X.: Storytelling from an image stream using scene graphs. In: Proc. 34th AAAI Conference on Artificial Intelligence & 32nd Innovative Applications of Artificial Intelligence Conference, pp. 9185–9192 (2020)
30. Wang, X., Chen, W., Wang, Y., Wang, W.Y.: No metrics are perfect: adversarial reward learning for visual storytelling. In: Proceedings of 56th Annual Meeting of the Association for Computational Linguistics, pp. 899–909 (2018)
31. Xie, S., Girshick, R., Dollár, P., Tu, Z., He, K.: Aggregated residual transformations for deep neural networks. In: Proceedings of 2017 IEEE Conference on Computer Vision and Pattern Recognition, pp. 1492–1500 (2017)
32. Xu, C., Yang, M., Li, C., Shen, Y., Ao, X., Xu, R.: Imagine, reason and write: visual storytelling with graph knowledge and relational reasoning. In: Proceedings of 35th AAAI Conference on Artificial Intelligence, pp. 3022–3029 (2021)
33. Yu, L., Bansal, M., Berg, T.L.: Hierarchically-attentive RNN for album summarization and storytelling. In: Proceedings of 2017 Conference on Empirical Methods in Natural Language Processing, pp. 966–971 (2017)
34. Zellers, R., Yatskar, M., Thomson, S., Choi, Y.: Neural Motifs: scene graph parsing with global context. In: Proceedings of 2018 IEEE Conference on Computer Vision and Pattern Recognition, pp. 5831–5840 (2018)
35. Zhang, M., Cui, Z., Neumann, M., Chen, Y.: An end-to-end deep learning architecture for graph classification. In: Proceedings of 32nd AAAI Conference on Artificial Intelligence, pp. 4438–4445 (2018)

TPS-YOLO: The Efficient Tiny Person Detection Network Based on Improved YOLOv8 and Model Pruning

Li Yao, Qianni Huang(✉), and Yan Wan

School of Computer Science and Technology, Donghua University,
Shanghai 201620, China
{yaoli,winniewan}@dhu.edu.cn, 2222823@mail.dhu.edu.cn

Abstract. Tiny Person detection in long-range scenes is a popular and challenging task. Current person detectors have two major issues. Firstly, their performance is poor in the case of tiny and heavily occluded persons. Secondly, they are computation-intensive and have large model sizes, which make them difficult to deploy on resource-limited devices. To solve the above issues, we proposed TPS-YOLO. Based on YOLOv8, we reconstruct the network structure by introducing shallow features of P2 into the feature fusion layers, which helps retain more spatial information important for tiny person detection. We design a fine-grained feature extraction module SPDCA to replace the standard convolution layer in the backbone network to enhance the feature representation of the network. In the feature fusion network, we use a weighted fusion method to fuse multi-scale features, which introduces learnable weights to learn the importance of different input features. We propose a lightweight module named C2f_Efficient, which integrates Depthwise Separable Convolution (DSC) to reduce the model parameters. Furthermore, we apply a model pruning method to further reduce the model's computational complexity. Experiments on the Tinypersonv2 and VisDrone-person datasets show that TPS-YOLO achieves satisfactory performance in terms of both efficiency and accuracy and has advantages on model lightweight.

Keywords: object detection · attention mechanism · feature fusion · YOLOv8 · tiny person detection

1 Introduction

Person/Pedestrian detection is an important topic in the field of computer vision with broad applications such as video surveillance, autonomous driving, and search-and-rescue operations. However, research on tiny person detection remains far from well explored. Directly applying previous detection models to scenes involving tiny persons has several challenges. Firstly, different from objects in proper scales, tiny persons are extreme small in size and has low signal noise ratio. Limited feature information about tiny person may disappear or mixed with the backgrounds during the down-sampling process in convolutional neural networks (CNNs) [1,2]. This complicates feature extraction for deep networks. Secondly, images taken from long distances often contain high-density objects, resulting in occlusion of persons [3]. Moreover, the computational limitations in real-world environments further increase the difficulty of model deployment [4]. Therefore, ensuring the performance and efficiency of tiny person detection under limited computational resources is a challenging issue.

In object detection, the YOLO series [5–7] are widely used in academic and industrial fields due to their good performance and efficiency. Shi et al. utilized YOLOv5 to detect tiny persons for emergency rescue [8]. Gao F et al. [9] improve the detection performance of occluded persons by introducing the self-attention module in the transformer block. While these one-stage detection methods demonstrate promising performance, their large model sizes make them not well suited for deployment on low-memory edge devices. Li G et al. [10] propose a real-time pedestrian detection network based on YOLO, which uses the MobileNet [11] network for feature extraction. The higher performance of this network is due to its improved detection head using Depthwise Separable Convolutions (DSC).

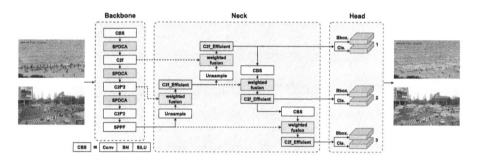

Fig. 1. The architecture of TPS-YOLO for tiny person detection. a) CSPDarknet53 backbone with fine-grained feature extraction module SPDCA. b) PANet neck with weighted feature fusion and C2f_Efficient. c) Three detection heads use feature maps from neck.

To improve detection performance while keeping the model lightweight, an efficient tiny person detection model TPS-YOLO is proposed in this work. The architecture is illustrated in Fig. 1. In order to retain more spatial information from tiny persons, we reconstruct the network architecture by introducing shallow features from P2 during feature fusion. For the backbone, SPDCA is introduced to enhance the feature representation by extracting fine-grained features and emphasising relative information. For the neck, weighted feature fusion is used to better balance the feature information of different scales, and the lightweight module C2f_efficient is designed which incorporates the lightweight module Depthwise Separable Convolutions (DSC) from MobileNet [11] to reduce the model size. Finally, we employ a channel pruning algorithm to further compress the model size. We conduct experiments on two publicly available datasets TinyPersonv2 and VisDrone-person to verify performance and efficiency of TPS-YOLO.

2 Method

2.1 Prediction Head for Tiny Objects

Tiny persons appear smaller size and have limited feature information. After many layers of downsampling, much of their feature information gets lost. In order to detect tiny persons effectively, smaller receptive field and higher resolution features need to be used [12]. Therefore, we use P2, P3, P4 of PAFPN instead of P3, P4, P5 for YOLOv8s. We named the adjusted version as YOLOv8s-tinyhead. This helps better perceive and capture the characteristics of tiny persons, thus enhancing the model's perception of tiny persons.

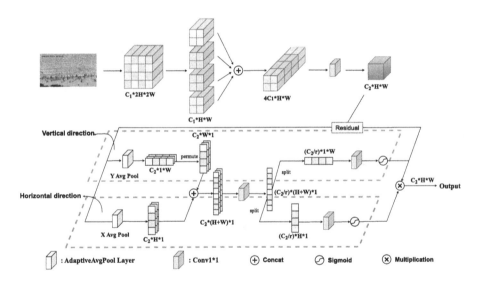

Fig. 2. The architecture of SPDCA, the Space-to-Depth (SPD) downsampling block combined with Coordinate Attention (CA) block.

2.2 Fine-Grained Feature Extraction

Since tiny persons have less pixel and feature information, spatial information maybe more important than deeper network model [1]. Strided convolution often loss detailed information during downsampling, thereby increasing the likelihood of false or missed detections [13]. To address this problem, we proposed a fine-grained feature extraction module, SPDCA, which is mainly composed of the

Space-to-Depth module (SPD) [13] and the Coordinate Attention mechanism (CA) [14]. The architecture is shown in Fig. 2. Firstly, we use the SPD module to reorganize spatial information. By slicing the feature maps into four submaps, each sub-map keeps the same number of channels but is half the original height and width. These sub-maps are conbined in the channel dimension, and we use a Pointwise convolution (PWConv) [15] to integrate information from different channels and reduce dimensionality. Compared to conventional strided convolution, SPD preserves more fine-grained information, although it might introduce some redundant features. The CA mechanism mitigates this by focusing on important areas, thus improving the accuracy of locating and identifying tiny objects. It decomposes the global pooling into two one-dimensional vector feature encodings, one for the horizontal and one for the vertical direction. This helps capture spatial relationships over long distances and maintain positional details, allowing the network to focus on important areas with low computational cost, thus improving the localization accuracy. In addition, the CA module has a lower computational cost compared to conventional attention mechanisms. By extracting fine-grained features and attention area, SPDCA helps focus on tiny persons and enhances the feature representation of the network.

2.3 Weighted Feature Fusion

When the PANet in the original YOLO fuses features of different resolutions, it adjusts them to the same resolution and then adds them directly. Since different input features have different resolutions, their effects on the fused output features are often unequal, and simple addition may lose semantic information. Inspired by BiFPN [16], we use the weighted feature fusion method to fuse features.

In this method, each input feature gets a learnable weight that allows the network to autonomously learn the significance of each of them. Unlike attention mechanisms that rely on the computation-intensive softmax function for normalization, we use a fast normalization fusion strategy. It directly divides the weight of each node into the sum of each node value to achieve normalization, which can speed up calculations. The process can be described as the following formula:

$$O = \sum_i \frac{w_i}{\varepsilon + \sum_j w_j} \cdot I_i. \quad (1)$$

where, w_i is the learnable weight, which is subsequently subjected to a ReLU function to ensure positivity. ε is a small value (0.0001) used to maintain numerical stability. I_i denotes the i-th input feature. The fast normalization fusion strategy is 30% faster than the softmax-based normalization, which ensures the model's real-time performance [16].

The feature fusion module of TPS-YOLO is shown in Fig. 3. For weighted feature fusion, features $P_2^{(1)}$, $P_3^{(1)}$ and $P_4^{(1)}$ extracted from the backbone at three different scales are used as inputs. Taking the node $P_3^{(2)}$ as an example, performing the fusion features is shown as the following formula. Where $P_3^{(2)}$ is the middle blue node in Fig. 3. Resize is the up-sampling or sub-sampling operation.

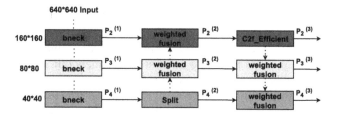

Fig. 3. Feature fusion module of TPS-YOLO.

By dynamically adjusts fusion weights, the weighted fusion method can filter out unimportant background information and better locate and recognize the tiny person in complex environments.

$$P_3^{(2)} = \frac{w_1 \cdot P_3^{(1)} + w_2 \cdot \text{Resize}(P_4^{(2)})}{w_1 + w_2 + \varepsilon} \quad (2)$$

2.4 C2f_Efficient

In order to improve the performance and efficiency of networks, research into lightweight models has received much attention. Models like MobileNet [11], ShuffleNet [17], and EfficientNet [16] achieve lightweight design through various techniques. Depthwise Separable Convolution (DSC) is a fundational element of many efficient network architectures due to its simplified computational design [18]. The architecture of DSC is shown in Fig. 4(a). Based on DSC, we propose a lightweight module, C2f_Efficient, which reduces the number of parameters and computations. Figure 4(b) shows the architecture of C2f_Efficient.

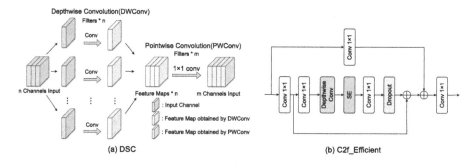

Fig. 4. (a) The architecture of DSC. (b) The architecture of C2f_Efficient.

DSC decomposes the Standard Convolution (SConv) operation into two steps: channel-by-channel convolution and point-by-point convolution. Deepthwise Convolution (DWConv) applies a single convolution for each input channel.

However, the number of feature maps obtained is the same as the number of channels in the input layer, and the feature maps cannot be extended. In addition, the deep convolution operation is performed independently for each channel, which cannot effectively use the feature information of different channels at the same spatial location. Therefore, Pointwise Convolution (PWConv) is required to combine these feature maps. PWConv is equivalent to a 1 × 1 standard convolution. The feature maps are combined in the depth direction to generate new feature maps to ensure that DSC has the same output dimension as the SConv. Compared to the Sconv, the number of parameters and computation of DSC is greatly reduced. Given input $I \in R^{C_{in} \times H \times W}$ and output $O \in R^{C_{out} \times H \times W}$, the FLOPs and parameters of SConv and DSC are as follows:

$$FLOPs\ SConv = C_{in} \times C_{out} \times h \times w \times k^2. \tag{3}$$

$$Params\ SConv = C_{in} \times C_{c_{out}} \times k^2. \tag{4}$$

$$FLOPs\ DSC = C_{in} \times h \times w \times (k^2 + C_{out}). \tag{5}$$

$$Params\ DSC = C_{in} \times (k^2 + C_{out}). \tag{6}$$

Comparative ratios of FLOPs and parameters are as follows:

$$\frac{C_{in} \times h \times w \times (k^2 + C_{out})}{C_{in} \times C_{out} \times w \times k^2} = \frac{1}{C_{out}} + \frac{1}{k^2}. \tag{7}$$

$$\frac{C_{in} \times (k^2 + C_{out})}{C_{in} \times C_{out} \times k^2} = \frac{1}{C_{out}} + \frac{1}{k^2}. \tag{8}$$

Using a 3 × 3 SConv as an example, the FLOPs and parameters of DSC are about 1/9 that of SConv. We replace the SConv in the original C2f module of YOLOv8 with the DSC, and propose a more lightweight module C2f_Efficient. As in the MobileNetv3 [11], we insert the Squeeze-and-Excitation (SE) [19] attention mechanism between the DWConv and PWConv to enhance the network's sensitivity to direction and location. By processing the information obtained by DWConv through the SE attention mechanism, the model can adaptively learn the importance of each channel. Additionally, a Dropout layer is included to help the network converge and reduce the risk of overfitting.

2.5 Channel Pruning

The advantages of the model pruning are that it reduces the number of model parameters and computation cost, and reduces hardware requirements in practical deployment. In this work, a simple yet effective model pruning method [20] is introduced to compress the improved TPS-YOLO model. The process is shown in Fig. 5, including three main steps.

The first step is sparse training, where less informative channels are identified for subsequent channel pruning. During training, L1 norm is applied to the scaling factor γ of the Batch Normalization (BN) layer to evaluate the importance of each channel. A lower scaling factor value means that the channel contributes

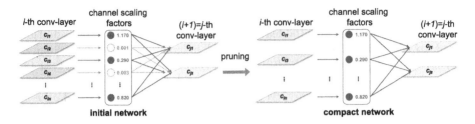

Fig. 5. Shows the process of model pruning. Channels with low scaling factor values (in red color) are pruned. After pruning, we obtain a compact network (right side). (Color figure online)

less to the output and thus can be removed. Specifically, the training objective is given by:

$$L = \sum_{(x,y)} l(f(x,W),y) + \lambda \sum_{\gamma \in \Gamma} f(\gamma). \quad (9)$$

The first term is the usual training loss. The second term is the sparsity penalty on the scaling factor, where $f(\gamma) = |\gamma|$ denotes L1-norm and λ denotes the penalty factor for balancing the loss of these two terms.

The second step is channel pruning. Channels with low scaling factor values are pruned according to a predefined pruning rate. After pruning, we can get a compact network. Comparative experiments for different pruning rates are shown in Table 3.

The third step is fine-tuning. A fine-tuning operation after channel pruning can compensate the temporarily drop in detection performance. The resulting model takes into account both accuracy and speed, and greatly reduces the model's computational and memory costs. This facilitates the deployment of the model on resource-constrained mobile or edge devices.

3 Experiments

3.1 Datasets and Details

Datasets. In this work, we use the publicly available datasets TinyPerosnv2 [21] and VisDrone-person [22]. TinyPerosnv2 is a dataset for tiny person detection collected via a seaside drone camera. The dataset contains 12,032 low-resolution images and 619,627 annotated persons. We use the same division as in [21], 5711 for training, 568 for validation and 5753 for testing. The image resolution is mainly 1920×1080 pixels with very small person size. The VisDrone dataset is a UAV image dataset containing images of different urban scenes, weather and lighting conditions. It contains 8629 images, of which 6471 are used for training, 548 for validation and 1610 for testing. We generate a single-category dataset Visdrone-person by extracting labels of two categories person and pedestrian from the original VisDrone dataset, and then merging the two categories into a single category named person.

Metrics. We use AP (Average Precision) for performance evaluation and set the IOU threshold to 0.5. In order to compare the detection accuracy of objects of different sizes, TinyPerson [23] benchmark divides the size range into several parts: in the range of 2 to 20 pixels is tiny, 20 to 32 is small, 32 to inf is reasonable, and 2 to inf is all. The corresponding evaluation metrics are AP_{50}^t, AP_{50}^s, AP_{50}^r, and AP_{50}. In addition, we use Parameters and FLOPs to validate the model size and computational consumption.

Details. Experiments are conducted on an Ubuntu 20.04 system using PyTorch version 1.11.0, CUDA 11.3, accelerated with a NVIDIA GeForce RTX 3090 GPU. The experimental parameters were set as follows: the input image size was 640×640, the initial learning rate was set to 0.01, and the Stochastic Gradient Descent (SGD) optimizer was used with a momentum of 0.9. The model pruning penalty factor λ is 0.01, and the pruning ratio is 3.0.

Table 1. Ablation Experiments on TinyPersonv2 dataset.

Model	T. H.	SPDCA	C2f_E.	W. F.	Prune	AP_{50}	AP_{50}^t	AP_{50}^s	AP_{50}^r	Params	FLOPs
YOLOv8s						54.2	29.5	67.0	84.1	11.2	28.6
(a)	✓					62.4	43.5	72.8	84.3	3.3	30.0
(b)	✓	✓				65.3	46.0	75.8	85.9	3.3	28.5
(c)	✓	✓	✓			62.7	44.0	73.0	84.5	3.3	26.6
(d)	✓	✓	✓	✓		65.2	46.2	75.5	86.2	3.3	26.6
(e)	✓	✓	✓	✓	✓	63.5	44.7	74.0	84.4	1.1	8.6

3.2 Ablation Experiments

In this section, we conduct ablation experiments to evaluate the effect of individual components of TPS-YOLO on both performance and computational efficiency. Table 1 sums up the results of the ablation experiments on TinyPersonv2 dataset. The first row of Table 1 shows the results of our benchmark YOLOv8s. Method (a) shows that by reconstructing the network architecture and introducing the shallow features for feature fusion, the performance of the model is significantly improved with an increase in AP_{50}^t of 14.0%. It verifies that the spatial information from shallow layers is important for tiny object detection, thus the adjusted structure is better able to perceive tiny persons. Method (b) verifies that the SPDCA moudule enhances the feature extraction of the network. The AP is improved by 2.9% compared with the previous step. Method (c) shows that when the lightweight module C2f_Efficient is introduced, the FLOPs of the network is reduced by 1.9G, albeit with a slight drop in AP. Method (d) shows that the introduction of weighted fusion improves the AP by 2.5% over the previous step, and the size of the network is almost constant. In order to further compress the network, the channel pruning method is utilized in method (e). It significantly reduces parameters and FLOPs with only a slight drop in accuracy. The model of TPS-YOLO selects Method (e) to ensure the optimal model performance. Compared to the benchmark model, the AP_{50}, AP_{50}^t, AP_{50}^s, AP_{50}^r of our method increased by 9.3%, 15.2%, 7.0%, and 0.3% respectively. It can be seen that the most improvement is on the AP_{50}^t metric, indicating that our model is effective in detecting very tiny persons. As for the model lightweight, the number of parameters is reduced by 90.2% and the FLOPs is reduced by 69.9% compared with the YOLOv8s benchmark, which verifies that our model is more lightweight while maintaining higher accuracy.

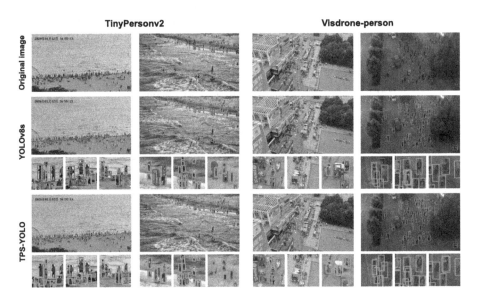

Fig. 6. Detection results for YOLOv8s and TPS-YOLO. Green marks the correctly detected ones and red marks the missed ones. (Color figure online)

Figure 6 shows the comparison of the detection results of the benchmark model YOLOv8s and TPS-YOLO on datasets TinyPersonv2 and Visdrone-person. Green marks the correctly detected persons, and red marks the missed persons. We takes some of the detail cases from the figure for a clearer comparison. It can be observed that compared with YOLOv8s, TPS-YOLO performs better, especially in tiny and heavy occlusion cases. This indicates that the introduction of tiny head and SPDCA module enhances the feature representation of the network, while the adoption of the weighted fusion method helps better focus on target areas and less on irrelevant environmental information. The experimental results verify the effectiveness of the improvement measures.

3.3 Comparation Experiments

Compared With Related Methods. We compare the detection performance of TPS-YOLO with the benchmark model YOLOv8s and other related models, including different YOLO models, one-stage detection model SSD512, two-stage detection model Faster-RCNN, and transformer-based method RT-DETR. Table 2 shows the experiment results on the two datasets. It can be observed that TPS-YOLO outperforms other methods by large margins. It is worth mentioning that our method shows the most significant improvement on the AP_{50}^t metric, outperforming the second-best mothod by 9.2% (44.7% vs 35.5%) and 8.1% (45.5% vs 37.4%) on the two datasets, which shows the effectiveness of our method in detecting very tiny cases. In the experiments, we also compare with the recently popular transformer-based method RT-DETR. Considering that transformer-based methods require a large amount of data for training, we use the model pre-trained on COCO dataset. It can be seen that RT-DETR achieves 30.7% AP on the TinyPersonv2 dataset and 37.2% on the Visdrone-person dataset, which is even lower than some anchor-based detectors. We think that the property of the transformer to capture long-range dependencies is not suitable for detecting tiny objects which have very small sizes. As for the model lightweight, compared with the lightweight models YOLOv3-Tiny, YOLOv7-Tiny, YOLOv5n, and YOLOv8n, TPS-YOLO has the least number of parameters (1.1M) and the FLOPs is only larger than YOLOv5n. The experimental results verify the model lightweight of TPS-YOLO.

Table 2. Comparison of Related Methods on TinyPersonv2 dataset and Visdrone-person dataset.

method	TinyPersonv2				Visdrone-person				FLOPs	Params
	AP_{50}	AP_{50}^t	AP_{50}^s	AP_{50}^r	AP_{50}	AP_{50}^t	AP_{50}^s	AP_{50}^r		
YOLOv8s	54.2	29.5	67.0	84.1	50.9	34.3	72.5	85.1	28.6	11.2
YOLOv3-Tiny	25.6	20.6	27.3	38.7	32.4	20.6	48.6	60.6	12.9	8.6
YOLOv5n	44.3	31.9	51.0	63.4	43.0	32.3	58.3	66.1	**4.1**	1.8
YOLOv7-Tiny	50.7	35.5	58.4	71.6	44.5	31.7	59.3	75.2	13.2	6.0
YOLOv8n	50.3	26.4	62.9	80.9	46.4	29.9	67.6	82.0	8.7	3.2
YOLOv8m	56.0	31.0	69.1	84.7	53.4	36.9	75.2	86.3	79.1	25.9
YOLOv9C	54.4	29.3	67.8	**85.1**	53.8	37.4	**75.5**	**87.4**	102.1	25.3
Faster-RCNN-FPN	37.8	10.2	53.5	76.8	49.0	32.7	73.5	78.6	207.9	41.3
SSD512	38.2	23.4	46.0	69.6	39.2	26.1	55.9	68.1	87.5	24.4
RT-DETR	30.7	15.1	38.4	54.9	37.2	22.7	53.9	73.8	108.0	32.8
TPS-YOLO(ours)	**63.5**	**44.7**	**74.0**	84.4	**57.2**	**45.5**	72.7	85.5	8.6	**1.1**

Comparative Pruning Experiments. The pruning rate affects the accuracy and model size after pruning, it represents the degree of compression of the network. For example, when the pruning rate is 2.0, the computation before pruning is 2.0 times the computation after pruning. We conduct experiments on our improved model to find the appropriate pruning rate. Specifically, we set the pruning rate from 2.0 to 5.0 and maintains same training parameters. Results on TinyPersonv2 dataset are shown in Table 3. As the pruning ratio increases, there is a slight drop in accuracy. However, it also reduces computational and memory costs significantly. It can be seen that when the pruning rate is between 2.0 and 3.0, the AP_{50}^t is still at a high level. When the pruning rate is beyond 3.0, AP_{50}^t decreases faster. After weighing the pruning rate and the loss of accuracy, we choose the model with a pruning rate of 3.0.

Table 3. Effect of Different Pruning Rates on TinyPersonv2 dataset.

Pruning Rate	AP_{50}	AP_{50}^t	AP_{50}^s	AP_{50}^r	FLOPs(G)	Params(M)
2.0	64.4	45.0	75.0	84.9	12.9	2.6
2.5	63.9	44.7	74.7	84.6	10.4	1.5
3.0	63.5	44.7	74.0	84.4	8.6	1.1
3.5	63.1	44.1	73.3	84.2	7.4	0.9
4.0	62.7	43.6	72.9	83.8	6.5	0.7
4.5	63.0	43.7	73.5	83.9	5.7	0.7
5.0	62.3	42.8	73.1	83.6	5.1	0.6

4 Conclusion

In order to improve the performance of distant tiny person detection, we propose a novel efficient network TPS-YOLO based on YOLOv8. We reconstruct the structure by introducing shallow features for feature fusion to retain more spatial information. We replace the SConv in the backbone network with the fine-grained feature extraction module SPDCA. The SPDCA utilizes space-to-depth downsampling module and coordinate attention mechanism to extract feature information effectively, significantly enhancing the feature representation of the network. In addition, we use a weighted fusion method during feature fusion to better focus on target areas and less on irrelevant environmental information. Moreover, to compress the model for practical deployment, we propose a lightweight module C2f_Efficient and introduce a model pruning method to trim redundant network parameters. Our experiments on the Tinypersonv2 and Visdrone-person datasets validate the effectiveness of TPS-YOLO. Compared to related methods, TPS-YOLO achieves superior performance with significantly lower computational and memory costs. In future work, we will further optimize the network and utilize super-resolution techniques to enhance features for tiny object detection.

References

1. Jiang, N., Yu, X., Peng, X., Gong, Y., Han, Z.: SM+: refined scale match for tiny person detection (2021)
2. Peng, G., Yang, Z., Wang, S., Zhou, Y.: AMFLW-YOLO: a lightweight network for remote sensing image detection based on attention mechanism and multi-scale feature fusion. IEEE Trans. Geosci. Remote Sens. **16** (2023)
3. Cao, J., Pang, Y., Xie, J., Khan, F.S., Shao, L.: From handcrafted to deep features for pedestrian detection: a survey. IEEE Trans. Pattern Anal. Mach. Intell. **44**(9) 4913–4934 (2021)
4. Khan, A.H., Nawaz, M.S., Dengel, A.: Localized semantic feature mixers for efficient pedestrian detection in autonomous driving. In: Proceedings of the IEEE/CVF Conference on Computer Vision and Pattern Recognition, pp. 5476–5485 (2023)
5. Redmon, J., Divvala, S., Girshick, R., Farhadi. A.: You only look once: unified, real-time object detection. In: Proceedings of the IEEE Conference on Computer Vision and Pattern Recognition, pp. 779–788 (2016)
6. Redmon .J., Farhadi. A.: YOLOv3: an incremental improvement. arXiv preprint
7. Wang, C.Y., Yeh, I.H., Liao, H.Y.M.: YOLOV9: learning what you want to learn using programmable gradient information. arXiv preprintarXiv:2402.13616 (2024)
8. Shi, Y., Li, S., Liu, Z., Zhou, Z., Zhou, X.: MTP-YOLO: you only look once based maritime tiny person detector for emergency rescue. J. Marine Sci. Eng. **12**(4) (2024)
9. Lin, T.-Y., Dollár, P., Girshick, R., He, K., Hariharan, B., Belongie, S.: Feature pyramid networks for object detection. In: Proceedings of the IEEE Conference on Computer Vision and Pattern Recognition, pp. 2117–2125 (2017)
10. Li, G., Yang, Y., Xingda, Q.: Deep learning approaches on pedestrian detection in hazy weather. IEEE Trans. Industr. Electron. **67**(10), 8889–8899 (2019)
11. Howard, A., et al.: Searching for MobileNetV3. In: Proceedings of the IEEE/CVF International Conference on Computer Vision, pp. 1314–1324 (2019)
12. Kim, B.J., Choi, H., Jang, H., Lee, D.G., Jeong, W., Kim, S.W.: Dead pixel test using effective receptive field. Pattern Recogn. Lett, **167**, 149–156 (2023)
13. Sunkara, R., Luo, T.: No more strided convolutions or pooling: a new CNN building block for low-resolution images and small objects. In: Joint European Conference on Machine Learning and Knowledge Discovery in Databases, pp. 443–459. Springer (2022)
14. Hou, Q., Zhou, D., Feng, J.: Coordinate attention for efficient mobile network design. In: Proceedings of the IEEE/CVF Conference on Computer Vision and Pattern Recognition, pp. 13713–13722 (2021)
15. Hua, B.-S., Tran, M.-K., Yeung, S.-K.: Pointwise convolutional neural networks. In: Proceedings of the IEEE Conference on Computer Vision and Pattern Recognition, pp. 984–993 (2018)
16. Tan, M., Pang, R., Le, Q.V.: EfficientDet: scalable and efficient object detection. In: Proceedings of the IEEE/CVF Conference on Computer Vision and Pattern Recognition, pp. 10781–10790 (2020)
17. Zhang, X., Zhou, X., Lin, M., Sun, J.: ShuffleNet: an extremely efficient convolutional neural network for mobile devices. In: Proceedings of the IEEE Conference on Computer Vision and Pattern Recognition, pp. 6848–6856 (2018)
18. Chollet, F.: Xception: deep learning with depthwise separable convolutions. In: Proceedings of the IEEE Conference on Computer Vision and Pattern Recognition, pp. 1251–1258 (2017)

19. Hu, J., Shen, L., Sun, G.: Squeeze-and-excitation networks. In: Proceedings of the IEEE Conference on Computer Vision and Pattern Recognition, pp. 7132–7141 (2018)
20. Liu, Z., Li, J., Shen, Z., Huang, G., Yan, S., Zhang, C.: Learning efficient convolutional networks through network slimming. In: Proceedings of the IEEE International Conference on Computer Vision, pp. 2736–2744 (2017)
21. Yu, X., et al.: Object localization under single coarse point supervision. In: Proceedings of the IEEE/CVF Conference on Computer Vision and Pattern Recognition, pp. 4868–4877 (2022)
22. Zhu, P., Wen, L., Bian, X., Ling, H., Hu, Q.: Vision meets drones: A challenge. arXiv preprint arXiv:1804.07437 (2018)
23. Yu, X., Gong, Y., Jiang, N., Ye, Q., Han, Z.: Scale match for tiny person detection. In: Workshop on Applications of Computer Vision (2020)

Uncertainty-Guided Joint Semi-supervised Segmentation and Registration of Cardiac Images

Junjian Chen and Xuan Yang[✉]

College of Computer Science and Software Engineering, Guangdong Provincial Key Laboratory of Popular High-Performance Computers, Shenzhen University, Shenzhen, China
yangxuan@szu.edu.cn

Abstract. Aleatoric uncertainty negatively impacts registration and segmentation results for medical image analysis. In this paper, we propose an uncertainty-guided framework for the joint semi-supervised segmentation and registration of cardiac images, aiming to take advantage of both tasks for each other. We propose a semi-supervised segmentation framework by predicting statistical shape models of the heart and generating uncertainty maps to guide anchor selection in pixel-level contrastive learning. Besides, we develop a registration network to predict the deformation vector field (DVF) and registration uncertainty, where the registration uncertainty ensures the registration model focuses on regions with high confidence. By employing estimated DVFs, additional constraints between segmentation results are embedded as losses to further improve segmentation and registration accuracy at the same time. The experimental results show that our proposed framework outperforms the state-of-the-art registration and segmentation networks.

Keywords: Image segmentation · Image registration · Contrastive learning · Statistical shape model · Aleatoric uncertainty

1 Introduction

Accurate delineation of cardiac structures through image segmentation and estimation of cardiac motion via image registration are both crucial for disease diagnosis and treatment planning. Cardiac image registration can benefit from anatomical labels of the heart by paying attention to tissue structure instead of image texture details. In this study, we use full images as templates for registration. Meanwhile, segmentation performance can be improved by constraining prediction results of image pairs to be consistent under the deformation vector field (DVF) provided by image registration. With the advent of deep learning, combining these two tasks has shown potential for performance improvement, leading to the development of various deep learning-based joint segmentation and registration models [13,18]. However, training data often contains inherent uncertainty, which refers to the intrinsic noise or unavoidable errors within the data, and these uncertainties can adversely affect predictions of both segmentation and registration models.

Pixel-level contrastive learning (PCL) has demonstrated promise in semi-supervised image semantic segmentation [6]. The core issue of PCL is the selection of samples based on pseudo labels or the spatial structures of unlabeled data, which inherently contain noise. In this context, pseudo labels refer to labels generated from model predictions, where each pixel is assigned a label indicating the category to which it belongs. To alleviate the challenges associated with this inherent noise, we employ prediction uncertainty to guide the sampling process and reduce the number of noisy samples.

In this context, the entropy of per-pixel probability distributions is frequently used as a measure of model uncertainty [23,25]. Nevertheless, these uncertainties often overlook the anatomical prior knowledge, potentially leading to noisy sample inclusion during pseudo-label construction.

Similarly, the uncertainty of deformable image registration (DIR) is essential for safe and reliable clinical deployment [20]. Recent deep learning-based models still face the challenge of accurate DIR with its proper uncertainty. There are two types of uncertainties in DIR, including aleatoric uncertainty and epistemic uncertainty. Aleatoric uncertainty is due to the image acquisition process, such as poor imaging quality. Epistemic uncertainty is model uncertainty, which arises from the limitations of the model to find a proper registration [20]. The Bayesian neural network is commonly used to estimate registration uncertainty using dropout [15] or estimating the posterior of network parameters using variational inference [9]. Another kind of approach estimated registration uncertainty directly from the network [12,21].

However, to the best of our knowledge, uncertainty-guided models that jointly perform segmentation and registration remain unstudied. The main motivation of this work is to propose an uncertainty-guided joint semi-supervised segmentation and registration framework for cardiac images. Our contributions include the following:

- We design a network to predict heart contours directly from images using statistical shape models. The anatomical shapes of the heart are leveraged as uncertainty to construct pseudo-labels in a pixel-level contrastive learning segmentation model, reducing noise sampling for cardiac segmentation. Moreover, we develop a cardiac image registration network to estimate the DVF and registration uncertainty simultaneously.
- A joint segmentation and registration framework is proposed. The segmentation results of the warped image and fixed image are used to guide the registration, and the estimated DVF is used to warp the segmentation results to optimize the pseudo-label in contrastive learning. Alternative segmentation and registration processes enhance the overall accuracy of both segmentation and registration.
- Experimental results on public datasets indicate that our proposed joint segmentation and registration framework outperforms the state-of-the-art reg-

2 Method

2.1 Semi-supervised Segmentation Using Statistical Shape Model

In semi-supervised cardiac segmentation, incorporating anatomical knowledge has been proven to enhance the performance of semi-supervised networks. Statistical Shape Models (SSMs) are effective tools for describing anatomical structures and are widely used in medical image analysis to represent organ structures [17]. One of the most established SSMs in the medical field is the Point Distribution Model (PDM) [10], which transforms shapes into point distribution patterns, capturing relationships between shapes and providing intuitive explanations.

Given labeled dataset $D_L = \{(x_i, y_i), i = 1, \ldots, N\}$ and unlabeled dataset $D_U = \{x_j, j = 1, \ldots, M\}$, where $M \gg N$. First, we establish the PDM using the labeled dataset D_L. For each labeled image x_i, the corresponding label y_i represents the ground truth segmentation. The shape Y_i is constructed from n points extracted from the contour of y_i after PDM processing, with these n points representing the boundary of a structure. The shape Y_i of x_i is n points extracted from label y_i uniformly. The shape set is $Y \in \mathbf{R}^{N \times 2n}$ with mean shape $\bar{Y} = \frac{1}{N} \sum_i^N Y_i$. The covariance matrix of Y are computed to obtain the eigenvectors matrix $\mathbf{P} = [p_1, p_2, \ldots, p_m]$, where p_i is the ith eigenvector. The shape parameter b_i of Y_i is a m-dimensional vector computed as $b_i = \mathbf{P}^T(Y_i - \bar{Y})$.

To predict the shape Y_j of unlabeled image $x_j \in D_U$, we design a ResNet to learn the shape parameters b_j end-to-end from cardiac images inspired by [16]. The ResNet architecture consists of 1 convolutional layer, 16 residual blocks, and 2 fully connected layers. After the ResNet is trained, the unlabeled image $x_j \in D_U$ is input to the ResNet to output the predictive parameter b_j, and the point set Y_j is obtained using $Y_j = \bar{Y} + \mathbf{P}b_j$. Contours of unlabeled images can be obtained by interpolating between adjacent points one by one.

Labeled images are used to train the ResNet in a fully supervised manner. The average Hausdorff distance is the loss function.

$$\mathcal{L}_{AHD} = \frac{1}{|Y_i|} \sum_{q \in Y_i} \min_{q' \in Y_i'} d(q, q') + \frac{1}{|Y_i'|} \sum_{q' \in Y_i'} \min_{q \in Y_i} d(q', q), \quad (1)$$

where, Y_i is the ground truth point set and Y_i' is the predicted point set. $d(q, q')$ is the Euclidean distance between points q and q'.

We embed the contours of unlabeled data into a semi-supervised contrastive learning semantic segmentation model. The key point of semi-supervised learning is to assign pseudo-labels to unlabeled images. These pseudo-labels are used to guide the selection of anchors and positive and negative samples in contrastive learning. Since incorrect pseudo-labels can negatively impact model performance, pseudo-labels are often combined with uncertainty assessments to filter out unreliable pseudo-labels. General uncertainty estimation evaluated the entropy of predictions. Nevertheless, this approach overlooks the inherent tissue structure information of objects. We propose an uncertainty assessment approach based on object shapes to generate pseudo-labels more accurately for unlabeled data. This method enhances the model's prediction reliability in boundaries and enfources boundary predictions consistent with anatomical structures. We hypothesize that pixels closer to the contour easily have higher label uncertainty. In contrast, pixels far from the contour are deemed to have lower uncertainty and can be used for pseudo-labels.

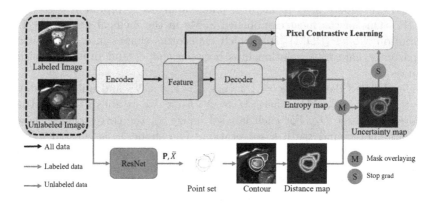

Fig. 1. The architecture of our semi-supervised segmentation model.

Suppose Y_j is the shape for unlabeled data x_j containing n points $Y_j = \{Y_{j_1}, Y_{j_2}, \ldots, Y_{j_n}\}$. To quantify uncertainty, we use the Manhattan distance between the pixel q in the prediction probability distribution p and the shape Y_j. The Manhattan distance is:

$$d(q, Y_j) = \min_k \left(|q - Y_{j_k}|_1 \right), \quad k \in \{1, 2, \ldots, n\}, \tag{2}$$

where $|\cdot|_1$ is the 1-norm. Each pixel q is assigned an uncertainty score based on this distance. Furthermore, we compute the entropy of the predicted probability $H(q)$ at q and exclude pixels with high uncertainty to estimate the pseudo-label y_p for unlabel data x_j.

$$y_p(q) = Argmax(q) \quad \text{if } d(q, Y_j) \geq \theta_1 \text{ and } H(q) \leq \theta_2 \tag{3}$$

where $Argmax$ is the label with the largest predictve probabilty; θ_1 and θ_2 are thresholds. The value of θ_1 ranges between 1 and 3 across different datasets, while θ_2 dynamically changes during training [23].

In contrastive learning segmentation, the features extracted from the encoder are embedded in a low-dimensional embedded representation space. Anchors, positive and negative samples are selected based on pseudo-labels, and the popular InfoNCE loss function is used:

$$\mathcal{L}_{cl} = -\sum_i \frac{1}{|P_i|} \sum_{v_i^+ \in P_i} \log \frac{e^{\cos(v_i, v_i^+)/\tau}}{e^{\cos(v_i, v_i^+)/\tau} + \sum_{v_i^- \in N_i} e^{\cos(v_i, v_i^-)/\tau}} \quad (4)$$

where P_i and N_i are positive and negative sample sets, respectively. v_i is the embedded representation of the ith anchor, v_i^+, v_i^- are positive and negative embedded representations for the ith anchor; τ is the temperature hyperparameter.

Besides $\mathcal{L}_{cl}$, other losses in [23] are used in the total loss $\mathcal{L}_{seg}$ to train the segmentation network.

The overall loss function of our segmentation network is:

$$\mathcal{L}_{seg} = \mathcal{L}_{sup} + \lambda_1 \mathcal{L}_{cl} + \lambda_2 \mathcal{L}_e \quad (5)$$

2.2 Uncertainty-Guided Image Registration

In deep learning models, providing higher uncertainty for incorrect predictions and lower uncertainty for correct ones is needed due to the inherent noise within the data. In cardiac image registration, the generated DVF also exhibits uncertainty. To reveal the inherent uncertainty in DVF, we designed a network to estimate the DVF and warped image uncertainty simultaneously.

Denote M and F as the moving and fixed images, respectively. Cardiac image registration is an unsupervised regression task to estimate the DVF, denoted as d, and the corresponding deformation function is $f(v) = v + d(v)$. The output of the registration network can be considered as a Gaussian distribution with mean $f(M)$ and a diagonal variance matrix Σ:

$$P(F|f(M)) = \mathcal{N}(f(M), \Sigma). \quad (6)$$

where $f(M) = M^w$ is the warped moving image. Each pixel of $f(M)$ can be viewed as a random variable, and the variance matrix Σ describes the variances of these random variables and their covariances, thereby quantifying the data uncertainty in the registered images.

We use two decoder branches in the registration network to predict the DVF and the uncertainty of warped images, respectively. The registration goal is to maximize the log-likelihood,

$$\log P(F|f(M)) = -\frac{1}{2}(F - M^w)^T \Sigma^{-1}(F - M^w) - \frac{1}{2} \log |\Sigma| + const. \quad (7)$$

In Equ.7, $(F - M^w)^T \Sigma^{-1}(F - M^w)$ indicates the registration error between F and M^w weighted by uncertainty. When the registration uncertainty is large, the contribution of the registration error decreases.

Furthermore, we use the deformation field constraint as VoxelMorph [3] to ensure the generated DVF is smooth:

$$\mathcal{L}_r = \sum_{q \in \Omega} \|\nabla d(q)\|^2, \tag{8}$$

where Ω represents the set of pixels in the DVF. The total loss for the registration network is:

$$\mathcal{L}_{reg} = \frac{1}{2}(F - M^w)^T \Sigma^{-1}(F - M^w) + \frac{1}{2}\log|\Sigma| + \lambda_3 \mathcal{L}_r, \tag{9}$$

where λ_3 is a balance coefficient.

2.3 Joint Segmentation and Registration Framework

To further improve both registration and segmentation, we combine the segmentation network with the registration network as a whole framework. The overall network structure is shown in Fig. 2. The source image M and fixed image F are input to the registration network and the segmentation network, resulting in the DVF, the uncertainty Σ, and segmentation results of M and F, denoted as S_M and S_F, respectively. Next, the warped image M^w is input to the segmentation to obtain S_{M^w}, and the segmentation result S_M is warped as $S_M{}^w$ using the estimated DVF.

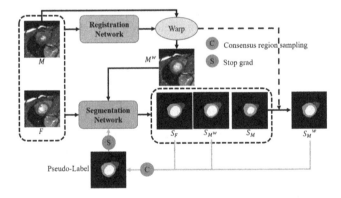

Fig. 2. The architecture of the proposed joint segmentation and registration framework.

The constraints between the registration and segmentation results are: 1) the segmentation mask S_{M^w} of the warped moving image is required to be similar to S_F of the fixed image; 2) the warped segmentation mask $S_M{}^w$ is required to be similar to S_F of the fixed image. These two constraints connect the DVF's predictions of segmentation and registration networks. Whether the segmentation or registration performance is poor, the above two constraints cannot be satisfied. The additional loss is defined as:

$$\mathcal{L}_w = \frac{1}{2}\left(\mathcal{L}_{Dice}(S_F, S_{M^w}) + \mathcal{L}_{Dice}(S_F, S_M{}^w)\right) \tag{10}$$

Moreover, the pixel consistency between $S_F, S_M{}^w, S_{M^w}$ can be used to guide the pseudo-label. Specifically, pixels of $S_F, S_M{}^w, S_{M^w}$ with the same label are marked as the pseudo-label, serving as candidate samples in contrastive learning. On the contrary, pixels with different labels are ignored.

$$y_p(q) = S_F(q) \quad \text{if } S_F(q) = S_{M^w}(q) = S_M{}^w(q) \tag{11}$$

Then, for the fixed image F, we use the new strategy given by Eq. 11 to construct pseudo-labels in joint segmentation and registration training. For the moving image M, we still use the original approach in Eq. 3 to construct pseudo-labels.

The total loss function of our joint segmentation and registration framework is:

$$\mathcal{L} = \mathcal{L}_{seg} + \mathcal{L}_{reg} + \lambda_4 \mathcal{L}_w \tag{12}$$

where λ_4 is a weight coefficient.

The training process of the joint model is divided into a pre-training phase and a joint training phase. In the pre-training phase, the segmentation network and the registration network are trained separately. During the joint training, the parameters of the registration network are fixed when training the segmentation network, and vice versa. This alternating training process maintained a ratio of 1:6 for the number of epochs required by the segmentation network to the registration network within a single round of joint training. This is because the registration task converges more slowly compared to the segmentation task.

3 Experiments

3.1 Datasets and Implement Details

Four public cardiac datasets were used to evaluate our framework, including MICCAI 2009 [19], York [1], ACDC [4], and M&Ms [5]. Details of four datasets are listed in Table 1. Experts provided end-diastolic (ED) and end-systolic (ES) contours for the left ventricle (LV), right ventricle (RV), or myocardium (MYO) for different datasets. We used the ES and ED images as moving and fixed images, respectively. All images were cropped to 224 × 224 with the heart centered in the image. The data augmentation techniques we used include random flipping, random rotation, random cropping, and color jittering. Semi-supervised training was performed with label amounts of 10% and 5%.

Our networks are trained using PyTorch on a computer equipped with an Intel(R) Core(TM) i7-7800X CPU @ 3.50GHz and an NVIDIA GeForce RTX 3080 GPU. The Adam optimizer is employed with a learning rate of 1e-4. U-Net

Table 1. Details of four datasets. Data format: number of patients (slices).

Dataset	Total Patients	Training		Validation	Test
		10% Labeled	5% Labeled		
MICCAI 2009	45 (666)	3 (56)	1 (18)	3(90)	15 (164)
York	33 (456)	2(28)	1 (16)	2(28)	11 (156)
ACDC	100 (1392)	7 (82)	3 (44)	10(154)	20 (290)
M&Ms	320 (3320)	15 (190)	7 (92)	34(438)	136 (1578)

is employed as the baseline architecture for image segmentation and registration, with the encoder portion of U-Net replaced by ResNet-50 for two networks. For the segmentation network, we initialized its parameters with weights pre-trained on ImageNet. The hyperparameters are $\lambda_1 = \lambda_2 = \lambda_3 = 0.1, \lambda_4 = 0.01$.

To compare the performance of our proposed method, we conducted a comparative analysis with 11 state-of-the-art semi-supervised image segmentation and registration deep learning algorithms. These include semi-supervised contrastive learning segmentation networks: UGPCL [23], U^{2}PL [25], DGCL [24], CSS [22], and SAMT [8]; registration networks: VoxelMorph [2], Dalcadiff [11], and TransMorph [7]; joint segmentation and registration networks: DeepAtlas [26], U-ReSNet [13], RSegNet [18], and Bi-JROS [14]. All methods were implemented with their official hyperparameter settings.

Among the joint segmentation and registration networks, U-ReSNet and RSegNet are fully supervised frameworks, while DeepAtlas operates as a semi-supervised framework. All three frameworks, along with our proposed method, were evaluated using 10% and 5% of the labeled ED and ES slices. Bi-JROS, a semi-supervised one-shot framework, is evaluated with labels provided for all ES slices but none for the ED slices, resulting in a total labeling rate of 50%. The Dice coefficient (Dice) and the Hausdorff distance (HD) are used as evaluation metrics.

3.2 Results

A comparison of segmentation results across various datasets using different networks is listed in Table 2 using 10% and 5% labeled data ("reg." and "seg." are abbreviations of registration and segmentation). Our method achieved the optimal Dice scores and lowest HD values compared to DGCL, U^{2}PL, SAMT, CSS, and UGPCL, demonstrating superior performance in segmentation accuracy and boundary precision. Notably, with only 5% labeled data, our method significantly outperformed other frameworks on smaller datasets like MICCAI 2009 and York due to its use of contour-guided pseudo-label generation in contrastive learning. Figure 3 illustrates the boxplots of Dice of the left ventricle (LV), right ventricle (RV), and myocardium (MYO), using different networks on various datasets. It can be observed that our segmentation network is robust for different tissues, especially for the RV with flexible shapes and MYO with thin structures.

Table 2. Comparison of segmentation results for different datasets with 10% and 5% labeled data. Dice data format: mean (standard deviation).

Labeled	Network	MICCAI 2009 and York		ACDC		M&Ms	
		Dice	HD	Dice	HD	Dice	HD
10%	DGCL	82.00(1.87)	8.20	88.01(0.65)	7.21	82.89(0.66)	9.77
	U^{2}PL	87.25(2.88)	8.26	88.78(0.75)	6.43	84.06(0.52)	9.32
	SAMT	84.96(1.80)	22.49	88.63(0.14)	14.29	84.45(0.58)	11.29
	CSS	84.47(1.71)	8.27	88.34(0.62)	7.38	81.32(0.87)	11.99
	UGPCL	86.77(1.90)	9.37	88.36(0.41)	9.14	85.08(0.41)	7.63
	Ours (wo. reg.)	87.53(1.73)	8.38	89.05(0.40)	8.71	85.45(0.39)	6.82
	Ours (w. reg.)	88.73(1.81)	5.15	89.90(0.66)	5.16	86.92(0.33)	3.70
5%	DGCL	45.56(2.92)	48.30	77.85(0.77)	23.63	80.94(0.69)	14.02
	U^{2}PL	49.84(4.56)	47.25	85.72(0.67)	8.36	78.07(1.34)	19.47
	SAMT	62.90(2.46)	42.98	85.66(0.24)	25.78	82.39(0.78)	17.76
	CSS	67.88(1.96)	14.36	80.07(0.47)	20.68	80.46(0.82)	14.98
	UGPCL	81.31(2.00)	22.44	83.88(0.43)	18.03	80.47(0.79)	14.56
	Ours (wo. reg.)	83.30(1.84)	14.06	87.13(0.65)	11.63	82.97(0.56)	12.06
	Ours (w. reg.)	83.37(1.92)	10.61	88.08(0.94)	8.26	85.09(0.44)	7.16

The comparison of registration results across various datasets using different networks is listed in Table 3. It is observed that TransMorph achieved the highest Dice scores and the lowest Hausdorff distances on the ACDC and M&Ms datasets. Our method performed competitively, slightly below TransMorph in both Dice scores and HD values, but still above average overall. The reason is that our method only uses simple network architecture, such as U-Net, instead of the Transformer-based network in TransMorph. However, the registration performance of our method can be improved greatly when joint segmentation and registration are performed, as listed in Table 4.

The comparison of joint segmentation and registration results across various datasets is shown in Table 4. After conducting joint segmentation and registration training, our method demonstrated notable improvements. With 10% labeled data, the average improvement in the Dice score of segmentation is 1.17% for all datasets compared with our only segmentation network, and the average registration accuracy increased by 5.54% compared to the only registration network. With 5% labeled data, the average segmentation Dice score improved by 1.04%, and the average registration accuracy increased by 3.25% compared to their respective standalone performances. Specifically, for the M&Ms dataset, our method's registration accuracy with 10% labeled data was 0.96% lower than

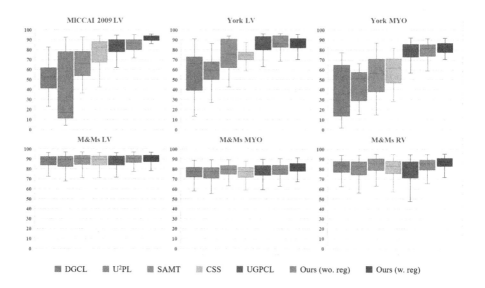

Fig. 3. Boxplots of Dice using different segmentation networks on various datasets.

Table 3. Comparison of registration results for different datasets. Dice data format: mean (standard deviation).

Network	MICCAI 2009 and York		ACDC		M&Ms	
	Dice	HD	Dice	HD	Dice	HD
VoxelMorph	75.46(3.49)	14.41	81.73(1.31)	10.59	79.63(0.56)	8.02
DalcaDiffNet	80.42(3.32)	9.65	84.01(1.38)	9.96	81.97(0.54)	7.61
TransMorph	78.40(3.49)	13.76	85.66(1.10)	9.90	82.70(0.56)	7.74
Ours (wo. seg.)	78.78(3.39)	13.77	83.53(0.89)	10.06	81.87(0.52)	7.64
Ours (w. seg. 5%)	82.75(3.44)	13.44	86.76(0.60)	11.65	84.44(0.42)	7.44
Ours (w. seg. 10%)	86.40(3.26)	12.86	88.83(0.61)	11.48	85.58(0.40)	7.13

Bi-JROS's accuracy with 50% labeled data but outperformed other methods on the remaining datasets. Overall, our method consistently provides excellent segmentation and competitive registration results, surpassing other methods in both tasks. The visualization of segmentation and registration results using our joint model are shown in Fig. 4 and Fig. 5.

3.3 Ablation Study

We conducted ablation experiments on the ACDC dataset with 10% labeled data to evaluate the impact of different components on the proposed method. The notation "Entropy" refers to constructing pseudo-labels using only the entropy of the predicted results in the segmentation network, while "Contour" denotes the

Table 4. Comparison of joint segmentation and registration results for different datasets with 50%, 10%, and 5% labeled data. Dice data format: mean (standard deviation).

Labeled	Network		MICCAI 2009 and York		ACDC		M&Ms	
			Dice	HD	Dice	HD	Dice	HD
50%	Bi-JROS	Seg.	82.25(2.04)	15.66	88.86(0.30)	11.22	83.76(0.64)	13.93
		Reg.	73.63(3.81)	7.38	88.57(0.98)	4.47	86.54(0.44)	3.43
10%	DeepAtlas	Seg.	70.89(2.86)	39.98	81.14(0.49)	31.57	80.79(0.75)	19.33
		Reg.	75.08(3.50)	14.58	82.02(1.15)	10.68	79.87(0.60)	8.07
	U-ReSNet	Seg.	49.91(5.44)	48.42	65.16(2.70)	31.14	67.67(1.87)	24.47
		Reg.	73.64(3.91)	14.48	83.12(0.95)	10.70	80.61(0.56)	8.25
	RSegNet	Seg.	70.03(2.83)	39.81	85.44(0.25)	21.87	81.89(0.61)	18.00
		Reg.	76.92(3.35)	13.85	84.90(0.87)	10.18	82.47(0.54)	7.48
	Ours	Seg.	88.73(1.81)	5.15	89.90(0.66)	5.16	86.92(0.33)	3.70
		Reg.	86.40(3.26)	12.86	88.83(0.61)	11.48	85.58(0.40)	7.13
5%	DeepAtlas	Seg.	43.95(3.12)	75.27	62.71(2.67)	52.08	75.33(1.19)	31.70
		Reg.	74.34(3.49)	14.59	81.53(1.32)	10.69	79.84(0.62)	8.08
	U-ReSNet	Seg.	41.78(4.40)	54.34	50.70(2.87)	42.80	57.54(3.08)	38.43
		Reg.	73.50(3.86)	14.56	81.58(0.94)	11.00	79.38(0.58)	8.33
	RSegNet	Seg.	40.43(3.53)	46.19	66.28(2.12)	52.70	76.91(1.12)	29.99
		Reg.	76.26(3.33)	13.86	84.77(0.84)	10.45	82.41(0.54)	7.55
	Ours	Seg.	83.37(1.92)	10.61	88.08(0.94)	8.26	85.09(0.44)	7.16
		Reg.	82.75(3.44)	13.44	86.76(0.60)	11.65	84.44(0.42)	7.44

use of contour uncertainty for pseudo-label generation. "Uncertainty" signifies the use of uncertainty in the registration network. "CP" represents constructing pseudo-labels for the fixed image using Eq. 3, and "WP" refers to using Eq. 11 for pseudo-label construction in the joint segmentation and registration network. Table 5 lists the results of the ablation experiments. From the first two rows, it is evident that contour-guided contrastive learning improves the Dice score by 1.37% and reduces the Hausdorff Distance (HD) by 2.02 pixels compared to using entropy-based uncertainty alone. This indicates that incorporating contour guidance results in segmentation outputs that better align with the anatomical structure of the heart. The results from the third and fourth rows show that incorporating uncertainty into the registration process increases the Dice score by 0.08% compared to using registration without uncertainty. The results from the last four rows demonstrate that using Eq. 11 for pseudo-label construction on the fixed image within the joint framework leads to a 0.23% improvement in segmentation Dice and a 1.30% improvement in registration Dice. It validates the effectiveness of combining segmentation results together using estimated DVF.

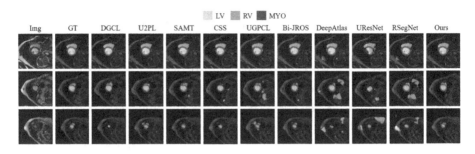

Fig. 4. Visualization of segmentation results using different networks and our joint model.

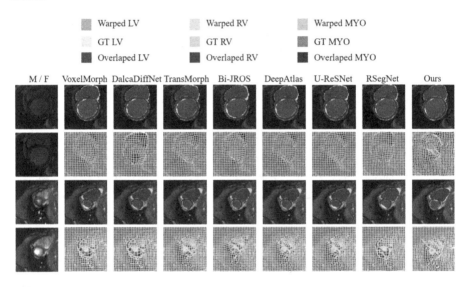

Fig. 5. Visualization of registration results using different networks and our joint model.

Table 5. Comparison of the influence of different parts in our method on ACDC dataset. Dice data format: mean (standard deviation).

		Entropy	Contour	Uncertainty	CP	WP	Dice	HD
	Segmentation	✓					87.68(0.67)	10.73
		✓	✓				89.05(0.40)	8.71
	Registration						83.45(0.94)	10.13
				✓			83.53(0.89)	10.06
Joint	Segmentation	✓	✓	✓	✓		89.67(0.60)	5.36
	Registration						87.53(0.71)	11.64
	Segmentation	✓	✓	✓		✓	89.90(0.66)	5.16
	Registration						88.83(0.61)	11.48

4 Conclusion

This paper presented an uncertainty-guided joint segmentation and registration framework of cardiac images. The heart contours are predicted to guide contrastive learning through uncertainty. For the registration task, we predicted the DVF and registration uncertainty at the same time. Within the joint framework, the segmentation results were used to guide the registration process, and the pseudo-labels within the segmentation model were refined using estimated DVF, further enhancing the accuracy of both segmentation and registration tasks. Experimental results on public datasets demonstrated that our joint framework outperforms existing networks both for segmentation and registration.

Acknowledgments. This paper has been supported by the Shenzhen Fundamental Research Program (JCYJ20220531102407018), Guangdong Province Key Laboratory of Popular High-Performance Computers 2017B030314073), the National Natural Science Foundation of China (Grant No.62201355, 62071303).

References

1. Andreopoulos, A., Tsotsos, J.K.: Efficient and generalizable statistical models of shape and appearance for analysis of cardiac MRI. Med. Image Anal. **12**(3), 335–357 (2008)
2. Balakrishnan, G., Zhao, A., Sabuncu, M.R., Guttag, J., Dalca, A.V.: An unsupervised learning model for deformable medical image registration. In: Proceedings of the IEEE Conference on Computer Vision and Pattern Recognition, pp. 9252–9260 (2018)
3. Balakrishnan, G., Zhao, A., Sabuncu, M.R., Guttag, J., Dalca, A.V.: VoxelMorph: a learning framework for deformable medical image registration. IEEE Trans. Med. Imag. **38**(8), 1788–1800 (2019)
4. Bernard, O., et al.: Deep learning techniques for automatic MRI cardiac multi-structures segmentation and diagnosis: is the problem solved? IEEE Trans. Med. Imag. **37**(11), 2514–2525 (2018)
5. Campello, V.M., et al.: Multi-centre, multi-vendor and multi-disease cardiac segmentation: the M&Ms challenge. IEEE Trans. Med. Imag. **40**(12), 3543–3554 (2021)
6. Chaitanya, K., Erdil, E., Karani, N., Konukoglu, E.: Contrastive learning of global and local features for medical image segmentation with limited annotations. Adv. Neural. Inf. Process. Syst. **33**, 12546–12558 (2020)
7. Chen, J., Frey, E.C., He, Y., Segars, W.P., Li, Y., Du, Y.: TransMorph: transformer for unsupervised medical image registration. Med. Image Anal. **82**, 102615 (2022)
8. Chen, Y., Chen, F., Huang, C.: Combining contrastive learning and shape awareness for semi-supervised medical image segmentation. Expert Syst. Appl. **242**, 122567 (2024)
9. Cheng, N., Malik, O.A., De, S., Becker, S., Doostan, A.: Bi-fidelity variational auto-encoder for uncertainty quantification. Comput. Methods Appl. Mech. Eng. **421**, 116793 (2024)
10. Cootes, T.F., Taylor, C.J., Cooper, D.H., Graham, J.: Active shape models-their training and application. Comput. Vis. Image Underst. **61**(1), 38–59 (1995)

11. Dalca, A.V., Balakrishnan, G., Guttag, J., Sabuncu, M.R.: Unsupervised learning for fast probabilistic diffeomorphic registration. In: Medical Image Computing and Computer Assisted Intervention–MICCAI 2018: 21st International Conference, Granada, Spain, September 16-20, 2018, Proceedings, Part I, pp. 729–738. Springer (2018)
12. Dalca, A.V., Balakrishnan, G., Guttag, J., Sabuncu, M.R.: Unsupervised learning of probabilistic diffeomorphic registration for images and surfaces. Med. Image Anal. **57**, 226–236 (2019)
13. Estienne, T., et al.: U-ReSNet: ultimate coupling of registration and segmentation with deep nets. In: Medical Image Computing and Computer Assisted Intervention–MICCAI 2019: 22nd International Conference, Shenzhen, China, October 13-17, 2019, Proceedings, Part III 22, pp. 310–319. Springer (2019)
14. Fan, X., Wang, X., Gao, J., Wang, J., Luo, Z., Liu, R.: Bi-level learning of task-specific decoders for joint registration and one-shot medical image segmentation. In: Proceedings of the IEEE/CVF Conference on Computer Vision and Pattern Recognition, pp. 11726–11735 (2024)
15. Gong, X., Khaidem, L., Zhu, W., Zhang, B., Doermann, D.: Uncertainty learning towards unsupervised deformable medical image registration. In: Proceedings of the IEEE/CVF Winter Conference on Applications of Computer Vision, pp. 2484–2493 (2022)
16. He, K., Zhang, X., Ren, S., Sun, J.: Deep residual learning for image recognition. In: Proceedings of the IEEE Conference on Computer Vision and Pattern Recognition, pp. 770–778 (2016)
17. Heimann, T., Meinzer, H.P.: Statistical shape models for 3D medical image segmentation: a review. Med. Image Anal. **13**(4), 543–563 (2009)
18. Qiu, L., Ren, H.: RSegNet: a joint learning framework for deformable registration and segmentation. IEEE Trans. Autom. Sci. Eng. **19**(3), 2499–2513 (2021)
19. Radau, P., Lu, Y., Connelly, K., Paul, G., Dick, A.J., Wright, G.A.: Evaluation framework for algorithms segmenting short axis cardiac MRI. MIDAS J. (2009)
20. Rivetti, L., Studen, A., Sharma, M., Chan, J., Jeraj, R.: Uncertainty estimation and evaluation of deformation image registration based convolutional neural networks. Phys. Med. Biol. **69**(11), 115045 (2024)
21. Smolders, A., Lomax, A., Weber, D.C., Albertini, F.: Deep learning based uncertainty prediction of deformable image registration for contour propagation and dose accumulation in online adaptive radiotherapy. Phys. Med. Biol. **68**(24), 245027 (2023)
22. Wang, C., Xie, H., Yuan, Y., Fu, C., Yue, X.: Space engage: collaborative space supervision for contrastive-based semi-supervised semantic segmentation. In: Proceedings of the IEEE/CVF International Conference on Computer Vision, pp. 931–942 (2023)
23. Wang, T., Lu, J., Lai, Z., Wen, J., Kong, H.: Uncertainty-guided pixel contrastive learning for semi-supervised medical image segmentation. In: Raedt, L.D. (ed.) Proceedings of the Thirty-First International Joint Conference on Artificial Intelligence, IJCAI-22, pp. 1444–1450. International Joint Conferences on Artificial Intelligence Organization (2022). https://doi.org/10.24963/ijcai.2022/201
24. Wang, X., Zhang, B., Yu, L., Xiao, J.: Hunting sparsity: density-guided contrastive learning for semi-supervised semantic segmentation. In: Proceedings of the IEEE/CVF Conference on Computer Vision and Pattern Recognition, pp. 3114–3123 (2023)

25. Wang, Y., et al.: Semi-supervised semantic segmentation using unreliable pseudo-labels. In: Proceedings of the IEEE/CVF Conference on Computer Vision and Pattern Recognition, pp. 4248–4257 (2022)
26. Xu, Z., Niethammer, M.: DeepAtlas: joint semi-supervised learning of image registration and segmentation. In: Medical Image Computing and Computer Assisted Intervention–MICCAI 2019: 22nd International Conference, Shenzhen, China, October 13–17, 2019, Proceedings, Part II 22, pp. 420–429. Springer (2019)

Understanding the Roles of Visual Modality in Multimodal Dialogue: An Empirical Study

Qian Cao, Ruihua Song[✉], and Xu Chen

Renmin University of China, Beijing, People's Republic of China
{caoqian4real,rsong}@ruc.edu.cn

Abstract. Multimodal dialogue systems aim to simulate human-world interactions by perceiving information from various modalities. By developing effective strategies to integrate textual and visual modalities, multimodal dialogue models have yielded significant advancements. However, limited research has systematically and comprehensively explored the impact of the visual modality on dialogue response generation. To address this, this paper presents empirical studies conducted on a prominent multimodal dialogue model named Maria. To facilitate our analysis, we design a diagnostic approach utilizing cross-modal input ablation experiments. Our methodology involves manipulating the visual content of images by substituting visual regions or removing textual visual concepts while keeping the remaining inputs unchanged. Through experiments conducted on three datasets, we observe that the enhancements achieved through visual region features exhibit varying levels of stability across datasets. However, carefully introducing textual visual concepts, particularly those providing supplementary information, yields positive effects. Furthermore, discrepancies arise when constructing multimodal dialogue datasets using different approaches, underscoring the attention needed for future research in this field.

Keywords: Multimodal Dialogue · Dialogue Systems · Empirical Study

1 Introduction

Multimodal dialogue has attracted much attention from the research community and emerged with great potential in real-world scenarios [3,30,36,45]. Compared with traditional conversational systems [1,5,41,52], multimodal dialogue systems aim to introduce additional modalities such as visual or audio information to textual conversations. Many promising models (especially visual ones) have been recently witnessed to achieve this, either paying attention to coarse or fine-grained visual features modeling [47] or designing attention mechanisms such as routing to handle the sparsity of image features [53]. Besides, other researchers [27] propose to employ not only visual region features but also visual concept labels identified from the image to enhance a response generator, providing us with another perspective on perceiving visual modalities. More recently, significant progress has been achieved owing to the advances in large language models (LLM) [6,11,46], further enhancing the capacity of multimodal models to comprehend both images and dialogues [26,32,54].

Despite the progress in multimodal dialogue models, most efforts focus on improving strategies for combining textual and visual features. However, there is limited systematic analysis of which types of visual information most enhance performance. Exploring this is essential to addressing key research questions in multimodal dialogue: (1) Whether image region features extracted from the visual modality are essential for the model? (2) Do visual concepts derived from images matter and what types of concepts have a greater impact on the model performance? (3) What are the characteristics of multimodal dialogue datasets that better leverage the complementary heterogeneous information (*i.e.*, visual regions and concepts) from different modalities? Addressing these questions could offer new insights and guide the development of more effective multimodal dialogue models and datasets.

Given the above research questions, the significance of different modalities has been explored and analyzed in other domains [7]. In multimodal machine translation (MMT), researchers have observed that visual features may not consistently provide reliable improvements in performance [44]. Other researchers [17] suggest that MMT models can exhibit insensitivity to irrelevant images, and their performance may not significantly deteriorate. These counterintuitive observations have led to different hypotheses, with researchers [49] believing that the performance improvements in MMT models are primarily driven by a regularization effect. On the other hand, another work [20] attributes the observed improvements to the higher quality of latent space representations. These findings highlight the complexity of understanding the impact of different modalities and emphasize the need to investigate the underlying mechanisms and factors influencing multimodal model performance.

Following the above discussions, this paper aims to thoroughly analyze the role of visual modality in multimodal dialogue systems. However, numerous unknown factors persist in large language models (LLMs), such as emergent capabilities [15,40,48] and hallucinations [19,21,31], highlighting the need for further investigation into multimodal attribution in large-scale models. As a result, our research focuses on the smaller-scale model Maria [27], which effectively integrates visual information as both visual features and visual concept labels. We conduct a series of cross-modal input ablation experiments to assess the significance of visual features (object regions) and textual inputs of visual concepts by actively modifying visual inputs while keeping other inputs constant. Specifically, for visual features, we replace the original image with an irrelevant one, a black/white image, or set visual region features to zero or noisy embeddings. For visual concepts, we either remove all or selectively eliminate those irrelevant to the dialogue context. Experiments reveal that the impact of visual region features varies across datasets, with added visual concepts absent

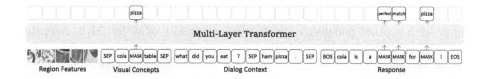

Fig. 1. The architecture of Maria [27]. We simplify it for better understanding.

from the dialogue context improving performance more than redundant ones. Additionally, datasets with native multimodal information are more effective in leveraging visual input for better dialogue responses.

2 Background

2.1 Multimodal Dialogue Systems

Images are frequently employed as a complementary modality to text in multimodal dialogue systems [9]. Previous researches concentrate on visual content understanding via question-answering tasks, like Visual Question Answering (VQA) [4] and Audio Visual Scene-Aware Dialog (AVSD) [2]. In addition, Visual Dialogue (VisDial) [12] extends dialogue turns and widens the scope of the conversation. Nevertheless, we focus on more natural open-domain multimodal dialogue, rather than these QA-style ones differing greatly from our daily communication. Image grounded conversation (IGC) [34] and Image-Chat [42] are more open-domain dialogue systems. Related works [33,35,43,53] capture visual information by an image encoder, which is then integrated into fusion modules to create responses while utilizing the context. Besides, owing to the lack of training data, other work [50] even considers recovering the image from a dialogue context. Recent advancements have significantly enhanced the ability of multimodal models to understand both images and dialogues [26,32,54], bringing multimodal dialogue systems into a new era.

2.2 Multimodal Dialogue Architecture

The utilization of visual modality is a crucial aspect of multimodal chitchat dialogue systems, which is the focus of our study. However, the existing unknown factors LLMs, such as emergent capabilities [15,40,48] and hallucinations [19,21,31], making it more challenging to conduct our research based on these uncertainties. Thus, we focus more on the smaller-scale models. In most of these models, the image encoder is either a multimodal encoder [18,53] or a relatively standalone visual module cascaded with the response generator [27,43], taking the concatenation of visual and word embeddings as input. Among them, researchers [27] propose one named Maria (See Fig. 1) which adopts an object detection module to perceive visual object regions in an image. It concatenates the visual objects, the visual concepts (*i.e.*, region labels), and the dialogue context as input to generate a response. In contrast to employing a coarse representation [42] of the whole image and utilizing the context as the only textual information [43], Maria adopts fine-grained visual information from objects. This fine-grained visual modeling [47], including visual object features and region labels, enables Maria to achieve effective improvements in multimodal dialogue response generation. Thus, we take it as the base model[1] to revisit the significance of visual objects and visual concepts.

Formally, given a multimodal dialogue dataset $\mathcal{S} = \{(I_i, C_i, R_i)\}_{i=1}^{N}$, where $i = 1, \ldots, N$, and I_i, C_i, R_i refer to an image, a dialogue context and a response respectively. Maria first gets a set of visual embeddings by an image encoder f_v:

$$\mathbf{E}_v = f_v(I_i), \qquad (1)$$

and then additional texts other than context like visual concepts are encoded the same way as each word in the context C_i:

$$\mathbf{E}_{od}, \mathbf{E}_t = h_t(x_{od}), h_t(x_{ctx}), \qquad (2)$$

where x_{od} and x_{ctx} represent the tokens in visual concepts and context sentences, and $x_{od} \in \mathcal{O}$ where $\mathcal{O}$ is the visual concept set extracted by an object detection module. $h_t(\cdot)$ is either a word embedding mapping matrix or an encoder. Finally, these embeddings are fed into the decoder to generate a response R_i:

$$R_i = g(concat(\mathbf{E}_v, \mathbf{E}_{od}, \mathbf{E}_t)), \qquad (3)$$

where $concat(\cdot)$ is concatenating operation and $g(\cdot)$ is a generator. The models are optimized in a Maximum Likelihood Estimate (MLE) manner to adapt multimodal representations to the dialogue response generation task.

3 Experimental Setup

To investigate how visual modality impacts response generation, we conduct experiments on representative datasets and evaluate results by widely used metrics. Implementation details can be found in the Appendix A.

3.1 Dataset

Current multimodal dialogue datasets are constructed from crowd-sourcing [12, 34,42], text-only dialogue datasets [24,27], movies&TVs [33,38], or crawling from social media [29,53]. However, datasets collected from movies and TV shows, such as OpenViDial [33], contain fewer vision-related responses [29], making them less suitable for our study. Therefore, we use three representative multimodal dialogue datasets collected in the other three ways, i.e., constructing conversations by crowd-sourcing, retrieving images for text conversations, and collecting conversations from social networks:

---
[1] Following the license at https://github.com/jokieleung/Maria.

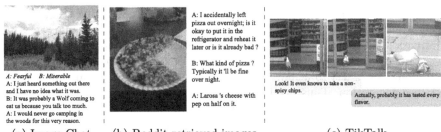

(a) Image-Chat   (b) Reddit-retrieved images   (c) TikTalk

Fig. 2. Examples of three multimodal dialogue datasets created in different ways. Cases in (a) and (c) are from [42] and [29] respectively, while (b) is sampled from our reconstruction of [27] of which retrieved images have not been public.

- **Image-Chat** [42] is a multimodal dialogue collection of 186K human-human conversations with grounded images, which is constructed using crowdworkers who are asked to converse based on an image in an engaging way. An example in Fig. 2a shows the role A plays as a fearful person worrying about a sound, which does not appear in the image but makes sense if the two people are in the woods. To preserve the essence of the multi-turn dialogue, we remove those data that only contain one utterance.

- **Reddit** [27] comprises 1 million high-quality text-only dialogues extracted from the Reddit Conversation Corpus [16]. For the multimodal dialogue task, [27] associated each session with a relevant image using a retriever. Images from the Open Images dataset [23] are retrieved, and the image I_i with the highest relevance score is assigned to each dialogue, as shown in Fig. 2b (*e.g.*, a pizza photo). Note that the context and response are taken as the text query in the training set while only using the context in validation and test set [27].

- **TikTalk** [29] is a Chinese multimodal dialogue dataset crawled from social media Douyin[2] (Chinese version of TikTok), containing 367,670 dialogues and 38,703 videos. As Fig. 2c shows, the dialogues come from the multi-level comments of each video, which are natural human-human conversations aroused by the video. To fit our setting, keyframes are fed into the model.

3.2 Evaluation Metric

We employ three types of widely used automatic metrics to assess the model performances: 1) **BLEU** [37] to measure the n-grams similarity degree of the ground-truth and generated ones; 2) **Rouge-L** [28] as a complement of the relevance to BLEU; and 3) **Distinct** [25] to consider the ratio of unique n-grams in the total to evaluate the diversity of generated responses. Metrics are calculated by MS COCO caption evaluation [10] for English and an NLG evaluation script[3] for Chinese. All the scores reported are on average.

---

[2] https://www.douyin.com/.
[3] https://github.com/Maluuba/nlg-eval.

4 RQ1: Do Region Features Matter?

In this section, we propose a probing method to verify the impact of visual region features on Maria's generation results, which is our first research question.

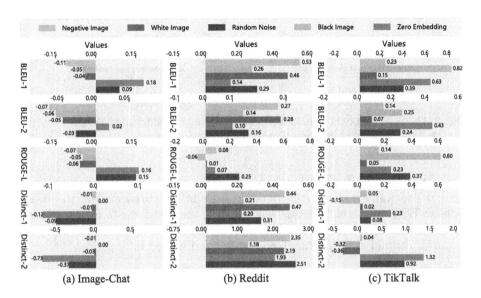

Fig. 3. The performance gap of Maria minus that of Maria with image region features replaced on three datasets. Best viewed in color.

4.1 Methodology: Probing Visual Regions

Despite the presence of images in multimodal dialogue data, it remains uncertain whether images provide complementary contributions to the textual information of visual concepts and dialogue context. Consequently, our study endeavors to explore this relationship by designing several methods of replacing visual region inputs to observe whether such replacements result in performance drops.

Negative Visual Input Replacement. We change the input image I_i with a randomly sampled irrelevant image I_i^{neg}. Accordingly, model generates new responses in Eq. 3 using $\mathbf{E}_v^{neg}$ rather than $\mathbf{E}_v$.

Black or White Visual Input Replacement. To explore the impact of a non-meaningful image on the model's preferences in the generation, we replace the input image I_i with a black or white image:

$$\mathbf{E}_v^{white}, \mathbf{E}_v^{black} = f_v(I_{white}), f_v(I_{black}) \qquad (4)$$

where I_{white} and I_{black} are a white or black image respectively, thus $\mathbf{E}_v$ in Eq. 3 is substituted by $\mathbf{E}_v^{white}$ and $\mathbf{E}_v^{black}$.

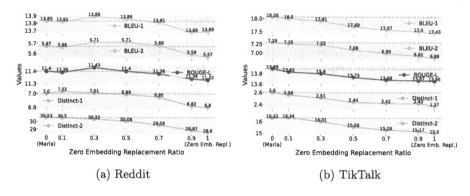

(a) Reddit (b) TikTalk

Fig. 4. The line charts of evaluation results of Maria on Reddit and TikTalk when the visual inputs are set to zero embeddings with different ratios.

Zero or Noise Visual Embedding Replacement. As non-zero features from white or black images may still contain a little information, we zero the visual input to $\mathbf{E}_v^{zero}$ in the embedding level. Besides, we follow [49] to inject different random noise $\mathbf{E}_v^{noise}$ to replace the visual input. In this way, the visual inputs no longer contain any structured, consistent, and unified information.

4.2 Result: It Matters on Reddit and TikTalk

The evaluation results of generated responses with different input image region features replaced are shown in Fig. 3. We subtracted the performance of the replaced image region features from that of Maria, resulting in positive values indicating a performance drop. These observations suggest that Maria indeed pays attention to image region features during response generation.

As Fig. 3 shows, negative image, zero embedding, and random noise replacements induce expected performance drops on both Reddit and TikTalk datasets, implying that Maria places attention on image region features during response generation. However, it is contrary to Image-Chat. For example, Maria's performance on Image-Chat improved when using randomly selected negative images instead of region features, indicating the latter may be unnecessary.

This means dataset characteristics may influence the model performance. For Image-Chat, the dataset is constructed with a focus on human engagement [42], potentially resulting in the divergence of image-based conversation and a decreased emphasis on visual information. Accounting for these factors, we conducted the remaining experiments using data from Reddit and TikTalk.

Substituting visual input with black or white images reduces response quality on Reddit and TikTalk, but increases diversity on TikTalk. This suggests that black or white images may contain unperceived implicit information while replacing visual inputs with zero embedding or random noise images affects the model's dependence on vision. Thus, replacing region features with zero embedding or random noise is a more suitable choice for future investigations.

4.3 Further Results On Model Variant

We further modified the Maria to verify model dependency on the inputs, where the region labels are removed during the training and inference phase, and we denote this model as Maria-V. To assess the model's performance with a complete image rather than noise as the visual input, we adopt NegVRep and calculate the median perplexity (PPL)[4] of ground truth responses in the test set on three datasets, and the results are illustrated in Fig. 5. Likewise, NegVRep can raise the median PPL of the ground truth, making the model more confused to generate it. From another perspective, Maria-V-NegVRep has more PPL growth against Maria-V than that of Maria-NegVRep against Maria on all three datasets. It may indicate that Maria-V focuses more on visual input than Maria does. As region labels contain useful information about the dialogue, the model may conduct shortcut learning during training, confusing the presence of irrelevant visual input.

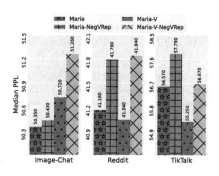

Fig. 5. Results on model variants, where Maria-V is the model trained without region labels and NegVRep is negative visual input replacement.

4.4 Further Analysis: TikTalk is More Stable

We examine how the model's performance varies as we incrementally replace visual inputs with zero embedding. By adjusting the replacement ratio from 0 to 1, we evaluate the model's generation capability on Reddit and TikTalk, depicted in Fig. 4. The line charts for TikTalk demonstrate a consistent decline in performance as the ratio increases. Relevance scores (BLEU and Rouge-L) and diversity score (Distinct) follow the same pattern, indicating a significant shift in the model's generation capability. This demonstrates that the decline in model performance is attributed to the gradual change in visual input rather than random factors. Thus, visual inputs not only contribute to generated responses but also enhance model robustness on the TikTalk dataset.

Replacing 30% of region features with zero embeddings on the Reddit dataset initially improves the BLEU and ROUGE-L lines, but they subsequently decline when 70% of the samples are replaced. Notably, the maximum values of all three metrics surpass Maria's performance, suggesting that visual inputs may not significantly contribute to generated responses on Reddit.

TikTalk gathers comment-reply pairs with visual context, while Reddit contains text-only utterances. Images are retrieved in the Reddit dataset but may have limited relevance or new information. This explains why image region features have more consistent contributions in TikTalk compared to Reddit.

---

[4] To avoid the effect of outliers on the mean, we adopt median values.

5 RQ2: Do Visual Concepts Matter?

In this section, we examine the contribution of visual concepts in Maria, focusing on determining the most influential types of visual concepts.

Table 1. Evaluation results of the generated responses on Reddit and TikTalk. Underline scores are higher than those of same-group methods. B, W, O, and R here denote black, white, zero, and noise visual input replacement respectively.

Methods \ Metrics	Reddit					TikTalk				
	B-1	B-2	R-L	D-1	D-2	B-1	B-2	R-L	D-1	D-2
Maria	13.83	5.67	11.40	7.00	30.53	18.06	7.29	13.89	2.60	16.32
Maria w/o all concepts	13.70	5.61	11.36	6.88	30.05	17.21	6.80	13.77	2.35	15.44
B - w/o additional ones	13.45	5.46	11.45	6.58	28.63	17.30	6.89	13.01	2.69	15.81
B - w/o context ones	13.57	5.53	11.46	6.79	29.30	17.32	7.08	13.29	2.75	16.59
W - w/o additional ones	13.12	5.25	11.44	6.09	26.39	17.26	6.90	13.82	2.43	16.09
W - w/o context ones	13.37	5.40	11.39	6.52	28.29	17.91	7.22	13.87	2.59	16.85
O - w/o additional ones	13.70	5.56	11.37	6.61	27.97	16.16	6.09	13.12	2.08	12.84
O - w/o context ones	13.68	5.58	11.33	6.79	28.59	17.29	6.77	13.60	2.35	14.85
R - w/o additional ones	13.57	5.49	11.18	6.74	28.58	16.67	6.40	13.18	2.22	13.71
R - w/o context ones	13.56	5.52	11.16	6.68	28.00	17.54	6.96	13.42	2.50	15.30

5.1 Methodology: Probing Visual Concepts

To assess the influence of additional textual inputs, we remove visual concept words. We further divide the concept set $\mathcal{O}$ into two subsets based on their presence or absence in the context:

$$\mathcal{O} = \mathcal{O}_{ctx} \cup \mathcal{O}_{addi} \tag{5}$$

where $\mathcal{O}_{ctx}$ and $\mathcal{O}_{addi}$ refer to visual concepts present or absent in the dialogue context. For example, in Fig. 1, "pizza" appears in context and leads to the retrieval of an image featuring both pizza and cola. Therefore, both "pizza" and "cola" are visual concepts for the model, with "cola" being an additional concept that co-occurred with "pizza" being a context concept.

Thus, we propose three probing methods: 1) removing all concepts, 2) removing additional concepts, and 3) removing context-related concepts. By analyzing Maria's performance drops, we can assess the importance of each concept type.

5.2 Results: Concepts Matter, Additional Ones Contribute More

Table 1 presents the results on Reddit and TikTalk. Removing all visual concepts consistently lowers Maria's performance on Image-Chat, Reddit, and TikTalk. The largest reduction occurs in TikTalk, while Image-Chat shows the smallest.

The varying impact of removing all visual concepts on TikTalk and Reddit datasets can be attributed to their distinct creation procedures. In TikTalk, images sampled from videos provide extra pertinent information, whereas Reddit dialogues lack visual context, resulting in potentially irrelevant images.

Table 1 indicates that removing additional visual concepts $\mathcal{O}addi$ results in larger performance drops than removing context ones $\mathcal{O}ctx$ for most settings on Reddit and all settings on TikTalk. This is intuitive since the visual concepts not presented in the dialogue context, as we called "additional", can offer the model extra-textual information from the corresponding images, which aids Maria in generating responses by copying labels as new knowledge during generation.

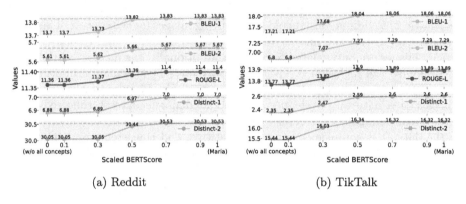

Fig. 6. Evaluation results of Maria on Reddit and TikTalk by removing visual concepts with varying BERTScore values. The BERTScores are scaled by min-max normalization to amplify differences among visual concepts.

5.3 Further Analysis: Moderately Relevant Concepts Matter Most

Dividing visual concepts based on their occurrence in the dialogue context can be imprecise, as some concepts may still be relevant. To enhance this, we propose a fine-grained approach that uses BERTScore [51] to differentiate the semantic relationship between a visual concept and the dialogue context. We vary the threshold BERTScore below which visual concepts are included and observe the resulting impact on model performance, which is illustrated in Fig. 6.

The steepest parts of the lines correlate to the most useful part of textual inputs for the model. The steepest ranges on Reddit and TikTalk datasets are from 0.3 to 0.5 and from 0.1 to 0.5, respectively. This means that only concepts with this medium relevance to dialogue context bring complementary benefits to dialogue models, while higher and lower relevance concepts are redundant or irrelevant. Table 2 presents examples of high and medium visual concepts. Although "carrot" and "sandwich" have more than 0.7 BERTScore with the context and the BERTScore of "cup" is between 0.3 to 0.5, the "cup" can bring new information to the dialogue, which may help response generation.

6 RQ3: Which Dataset is More Sensitive to Vision?

The above experiments confirm the importance of visual and textual representations from images in multimodal dialogue systems on three datasets. However, result variations are observed due to differences in the datasets, with TikTalk being the most sensitive to visual modality and Image-Chat being the least.

This could result from their construction and target scenarios: 1) Image-Chat is manually annotated, but the predefined engaging style may lead to less focus on the image. While this scenario is closest to face-to-face visual chatting, the dialogues may not necessarily contain the objects or scenes in the image. Figure 3 reveals Maria trained on Image-Chat pays little attention to visual features, indicating a mismatch between visual knowledge and human-annotated responses. 2) Reddit is constructed by retrieving images for the context. This may cause a gap, as training images are related to both context and response while testing images are only related to context. Additionally, the dataset is originally text-only, so the inclusion of visual modality adds little semantic information and may be redundant. Results in Table 1 and Fig. 4a support this view. 3) TikTalk is a naturally occurring multimodal dialogue dataset collected from social media comments and replies to videos. The visual and textual information in the dataset is complementary, not repetitive, and consistently enhances the training of multimodal dialogue models, as shown in Fig. 3c and Table 1.

Thus, careful selection of suitable multimodal dialogue datasets is crucial for research in this field, as different datasets cover a wide range of scenarios.

Table 2. Examples of visual concepts and their dialogue contexts from Reddit, categorized as high (BERTScore > 0.7) and medium (0.3 < BERTScore < 0.5).

#	Context	Medium	High
1	"Homemade aged cheeses and cured meat platter.", "Recipe for crackers please."	'cup'	'carrot' 'sandwich'
2	"On the plus side, somebody will come before the pizzas do." "All over your face."	'fork'	'pizza'
3	"What's your favorite pizza toppings?" "Chicken, pepperoni, and BBQ sauce (on the side)."	'bottle'	'pizza'

7 Conclusion and Future Work

In this paper, we examine the role of visual and textual inputs from images in multimodal dialogue through an analysis of the effects of their removal from the Maria model. Our experimental results demonstrate that visual region features enhance the model performance on Reddit and TikTalk datasets, while their impact on the Image-Chat dataset is negligible. Notably, the TikTalk dataset exhibits more consistent and substantial performance improvements when utilizing visual features. Additionally, our findings indicate that visually represented concepts that are moderately relevant to the context contribute new information and offer more assistance compared to redundant or similar ones.

In the future, we aim to explore the editing of different regions to enhance and explain multimodal dialogue models. Despite recent advancements in large language models and large vision-language models, which have greatly improved image-based question answering, the optimal approach for selectively incorporating visual modality in chitchat scenarios, particularly when visual modality is abundant or redundant, remains unclear. For example, in video-based chitchat dialogues like TikTalk, determining which frames to use or whether to include visual information in generating a response is still unknown. This is an area of research that greatly interests us for our upcoming studies.

Acknowledgments. This work is supported by the National Natural Science Foundation of China (No. 62276268). We acknowledge the anonymous reviewers for their helpful comments.

A Appendix

Implement Details. We implement the retrieval model in Maria by CLIP [39] to assign the most relevant image to a dialogue. All the candidates from Open Images and text queries are encoded into embeddings, using the FAISS [22] library for indices building and retrieval. We adopt DETR [8] to extract the object regions and their corresponding labels. The response generation module is a 12-layer transformer with hidden size 768. We use UniLM [14] to initialize the parameters on Image-Chat and Reddit, and $BERT_{base-Chinese}$ [13] for TikTalk. The model is trained using maximum sequence length 120, batch size 128, and learning rate 3e-5 over 20 epochs on 8 Nvidia A100 40G GPUs, until validation loss convergence. During inference, we used beam size of 5, a length penalty of 0.3, and a repetition penalty of 3. Scores are averaged for each test sample.

References

1. Adiwardana, D., et al.: Towards a human-like open-domain chatbot. arXiv preprint arXiv:2001.09977 (2020)
2. Alamri, H., et al.: Audio visual scene-aware dialog. In: Proceedings of the IEEE/CVF Conference on Computer Vision and Pattern Recognition, pp. 7558–7567 (2019)
3. Alayrac, J., et al.: Flamingo: a visual language model for few-shot learning. In: Advances in Neural Information Processing Systems 35: Annual Conference on Neural Information Processing Systems 2022, NeurIPS 2022 (2022)
4. Antol, S., Agrawal, A., Lu, J., Mitchell, M., Batra, D., et al.: VQA: visual question answering. In: International Conference on Computer Vision (ICCV) (2015)

5. Bao, S., He, H., Wang, F., Wu, H., Wang, H.: Plato: pre-trained dialogue generation model with discrete latent variable. arXiv preprint arXiv:1910.07931 (2019)
6. Brown, T.B., et al.: Language models are few-shot learners. In: Larochelle, H., Ranzato, M., Hadsell, R., Balcan, M., Lin, H. (eds.) Advances in Neural Information Processing Systems 33: Annual Conference on Neural Information Processing Systems 2020, NeurIPS 2020, December 6-12, 2020 (2020)
7. Cao, Q., Chen, X., Song, R., Wang, X., Huang, X., Ren, Y.: See or guess: counterfactually regularized image captioning. arXiv preprint arXiv:2408.16809 (2024)
8. Carion, N., et al.: End-to-end object detection with transformers. In: European Conference on Computer Vision, pp. 213–229. Springer (2020)
9. Chen, X., et al.: Multimodal dialog systems with dual knowledge-enhanced generative pretrained language model. ACM Trans. Inf. Syst. **42**(2), 53:1-53:25 (2024)
10. Chen, X., Fang, H., Lin, T.Y., Vedantam, R., et al.: Microsoft coco captions: data collection and evaluation server. arXiv preprint arXiv:1504.00325 (2015)
11. Chowdhery, A., et al.: Palm: scaling language modeling with pathways. J. Mach. Learn. Res. **24**, 240:1–240:113 (2023). http://jmlr.org/papers/v24/22-1144.html
12. Das, A., et al.: Visual dialog. In: Proceedings of the IEEE Conference on Computer Vision and Pattern Recognition, pp. 326–335 (2017)
13. Devlin, J., et al.: BERT: pre-training of deep bidirectional transformers for language understanding. arXiv preprint arXiv:1810.04805 (2018)
14. Dong, L., et al.: Unified language model pre-training for natural language understanding and generation. Adv. Neural Inf. Process. Syst. (2019)
15. Du, Z., Zeng, A., Dong, Y., Tang, J.: Understanding emergent abilities of language models from the loss perspective. arXiv preprint arXiv:2403.15796 (2024)
16. Dziri, N., Kamalloo, E., et al.: Augmenting neural response generation with context-aware topical attention. arXiv preprint arXiv:1811.01063 (2018)
17. Elliott, D.: Adversarial evaluation of multimodal machine translation. In: Proceedings of the 2018 Conference on Empirical Methods in Natural Language Processing, pp. 2974–2978 (2018)
18. Hu, R., Singh, A.: Unit: multimodal multitask learning with a unified transformer. In: Proceedings of the International Conference on Computer Vision (2021)
19. Huang, L., et al.: A survey on hallucination in large language models: principles, taxonomy, challenges, and open questions. arXiv preprint arXiv:2311.05232 (2023)
20. Huang, Y., et al.: What makes multi-modal learning better than single (provably). Adv. Neural. Inf. Process. Syst. **34**, 10944–10956 (2021)
21. Ji, Z., et al.: Survey of hallucination in natural language generation. ACM Comput. Surv. **55**(12), 248:1–248:38 (2023). https://doi.org/10.1145/3571730
22. Johnson, J., Douze, M., Jégou, H.: Billion-scale similarity search with GPUs. IEEE Trans. Big Data **7**(3), 535–547 (2019)
23. Kuznetsova, A., et al.: The open images dataset V4. Int. J. Comput. Vis., pp. 1956–1981 (2020)
24. Lee, N., et al.: Constructing multi-modal dialogue dataset by replacing text with semantically relevant images. Association for Computational Linguistics (2021)
25. Li, J., Galley, M., Brockett, C., Gao, J., Dolan, B.: A diversity-promoting objective function for neural conversation models. arXiv preprint arXiv:1510.03055 (2015)
26. Li, J., et al.: Blip-2: bootstrapping language-image pre-training with frozen image encoders and large language models. arXiv preprint arXiv:2301.12597 (2023)
27. Liang, Z., Hu, H., Xu, C., Tao, C., Geng, X., et al.: Maria: a visual experience powered conversational agent. arXiv preprint arXiv:2105.13073 (2021)
28. Lin, C.Y.: Rouge: a package for automatic evaluation of summaries. In: Text Summarization Branches Out, pp. 74–81 (2004)

29. Lin, H., Ruan, L., Xia, W., Liu, P., Wen, J., et al.: Tiktalk: a multi-modal dialogue dataset for real-world chitchat (2023)
30. Liu, G., Wang, S., Yu, J., Yin, J.: A survey on multimodal dialogue systems: recent advances and new frontiers. In: 2022 5th International Conference on Advanced Electronic Materials, Computers and Software Engineering (AEMCSE) (2022)
31. Liu, H., et al.: A survey on hallucination in large vision-language models. arXiv preprint arXiv:2402.00253 (2024)
32. Liu, H., et al.: Visual instruction tuning. In: Advances in Neural Information Processing Systems 36: Annual Conference on Neural Information Processing Systems 2023, NeurIPS 2023, New Orleans, LA, USA, December 10 - 16, 2023 (2023)
33. Meng, Y., Wang, S., Han, Q., et al.: OpenViDial: a large-scale, open-domain dialogue dataset with visual contexts. arXiv preprint arXiv:2012.15015 (2020)
34. Mostafazadeh, N., et al.: Image-grounded conversations: multimodal context for natural question and response generation. In: Proceedings of the Eighth International Joint Conference on Natural Language Processing (2017)
35. Nie, L., Wang, W., et al.: Multimodal dialog system: generating responses via adaptive decoders. In: Proceedings of the 27th ACM International Conference on Multimedia, pp. 1098–1106 (2019)
36. OpenAI: GPT-4 technical report (2023)
37. Papineni, K., Roukos, S., Ward, T., Zhu, W.J.: Bleu: a method for automatic evaluation of machine translation. In: Proceedings of the 40th annual meeting of the Association for Computational Linguistics, pp. 311–318 (2002)
38. Poria, S., et al.: MELD: a multimodal multi-party dataset for emotion recognition in conversations, pp. 527–536. Association for Computational Linguistics (2019)
39. Radford, A., et al.: Learning transferable visual models from natural language supervision. In: International Conference on Machine Learning. PMLR (2021)
40. Schaeffer, R., et al.: Are emergent abilities of large language models a mirage? In: NeurIPS 2023 (2023)
41. Serban, I., et al.: Building end-to-end dialogue systems using generative hierarchical neural network models. In: Proceedings of the AAAI Conference (2016)
42. Shuster, K., Humeau, S., Bordes, A., Weston, J.: Image chat: engaging grounded conversations. arXiv preprint arXiv:1811.00945 (2018)
43. Shuster, K., Smith, E.M., Ju, D., Weston, J.: Multi-modal open-domain dialogue. arXiv preprint arXiv:2010.01082 (2020)
44. Specia, L., Frank, S., Sima'An, K., et al.: A shared task on multimodal machine translation and crosslingual image description. In: Proceedings of the First Conference on Machine Translation: Volume 2, pp. 543–553 (2016)
45. Sundar, A., Heck, L.: Multimodal conversational AI: a survey of datasets and approaches. arXiv preprint arXiv:2205.06907 (2022)
46. Touvron, H., et al.: Llama: open and efficient foundation language models. arXiv preprint arXiv:2302.13971 (2023)
47. Wang, S., Meng, Y., et al.: Modeling text-visual mutual dependency for multimodal dialog generation. arXiv preprint arXiv:2105.14445 (2021)
48. Wei, J., et al.: Emergent abilities of large language models. Trans. Mach. Learn. Res. (2022)
49. Wu, Z., et al.: Good for misconceived reasons: an empirical revisiting on the need for visual context in multimodal machine translation. arXiv:2105.14462 (2021)
50. Yang, Z., et al.: Open domain dialogue generation with latent images. In: Proceedings of the AAAI Conference on Artificial Intelligence, pp. 14239–14247 (2021)
51. Zhang, T., Kishore, V., Wu, F., Weinberger, K.Q., Artzi, Y.: BERTScore: evaluating text generation with BERT. arXiv preprint arXiv:1904.09675 (2019)

52. Zhang, Y., et al.: DialoGPT: large-scale generative pre-training for conversational response generation. arXiv preprint arXiv:1911.00536 (2019)
53. Zheng, Y., Chen, G., Liu, X., Lin, K.: MMChat: multi-modal chat dataset on social media. arXiv preprint arXiv:2108.07154 (2021)
54. Zhu, D., et al.: MiniGPT-4: enhancing vision-language understanding with advanced large language models. In: ICLR 2024 (2024)

Vision-Language Pretraining for Variable-Shot Image Classification

Sotirios Papadopoulos[1,2(✉)], Konstantinos Ioannidis[1], Stefanos Vrochidis[1], Ioannis Kompatsiaris[1], and Ioannis Patras[2]

[1] Information Technologies Institute, Thessaloniki, Greece
{papasoti,kioannid,stefanos,ikom}@iti.gr
[2] Queen Mary University of London, London, UK
{s.papadopoulos,i.patras}@qmul.ac.uk

Abstract. Contrastively pretrained vision-language models (VLMs) such as CLIP have shown impressive zero-shot classification performance without any classification-specific training. They create a common embedding space by contrastively pretraining an image and a text encoder to align positive image-text pairs and repel negative pairs. Then zero-shot classification of an image can be performed by measuring the cosine similarities between the image embedding and embeddings of texts that describe the classes. However, relevant works do not address the scenario in which few image examples for some (not all) classes are available. In this novel task which we term variable-shot (v-shot) classification, these models fail due to the embedding space modality gap, i.e. the fact that image-to-image similarities are higher than image-to-text ones. To this end, we propose to enable v-shot capabilities in pre-trained VLMs with minimal training complexity by re-projecting embeddings of frozen pre-trained image encoders using a shallow network, RectNet, which we train both with the standard CLIP contrastive loss function, as well as a novel modality alignment loss function specifically constructed to bridge the modality gap. Finally, we introduce three v-shot classification benchmarks, on which the proposed architecture achieves 32.22%, 29.58% and 45.15% increases in top-1 classification accuracy respectively.

Keywords: variable-shot classification · v-shot classification · vision-language pretraining · computer vision

1 Introduction

The rise of joint image-text contrastive pretraining has transformed how computers understand the link between visual content and language. Models like CLIP [21] utilize vast web-collected image-text datasets to achieve impressive zero-shot classification abilities, excelling in tasks like image captioning and retrieval without fine-tuning, thus demonstrating remarkable versatility. Additionally, with minor adaptations, they can perform tasks such as open-vocabulary object detection [13,15,16,30] and open-vocabulary image semantic segmentation [11,27,28]. In all tasks, zero-shot classification (image, object or pixel classification) is achieved by comparing visual embeddings with embeddings of text descriptions of the classes.

However, there can be classification scenarios where purely textual descriptions are insufficient for certain fine-grained classes, whether due to the ambiguity of language or due to the need for excessively long descriptions to adequately define the class. In such cases, combining textual descriptions for some classes with a few image examples for others (few-shot) could be beneficial. We propose this new paradigm, termed "v-shot classification", in order to benchmark classification generalizability. In this setting, query image embeddings are compared to the text embeddings for some classes and the image embeddings for others.

Nevertheless, a recently studied phenomenon called the embedding space *modality gap* [12,23,25,26] causes image and text embeddings to occupy distinct clusters within the latent space, leading to higher intra-modality similarities (image-to-image) compared to inter-modality similarities (image-text). Consequently, when using existing contrastively pretrained image/text encoders, test images are overwhelmingly classified into image-based classes, ignoring text-based ones, making v-shot classification infeasible.

To enable v-shot classification in pretrained CLIP models while retaining their strong zero-shot capabilities, we propose to better align image embeddings with the text embeddings of zero-shot classes with minimal training complexity and inference overhead using RectNet, a 2-layer embedding rectification MLP network placed after a frozen pretrained CLIP image encoder. Additionally, we propose a Contrastive Modality Alignment loss ($\mathcal{L}_{CMA}$) to bridge the modality gap by pushing same-modality embeddings apart and bringing positive image-text pairs closer, complementing the baseline CLIP contrastive loss. Finally, we introduce a set of v-shot classification datasets and experimentally demonstrate that RectNet achieves state-of-the-art v-shot classification performance. Although $\mathcal{L}_{CMA}$ helps bridge the gap, the best v-shot classification results are achieved by RectNet trained only on the baseline CLIP contrastive loss. In summary,

- We propose "v-shot classification", a novel classification setting with classes of mixed zero and few image availability, as well as set of benchmarks based on common image classification datasets.
- To enable v-shot classification capabilities in pre-trained VLM architectures, we introduce a shallow neural network, RectNet, that projects pretrained image embeddings to the text embedding space with trivial computational overhead and significant training efficiency.
- We propose a novel contrastive loss function to reduce the modality gap that causes CLIP to fail in v-shot scenarios.
- We show that RectNet achieves the best top-1 and top-5 v-shot classification accuracy across all datasets.

2 Related Work

Our proposed strategy is a VLM pretraining scheme to minimize the modality gap and enable v-shot classification in VLMs. Therefore, the related work in this section is categorized into methods that do contrastive VLM pretraining and methods that study the modality gap phenomenon.

Contrastive VLM Pretraining. Contrastive language-image pre-training has gained popularity since the introduction of CLIP [21] and ALIGN [7], which applied contrastive learning [2,8] to large-scale image-text datasets. Both models excel in zero-shot transfer tasks, including classification and retrieval. Following their success, there has been a surge of research in contrastive VLP. DeCLIP [10], SLIP [17] and MaskCLIP [5] reinforce the baseline CLIP training with per-modality self supervision to achieve superior performance, with SimSiam [3], SimCLR [2] or masked image modeling [1] for self-supervision of the visual encoder, and Masked Language modeling [4] for the text encoder. CyCLIP [6] added geometric consistency regularizations to the CLIP objective. SigLIP [29] substituted the softmax loss with a sigmoid loss, allowing for efficient training with much larger batch sizes, while simultaneously enhancing classification performance at smaller batch sizes. However none of the above methods evaluate their performance on the v-shot classification task. We show that due to the modality gap in CLIP-based methods, v-shot classification cannot be accomplished.

CLIP Modality Gap. Methods like CLIP bring positive image-text pairs close in the embedding space while pushing negative pairs apart. However, a noticeable "modality gap" exists between the embedding distribution means of the two modalities (see fig. 2a), a phenomenon highlighted by several works [12,19,22,23,25,26]. Liang et al. [12] first introduced this phenomenon, attributing it to factors like the narrow cone effect, different parameter initializations for two different encoders (image/text), the nature of the contrastive loss itself, non-linear activation functions, the value of the softmax temperature, and mismatched image-text pairs in pretraining datasets . Interestingly, they demonstrate that increasing the modality gap can sometimes improve specific downstream tasks like zero-shot classification. Udandarao [26] found that the spread of image embeddings influences the gap, showing that as the mean distance between image embeddings decreases, the text embeddings yielding the lowest contrastive loss are further away from their respective image embeddings. Shi et al. [25] demonstrated that the contrastive loss comprises conflicting uniformity and alignment terms, trapping CLIP in a local minimum with a pronounced modality gap. Oh et al. [19] reiterated that high uniformity and alignment are essential for quality embeddings. They blend images with corresponding text to create hard negative samples, incentivizing further alignment and minimizing the gap. Schrodi et al. [23] showed that only a few embedding dimensions form the gap, which arises from the information imbalance between modalities: images provide detailed appearance information, while texts offer sparse descriptions. More detailed captions reduce the gap. In contrast to all these works, we study the effect of the gap on the v-shot classification task.

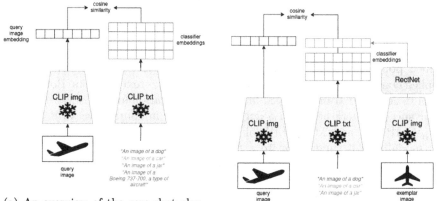

(a) An overview of the zero-shot classification scheme with a baseline VLM [21].

(b) An overview of the VSC scheme with our RectNet.

Fig. 1. The v-shot classification scheme (VSC). In fig. 1a a query image embedding is compared to text embeddings of class descriptions (standard zero-shot classification). In fig. 1b a query image embedding is compared to text embeddings of class descriptions for some classes and to RectNet image exemplar embeddings for other classes (VSC).

3 Proposed Method

In this section, we first review the contrastive vision-language pretraining strategy. We then formulate the novel v-shot classification scenario and lastly, we introduce RectNet, our image embedding re-projection network, as well as our novel gap-closing loss. Refer to Fig. 1 for the outline of the proposed classification setting with RectNet.

3.1 VLM Contrastive Pretraining

Vision-language pretraining involves a large dataset $\mathcal{D}_{pre} = (I_i, t_i)|i = 1, .., r$ of r image-caption pairs, where I_i is an image and t_i a corresponding text prompt. The goal is to create a common representation space where cosine similarities reflect semantic similarity between these different modalities. This space can be realised using two encoding functions:

$$\begin{aligned} \mathbf{e}_i^v &= E^v(I_i), \\ \mathbf{e}_i^t &= E^t(t_i), \end{aligned} \quad (1)$$

where E^t, E^v are usually represented by deep neural networks.

The common multi-modal embedding space is achieved by jointly training the two modality-specific encoders on a contrastive loss function [21]:

$$\mathcal{L}_{\text{CLIP}} = -\frac{1}{2B} \left(\sum_i \log \frac{exp(\mathbf{e}_i^v \cdot \mathbf{e}_i^t/\tau)}{\sum_j exp(\mathbf{e}_i^v \cdot \mathbf{e}_j^t/\tau)} + \sum_i \log \frac{exp(\mathbf{e}_i^v \cdot \mathbf{e}_i^t/\tau)}{\sum_j exp(\mathbf{e}_i^t \cdot \mathbf{e}_j^v/\tau)} \right), \quad (2)$$

where τ is the softmax temperature parameter, B the batch size and $\cdot$ the dot product.

3.2 V-Shot Classification (VSC)

In the zero-shot image classification task, each class c is represented by the embedding $\mathbf{e}_c^t$ of some text t_c that describe this class, e.g. *"an image of a {class}"*. An image I_i can be then classified by assigning to it a class c_i using the following:

$$c_i = \arg\max_c (\mathbf{e}_i^v \cdot \mathbf{e}_c^t), \quad (3)$$

where the embeddings come from encoders that are pretrained on the contrastive objective.

In cases where a subset of the class set contains classes with no sufficient textual description available, the classification task is modified in order to incorporate k example images I_c^i, $i = 1, .., k$ for such object classes (image-described classes) instead of text descriptions. In that case, which we term Variable-shot Classification (VSC), the classification task can be summarised in the following formulation:

$$c_i = \arg\max_c (\mathbf{e}_i^v \cdot \mathbf{e}_c),$$
$$\text{where } \mathbf{e}_c = \begin{cases} E^t(t_c), \text{ for text-described classes} \\ \frac{1}{k}\sum_i^k E^v(I_c^i), \text{ for image-described classes} \end{cases} \quad (4)$$

Visualization of CLIP embeddings from image-text pairs (Fig. 2a) of CC3M [24] shows distinct clusters for each modality. This separation makes vanilla CLIP embeddings unsuitable for VSC, as input image embeddings $\mathbf{e}_i^v$ are more similar to other image embeddings than to any text embedding, regardless of the true class. Consequently, images are assigned to classes represented by example images, neglecting textually described classes.

3.3 Embedding Rectification Network (RectNet)

Retraining VLM encoders for VSC requires substantial GPU resources and energy, and might degrade zero-shot classification due to limited training data. Instead, we propose RectNet, a shallow neural network that aligns modalities by re-projecting pretrained CLIP image embeddings into the text embedding space. This allows us to keep the VLM image encoder frozen and only optimize RectNet weights, ensuring efficiency.

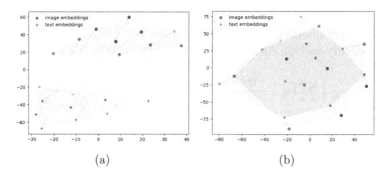

Fig. 2. t-SNE visualisation of embeddings of 10 random image-text pairs from the CC3m [24] dataset using: i) vanilla CLIP (Fig. 2a), and ii) our RectNet+$\mathcal{L}_{CMA}$ (Fig. 2b). Different shapes indicate different modalities and colours represent correspondences. Clearly, RectNet is able to eliminate the two-cluster structure while keeping positive image-text pairs close together.

Given the strength of pretrained CLIP encoders in vision-language reasoning, RectNet is placed directly after the CLIP image encoder:

$$\mathbf{e}_c = \begin{cases} E^t(t_c), \text{ for text-described classes} \\ \frac{1}{k}\sum_i^k R(E^v(I_c)), \text{ for image-described classes,} \end{cases} \quad (5)$$

where $R(\cdot)$ denotes RectNet. Both CLIP encoders remain frozen during training, optimizing only the RectNet parameters. A visualization of RectNet is shown in Fig. 3.

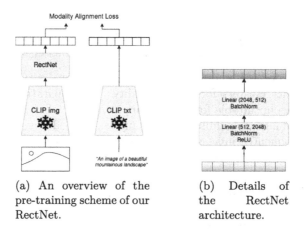

(a) An overview of the pre-training scheme of our RectNet.

(b) Details of the RectNet architecture.

Fig. 3. RectNet architecture and training. Figure 3a shows where RectNet is placed in the CLIP framework as well as its training scheme. Figure 3b shows a detailed overview of RectNet architecture and parameters.

3.4 Contrastive Modality Alignment Loss

For successful v-shot classification, positive image-text pairs must exhibit higher similarities than both negative image-text pairs and negative image-image pairs (when comparing a query image with an image exemplar). To achieve this within a CLIP-style classification framework, we propose to introduce same-modality negatives in the contrastive objective, creating the following contrastive modality alignment (CMA) loss function:

$$\mathcal{L}_{\text{CMA}} = -\frac{1}{2B} \left(\sum_i \log \frac{exp(\mathbf{e}_i^v \cdot \mathbf{e}_i^t / \tau)}{\sum_{j \neq i} exp(\mathbf{e}_i^v \cdot \mathbf{e}_j^v / \tau)} + \sum_i \log \frac{exp(\mathbf{e}_i^v \cdot \mathbf{e}_i^t / \tau)}{\sum_{j \neq i} exp(\mathbf{e}_i^t \cdot \mathbf{e}_j^t / \tau)} \right). \tag{6}$$

In order to maintain the powerful zero-shot classification capability and at the same time align the two modality-specific clusters, the final loss function is updated as:

$$\mathcal{L}_f = \alpha \mathcal{L}_{\text{CLIP}} + \beta \mathcal{L}_{\text{CMA}}, \tag{7}$$

where α and β are weighting coefficients, both set to 1.

4 Experiments

4.1 Experimental Setup

RectNet Pre-Training. RectNet comprises two blocks, each with a linear and a batch normalization layer, a hidden space of 2048, and a ReLU activation function after the first layer (see Fig. 3b). We train RectNet on top of a frozen pre-trained CLIP ViTB/32 image encoder on the proposed loss function (eq. 7) using Adam optimizer with a batch size of 2048 on the CC3M dataset [24], with 2,334,451 image-text pairs available. The weights are initialized with the identity matrix to leverage the robust baseline CLIP weights. Training for 30 epochs took less than 7 h on a single NVIDIA RTX 3090 GPU.

VSC Datatasets. To empirically validate the VSC capability of RectNet, we conduct extensive evaluations on custom classification tasks combining a general-purpose image classification dataset for the zero-shot classes with a dataset containing specialised (fine-grained) classes as the few-shot classes, as text descriptions of fine-grained classes are generally inadequate. We create three such classification benchmarks, by combining CIFAR100 [9] (containing 100 general-purpose classes, such as "shark", "telephone", "lobster", etc.) and each of the following fine-grained image classification datasets:

- FGVC-Aircraft [14], which is a dataset that contains photos of aircraft, categorized into aircraft models.
- Oxford-IIIT Pet [20], which contains photos of pets, classified into dog/cat breeds, and
- Flowers102 [18], which contains flower species commonly occuring in the United Kingdom.

4.2 Performance on VSC Datasets

For all three tasks, we compare the baseline CLIP classification performance with RectNet pretrained on the baseline CLIP loss (RectNet-ft) and RectNet trained on the proposed loss (eq. 7). In the rest of the section, zero-shot and v-shot performance on the VSC benchmarks is analyzed and finally, the relationship between the modality gap and VSC performance is discussed.

Zero-Shot Classification. We first evaluate on the baseline zero-shot classification, where all classes are text-described (eq. 3). These evaluations can be seen in Tables 2, 1 and 3 in the blocks titled "zero-shot". While it is evident that the baseline CLIP model still has the top zero-shot classification accuracy figures, this metric is only presented for the sake of completeness and is not the main focus of this work.

V-Shot Classification. VSC is evaluated on pretrained VLM encoders by comparing a query image embedding with text description embeddings for regular classes (CIFAR) and image exemplar embeddings for fine-grained classes (FGV-CAircraft, OxfordPets and Flowers102). Due to non-alignment between the two modalities in the baseline CLIP embedding space, as described in the previous section, the baseline pretrained CLIP encoders present a 0.0% accuracy in zero-shot classes, as it can be observed in Tables 2, 1 and 3. Note that we create different variants of classifier image embeddings, each time using a different number (k) of image exemplars for each few-shot class (see eq. 5). It can be also observed that, in general, accuracy rises in proportion to k.

By incorporating RectNet into the CLIP VSC scheme, VSC becomes feasible even with just one image exemplar per class (see rows labeled 'v-shot: k=1' in Tables 2, 1 and 3). Although for OxfordPets the zero-shot performance of vanilla CLIP cannot be surpassed with v-shot (Table 1), probably due to the high availability of pet photographs in the CLIP pretraining dataset, with just five example images for the fine-grained classes of the other two datasets (Tables 2, 3) the zero-shot performance can be surpassed. Using all available image exemplars from each fine-grained dataset's training set, all v-shot experiments have shown that RectNet outperforms the baseline. The following accuracy gains can be observed for each dataset compared to CLIP's zero-shot performance:

- CIFAR+FGVC-Aircraft: a 6.55% and 6.92% increases can be observed for overall top-1 and top-5 accuracy respectively, while a 13.1% and 13.61% increase is observed in top-1 and top-5 accuracy for the fine-grained classes.
- CIFAR+Flowers102: a 7.74% and 5.39% increases can be observed for overall top-1 and top-5 accuracy respectively, while a 13.92% and 9.41% increase is observed in top-1 and top-5 accuracy for the fine-grained classes.
- CIFAR+Oxford-IIIT Pet: Both the overall and fine-grained classes' top-1 and top-5 accuracies are slightly reduced, probably due to the strength of the text embeddings of classes that relate to pets, as they are prominent in the pretraining dataset. However, accuracy on CIFAR classes is still higher than the baseline.

Table 1. Zero-shot and V-shot classification performance on CIFAR100 + OxfordPets. "zs" and "fs" means zero-shot and few-shot respectively. Best results are highlighted with **bold**, while second best are underlined.

	Model	Overall		zs classes		fs classes	
		top1	top5	top1	top5	top1	top5
zero-shot	CLIP	70.42	90.91	64.31	88.01	86.95	98.76
	RectNet-ft	62.55	88.40	60.52	85.41	68.03	96.48
	RectNet-$\mathcal{L}_f$	39.97	57.69	52.95	74.66	4.88	11.82
v-shot: k=1	CLIP	9.09	18.92	0	0	<u>33.64</u>	<u>70.06</u>
	RectNet-ft	**55.45**	<u>82.21</u>	**65.17**	**89.22**	29.19	63.26
	RectNet-$\mathcal{L}_f$	<u>53.59</u>	**82.28**	<u>60.10</u>	<u>84.98</u>	**35.98**	**75.00**
v-shot: k=5	CLIP	16.69	25.40	0	0	**61.79**	**94.04**
	RectNet-ft	**61.45**	**88.90**	**65.17**	**89.22**	51.41	88.02
	RectNet-$\mathcal{L}_f$	<u>58.66</u>	<u>84.11</u>	<u>58.39</u>	<u>80.51</u>	<u>59.39</u>	<u>93.85</u>
v-shot: k=10	CLIP	18.68	25.96	0	0	**69.17**	<u>96.11</u>
	RectNet-ft	**63.56**	**90.05**	**65.17**	**89.22**	59.21	92.29
	RectNet-$\mathcal{L}_f$	<u>60.88</u>	<u>84.83</u>	<u>58.54</u>	<u>80.36</u>	<u>67.23</u>	**96.93**
v-shot: k=20	CLIP	20.15	26.31	0	0	**74.61**	<u>97.44</u>
	RectNet-ft	**64.92**	**90.64**	**65.17**	**89.22**	64.26	94.47
	RectNet-$\mathcal{L}_f$	<u>62.59</u>	<u>85.03</u>	<u>58.83</u>	<u>80.22</u>	<u>72.74</u>	**98.04**
v-shot: k=50	CLIP	20.92	26.39	0	0	**77.45**	<u>97.71</u>
	RectNet-ft	**65.79**	**90.84**	**65.17**	**89.22**	67.45	95.21
	RectNet-$\mathcal{L}_f$	<u>63.97</u>	<u>85.43</u>	<u>59.38</u>	<u>80.53</u>	<u>76.37</u>	**98.65**
v-shot: k=100	CLIP	21.54	26.49	0	0	**79.76**	<u>98.09</u>
	RectNet-ft	**66.69**	**90.93**	**65.17**	**89.22**	70.80	95.55
	RectNet-$\mathcal{L}_f$	<u>64.77</u>	<u>85.24</u>	<u>59.31</u>	<u>80.18</u>	<u>79.54</u>	**98.91**

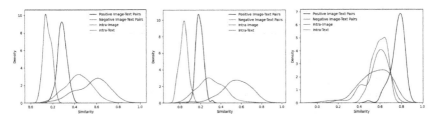

(a) KDE of intra and inter modality similarities for the baseline CLIP. (b) KDE of intra and inter modality similarities for RectNet-ft. (c) KDE of intra and inter modality similarities for RectNet-$\mathcal{L}_f$.

Fig. 4. Kernel density estimation (KDE) visualization of similarities between positive and negative image-text pairs, as well as intra-modal similarities. The held-out validation set of CC3m is used.

Table 2. Zero-shot and V-shot classification performance on CIFAR100 + FGVCAircraft. "zs" and "fs" means zero-shot and few-shot respectively. Best results are highlighted with **bold**, while second best are underlined.

	Model	Overall		zs classes		fs classes	
		top1	top5	top1	top5	top1	top5
zero-shot	CLIP	42.02	69.91	65.05	88.97	18.99	50.84
	RectNet-ft	38.99	64.93	61.38	86.20	16.62	43.66
	RectNet-$\mathcal{L}_f$	29.25	47.31	52.95	74.65	5.30	19.97
v-shot: k=1	CLIP	7.81	19.04	0	0	15.63	<u>38.10</u>
	RectNet-ft	**40.07**	**63.85**	65.15	**89.20**	<u>15.00</u>	**38.50**
	RectNet-$\mathcal{L}_f$	<u>38.74</u>	<u>62.16</u>	63.04	87.04	14.43	37.28
v-shot: k=5	CLIP	12.60	27.34	0	0	**25.20**	**54.68**
	RectNet-ft	**44.19**	**70.88**	65.14	**89.20**	<u>23.24</u>	<u>52.55</u>
	RectNet-$\mathcal{L}_f$	<u>42.57</u>	<u>68.57</u>	63.21	86.46	21.93	50.68
v-shot: k=10	CLIP	14.17	29.60	0	0	**28.34**	**59.20**
	RectNet-ft	**45.57**	**73.05**	65.14	**89.20**	<u>25.99</u>	<u>56.90</u>
	RectNet-$\mathcal{L}_f$	<u>43.96</u>	<u>70.71</u>	63.58	86.70	24.35	54.73
v-shot: k=20	CLIP	15.52	31.41	0	0	**31.05**	**62.81**
	RectNet-ft	**46.84**	**74.73**	65.15	**89.20**	<u>28.53</u>	<u>60.26</u>
	RectNet-$\mathcal{L}_f$	<u>45.17</u>	<u>72.29</u>	63.76	86.75	26.58	57.83
v-shot: k=33	CLIP	17.35	33.48	0	0	**34.70**	**66.96**
	RectNet-ft	**48.57**	**76.83**	65.15	**89.20**	<u>32.00</u>	<u>64.45</u>
	RectNet-$\mathcal{L}_f$	<u>47.40</u>	<u>74.56</u>	63.90	86.86	30.89	62.25

It must be noted that VSC with RectNet does not require any further training on classification datasets than its pre-training on the CLIP loss (eq. 2) or the proposed loss (eq. 7). Moreover, the exemplar embeddings can be computed in a pre-testing phase and saved, therefore during the testing v-shot classification stage, no overhead is induced to the baseline CLIP model.

Modality Gap Size and Performance. We evaluated the modality gap for each method as proposed in [12] using the held-out evaluation set of CC3M. The results, shown in Table 4, reveal an interesting trend: while we expected VSC performance to improve as the gap decreased, the opposite occurred. Although $\mathcal{L}_f$ reduces the modality gap and achieves better VSC accuracies than the baseline CLIP model, RectNet trained only on the regular CLIP loss function (eq. 2) yields the best overall results despite increasing the gap. This suggests a more complex relationship between the gap and VSC performance. As illustrated in Fig. 4, $\mathcal{L}_f$ (subfig. 4c) enhances positive image-text pair similarities over intra-image similarities. Conversely, RectNet-ft (subfig. 4b) does not make positive inter-modal similarities stronger than intra-image similarities. However,

Table 3. Zero-shot and V-shot classification performance on CIFAR100 + Flowers102. "zs" and "fs" means zero-shot and few-shot respectively. Best results are highlighted with **bold**, while second best are <u>underlined</u>.

	Model	Overall		zs classes		fs classes	
		top1	top5	top1	top5	top1	top5
zero-shot	CLIP	65.95	87.40	63.82	87.97	67.94	86.86
	RectNet-ft	57.65	83.83	60.18	84.68	55.29	83.04
	RectNet-$\mathcal{L}_f$	30.16	52.25	52.23	73.93	9.61	32.06
v-shot: k=1	CLIP	30.62	43.02	0	0	**59.51**	**83.25**
	RectNet-ft	**60.87**	**85.42**	**64.91**	**89.04**	57.11	<u>82.04</u>
	RectNet-$\mathcal{L}_f$	<u>58.65</u>	<u>81.91</u>	<u>59.02</u>	<u>82.72</u>	<u>58.30</u>	81.16
v-shot: k=5	CLIP	41.06	47.41	0	0	**83.55**	**96.28**
	RectNet-ft	**70.01**	**91.57**	**64.91**	**89.04**	74.76	<u>93.93</u>
	RectNet-$\mathcal{L}_f$	<u>68.37</u>	<u>86.76</u>	<u>58.83</u>	<u>81.35</u>	<u>77.25</u>	91.79
v-shot: k=10	CLIP	44.11	48.48	0	0	**89.71**	**98.43**
	RectNet-ft	**73.69**	**92.79**	**64.92**	**89.04**	81.86	<u>96.27</u>
	RectNet-$\mathcal{L}_f$	<u>71.77</u>	<u>87.64</u>	<u>59.46</u>	<u>81.21</u>	<u>83.24</u>	93.63

Table 4. Size of the modality gap (as per [12]), evaluated on CC3m. The lower the size, the more aligned the two modalities are.

Model	Modality gap
CLIP	0.8237
RectNet-ft	0.8540
RectNet-$\mathcal{L}_f$	0.1553

compared to the baseline CLIP (subfig. 4a), RectNet-ft brings the peaks of the probability density functions for inter-modal and intra-image similarities closer together, which may explain its VSC accuracy gains over the baseline CLIP.

4.3 Influence of Embedding Dimensions in VSC Performance

As noted in [23], only a handful of embedding dimensions form the modality gap. We evaluate VSC performance on the CIFAR100+Flowers102 dataset by gradually removing the most influential embedding dimensions (dimensions with the highest absolute difference between the mean image embedding and the mean text embedding, following [23]) of baseline CLIP image and text embeddings. The results are shown in Table 5. It can be seen that in v-shot settings, by removing the most influential dimension the performance in image-described classes drops significantly, while performance in text-described classes increase sharply (therefore enabling VSC capability), although not reaching the zero-shot performance levels of the full 512 dimensions. Then, with each removed

Table 5. V-shot classification performance on CIFAR100+Flowers102 with progressive removal of gap-forming dimensions of baseline CLIP embeddings.

Model	emb. dim	Overall		CIFAR100		Flowers102	
		top1	top5	top1	top5	top1	top5
zero-shot	512	**65.95**	**87.40**	**63.82**	**87.97**	**67.94**	**86.86**
	511	59.38	83.06	55.87	79.28	62.65	86.57
	510	56.94	77.87	48.82	70.96	64.51	84.31
	509	49.58	70.05	36.28	56.83	61.96	82.35
	508	49.88	70.44	36.91	57.23	61.96	82.75
	507	52.88	73.41	42.19	63.48	62.84	82.65
v-shot: k=1	512	30.62	43.02	0	0	**59.15**	**83.09**
	511	**48.85**	**74.49**	**42.45**	**67.49**	54.81	81.01
	510	35.74	56.75	15.41	30.76	54.67	80.95
	509	29.22	44.44	2.08	5.30	54.51	80.89
	508	29.52	45.11	2.57	6.58	54.63	81.00
	507	30.13	46.46	3.90	9.47	54.56	80.91
v-shot: k=5	512	43.35	49.82	0	0	**83.72**	**96.23**
	511	**61.93**	**82.85**	**46.71**	**70.41**	76.11	94.44
	510	49.18	65.85	20.44	35.21	75.95	94.38
	509	40.83	52.24	3.31	7.02	75.78	94.36
	508	41.26	53.05	4.12	8.69	75.85	94.36
	507	42.13	54.92	6.05	12.61	75.73	94.33
v-shot: k=10	512	46.50	50.86	0	0	**89.80**	**98.24**
	511	**66.79**	**84.68**	**47.97**	**71.38**	84.31	97.06
	510	54.13	67.85	22.03	36.49	84.02	97.06
	509	45.32	53.95	3.78	7.67	84.02	97.06
	508	45.77	54.77	4.69	9.36	84.02	97.06
	507	46.77	56.83	6.88	13.63	83.92	97.06

dimension the performance on all classes is diminishing, but gradually coming up again after removing 4 dimensions. Here, k is the number of image examples for the few-shot classes.

5 Conclusion

In conclusion, the proposed approach introduces variable shot (v-shot) image classification, a novel classification setting for vision-language models, along with a comprehensive set of v-shot evaluation benchmarks. Through our experiments, it was demonstrated that CLIP struggles with v-shot classification due to the modality gap between visual and textual representations. To tackle this issue, we proposed a novel contrastive modality alignment loss, which is specifically designed to reduce this gap. This loss function was used to train our image embedding re-projection network, RectNet. RectNet rectifies the pretrained CLIP image embeddings, retaining the strong zero-shot capabilities of pretrained CLIP models while significantly reducing the modality gap. This enables efficient adaptation of pretrained VLMs for v-shot classification. Our results showed a marked improvement in v-shot classification accuracy across all proposed datasets As next steps, we plan to extend our research to address the v-shot object detection problem. Additionally, we aim to further explore the structure of the modality gap and its impact on v-shot performance, which may provide deeper insights and lead to further advancements in vision-language model capabilities.

Acknowledgments.. This project has received funding from the European Defence Fund programme under grant agreement No 101103386. Views and opinions expressed are, however, those of the authors only and do not necessarily reflect those of the European Union or the European Commission. Neither the European Union nor the granting authority can be held responsible for them.

Disclosure of Interests. The authors have no competing interests to declare that are relevant to the content of this article.

References

1. Bao, H., Dong, L., Piao, S., Wei, F.: BeIT: BERT pre-training of image transformers. arXiv preprint arXiv:2106.08254 (2021)
2. Chen, T., Kornblith, S., Norouzi, M., Hinton, G.: A simple framework for contrastive learning of visual representations. In: International Conference on Machine Learning, pp. 1597–1607. PMLR (2020)
3. Chen, X., He, K.: Exploring simple Siamese representation learning. In: Proceedings of the IEEE/CVF Conference on Computer Vision and Pattern Recognition, pp. 15750–15758 (2021)
4. Devlin, J., Chang, M.W., Lee, K., Toutanova, K.: BERT: pre-training of deep bidirectional transformers for language understanding. arXiv preprint arXiv:1810.04805 (2018)
5. Dong, X., et al.: MaskCLIP: masked self-distillation advances contrastive language-image pretraining. In: Proceedings of the IEEE/CVF Conference on Computer Vision and Pattern Recognition, pp. 10995–11005 (2023)
6. Goel, S., Bansal, H., Bhatia, S., Rossi, R., Vinay, V., Grover, A.: CyCLIP: cyclic contrastive language-image pretraining. Adv. Neural. Inf. Process. Syst. **35**, 6704–6719 (2022)
7. Jia, C., et al.: Scaling up visual and vision-language representation learning with noisy text supervision. In: International Conference on Machine Learning, pp. 4904–4916. PMLR (2021)
8. Khosla, P., et al.: Supervised contrastive learning. Adv. Neural. Inf. Process. Syst. **33**, 18661–18673 (2020)
9. Krizhevsky, A., Hinton, G., et al.: Learning multiple layers of features from tiny images (2009)

10. Li, Y., et al.: Supervision exists everywhere: a data efficient contrastive language-image pre-training paradigm. arXiv preprint arXiv:2110.05208 (2021)
11. Liang, F., et al.: Open-vocabulary semantic segmentation with mask-adapted clip. In: Proceedings of the IEEE/CVF Conference on Computer Vision and Pattern Recognition, pp. 7061–7070 (2023)
12. Liang, V.W., Zhang, Y., Kwon, Y., Yeung, S., Zou, J.Y.: Mind the gap: understanding the modality gap in multi-modal contrastive representation learning. Adv. Neural. Inf. Process. Syst. **35**, 17612–17625 (2022)
13. Ma, C., Jiang, Y., Wen, X., Yuan, Z., Qi, X.: CoDet: co-occurrence guided region-word alignment for open-vocabulary object detection. Adv. Neural Inf. Process. Syst. **36** (2024)
14. Maji, S., Rahtu, E., Kannala, J., Blaschko, M., Vedaldi, A.: Fine-grained visual classification of aircraft. arXiv preprint arXiv:1306.5151 (2013)
15. Minderer, M., Gritsenko, A., Houlsby, N.: Scaling open-vocabulary object detection. Adv. Neural Inf. Process. Syst. **36** (2024)
16. Minderer, M., et al.: Simple open-vocabulary object detection. In: European Conference on Computer Vision, pp. 728–755. Springer (2022)
17. Mu, N., Kirillov, A., Wagner, D., Xie, S.: SLIP: self-supervision meets language-image pre-training. In: European Conference on Computer Vision, pp. 529–544. Springer (2022)
18. Nilsback, M.E., Zisserman, A.: Automated flower classification over a large number of classes. In: 2008 Sixth Indian Conference on Computer Vision, Graphics and Image Processing, pp. 722–729. IEEE (2008)
19. Oh, C., et al.: Geodesic multi-modal mixup for robust fine-tuning. Adv. Neural Inf. Process. Syst. **36** (2024)
20. Parkhi, O.M., Vedaldi, A., Zisserman, A., Jawahar, C.: Cats and dogs. In: 2012 IEEE Conference on Computer Vision and Pattern Recognition, pp. 3498–3505. IEEE (2012)
21. Radford, A., et al.: Learning transferable visual models from natural language supervision. In: International Conference on Machine Learning, pp. 8748–8763. PMLR (2021)
22. Ramasinghe, S., Shevchenko, V., Avraham, G., Thalaiyasingam, A.: Accept the modality gap: an exploration in the hyperbolic space. In: Proceedings of the IEEE/CVF Conference on Computer Vision and Pattern Recognition, pp. 27263–27272 (2024)
23. Schrodi, S., Hoffmann, D.T., Argus, M., Fischer, V., Brox, T.: Two effects, one trigger: on the modality gap, object bias, and information imbalance in contrastive vision-language representation learning. arXiv preprint arXiv:2404.07983 (2024)
24. Sharma, P., Ding, N., Goodman, S., Soricut, R.: Conceptual captions: a cleaned, hypernymed, image alt-text dataset for automatic image captioning. In: Proceedings of the 56th Annual Meeting of the Association for Computational Linguistics, Volume 1: Long Papers, pp. 2556–2565 (2018)
25. Shi, P., Welle, M.C., Björkman, M., Kragic, D.: Towards understanding the modality gap in clip. In: ICLR 2023 Workshop on Multimodal Representation Learning: Perks and Pitfalls (2023)
26. Udandarao, V.: Understanding and fixing the modality gap in vision-language models, Master's thesis, University of Cambridge (2022)
27. Xie, B., Cao, J., Xie, J., Khan, F.S., Pang, Y.: SED: a simple encoder-decoder for open-vocabulary semantic segmentation. In: Proceedings of the IEEE/CVF Conference on Computer Vision and Pattern Recognition, pp. 3426–3436 (2024)

28. Xu, M., Zhang, Z., Wei, F., Hu, H., Bai, X.: Side adapter network for open-vocabulary semantic segmentation. In: Proceedings of the IEEE/CVF Conference on Computer Vision and Pattern Recognition, pp. 2945–2954 (2023)
29. Zhai, X., Mustafa, B., Kolesnikov, A., Beyer, L.: Sigmoid loss for language image pre-training. In: Proceedings of the IEEE/CVF International Conference on Computer Vision, pp. 11975–11986 (2023)
30. Zhong, Y., et al.: RegionCLIP: region-based language-image pretraining. In: Proceedings of the IEEE/CVF Conference on Computer Vision and Pattern Recognition, pp. 16793–16803 (2022)

Visual Anomaly Detection on Topological Connectivity Under Improved YOLOv8

Yu Li[1,2] and Zhenping Xie[1,2(✉)]

[1] School of Artificial Intelligence and Computer Science, Jiangnan University, Wuxi 214122, China
6223115002@stu.jiangnan.edu.cn, xiezp@jiangnan.edu.cn
[2] Jiangsu Key University Laboratory of Software and Media Technology under Human-Computer Cooperation, Wuxi 214122, China

Abstract. With the advancement of AI-Generated Content (AIGC), generated unrealistic images increasingly distort visual perception, posing serious challenges to distinguish between reality and fabrication. To alleviate this challenge, we propose a topological anomaly detection method. Firstly, topological connectivity anomaly phenomenon refers to regions that appear continuous on the plane but actually disconnected. Based on this phenomenon, we construct an innovative dataset containing plentiful scene images of both topological anomalous and topological normal objects and corresponding depth maps in Unity3D. Secondly, we construct an improved version of YOLOv8 integrated with depth estimation module, enabling more efficient in detecting breakpoints of pseudo-connected objects. Finally, our method is evaluated comprehensively in different experimental settings, achieving a final mean average precision(mAP) of 89.2% that is superior to the latest general YOLOv8 models. This research breaks through the ability of visual models to recognize situations that violate physical laws and provides a feasibility foundation for the novel field of image anomaly detection.

Keywords: impossible objects · optical illusion · topological visual cognition · monocular depth estimation · anomaly detection

1 Introduction

AI-Generated Content (AIGC) is any form of content that is created with the help of artificial intelligence technology [1], such as text, images, audio, or video. With the flourishing development of AIGC, AIGC Images (AIGIs), as the representative research [2,3], become more prevalent in areas such as culture, education, social media, etc. However, due to the hardware limitations and technical proficiency, the quality of AIGIs is inconsistent and various, struggling to align accurately with human perception. When people are exposed to unrealistic images, they may develop false perceptions, weaken information processing ability and even cause misunderstandings in knowledge memory. Thus, detecting anomalmous regions that defy human perception in AIGIs is urgently needed.

In recent years, topological perception theory have attracted increasing attention in exploring human visual cognition. This theory, based on the perspective of holistic priority, proves that topological properties are perceived first in the early visual perception of objects [4]. Topological properties are those that remain unchanged under any topological transformation, such as connectivity, the number of holes and the inside-outside relationship [5]. One of the most important basis of topological properties is topological connectivity, which provides a stable and reliable foundation for object perception. Moreover, topological connectivity analysis can help identify whether there are any anomalies in images and evaluate the logical consistency of the image.

In addition, as an optical illusion formed by the instantaneous conscious projection of the human visual system, impossible objects consist of locally consistent possible parts that join together to form an impossible loop, such as the Penrose triangle [6]. From certain perspectives, they appear as an entirety, i.e., connected. However, when the perspective changes, the hidden disconnected parts of the object become apparent, i.e., disconnected, as shown in Fig. 1. This pseudo-connected property reflects precisely the topological connectivity anomalies, disrupting sensory processing mechanisms and leads to confusion in object recognition. Therefore, based on impossible objects with topological connectivity anomaly, we can train models to recognize whether images exist topological anomalous regions, thereby comprehensively assessing the authenticity of AIGIs.

Fig. 1. Impossible objects from different angles.

In this paper, we construct a topological anomaly image dataset called 'Scenes of Topological Anomaly' (STA), and propose a topological connectivity anomaly detection model. To the best of our knowledge, what we do is the first attempt to explicitly introduce topological connectivity into images anomaly detection. Our contributions are summarized as follows:

1. We propose a novel dataset STA which includes scene images and depth maps containing both topological anomalous and topological normal objects. It provides the necessary training samples for models to deal with topological connectivity anomalies.
2. We construct an improved YOLOv8 model by using depth information to guide detection of anomaly regions, which significantly improves the accuracy of topological connectivity anomaly detection compared to the latest general YOLOv8 models, with a final mean average precision of 89.2%.
3. We prove that our method detects topological abnormal regions successfully in distinct image source, providing a feasible solution for evaluating AIGIs authenticity.

2 Related Work

2.1 Topological Visual Cognition

In the field of computer vision, many studies explore human visual cognition based on topological properties. [7] input binary images generated from hole filtering as a topological channel into a quaternion model, yielding more precise attention saliency maps. [8] calculated the topological map based on regional contrast and spatial features, solving the scaling problem caused by advanced feature extraction in salient object detection. [9] assigned corresponding topological complexity values based on the number of holes in ultrametric contour maps to images, achieving more accurate salient object detection.

2.2 Monocular Depth Estimation

Monocular Depth Estimation (MDE) is a task that attempts to estimate the distance from each pixel point in the scene to the camera by using only a single two-dimensional image. It has two distinct concepts. On the one hand, absolute depth estimation [10] directly calculates the actual physical distance from each pixel in an image to the camera. Many studies focus on innovative improvements to continuously improve the accuracy of absolute depth estimation, such as data integration strategies [11], multitask [12,13] and loss function [14,15]. On the other hand, relative depth estimation [16,17] focuses on the relationship of depth order between pixels in an image. It only needs to determine which objects are closer to or farther away from the camera than others. Researches on this aspect mainly center on construction of relative depth dataset and balance ordinal relationships between depths [18,19]. In the latest research, [20] developed a general model for MDE, which can generate high-quality depth information under any circumstances.

2.3 Impossible Objects

Most of the current research based on impossible objects remains in modeling and rendering [21–24] and visual experimentation [25–28], which usually create a possible representation for impossible objects followed by realizing the illusion of impossibility through deformation and other operations to continue further research. [29] first created an interactive modeling environment to generate impossible objects by applying special deformations to regular 3D geometric objects. [30] reduced the unnaturalness of rendering impossible objects by changing the connected prisms' order and introducing corner objects. [31] proposed a generic method for modeling impossible objects in 3D space, rendered their shadows to increase realism and investigated how the introduction of shadows affects the perception on impossible objects.

2.4 Discussion

Although most studies have used topological properties to guide the visual tasks, these researches just utilize topological properties as auxiliary features and do not effectively demonstrate the key role of topological properties in human perception. Due to the diversity of AIGIs, we extract the topological properties shared by things for experiments to determine whether AIGIs cause wrong perception to human eyes. However, since the rarity of impossible objects with topological anomaly property in the three-dimensional world, there is also no suitable model for this scenario. Therefore, inspired by the pseudo-connectivity and depth ambiguity reflected in impossible objects, we construct a topological anomaly scene dataset and establish a novel topological connectivity anomaly method by combination of impossible illusion and MDE. To our knowledge, there is currently no scientific research specifically targeted on this aspect.

3 Method

3.1 Modeling of Impossible Objects

Impossible objects' illusion depends on the overlapping parts of the objects' geometrical structure. Through the occlusion of these overlapping parts, we can get the illusion that one part of the object appears to be in front while at the same time behinds. Traditional methods for modeling impossible objects often create the illusion of impossibility by concealing the deformation of objects' geometry along the certain viewpoints [29,31]. This approach distorts the object's geometric part severely and even leads to its width approaching zero, inconsistencies in texture. In order to avoid above problems, we use 3D modeling technology to directly adjust the geometric structure of the object.

Taking the Penrose triangle [6] as an example, its original shape is assembled with three prisms, only two of which are connected to other prisms at one end. Firstly, we determine the number of prisms seen under the illusion viewpoint and construct the base form of the object. Secondly, we designate the two prisms, which connect to other prisms at one end, as P_{front} and P_{back} respectively to form the illusion. Thirdly, we build a prism P_{hide} with the same properties as other prisms, and place it in the position connected to P_{back} while maintaining the same orientation as P_{front} in space. Then, we adjust the whole object so that P_{front} can completely cover the P_{hide} in current viewpoint. Finally, we adjust the prism surfaces of P_{front} to make P_{back} naturally stack on top of the P_{front}, resulting in a pseudo-connected object. Complete process is illustrated in Fig. 2.

In this way, we construct twenty topological anomalous objects. To ensure that the model can distinguish different objects, we also construct twenty topological normal objects. All the topological anomalous objects and topological normal objects are shown in Fig. 3.

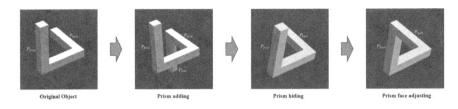

Fig. 2. Pipeline of modeling impossible objects in Unity3D.

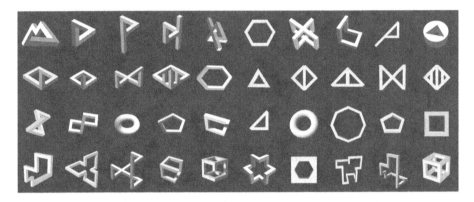

Fig. 3. All topological anomalous objects and topological normal objects built for experiment in Unity3D. The left five columns are topological anomalous objects and the right five columns are topological normal objects.

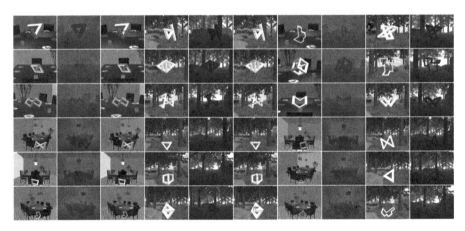

Fig. 4. STA Dataset. The left six columns are topological anomalous images with corresponding depth maps and annotations. The right four columns are topological normal images with corresponding depth maps.

3.2 Dataset Construction

In this section, we propose an innovative dataset named 'Scenes of Topological Anomaly' (STA) with 34,280 scene images and corresponding depth maps totally. There are 17,500 abnormal images containing topological anomalous objects and 16,780 normal images containing topological normal objects in STA, among which there are 7200 images rendering objects with different color. Part of STA is shown in Fig. 4.

To construct the dataset, firstly we model indoor as well as outdoor scenes in Unity3D and integrate topological anomalous objects or topological normal objects with every scene map. Viewpoints, lighting and other conditions are adjusted to make sure that our dataset can cover as many situations as possible. In particular, in preliminary experiments, we find that the color attributes of objects have an impact on the robustness of identifying topological anomalous regions. Therefore, we randomly select ten categories of objects with topological anomalous or normal property in both indoor and outdoor scenes, and render them using ten common color: red, orange, yellow, green, cyan, blue, purple, pink, grey and black, to enrich the training data. Specific modeling conditions for each object are shown in Table 1. Secondly, we generate corresponding depth maps alongside the collection of scene maps by utilizing the vertex shared to realize depth texture rendering point by point. Finally, for the abnormal image set, we use the Labelme annotation tool to label the breakpoint regions with rectangular boxes and assign its category as 'impossible'. No additional operations are needed for the normal image set because of its absence of topological anomalous regions.

Table 1. Modeling conditions of indoor scene and outdoor scene for each object.

Condition	Angle of viewpoint	Clipping Planes	Indoor scenes			Outdoor scenes		
			Field of View	Position of the object	Number of depth maps	Field of View	Position of the object	Number of depth maps
1	Left	0.3/1000	15	Middle	36	30/40	Middle	36
2	Left	0.3/1000	30	Middle	36	60	Middle	36
3	Left	0.3/1000	50	Left	36	60	Left	36
4	Left	0.3/1000	50	Corner	5	60	Corner	5
5	Right	0.3/1000	15	Middle	36	30/40	Middle	36
6	Right	0.3/1000	30	Middle	36	60	Middle	36
7	Right	0.3/1000	50	Right	36	60	Right	36
8	Right	0.3/1000	50	Corner	5	60	Corner	5
9	Center	0.3/1000	15	Middle	36	30/40	Middle	36
10	Center	0.3/1000	30	Middle	36	60	Middle	36
11	Center	0.3/1000	50	Middle	36	60	Bottom	36
12	Center	0.3/1000	50	Corner	4	60	Corner	5
	Total				338			339

3.3 Proposed Model

YOLOv8 [32], based on the anchor-free idea, has been used to achieve state-of-the-art results on end-to-end target detection. Considering the performance and flexibility of the model, we use YOLOv8 as our method's backbone. Additionally, due to the differences between indoor and outdoor scenes, most of MDE studies focus only on indoor or outdoor scenes to grasp accurate depth features. In order to generalize the MDE task for both scenes, inspired by [13], we use the main architecture of Vision Transformer as our depth estimation module (DEM)

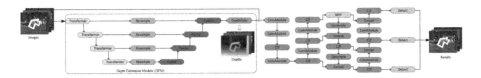

Fig. 5. The main framework of our model.

to provide finer-grained and more globally relevant predictions. The resulting model is summarized in Fig. 5, which utilizes entire images as input and achieves topological anomalous detection.

It is worth noting that our novelty lies in the combination of target detection and MDE, using depth information to guide detection of anomaly regions. Target detection is not new, nor is the MDE. But to the best of our knowledge, such a combination has not been proposed before, and in particular not for topological anomalous detection and analysis of image authenticity.

During training, we use the function weighted by BCE Loss as classification loss and CIOU (Complete-IOU) + DFL (Deep Feature Loss) as regression loss, as shown in Eqs. (1)(2)(3).

$$L_{BCE} = \frac{1}{N}\sum_i L_i = \frac{1}{N}\sum_i -[y_i \log(p_i) + (1-y_i)\log(1-p_i)] \qquad (1)$$

where N is the number of samples, and y_i and p_i respectively is the true label and predicted probability of the ith sample.

$$L_{CIou} = 1 - IoU + \frac{\rho^2(b, b^g)}{c^2} + \alpha v \qquad (2)$$

where IoU is the Intersection over Union between the predicted and ground truth boxes, b and b^g are the centroids of the two rectangular boxes. ρ is the Euclidean distance between the two boxes, c is the diagonal distance between their closed regions, v measures the concordance of their relative proportions, and α is the weighting factor.

$$DFL(S_i, S_{i+1}) = -\big((y_{i+1} - y)\log(S_i) + (y - y_i)\log(S_{i+1})\big) \qquad (3)$$

where y_i and y_{i+1} are the values of discrete points i and i+1, respectively. S_i and S_{i+1} are the predicted probability values of these two points, and y is the true value, from which more accurate bounding box prediction results are generated.

4 Experiments

4.1 Experimental Settings

For the dataset, we first randomly select 8 types of scenes from both the indoor and outdoor sets, with an equal ratio of abnormal images to normal images to validate our model's performance on untrained images. Then, we divide the remaining dataset into training set, validation set and test set according to 8:1:1. Unless otherwise specified, the YOLOv8n network is used as the detection framework, with the input image resolution of 640 × 640. The experiments are conducted on NVIDIA GeForce RTX 3060 Ti, with epoch set to 200, and the learning rate is adjusted using the SGD algorithm, starting with an initial learning rate of 0.01.

To evaluate the performance of the model, we use the following metrics:

- Accuracy: $\frac{TP+TN}{TP+TN+FP+FN}$
- Precision: $\frac{TP}{TP+FP}$
- Recall: $\frac{TP}{TP+FN}$
- F1-score: $\frac{2*Precision*Recall}{Precision+Recall}$
- Mean Average Precision (mAP): $\frac{1}{N}\sum_{i=1}^{N} AP_i$

where TP represents true positives, TN represents true negatives, FP represents false positives, FN represents false negatives, and AP stands for average precision for a single label, which is the area under the Precision-Recall curve.

4.2 Ablation Study

To further demonstrate the effectiveness of our proposed method, we conduct an ablation study about the DEM and investigate the impact of rendering objects with different color in STA on the experiments.

Contribution of DEM. As shown in Table 2, the overall performance of the model drops sharply without DEM. When the model integrates the DEM, it outperforms all other ablation models, proving the effectiveness of this module. Visual comparisons in Fig. 6 also support the idea that integration of DEM achieves more accurate detection. Although the model without DEM can recognize topological anomalous regions correctly, it makes wrong judgments more frequently.

Fig. 6. Impact of DEM on experimental results. Without DEM, the general model will incorrectly recognize or label normal and abnormal regions.

Table 2. Results of ablation experiments on DEM and color attributes.

DEM	Color	Accuracy	Precision	Recall	F1-Score	mAP0.5	mAP0.5:0.95
✗	✗	0.585	0.791	0.383	0.516	0.531	0.228
✔	✗	0.709	0.811	0.488	0.609	0.748	0.368
✗	✔	0.816	0.887	0.618	0.728	0.852	0.48
✔	✔	**0.827**	**0.924**	**0.63**	**0.749**	**0.892**	**0.513**

Table 3. Experimental results under different color settings.

Dataset	Accuracy	Precision	Recall	F1-Score	mAP0.5	mAP0.5:0.95
Grayscale	0.754	0.828	0.54	0.654	0.78	0.385
Single-Color	0.709	0.811	0.488	0.609	0.748	0.368
Colorful	**0.847**	**0.902**	**0.667**	**0.767**	**0.889**	**0.483**

Contribution of Color Attributes. In preliminary experiments without the inclusion of color attributes, we compare experimental results and find that even with data augmentation, the model is still unable to detect properly images with different color attributes given to the objects. Considering that the initial dataset only rendering objects with a single material, we not only just train with grayscale images to ignore the effect of color, but also render objects with different color. The results in Table 3 demonstrate that models trained on images featuring objects in varying colors achieve higher detection accuracy compared to other models, which is consistent with the findings in Table 2.

4.3 Qualitative Comparisons

Based on the results, we provide a visual demonstration of our method in Fig. 7. As shown in these examples, our model can produce accurate detection results in different situations, including low contrast (e.g., row 2), center bias (e.g., rows 2, 4, 5), multiple topological anomalous regions (e.g., row 5), large-scale salient objects (e.g., rows 1,2), and small-scale objects (e.g., rows 1,4). In addition, we test images from AIGIs to verify the substantial potential of topological connectivity in the assessment of images authenticity. Figure 8 shows the detection in these cases. The results demonstrate that our model can effectively handle with generated images with different backgrounds and lighting renderings, guide humans to discover the abnormal areas and provide a necessary basis for evaluating the authenticity of AIGIs.

4.4 Quantitative Comparisons

We conduct same experiments of topological anomaly region detection under different versions of YOLOv8. Input size of images here is uniformly adjusted to 160 × 160, and average inference time is added as a performance metric.

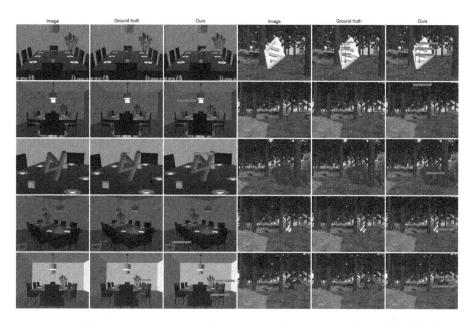

Fig. 7. Topological anomaly detection results in the test set of STA.

Fig. 8. Results of topological anomaly detection in AIGIs.

According to Table 4, experimental results prove that with the assistance of DEM, all the versions of YOLOv8 perform better than usual. With the deepening of the network layers, the models' accuracy and other metrics almost continuously improve with increasing inference time. Under the condition that the resolution of the input image size is limited, although YOLOv8n has the simplest structure and all metrics are at the bottom, it still achieves an accuracy approaching 60%. On the other hand, YOLOv8x can extract more accurate image features, and excel in mAP metrics with the slowest inference speed due to its deepest layers. Moreover, we can see that YOLOv8m(+DEM) may be better suited for the topological anomaly detection task, because it outperforms the other hierarchies in several metrics, and performs almost as well as the best-performing YOLOv8x in $mAP^{0.5}$.

Table 4. Comparison of performance in different YOLOv8 version.

Model	Accuracy	Precision	Recall	F1-Score	mAP0.5	mAP0.5:0.95	Average inference time(millisecond)
YOLOv8n	0.572	0.895	0.385	0.538	0.578	0.266	2.3
YOLOv8n(+DEM)	0.599	0.898	0.403	0.556	0.619	0.283	3.2
YOLOv8s	0.641	0.887	0.431	0.58	0.653	0.312	3.3
YOLOv8s(+DEM)	0.656	0.912	0.443	0.596	0.702	0.34	3.8
YOLOv8m	0.637	0.93	0.431	0.589	0.673	0.342	3.4
YOLOv8m(+DEM)	**0.686**	**0.946**	**0.469**	**0.627**	0.722	0.367	4.0
YOLOv8l	0.668	0.924	0.454	0.609	0.73	0.377	4.1
YOLOv8l(+DEM)	0.667	0.93	0.453	0.609	0.728	0.385	4.6
YOLOv8x	0.662	0.927	0.449	0.605	**0.737**	0.394	4.7
YOLOv8x(+DEM)	0.662	0.931	0.449	0.606	0.736	**0.397**	5.0

5 Conclusion

In this paper, based on the topological connectivity properties, we construct an innovative dataset STA by modeling impossible objects and an improved YOLOv8 with DEM to achieve end-to-end anomalous region detection. Experimental results show that our model can extract visual features of different scenes and detect topological abnormal regions successfully with a significant improvement compared to the latest general YOLOv8 models, which relieves the negative influence of unrealistic images on human perception. Although this research provides a foundation for anomaly detection of AIGIs, we still need to keep on exploring to make anomaly detection more robust and promoting the development of models in assessing as well as optimizing the authenticity of AIGC.

Disclosure of Interests. The authors have no competing interests to declare that are relevant to the content of this article.

References

1. Wu, J., Gan, W., Chen, Z., Wan, S. and Lin, H.: AI-generated content (AIGC): a survey. arXiv preprint arXiv:2304.06632 (2023)
2. Peng, F., et al.: AIGC image quality assessment via image-prompt correspondence. In: Proceedings of the IEEE/CVF Conference on Computer Vision and Pattern Recognition Workshops, vol. 6 (2024)
3. Zhang, Z., Li, C., Sun, W., Liu, X., Min, X., Zhai, G.: A perceptual quality assessment exploration for AIGC images. In: 2023 IEEE International Conference on Multimedia and Expo Workshops (ICMEW), pp. 440–445. IEEE (2023). https://doi.org/10.1109/ICMEW59549.2023.00082
4. Chen, L.: Topological structure in visual perception. Science **218**(4573), 699–700 (1982). http://dx.doi.org/10.1126/science.7134969
5. Chen, L.: The topological approach to perceptual organization. Vis. Cogn. **12**(4), 553–637 (2005). http://dx.doi.org/10.1080/13506280444000256
6. Penrose, L.S., Penrose, R.: Impossible objects: a special type of visual illusion. Br. J. Psychol. (1958). http://dx.doi.org/10.1111/j.2044-8295.1958.tb00634.x

7. Fang, Y., Gu, X., Wang, Y.: Attention selection model using weight adjusted topological properties and quantification evaluating criterion. In: The 2011 International Joint Conference on Neural Networks, pp. 328–335. IEEE (2011). http://dx.doi.org/10.1109/IJCNN.2011.6033239
8. Zhou, L., Gu, X.: Embedding topological features into convolutional neural network salient object detection. Neural Netw. **121**, 308–318 (2020). http://dx.doi.org/10.1016/j.neunet.2019.09.009
9. Peng, P., Yang, K.F., Luo, F.Y., Li, Y.J.: Saliency detection inspired by topological perception theory. Int. J. Comput. Vis. **129**(8), 2352–2374 (2021). http://dx.doi.org/10.1007/s11263-021-01478-4
10. Eigen, D., Puhrsch, C., Fergus, R.: Depth map prediction from a single image using a multi-scale deep network. Adv. Neural Inf. Process. Syst. **27** (2014). https://doi.org/10.48550/arXiv.1406.2283
11. Ranftl, R., Lasinger, K., Hafner, D., Schindler, K., Koltun, V.: Towards robust monocular depth estimation: mixing datasets for zero-shot cross-dataset transfer. IEEE Trans. Pattern Anal. Mach. Intell. **44**(3), 1623–1637 (2020). https://doi.org/10.1109/TPAMI.2020.3019967
12. Zhang, Z., Cui, Z., Xu, C., Yan, Y., Sebe, N., Yang, J.: Pattern-affinitive propagation across depth, surface normal and semantic segmentation. In: Proceedings of the IEEE/CVF Conference on Computer Vision and Pattern Recognition, pp. 4106–4115 (2019). https://doi.org/10.1109/CVPR.2019.00423
13. Ranftl, R., Bochkovskiy, A., Koltun, V.: Vision transformers for dense prediction. In: Proceedings of the IEEE/CVF International Conference on Computer Vision, pp. 12179–12188 (2021). https://doi.org/10.48550/arXiv.2103.13413
14. Laina, I., Rupprecht, C., Belagiannis, V., Tombari, F., Navab, N.: Deeper depth prediction with fully convolutional residual networks. In: 2016 Fourth International Conference on 3D Vision (3DV), pp. 239–248. IEEE (2016). https://doi.org/10.1109/3DV.2016.32
15. Patil, V., Sakaridis, C., Liniger, A., Van Gool, L.: P3Depth: monocular depth estimation with a piecewise planarity prior. In: Proceedings of the IEEE/CVF Conference on Computer Vision and Pattern Recognition, pp. 1610–1621 (2022). https://doi.org/10.48550/arXiv.2204.02091
16. Zoran, D., Isola, P., Krishnan, D., Freeman, W.T.: Learning ordinal relationships for mid-level vision. In: Proceedings of the IEEE International Conference on Computer Vision, pp. 388–396 (2015). https://doi.org/10.1109/ICCV.2015.52
17. Birkl, R., Wofk, D., Müller, M.: MiDaS v3. 1-a model zoo for robust monocular relative depth estimation. arXiv preprint arXiv:2307.14460 (2023)
18. Chen, W., Fu, Z., Yang, D., Deng, J.: Single-image depth perception in the wild. Adv. Neural Inf. Process. Syst. **29** (2016). https://doi.org/10.48550/arXiv.1604.03901
19. Xian, K., et al.: Monocular relative depth perception with web stereo data supervision. In: Proceedings of the IEEE Conference on Computer Vision and Pattern Recognition, pp. 311–320 (2018). http://doi.org/10.1109/CVPR.2018.00040
20. Yang, L., Kang, B., Huang, Z., Xu, X., Feng, J., Zhao, H.: Depth anything: unleashing the power of large-scale unlabeled data. In: Proceedings of the IEEE/CVF Conference on Computer Vision and Pattern Recognition, pp. 10371–10381 (2024). https://doi.org/10.48550/arXiv.2401.10891
21. Sugihara, K.: Evolution of impossible objects. In: 9th International Conference on Fun with Algorithms (FUN 2018). Schloss Dagstuhl-Leibniz-Zentrum fuer Informatik (2018). https://dx.doi.org/10.4230/LIPIcs.FUN.2018.2

22. Sánchez-Reyes, J., Chacón, J.M.: How to make impossible objects possible: anamorphic deformation of textured NURBs. Comput. Aided Geom. Des. **78**, 101826 (2020). http://dx.doi.org/10.1016/j.cagd.2020.101826
23. Kanayama, H., Hidaka, S.: Impossible objects of your choice: designing any 3D objects from a 2D line drawing. In: 2022 Nicograph International (NicoInt), pp. 37–43. IEEE (2022). http://dx.doi.org/10.1109/NicoInt55861.2022.00015
24. Sugihara, K.: Computer-aided creation of impossible objects and impossible motions. In: Kyoto International Conference on Computational Geometry and Graph Theory, pp. 201–212. Springer (2007). http://dx.doi.org/10.1007/978-3-540-89550-3_22
25. Freud, E., Ganel, T., Avidan, G.: Representation of possible and impossible objects in the human visual cortex: evidence from fMRI adaptation. NeuroImage **64**, 685–692 (2013). https://doi.org/10.1016/j.neuroimage.2012.08.070
26. Schacter, D.L., Reiman, E., Uecker, A., Roister, M.R., Yun, L.S., Cooper, L.A.: Brain regions associated with retrieval of structurally coherent visual information. Nature **376**(6541), 587–590 (1995). http://dx.doi.org/10.1038/376587a0
27. Friedman, D., Cycowicz, Y.M.: Repetition priming of possible and impossible objects from ERP and behavioral perspectives. Psychophysiology **43**(6), 569–578 (2006). http://dx.doi.org/10.1111/j.1469-8986.2006.00466.x
28. Habeck, C., Hilton, H.J., Zarahn, E., Brown, T., Stern, Y.: An event-related fMRI study of the neural networks underlying repetition suppression and reaction time priming in implicit visual memory. Brain Res. **1075**(1), 133–141 (2006). http://dx.doi.org/10.1016/j.brainres.2005.11.102
29. Elber, G.: Modeling (seemingly) impossible models. Comput. Graph. **35**(3), 632–638 (2011). http://dx.doi.org/10.1016/j.cag.2011.03.015
30. Chiba, T., Moriya, T., Takahashi, T.: An extended modeling method of optical illusion objects in general rendering environments. In: 2018 International Workshop on Advanced Image Technology (IWAIT), pp. 1–4. IEEE (2018). https://doi.org/10.1109/IWAIT.2018.8369719
31. Taylor, B.A.: Modeling and Rendering Three-Dimensional Impossible Objects. Bangor University, United Kingdom (2020)
32. Sohan, M., Sai Ram, T., Reddy, R., Venkata, C.: A review on yolov8 and its advancements. In: International Conference on Data Intelligence and Cognitive Informatics, pp. 529–545. Springer (2024). https://doi.org/10.1007/978-981-99-7962-2_39

Wavelet Integrated Convolutional Neural Network for ECG Signal Denoising

Takamasa Terada and Masahiro Toyoura[✉]

University of Yamanashi, 4-3-11 Takeda, Kofu, Yamanashi 400-8511, Japan
mtoyoura@yamanashi.ac.jp

Abstract. Wearable electrocardiogram (ECG) measurement using dry electrodes has a problem with high-intensity noise distortion. Hence, a robust noise reduction method is required. However, overlapping frequency bands of ECG and noise make noise reduction difficult. Hence, it is necessary to provide a mechanism that changes the characteristics of the noise based on its intensity and type. This study proposes a convolutional neural network (CNN) model with an additional wavelet transform layer that extracts the specific frequency features in a clean ECG. Testing confirms that the proposed method effectively predicts accurate ECG behavior with reduced noise by accounting for all frequency domains. In an experiment, noisy signals in the signal-to-noise ratio (SNR) range of -10—10 are evaluated, demonstrating that the efficiency of the proposed method is higher when the SNR is small.

Keywords: Electrocardiogram · Wavelet transform · Denoising autoencoder · Convolutional neural network

1 Introduction

Studies on machine-learning-supported detection and recognition of electrocardiogram (ECG) abnormalities have been conducted to improve the early prognosis of heart disease and to make evaluation easier for clinicians. To train these neural networks (NNs), clean ECG data with little-to-no noise are required to improve identification accuracy. Hence, preprocessing noise is a vital step when preparing the model.

In recent years, wearable devices have been used to acquire ECG data [11]. Although these devices make data acquisition easy, ECG noise caused by patient activity becomes a problem. For such devices, non-invasive dry electrodes are often used for comfort and convenience; however, they are more susceptible to noise. Hence, data preprocessing is unavoidable. This study provides a novel noise reduction method that effectively reduces high-intensity ECG noise to acceptable levels.

Conventional methods, such as finite input response (FIR) filters and wavelet and thresholding techniques, have been proposed to remove various types of noise in advance, according to the method. Deep-learning autoencoder-based models also have been demonstrated to remove noise with high accuracy. However, these and other methods continue to have problems removing noise when multiple frequencies overlap. Notably, the accuracy of the algorithm varies depending on the preset parameters. It has also been reported that the accuracy of feature extraction decreases with noise intensity [8].

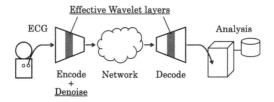

Fig. 1. The proposed convolutional neural network (CNN) model with an additional wavelet transform layer in which the features are separated into high and low components.

With overlapping ECG noise frequencies, fixed-parameter filters tend to lose original ECG information, which degrades accuracy. To improve this, a mechanism is needed that changes the parameter characteristics according to noise intensity and type. Thus, this study proposes *a convolutional neural network (CNN) model with an additional wavelet transform layer in which the features are separated into high and low components* as shown in Fig. 1. Then, the CNN is trained using parameters based on specific frequency bands. The combination of high feature extraction separation by the wavelet transform makes it possible for the network to learn changing filter behaviors based on a clean ECG, even when the frequency bands overlap. In our experiment, noisy signals in a signal-to-noise ratio (SNR) range of -10—10 are evaluated, demonstrating that the efficiency of the proposed method is higher when high-intensity noise is corrupted.

The remainder of this paper is organized as follows: Sect. 2 provides an overview of the related research. Section 3 explains the proposed CNN model with an integrated wavelet transform layer. Section 4 describes the experimental conditions, and Sect. 5 presents the results. Section 6 presents a discussion, and Sect. 7 concludes the paper.

2 Related Work

In ECG measurements, frequency noise is known to be a problem. The types of noise can be divided into low- and high-frequency, and the low-frequency type includes baseline wandering (BW), which is caused by breathing and other movements. High-frequency noise includes muscle artifacts (MAs) during electromyography (EMG) and electrode motions (EMs) from electrode misalignments. There can also be commercial powerline interference (PLI) at specific frequencies of 50 and 60 Hz and thermal noise from electronics, which is normally treated as additive white Gaussian noise. To remove noise, two types of denoising algorithms have been proposed: machine learning (e.g., deep learning) and others (e.g., filtering and wavelet transform). Notably, there is a trade-off between computational accuracy and computational complexity.

The computational cost of denoising without machine learning is low, and it is possible to remove noise in real time. FIR filters are typically used, and low-pass filters (LPFs) and high-pass filters (HPFs) are widely employed to remove high- and low-frequency noise, respectively.

Jenkal et al. [5] proposed a method that combines a wavelet transform with an adaptive dual-threshold filter. Focusing on the fact that different types of noise are mixed in different frequency bands, high-frequency signals are removed by wavelet transform in advance, and an adaptive dual-threshold filter is then applied. This combination of techniques successfully removes high-frequency noise, including that of EMG, PLI, and BW. Prashar et al. [13] proposed a method that applies a dual-tree complex wavelet transform to produce a noise-robust method with a threshold determined by eight different rules. Other wavelet transform and thresholding methods are used, but the parameters are determined in different ways (e.g., S-median [12] and improved thresholding [14]).

When predetermined parameters are used, the types of noise that can be removed are limited. In other cases, if the intensity of the unintended noise is large, it may not be sufficiently removed. If the noise is complex and has a large intensity, a more powerful method is needed.

In recent years, deep-learning-based methods have been proposed to automatically extract features. For example, Xiong et al. [18] proposed a method that combines wavelet transform and an NN. In this method, a deep-learning model is trained on data that has been denoised by wavelet transform and thresholding. By performing the wavelet transform in advance, the denoising effect is higher than when training the ECG with noise. On the other hand, Birok et al. [2] proposed a method that combines EMD and NN, with which ECG signals are first converted to clean and noisy IMFs. Then, the noisy IMFs are denoised by an NN. As a result, denoising performance is improved over the EMD and NN methods, separately.

CNNs are known to improve denoising performance better than NNs that use fully connected layers. CNNs require fewer parameters to be trained than do full-connected models. Thus, they can be used to build lightweight models. Yildirim et al. [19] proposed an autoencoder that extracts features with convolutional layers and pooling functions. Chiang et al. [3] showed that ECG denoising performance could be improved by giving the convolutional filter a stride of two. The model without pooling performs better.

Wang et al. showed that generative adversarial networks (GANs) could also improve denoising performance [17]. The GAN has a generator that removes noise from the ECG and a discriminator that judges whether the ECG is ground truth or fake. This adversarial learning method greatly improves denoising performance by using a generator and a discriminator with all fully connected layers. On the other hand, Singh et al. [16] proposed a GAN with fully convolutional layers, proving effective in denoising ECGs.

In recent years, edge-terminal processing (e.g., on wearable devices) has been considered. Hence, it is even more crucial to reduce data traffic by increasing the compression ratios. To do so, the encoder and decoder are placed at different locations, and the encoder sends compressed features to the decoder. For example, in the model discussed in [19], a 2,000-dimension vector was compressed to 62 dimensions by the encoders. Chiang et al. [3] successfully compressed a 1,024-dimension vector to 32 dimensions. On the other hand, the encoder of GAN in [16] expands 1,024-dimensional vectors to 8 × 1,024 dimensions, also using the encoder. Hence, a contraction path [15] is needed to concatenate the output of each encoder and the input of each decoder to improve performance. When using this model, data traffic increases in the contraction path and the output of the encoder, which then requires parameter reduction.

In another direction, Liu et al. [7] have proposed a method to find periods with low noise levels and perform the necessary recognition at these periods. If the signal contains such periods, it would be useful to use such methods together.

In the field of image processing, Li et al. [6] proposed a network that integrates wavelet transforms into a CNN for denoising. This is also useful for extracting ECG features in conventional methods and it demonstrates excellent feature representation. We integrate a wavelet transform into an ECG denoising network and train the denoising parameters to improve the denoising performance of existing models.

The difficulty in denoising ECGs lies in the overlap between the noise and the ECG frequency bands, and when using a handmade threshold in a wavelet transform, there is a possibility of erasing valid information that should be kept. There is overlap in all frequency bands of ECG and noise in the dataset we use in the experiment. In the ECG, there are frequency bands with large spike-like power, and by changing the filter strength for each frequency band, it is possible to remove the noise appropriately. In the range up to about 100 Hz, which is often the target of observation, the signal power in the low and high frequency bands is relatively small, so it may be possible to increase the filter strength in these frequency bands.

3 Proposed Method

3.1 Wavelet Layer

An overview of the conventional discrete wavelet transform is shown in Fig. 2. As mentioned in Sect. 2, wavelet transform denoising is designed for setting the filter after decomposing it into its components for each frequency band in order to give different filter strengths. First, the signal obtained by the HPF and LPF is down-sampled. Then, the HPF and LPF are applied to the output of the LPF among the down-sampled signals, and down-sampling is performed again. When repeated, the high-frequency component of each level is decomposed as Detail, and the low-frequency component of the last level is decomposed as Approximate. In conventional denoising, noise is removed by setting a threshold according to the frequency of each level and attenuating the component corresponding to noise. However, it is difficult to deal with complex or high-intensity noise with only a threshold value; additionally, the behavior of the filter must be changed according to the characteristics of the ECG. Therefore, we add a wavelet transform layer to the CNN and train different filters for each frequency to improve denoising performance.

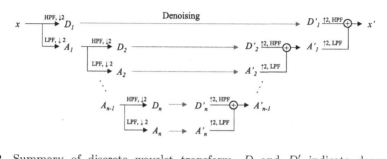

Fig. 2. Summary of discrete wavelet transform. D and D' indicate decomposed denoised component details. D and A are high- and low-frequency components, respectively.

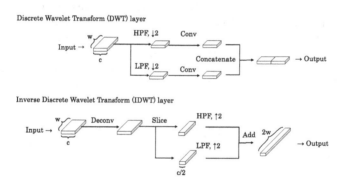

Fig. 3. Discrete Wavelet Transform (DWT) and Inverse Discrete Wavelet Transform (IDWT) layers.

Therefore, we introduced a CNN architecture to improve the denoising performance. For each level of the signal to be downsampled, the CNN can learn the optimal thresholds obtained from the training data. In addition, to make the network learn the behavior at each frequency level, we replace some convolutional layers in the encoder with Discrete Wavelet Transform (DWT) layers and some convolutional layers in the decoder with Inverse Discrete Wavelet Transform (IDWT) layers. By adding wavelet transform layers to the CNN and training

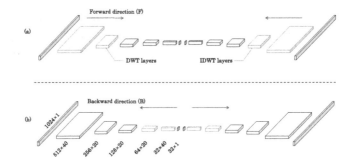

Fig. 4. Wavelet integrated convolutional neural network: (a) Forward(F)-type; (b) Backward(B)-type introduction of DWT and IDWT layers.

different filters for each frequency level, we can expect to improve the denoising performance.

The proposed wavelet and inverse wavelet layers are shown in Fig. 3. The wavelet layer first performs a wavelet transform to decompose the input features into high- and low-frequency components. Then, convolutions are performed on each feature, and these are concatenated in the channel direction. By explicitly specifying the convolution layer weights for the HPF and LPF and optimizing them, we get different optimized filters for different frequency bands.

Meanwhile, the inverse wavelet layer performs the transpose convolution symmetrically with the wavelet layer of the encoder. Then, an HPF and an LPF are applied to each feature map for up-sampling, and each feature map is added to form the output. During training, we aim to improve the modelâĂŹs robustness to complex noise by learning the parameters of the convolution layer for each frequency level.

The input and output dimensions are the same so that the convolutional layer, DWT, and IDWT are compatible. On the encoder side, the output dimension is half of the input dimension, and on the decoder side, the output dimension is twice the input dimension. It may be possible to further optimize these parameters, but we did not consider this part in this study to avoid making the discussion too complicated. This is an issue for future work.

3.2 CNN Model with Wavelet Layer

One possibility is to replace all convolutional layers with DWT and IDWT, but experiments have shown that this is not the best choice. In deep-learning models, it is known that deeper models better perform high feature extractions [4]. However, if the model is separated by the frequency at a stage where the extraction of features is insufficient, accurate extraction ability may be reduced. In our experiment, we show that the ECG feature extraction capability can be improved without increasing the model parameters by retaining the CNN in its shallow layers and performing frequency separation using DWT and IDWT in the deep layers. To clarify the effect of DWT and IDWT layers, we prepared models with different numbers of layers and ablation studies are conducted.

Figure 4(a) shows a network that replaces some layers from the shallow layer with DWTs in the forward direction, and also replaces the corresponding IDWTs just before the output. Figure 4(b) shows the opposite model, where some layers from the deep layer are replaced by DWTs in the backward direction, and the corresponding IDWTs in the deep layer are also replaced just before the output. The replacement from the shallow layer is called forward (F), and the replacement from the deep layer is called backward (B). We also prepared models with all convolutional layers and all wavelet layers, which correspond to conventional networks. Through experiments, we will verify how many layers can be replaced with DWTs and IDWTs to obtain optimal noise denoising performance.

4 Experimental Condition

The dataset and network parameters used in the experiments are described in this section.

4.1 Dataset

To clarify the effectiveness of the proposed method, the MIT-BIH arrhythmia dataset [9] and the MIT-BIH noise stress test dataset (MIT-NST) [10] were used as the ECG dataset and the noise dataset, respectively. The MIT-BIH dataset contains ECG data from 48 patients, each with a sampling frequency of 360 Hz and a measurement time of 30 min. The MIT-NST contains BW, EM, and MA noise, and all were used for evaluation. In the experiments, the first 90% of the data was used for training, and the remainder was for evaluation. To preserve the most accurate model, the last 20% of the training data was used for validation. For training, patients 102 and 104 were omitted because they included paced peaks. To apply the CNN model, 3 s of data was extracted with a window width of 1,024, as in the conventional method. For each patient, we randomly selected 160 samples for training and 40 for validation for a total of 200 samples. There were 9,200 total training samples, including validation.

The MIT-BIH dataset is band-pass filtered in the range of 0.1—100 Hz [9], but because the correct data also contain noise, a fifth-order HPF with a cutoff frequency of 0.67, an LPF at 100 Hz, and a smoothing filter with a kernel size of five were applied before dividing the data by window size. After dividing the data into windows, power per window was calculated to exclude outliers, and only samples within the upper 95th and lower 5th percentiles were used. As the amplitude of the signal differed for each patient, normalization was performed for each patient. The SNRs for the training and validation data were -2.5, 0, 2.5, 5.0, and 7.5, and those for evaluation were -10, -7.0, -3.0, -1.0, 3.0, 7, and 10. In order to verify whether the system can handle noise at levels not used in training, the test was conducted using a wider range of noise than in training.

4.2 Model Parameters

In this experiment, the denoising performance of the proposed wavelet layer was evaluated. The fully convolutional network model [3] was used as the conventional model. This model is an autoencoder that uses only convolution operations, making it possible to compress features with lengths of 1,024 to 32.

The parameters in each layer are shown in Table I. The stride of the convolution of each layer was two, and the dimension was compressed during convolution. The number of feature maps was set to 40 for the first and second-to-last convolutional layers and 20 for the other layers to reduce the number of parameters during training while increasing accuracy by extracting more input and output features. The parameters of a window width of 1,024, as in the conventional method. For the decoder were in reverse order of the encoder. There was no contraction path between the encoder and decoder; thus, it can also be employed in systems where the encoder and decoder are placed in different locations to perform compression and reconstruction. To apply the wavelet transform in the wavelet layer, the wavelet Daubechies 6 (db6) coefficient was used as the mother wavelet. This coefficient is similar to an ECG morphology; thus, it was used in the conventional methods [1,5].

For the convolutional and wavelet layers, apart from the last outputs of the encoder and decoder, a batch normalization layer, an exponential linear unit for the activation function, and a dropout layer with 10% probability were applied. An Adam optimizer with a learning rate of 0.0001 was used to optimize training. The batch size was 200, and 200 epochs were trained. Training and evaluation were performed 10 times, and the average value was used as the result. The implementation was performed in Python v.3.8 and TensorFlow v.2.4.

5 Experimental Results

5.1 Quantitative Evaluation

For evaluation metrics, root mean squared error ($RMSE$) and SNR improvement (SNR_{imp}) were applied. The experimental results of RMSE and SNR improvement are shown in Tables 2 and 3, respectively. A smaller value is better for RSME, and a smaller value is better for SNR improvement. Each ID in the tables shows different experimental conditions. ID1 represents the conventional CNN model, ID2–ID5 represent the proposed model that increases the wavelet layer in the forward direction (Fig. 4(a)), ID6–ID9 represent the proposed model that increases the wavelet layer in the backward direction (Fig. 4(b)), and ID10 shows the model with all wavelet layers. These different models are forward-, backward-, and all-wavelet-layer-type fully convolutional networks (FCNs), respectively.

In Table 2, ID6 shows the best RMSE when SNRs are in the range of −3–10, and ID9 shows the best RMSE when SNRs are −10 and −7. ID6 and ID9 are backward-type models with one wavelet layer and three wavelet layers, respectively. The RMSE differences between the FCN and the best models are 0.0293 and 0.0034 when SNRs are −10 and 10. The backward-type model shows the

Table 1. Parameters of training model. The numbers in Conv and Deconv represent the number of filters, kernel size, and stride, in order. This represents a backward-type model comprising three wavelet layers.

	No	Output	FCN [3]	Proposed
Input	-	1024 × 1		
Encoder	1	512 × 40	Conv(40, 16, 2)	Conv(40, 16, 2)
	2	256 × 20	Conv(20, 16, 2)	Conv(20, 16, 2)
	3	128 × 20	Conv(20, 16, 2)	HPF, Conv(20, 8, 2)
				LPF, Conv(20, 8, 2)
	4	64 × 20	Conv(20, 16, 2)	HPF, Conv(20, 8, 2)
				LPF, Conv(20, 8, 2)
	5	32 × 40	Conv(20, 16, 2)	HPF, Conv(20, 8, 2)
				LPF, Conv(20, 8, 2)
	6	32 × 1	Conv(1, 16, 1)	Conv(1, 16, 1)
Decoder	7	32 × 1	Conv(1, 16, 1)	Conv(1, 16, 1)
	8	64 × 40	Conv(40, 16, 2)	Conv(40, 16, 2)
	9	128 × 20	Conv(20, 16, 2)	Conv(20, 16, 2)
	10	256 × 20	Conv(20, 16, 2)	Conv(20, 16, 2)
	11	512 × 20	Conv(20, 16, 2)	Conv(20, 16, 2)
	12	1024 × 40	Conv(40, 16, 2)	Conv(40, 16, 2)
Output	13	1024 × 1	Conv(1, 16, 1)	Conv(1, 16, 1)

Table 2. Calculation results for RMSE. B and F are backward- and forward-types (Fig. 4), respectively. FCN: fully convolutional network; RMSE: root mean-square error; SNR: signal-to-noise ratio.

ID	Model	Wavelet	Type	Input SNR						
				−10	−7	−3	−1	3	7	10
1	FCN [3]	-	-	0.2104	0.1580	0.1141	0.0997	0.0806	0.0707	0.0670
2		1	F	0.2094	0.1575	0.1140	0.0996	0.0803	0.0701	0.0664
3		2	F	0.2011	0.1533	0.1119	0.0977	0.0785	0.0686	0.0651
4		3	F	0.1935	0.1489	0.1090	0.0953	0.0771	0.0679	0.0646
5		4	F	0.1814	0.1423	0.1063	0.0936	0.0766	0.0683	0.0654
6	Proposed	1	B	0.1811	0.1399	**0.1031**	**0.0907**	**0.0743**	**0.0663**	**0.0636**
7		2	B	0.1817	0.1403	0.1036	0.0911	0.0749	0.0670	0.0644
8		3	B	0.1774	0.1393	0.1039	0.0915	0.0752	0.0673	0.0648
9		4	B	**0.1772**	**0.1392**	0.1045	0.0922	0.0761	0.0683	0.0657
10		5	-	0.1822	0.1417	0.1049	0.0922	0.0758	0.0680	0.0655

Table 3. Calculation results for SNR improvement. B and F are backward- and forward-types (Fig. 4), respectively. FCN: fully convolutional network; SNR: signal-to-noise ratio.

ID	Model	Wavelet	Type	Input SNR						
				-10	-7	-3	-1	3	7	10
1	FCN [3]	-	-	18.51	18.10	17.08	16.30	14.14	11.26	8.72
2		1	F	18.55	18.13	17.10	16.32	14.19	11.33	8.80
3		2	F	18.89	18.36	17.26	16.49	14.40	11.54	8.99
4		3	F	19.20	18.62	17.52	16.73	14.57	11.66	9.08
5		4	F	19.77	19.04	17.76	16.91	14.64	11.63	8.99
6	Proposed	1	B	19.82	19.23	**18.07**	**17.23**	**14.95**	**11.92**	**9.28**
7		2	B	19.80	19.20	18.03	17.19	14.89	11.83	9.17
8		3	B	19.96	**19.24**	17.99	17.14	14.84	11.79	9.13
9		4	B	**19.97**	**19.24**	17.94	17.07	14.72	11.65	8.99
10		5	B	19.74	19.09	17.90	17.06	14.77	11.70	9.04

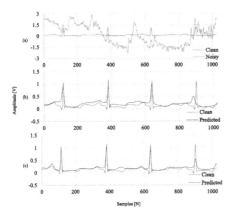

Fig. 5. Outputs of the convolutional neural network when patient is 100 and signal-to-noise ratio is −10: (a) Noisy electrocardiogram; (b) Fully convolutional network (FCN); (c) Backward-type discrete wavelet transform (DWT) FCN with one DWT layer.

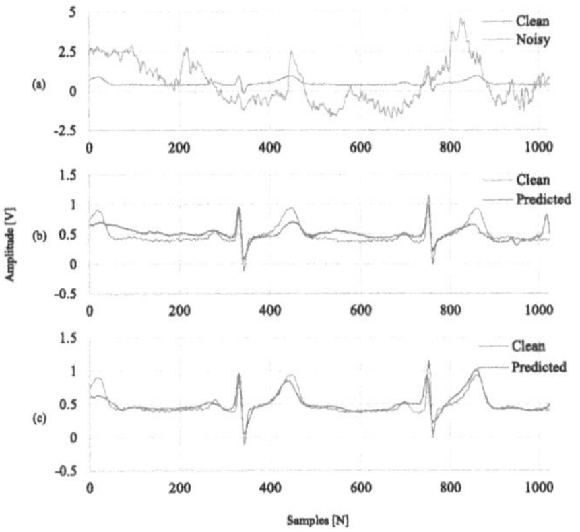

Fig. 6. Outputs of convolutional neural network when patient is 117 and signal-to-noise ratio is −3; (a) Noisy electrocardiogram; (b) Fully convolutional network (FCN); (c) Backward-type discrete wavelet transform (DWT) FCN with one DWT layer.

best RMSE in all SNRs, and there is more improvement when high-intensity noise is infected.

When focusing on the order of wavelet layers, there is no significant difference between the FCN and ID2. In the case of the forward-type model, the RMSE improves when the wavelet layer increases, as shown in ID2–ID5. However, the

RMSE of ID5 is not better than that of ID6, which is a backward-type model. Therefore, the performance of the forward-type model is inferior to that of the backward-type model. Furthermore, in the case of all wavelet models, the RMSE differences between FCN and ID10 are 0.0282 and 0.0015, respectively, when SNRs are -10 and 10. The performance of all wavelet models is better than that of the FCN, but it is inferior to the backward-type model. From these results, it is confirmed that the backward-type model shows the best performance.

As shown in Table 3, SNR improvements show similar trends to the RMSE. Specifically, ID6 is the best when the SNRs are in the range of -3–10, and ID9 shows the best when the SNRs are -10 and -7. When comparing the PRD of ID1 and ID6 with and without the wavelet layer, the PRDs improve by 10.19 and 1.04 when the SNR is -10 and 10. On the other hand, the SNR improvement of ID1 and ID6 improves 1.32 and 0.56, respectively, when the SNR is -10 and 10. Because the best model is the same as the RMSE, it can be concluded that the backward-type model improves denoising performance.

5.2 Qualitative Evaluation

For qualitative evaluation, the output results of samples 100 and 117 as different ECG waveforms are shown in Figs. 5 and 6, respectively. In Fig. 5, the proposed method extracts the signal more faithfully, and the shape before the first R-wave is closer to the clean signal. In the conventional method, the QRS waveform is distorted due to low-frequency noise in the second and third R-waves. In addition, the peak of the fourth R-wave is misaligned, whereas the peak of the proposed method is correct.

In Fig. 6, there is no difference in the shape of the R-wave between the two methods, but the ST-wave waveform of the proposed method is more like the original. Furthermore, in this sample, the conventional method incorrectly recovers a peak that does not exist in the proposed method. On the other hand, the output of the proposed method does not show such a peak, indicating that the proposed method is superior to the original method.

Furthermore, the wavelet transform is applied to the reconstruction results of ID1, ID2, ID6, and ID10 to clarify which frequency bands are recovered well, depending on the type of wavelet layer. The results of the wavelet transform are shown in Fig. 7. When comparing the results of ID1 and ID6, there is a difference in D3, which is the frequency band between 22.5 and 45 Hz. For example, the original ECG has four peaks, but the output of ID1 does not have a peak at the fourth location. ID2 and ID10, which are less accurate than ID6, differ in D1 from 45–90 Hz, indicating that they contain many frequency components that are not present in the original ECG signal. Therefore, it can be concluded that the noise of the high-frequency components is not removed, including ID1 and ID6.

From these results, it is confirmed that the wavelet layer improves the ECG denoising performance on the CNN, both quantitatively and qualitatively.

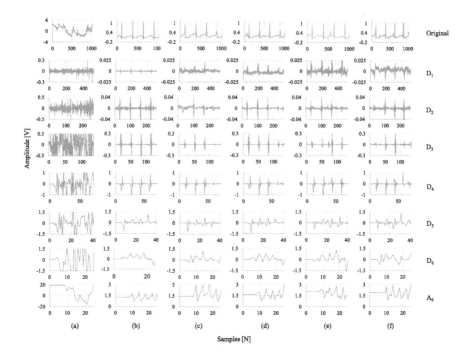

Fig. 7. Result of discrete wavelet transform (DWT) decomposition when patient is 100 and signal-to-noise ratio is -10: (a) Noisy electrocardiogram (ECG); (b) Clean ECG; (c) Output of fully convolutional network (FCN) for ID1; (d) Backward DWT FCN (ID6); (e) Forward DWT FCN (ID2); (f) All DWT-layer-type FCN (ID10).

6 Discussion

In our experiments, a comparison of all convolutional layers, a forward-type model, a backward-type model, and an all-wavelet-layers model was performed. It was shown that the backward-type model with one wavelet layer had the best accuracy and robustness to noise.

Although the wavelet layer improved the denoising performance of the conventional CNN, the accuracy of models with more wavelets in the forward direction and those with all wavelet layers was degraded. This is thought to be related to the frequency range of the ECG. Many theories suggest that an ECG contains important features between 0.5 and 40 Hz. The first layer of our model represents the components in the range of 90–180 Hz in the sampling frequency. When we separated the low- and high-frequency components in the shallow layer, the noise-removal capability of the high-frequency components was insufficient, and the accuracy was reduced. Thus, as the layers of the CNN became deeper in the feature extraction stage, they became closer to the frequency band that originally represented the ECG. Thus, replacing the deeper layers with wavelet layers may have led to improved accuracy. Additionally, the proposed model did not use a contraction path, which is used in existing methods to improve accuracy

[16]; hence, recovery accuracy is insufficient and needs further improvement. A related future challenge is to modify the network structure and parameters to create a model with higher accuracy, even under severe hardware constraints, such as at edge terminals.

7 Conclusion

This paper proposed a model with a wavelet layer for ECG denoising that is more robust to noise. The wavelet layer separates high and low frequencies, which is expected to improve the denoising performance for frequency bands where noise and ECG overlap.

In the experiments, the different model types were compared on different SNRs from -10 to 10, and the model with one wavelet layer in the backward direction showed the highest accuracy compared with the conventional model. The improvement was observed qualitatively and quantitatively; thus, it was confirmed that the wavelet layer was effective for ECG denoising. In the future, we aim to propose a model that is lighter and has higher denoising accuracy.

Acknowledgments.. This work was supported by JST PRESTO Grant Number JPMJPR2135 and JSPS KAKENHI Grant Number JP20H04472.

References

1. Banerjee, S., Gupta, R., Mitra, M.: Delineation of ECG characteristic features using multiresolution wavelet analysis method. Measurement **45**(3), 474–487 (2012)
2. Birok, R., et al.: ECG denoising using artificial neural networks and complete ensemble empirical mode decomposition. Turk. J. Comput. Math. Educ. (TURCOMAT) **12**(2), 2382–2389 (2021)
3. Chiang, H.T., Hsieh, Y.Y., Fu, S.W., Hung, K.H., Tsao, Y., Chien, S.Y.: Noise reduction in ECG signals using fully convolutional denoising autoencoders. IEEE Access **7**, 60806–60813 (2019)
4. He, K., Zhang, X., Ren, S., Sun, J.: Deep residual learning for image recognition. In: Proceedings of the IEEE Conference on Computer Vision and Pattern Recognition, pp. 770–778 (2016)
5. Jenkal, W., Latif, R., Toumanari, A., Dliou, A., El Bfcharri, O., Maoulainine, F.M.: An efficient algorithm of ECG signal denoising using the adaptive dual threshold filter and the discrete wavelet transform. Biocybernetics Biomed. Eng. **36**(3), 499–508 (2016)
6. Li, Q., Shen, L., Guo, S., Lai, Z.: WaveCNet: wavelet integrated CNNs to suppress aliasing effect for noise-robust image classification. IEEE Trans. Image Process. **30**, 7074–7089 (2021)
7. Liu, F., et al.: Wearable electrocardiogram quality assessment using wavelet scattering and LSTM. Front. Physiol. **13** (2022)
8. Mohd Apandi, Z.F., Ikeura, R., Hayakawa, S., Tsutsumi, S.: An analysis of the effects of noisy electrocardiogram signal on heartbeat detection performance. Bioengineering **7**(2), 53 (2020)

9. Moody, G.B., Mark, R.G.: The impact of the MIT-BIH arrhythmia database. IEEE Eng. Med. Biol. Mag. **20**(3), 45–50 (2001)
10. Moody, G.B., Muldrow, W., Mark, R.G.: A noise stress test for arrhythmia detectors. Comput. Cardiol. **11**(3), 381–384 (1984)
11. Nigusse, A.B., Mengistie, D.A., Malengier, B., Tseghai, G.B., Langenhove, L.V.: Wearable smart textiles for long-term electrocardiography monitoring-a review. Sensors **21**(12), 4174 (2021)
12. Poornachandra, S.: Wavelet-based denoising using subband dependent threshold for ECG signals. Digital Sig. Process. **18**(1), 49–55 (2008)
13. Prashar, N., Sood, M., Jain, S.: Design and implementation of a robust noise removal system in ECG signals using dual-tree complex wavelet transform. Biomed. Sig. Process. Control **63**, 102212 (2021)
14. Reddy, G.U., Muralidhar, M., Varadarajan, S.: ECG de-noising using improved thresholding based on wavelet transforms. Int. J. Comput. Sci. Netw. Secur. **9**(9), 221–225 (2009)
15. Ronneberger, O., Fischer, P., Brox, T.: U-Net: convolutional networks for biomedical image segmentation. In: International Conference on Medical Image Computing and Computer-assisted Intervention, pp. 234–241. Springer (2015)
16. Singh, P., Pradhan, G.: A new ECG denoising framework using generative adversarial network. IEEE/ACM Trans. Comput. Biol. Bioinf. **18**(2), 759–764 (2020)
17. Wang, J., et al.: Adversarial de-noising of electrocardiogram. Neurocomputing **349**, 212–224 (2019)
18. Xiong, P., Wang, H., Liu, M., Zhou, S., Hou, Z., Liu, X.: ECG signal enhancement based on improved denoising auto-encoder. Eng. Appl. Artif. Intell. **52**, 194–202 (2016)
19. Yildirim, O., San Tan, R., Acharya, U.R.: An efficient compression of ECG signals using deep convolutional autoencoders. Cogn. Syst. Res. **52**, 198–211 (2018)

WavFusion: Towards Wav2vec 2.0 Multimodal Speech Emotion Recognition

Feng Li[1,2(✉)], Jiusong Luo[1], and Wanjun Xia[1]

[1] Department of Computer Science and Technology, Anhui University of Finance and Economics, Anhui, China
lifeng@aufe.edu.cn
[2] School of Information Science and Technology, University of Science and Technology of China, Anhui, China

Abstract. Speech emotion recognition (SER) remains a challenging yet crucial task due to the inherent complexity and diversity of human emotions. To address this problem, researchers attempt to fuse information from other modalities via multimodal learning. However, existing multimodal fusion techniques often overlook the intricacies of cross-modal interactions, resulting in suboptimal feature representations. In this paper, we propose WavFusion, a multimodal speech emotion recognition framework that addresses critical research problems in effective multimodal fusion, heterogeneity among modalities, and discriminative representation learning. By leveraging a gated cross-modal attention mechanism and multimodal homogeneous feature discrepancy learning, WavFusion demonstrates improved performance over existing state-of-the-art methods on benchmark datasets. Our work highlights the importance of capturing nuanced cross-modal interactions and learning discriminative representations for accurate multimodal SER. Experimental results on two benchmark datasets (IEMOCAP and MELD) demonstrate that WavFusion succeeds over the state-of-the-art strategies on emotion recognition.

Keywords: Speech emotion recognition · multimodal · wav2vec 2.0 · A-GRU · A-GRU-LVC

1 Introduction

Recently, speech emotion recognition (SER) is a fascinating field that utilizes technology to analyze and identify different emotions present in human speech [1]. This technology has various applications, including in customer service and market research [2], learning and education [3], mental health [4], and social media analytics [5]. In real-life scenarios, humans express emotions not only through speech but also through alternative modalities, such as text and visuals [6,7]. Previous studies on SER typically rely on speech information. However, different modalities provide complementary information for emotion recognition, and emotion recognition of the single modality is not inadequate to meet real-world demands. To address this problem, researchers utilize multimodal information to identify emotional states [8]. In the domain of Multimodal Emotion Recognition (MER), the information of diverse modalities is complementary, providing additional cues to mitigate semantic and emotional ambiguities.

In addition to multimodality, another challenge of SER is achieving better interaction during the fusion of different modalities. Firstly, multimodal data often exhibit asynchrony [9]. For instance, visual signals typically precede audio signals by approximately 120 ms in emotional expressions [10]. This asynchronicity poses a challenge to feature fusion and model design, necessitating methods to address temporal alignment and matching issues. To address this issue, Tsai et al. [11] have proposed specific asynchronous models and cross-modal attention mechanisms. Zheng et al. [12] solved heterogeneity among different encoder output features by employing unsupervised training of a multi-channel weight-sharing autoencoder. This approach minimizes the differences among features extracted from different modalities. Additionally, the interactions are simulated by supervised training of cascaded multi-head attention mechanisms. However, most methods with cross-modal attention mechanisms ignore redundant information during the fusion process, thus restricting the performance of MER. Additionally, samples with the same emotion in multimodal data may exhibit differences across modalities, referred to as homogeneous feature differences. For instance, some features in speech and text may exhibit formal similarity but convey different emotional states [13]. Hazarika et al. [14] projected each modality into two different subspaces capturing modality-invariant and modality-specific features. However, they only considered the differences between the different emotion of same modalities and ignored the differences between different modalities with the same emotion. DialogueTRM explores intra- and inter-modal emotional behaviors in conversations, using Transformers to model the context [15]. MMGCN proposes a multimodal fusion approach via a deep graph convolution network, modeling the interactions between different modalities using a graph [16]. MM-DFN introduces a dynamic fusion network that leverages intra- and inter-modal information at different levels of representation [17]. M2FNet proposes a multi-modal fusion network that learns and fuses complementary information from audio, visual, and textual modalities [18].

Therefore, in this paper, we propose a novel architecture called WavFusion for emotion recognition. Unlike DialogueTRM, WavFusion specifically focuses on incorporating wav2vec2.0 [19] with a gated cross-modal attention mechanism to dynamically fuse multimodal features. Additionally, WavFusion introduces multimodal homogeneous feature discrepancy learning to distinguish between same-emotion but different-modality representations. WavFusion does not rely on graph-based modeling but instead uses a transformer architecture with a modified cross-modal attention mechanism. WavFusion also emphasizes capturing both global and local visual information through the A-GRU-LVC module. While MM-DFN focuses on dynamic fusion strategies, WavFusion emphasizes the use of wav2vec2.0 pre-trained representations and a gated cross-modal attention mechanism to mitigate redundant information during fusion. Additionally, WavFusion incorporates multimodal homogeneous feature discrepancy learning to distinguish between representations of the same emotion across different modalities.

The main contributions of this paper can be summarized as follows:

- We propose a multimodal speech emotion recognition model (WavFusion) that leverages the power of wav2vec 2.0 and incorporates textual and visual modalities to enhance the performance of audio-based emotion recognition.
- We integrate the designed gated cross-modal attention mechanism into the wav2vec 2.0 model to mitigate redundant information during the fusion process. Meanwhile, we employ multimodal homogeneous feature discrepancy learning to enhance the discriminative capability of the model.
- Experimental results on two benchmark datasets demonstrate the effectiveness of the proposed method. Our WavFusion succeeds over existing state-of-the-art methods.

2 Proposed Method

2.1 Problem Statement

Given a multimodal signal $S_j = \{S_j^a, S_j^t, S_j^v\}$, we can represent the unimodal raw sequence extracted from the video fragment j as $S_j^m, m \in \{a, t, v\}$. Here, the modalities are denoted by $\{a, t, v\}$, which refer to audio, text, and visual modalities.

In WavFusion, we aim to predict the emotion category for each utterance. It focuses on categorizing the emotion conveyed in each utterance, assigning it to a specific emotion class or category, $y_j \in R^c$. c is the number of emotion categories. Figure 1 illustrates the overall structure of WavFusion, including an auxiliary modal encoder, a primary modal encoder, and multimodal homogeneous feature discrepancy learning. The orange color represents positive emotions and the green color represents negative emotions.

2.2 Auxiliary Modality Encoder

Video Representation. For the visual modality, we use the EfficientNet pre-trained model as a feature extractor to obtain visual features $\mathbf{e}_j^v$. This model is a self-supervised framework for visual representation learning. In this paper, we attempt to extend EfficientNet to emotion recognition. $\mathbf{e}_j^v$ can be formulated as:

$$\mathbf{e}_j^v = \Phi_{visual}\left(S_j^v\right) \tag{1}$$

where Φ_{visual} denotes the function of EfficientNet model.

On the other hand, we consider the context and situation conveyed by the global information in the visual modality, along with the specific details of actions

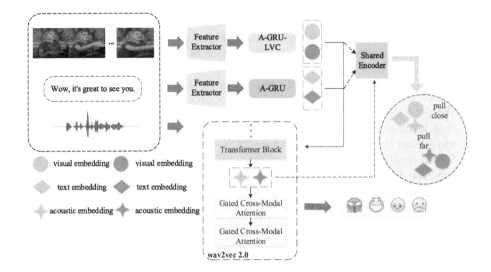

Fig. 1. The overview of WavFusion.

and expressions from local information. The visual feature is fed into the proposed A-GRU-LVC module, which aims to extract both global and local features.

$$X_j^{v1} = F_{SA}\left(F_{GRU}\left(\mathbf{e}_j^v\right)\right) \tag{2}$$

where F_{SA} and F_{GRU} denote the learning functions of GRU and self-attentive mechanism, respectively.

Simultaneously, to preserve local corner point regions and extract local information, a learnable visual center (LVC) is implemented on the visual features [20]. This LVC aggregates features from local areas, ensuring that important local information is retained. In contrast to the approach, we utilize one-dimensional convolution instead of two-dimensional convolution.

$$X_j^{v2} = F_{LVC}\left(e_j^v\right) \tag{3}$$

where F_{LVC} denotes the learning functions of the LVC block.

Finally, the output of the A-GRU-LVC block is obtained by connecting the output of the self-attention module X_j^{v1} and the output of the LVC block X_j^{v2} along the last dimension.

$$X_j^v = X_j^{v1} \oplus X_j^{v2} \tag{4}$$

Contextualized Word Representation. To capture rich contextual information from textual data, we utilize the RoBERTa-base model, which belongs to the transformer family, as a contextual encoder. The architecture of RoBERTa consists of multiple Transformer layers, including a stack of encoders. Each encoder layer contains a multi-head self-attention mechanism and a feed-forward neural network. RoBERTa is designed to capture contextualized representations of words in a sentence, allowing it to understand the meaning and relationships between different words. e_j^t can be formulated as:

$$e_j^t = \Phi_{text}\left(S_j^t\right) \quad (5)$$

where Φ_{text} denotes the function of the RoBERTa pre-train model.

To further consider context-sensitive dependence for text features, we feed it into the GRU and the self-attention mechanism to obtain global features of the text information. Due to the strong temporal continuity present in textual information, we opted not to employ the LVC mechanism to capture local feature.

$$X_j^t = F_{SA}\left(F_{GRU}\left(e_j^i\right)\right) \quad (6)$$

where F_{SA} and F_{GRU} denote the learning functions of GRU and self-attentive mechanism, respectively.

Major Modality Encoder. In WavFusion, we encode low-level audio features through the shallow transformer layer, followed by combining text and visual features through the deep transformer layer to form a comprehensive multimodal representation. We define the original transformer layer as a shallow transformer layer and the modified transformer layer as a deep transformer. The incorporation of text and vision into wav2vec 2.0 detects relevant information within the extensive pre-trained audio knowledge, thereby enhancing emotional information within the multimodal fusion representation. The low-level acoustic features X_j^a extracted by the shallow transformer block are calculated as follows:

$$X_j^a = F_{ST}\left(S_j^a\right) \quad (7)$$

where F_{ST} is the learning function of shallow transformer layers.

$$X_j^{F1} = CM_{A-T}\left(X_j^a, X_j^t\right) \quad (8)$$

$$X_j^{F2} = CM_{A-V}\left(X_j^a, X_j^v\right) \quad (9)$$

Finally, the augmented features X_j^{F1} and X_j^{F2} are processed through the following gated filtering mechanism. The ratio of each channel can be dynamically defined by a learnable parameter that filters out misinformation generated during cross-modal interactions.

$$P_* = sigmoid\left(FC\left(X_j^{F1} \oplus X_j^{F1}\right)\right) \quad (10)$$

$$X_j^F = P_* \odot X_j^{F1} + (1 - P_*) \odot X_j^{F2} \quad (11)$$

Multimodal Homogeneous Feature Discrepancy Learning. Multimodal homogeneous feature discrepancy learning has made significant progress in multimodal emotion recognition. It can optimize the modal representation ability and extract richer and more accurate emotional information by learning the relationships and differences between homogeneous features. First, we feed unfused audio features X_j^a, text features X_j^t and visual features X_j^v into a shared encoder to obtain homogeneous features. It minimizes the feature gap from different modalities and contributes to multimodal alignment.

$$X_{com}^{m[i]} = SD\left(X_j^m\right), m \in (a, t, v) \tag{12}$$

where SD is the shared encoder learning function that consists of a simple linear layer.

In this study, we perform multimodal homogeneous feature discrepancy learning to enhance the interactions between the same emotions but different modalities, and amplify the differences between the same modalities but different emotions. We define this loss function as margin loss.

$$L_{mar} = \frac{1}{M} \sum_{(i,j,k) \in M} max(0, \alpha - cos(X_{com}^{m[i],c[i]}, X_{com}^{m[j],c[j]}))$$
$$+ cos(X_{com}^{m[i],c[i]}, X_{com}^{m[k],c[k]})) \tag{13}$$

where

$$M = \{(i, j, k) \mid m[i] \neq m[j], m[i] = m[k], c[i] = c[j], c[i] \neq c[k]\}$$

is the modality of the sample i, and the $c[i]$ is the label of sample i. cos denotes the cosine similarity between two feature vectors. By applying a distance margin α, we ensure that the distance between positive samples is smaller than the distance between negative samples. Here, positive samples refer to the same emotion but different emotions, and negative samples refer to the same modality but different emotions.

Similarly, cross-entropy serves as a commonly employed loss function for optimizing model parameters and enhancing classification accuracy during training.

$$L_{task}^{emotion} = -\frac{1}{N_D} \sum_{j=0}^{N_D} y_j \cdot \log \hat{y}_j \tag{14}$$

where y_j is the true label of the sample, $\hat{y}_j$ is the prediction of the sample, and N_D is the number of samples in the dataset D.

$$L_{total}^{emotion} = L_{task}^{emotion} + \lambda L_{mar} \tag{15}$$

where λ is the balance factor.

3 Evaluation

3.1 Dataset

We evaluate our proposed method on two prevalent benchmark datasets for ERC, including IEMOCAP [21] and MELD [22], respectively. The IEMOCAP dataset consists of 12 h of improvised and scripted audio-visual data from 10 UC theatre actors (five males and five females). The dataset is divided into five binary sessions, and each conversation is annotated with emotional information in four modalities: video, audio, transcription, and motion capture of facial movements. We evaluate our model using audio, transcribed, and video data. The dataset contains a total of 7380 data samples. E.g., happy, neutral, angry, excited, sad, and frustrated. For evaluation, we employ a five-fold cross-validation approach. The first four sessions are utilized as the training set and the validation set, and the last session is utilized as the testing set.

The MELD dataset is derived from over 1,400 dialogues and 13,000 utterances extracted from the TV series Friends. Each utterance in the dataset is annotated with one of seven emotion labels: neutral, surprise, fear, sadness, joy, disgust, and anger. The dataset includes multimodal scenes, making it suitable for studying multimodal emotion recognition tasks. For our experiments, we utilize the predefined training/validation splits provided with the MELD dataset. This ensures consistency with existing approaches and allows for a fair comparison with other models.

3.2 Setting

For text and visual modalities, we freeze the parameters in the RoBERTa and EfficientNet pre-trained models and treat them as a feature extractor. The last dimension of the text and visual features is 768 and 64. For speech modalities, we unfreeze the parameters of the deep transformer layer in the wav2vec 2.0 pre-train model. These parameters are updated during model training, while the parameters of the other layers are freezing. The last dimension of the speech features are 768 and 64.

3.3 Comparative Analysis

In Tables 1 and 2, we show the performance of different approaches on the IEMOCAP and MELD datasets. The evaluation metrics are Accuracy (ACC) and Weighted F1 score (WF1). On the IEMOCAP dataset, our method outperformed the state-of-the-art by 0.84% in ACC and 0.74% in WF1. Similarly, on the MELD dataset, our method surpassed the state-of-the-art by 0.43% in ACC and 0.44% in WF1. The reasons are probably twofold. Firstly, we argue that this is because most of these models do not explicitly consider redundant information in the cross-modal fusion process, but our proposed method considers these through a gated cross-modal attention mechanism. Secondly, most of them only take into account the distances of different emotion samples of the

Table 1. The results of different methods on the IEMOCAP database.

Method	ACC(%)	WF1(%)	Year
DialogueTRM [15]	68.92	69.23	2020
HiTrans [23]	-	64.5	2020
DialogXL [24]	-	65.94	2021
MMGCN [16]	-	66.22	2021
COGMEN [25]	68.2	67.63	2022
MM-DFN [17]	68.21	68.18	2022
M2FNet [18]	69.69	69.86	2022
HAAN-ERC [26]	69.48	69.47	2023
Ours	**70.53**	**70.6**	**2024**

Table 2. The results of different methods on the MELD database.

Method	ACC(%)	WF1(%)	Year
DialogueTRM [15]	65.66	63.55	2020
MMGCN [16]	-	58.65	2021
MM-DFN [17]	62.49	59.46	2022
UniMSE [27]	65.09	65.51	2022
HAAN-ERC [26]	66.5	65.66	2023
Ours	**66.93**	**66.1**	**2024**

same modality, but not the distances of the same emotion samples of different modalities.

3.4 Ablation Studies

To verify the effectiveness of WavFusion model, we conduct ablation studies on the IEMOCAP dataset. First, we reveal the importance of each modality in this section. Specifically, when utilizing a single modality, we omitted the gated cross-modal attention and multimodal homogeneous feature discrepancy learning. The results in Table 3 illustrate that the highest accuracy and weighted average F1 scores are attained when incorporating all three modalities. Due to the complexity of emotion recognition, recognizing emotions using a single modality is challenging to meet the demands of reality. We can achieve better recognition performance by integrating multimodal information.

Additionally, we introduce LVC blocks to capture local information related to visual features. To assess the significance of LVC blocks, we conducted an experiment where we omitted the LVC blocks from the model, thus failing to capture local information about visual features. From Table 4, we observe that the model with the LVC block outperforms the model without the LVC block. The inclusion of LVC blocks improves ACC by 0.63% and the WF1 by 0.76%.

WavFusion: Towards Wav2vec 2.0 Multimodal Speech Emotion Recognition 333

Table 3. Experiment results on the different modalities.

Modality	ACC(%)	WF1(%)
A	66.06	65.59
T	58.74	58.63
V	29.88	26.31
A+T	67.75	67.45
A+V	66.33	64.14
A+V+T	70.53	70.6

The experiment demonstrates that the LVC blocks are beneficial for capturing relevant contextual details and spatial dependencies.

Table 4. Experiment results on LVC BLOCK.

Models	ACC(%)	WF1(%)
w/o LVC block	69.90	69.84
w/ LVC block	70.53	70.60

We also investigate the impact of multimodal homogeneous feature discrepancy learning in our framework. In this work, we assigned weights to the balance factor λ for margin loss and observed its effects across various weight values. The corresponding results are presented in Table 5. The results indicate that the optimal performance on the IEMOCAP dataset is achieved. The model shows a significant improvement by 2.64% and WF1 by 2.94% compared to the absence of margin loss ($\lambda = 0$). This demonstrates the effectiveness of multimodal homogeneous feature difference learning in enhancing the model's capacity to discern emotions across diverse modalities. However, we also observe that the performance deteriorates when the balance factor is excessively large ($\lambda = 10$). This suggests that an excessive emphasis on margin loss might have a detrimental effect on the original classification task.

Table 5. Experiment results on the different λ.

λ	ACC(%)	WF1(%)
0	67.89	67.66
0.01	68.63	68.39
0.1	69.11	68.96
1	70.53	70.6
10	64.43	64.19

Table 6. Experiment results on the transformed layers in wav2vec 2.0.

method	Shallow transformer	Deep transformer	ACC(%)	WF1(%)
concat	12	0	66.67	66.78
Attention	11	1	68.61	68.55
	10	2	68.54	68.32
	9	3	70.53	70.6
	8	4	69.29	69.06

We also observe the effect of gated cross-modal attention mechanism in the proposed framework. In our experiments, we define the original transformer layer the shallow transformer and the modified transformer layer as deep transformer, and observe their effect on the different numbers. In the first line, we omit the proposed gated cross-modal attention mechanism and solely conduct a basic concatenation of the three modal features at the last dimension. The corresponding results are shown in Table 6 where it is observed that 9 shallow transformer layers, 3 deep transformer layers yield the optimal performance for the IEMOCAP dataset. Moreover, from the first and second lines, we can discern the significance of the gated cross-modal attention mechanism for fusion.

4 Conclusion

In this paper, we propose a novel SER approach, which is designed a gated cross-modal attention alternative to self-attention in the wav2vec 2.0 pre-trained model to dynamically fuse features from different modalities. Additionally, we introduce a novel LVC block to efficiently capture the local information of visual features. The model can more effectively utilize the spatial characteristics of visual data, resulting in more comprehensive representations. Finally, we design the concept of multimodal homogeneous feature discrepancy learning, which helps the model to effectively learn and distinguish representations of the same modalities but different emotions. The effectiveness of the proposed model is demonstrated on the IEMOCAP and MELD datasets. The results show promising performance compared to state-of-the-art methods. In the future, we plan to utilize the leveraging large amounts of unlabeled audio and video data available to recognize the different emotion.

Acknowledgments.. This work was supported in part by the Natural Science Foundation of the Higher Education Institutions of Anhui Province under Grant Nos. 2024AH050018 and KJ2021A0486, Excellent Research and Innovation Team of Universities at Anhui Province under Grant Nos. 2024AH010001 and 2023AH010008, and Science Research Fund of Anhui University of Finance and Economics under Grant No. ACKYB23016.

References

1. Ayadi, M.E., Kamel, M.S., Karray., F.: Survey on speech emotion recognition: features, classification schemes, and databases. Pattern Recogn. **44**(3), 572–587 (2011)
2. Li, X., Lin, R.: Speech emotion recognition for power customer service. In: 2021 7th International Conference on Computer and Communications (ICCC), pp. 514–518 (2021)
3. Li, W., Zhang, Y., Fu, Y.: Speech emotion recognition in e-learning system based on affective computing. In: Third International Conference on Natural Computation (ICNC 2007), vol. 5, pp. 809–813 (2007)
4. Elsayed, E., ElSayed, Z, Asadizanjani, N., et al.: Speech emotion recognition using supervised deep recurrent system for mental health monitoring. In: 2022 IEEE 8th World Forum on Internet of Things (WF-IoT), pp. 1–6 (2022)
5. Ahire, V., Borse, S.: Emotion detection from social media using machine learning techniques: a survey. In: Iyer, B., Ghosh, D., Balas, V.E. (eds.) Applied Information Processing Systems. AISC, vol. 1354, pp. 83–92. Springer, Singapore (2022). https://doi.org/10.1007/978-981-16-2008-9_8
6. Calefato, F., Lanubile, F., Novielli, N.: EmoTxt: a toolkit for emotion recognition from text. In: 2017 Seventh International Conference on Affective Computing and Intelligent Interaction Workshops and Demos (ACIIW), pp. 79–80 (2017)
7. You, Q., Luo, J., Jin, H., et al.: Building a large scale dataset for image emotion recognition: the fine print and the benchmark. In: Proceedings of the AAAI conference on artificial intelligence, vol. 30 (2016)
8. Abdullah, S.S, Ameen, S.A., Sadeeq, A., Zeebaree, S.: Multimodal emotion recognition using deep learning. J. Appl. Sci. Technol. Trends **2**(02), 52–58 (2021)
9. Wu, W., Zhang, C, Woodland, P.: Emotion recognition by fusing time synchronous and time asynchronous representations. In: ICASSP 2021-2021 IEEE International Conference on Acoustics, Speech and Signal Processing (ICASSP), pp. 6269–6273 (2021)
10. Grant, K.W., Greenberg, S.: Speech intelligibility derived from asynchronous processing of auditory-visual information. In: AVSP 2001-International Conference on Auditory-Visual Speech Processing (2001)
11. Tsai, Y.H.H., Bai, S.J., Liang, P.P., et al.: Multimodal transformer for unaligned multimodal language sequences. In: Proceedings of the conference. Association for Computational Linguistics. Meeting, vol. 2019, p. 6558. NIH Public Access (2019)
12. Zheng, J., Zhang, S., Wang, Z., et al.: Multi-channel weight-sharing autoencoder based on cascade multi-head attention for multimodal emotion recognition. IEEE Trans. Multimedia **22**, 2213–2225 (2022)
13. Chen, B., Cao, Q., Hou, M., et al.: Multimodal emotion recognition with temporal and semantic consistency. IEEE/ACM Trans. Audio Speech Lang. Process. **29**, 3592–3603 (2021)
14. Hazarika, D., Zimmermann, R., Poria, S.: MISA: modality-invariant and-specific representations for multimodal sentiment analysis. In Proceedings of the 28th ACM International Conference on Multimedia, pp. 1122–1131 (2020)
15. Mao, Y., Sun, Q., Liu, G., et al.: DialogueTRM: exploring the intra-and inter-modal emotional behaviors in the conversation. arXiv preprint arXiv:2010.07637 (2020)
16. Hu, L., Liu, Y., Zhao, J., et al.: MMGCN: multimodal fusion via deep graph convolution network for emotion recognition in conversation. arXiv preprint arXiv:2107.06779 (2021)

17. Hu, D., Hou, X., Wei, L., et al.: MM-DFN: multimodal dynamic fusion network for emotion recognition in conversations. In: ICASSP 2022-2022 IEEE International Conference on Acoustics, Speech and Signal Processing (ICASSP), pp. 7037–7041 (2022)
18. Chudasama, V., Kar, P., Gudmalwar, A., et al.: M2FNET: multi-modal fusion network for emotion recognition in conversation. In Proceedings of the IEEE/CVF Conference on Computer Vision and Pattern Recognition, pp. 4652–4661 (2022)
19. Baevski, A., Zhou, Y., Mohamed, A., et al.: wav2vec 2.0: a framework for self-supervised learning of speech representations. In: Advances in Neural Information Processing Systems, vol. 33, pp. 12449–12460 (2020)
20. Quan, Y., Zhang, D., Zhang, L., et al.: Centralized feature pyramid for object detection. IEEE Trans. Image Process. **32**, 4341–4354 (2023)
21. Busso, C., Bulut, M., Le, C., et al.: IEMOCAP: interactive emotional dyadic motion capture database. Lang. Resour. Eval. **42**, 335–359 (2008). https://doi.org/10.1007/s10579-008-9076-6
22. Poria, S., Hazarika D., Majumder, N., et al.: MELD: a multimodal multi-party dataset for emotion recognition in conversations. arXiv preprint arXiv:1810.02508 (2018)
23. Li, J., Ji, D., Li, F., et al.: HiTrans: a transformer-based context-and speaker-sensitive model for emotion detection in conversations. In Proceedings of the 28th International Conference on Computational Linguistics, pp. 4190–4200 (2020)
24. Shen, W., Chen, J., Quan, X., et al.: DialogXL: All-in-one XLNet for multi-party conversation emotion recognition. In Proceedings of the AAAI Conference on Artificial Intelligence, vol. 35, pp. 13789–13797 (2021)
25. Joshi, A., Bhat, A., Jain, A., et al.: COGMEN: COntextualized GNN based multimodal emotion recognition. arXiv preprint arXiv:2205.02455 (2022)
26. Zhang, T., Tan, Z., Wu, X.: HAAN-ERC: hierarchical adaptive attention network for multimodal emotion recognition in conversation. Neural Comput. Appl., 1–14 (2023). https://doi.org/10.1007/s00521-023-08638-2
27. Hu, G., Lin, T., Zhao, Y., et al.: UniMSE: towards unified multimodal sentiment analysis and emotion recognition. arXiv preprint arXiv:2211.11256 (2022)

Zero-Shot Sketch-Based Image Retrieval with Hybrid Information Fusion and Sample Relationship Modeling

Weijie Wu[1], Jun Li[1(✉)], Zhijian Wu[2], and Jianhua Xu[1]

[1] School of Computer and Electronic Information, Nanjing Normal University, Nanjing 210023, China
lijuncst@njnu.edu.cn
[2] School of Data Science and Engineering, East China Normal University, Shanghai 200062, China

Abstract. Despite achieving dramatic advances, current zero-shot sketch-based image retrieval (ZS-SBIR) models are mainly challenged by two issues. They must overcome enormous domain gap between sketches and images along with significant intra-class variations. In addition, the scarcity of sketch samples may result in degraded feature representation learning from the training data, thereby limiting the model's generalization ability of recognizing different categories or instances. To address these issues, we propose a novel ZS-SBIR model in this study. On the one hand, a hybrid information fusion module is introduced to combine both the structured information (e.g., shapes and contours) and sequential information (e.g., stroke order) of sketches and images for a comprehensive representation. Furthermore, this module is capable of capturing multi-level feature representations of sketches, such that the model is better adapted to different types of sketches. On the other hand, both intra- and inter-sample correlation between sketches and images are modeled via a dual-sample relationship modeling module. To be specific, it can fully take advantage of each sample in data-scarce scenarios by extracting complementary semantic information among different images in a mini-batch, which allows the model to learn more robust feature representations and improve the discrimination of unseen class features. Extensive experiments conducted on three public benchmark datasets demonstrate the superiority of our proposed method to the state-of-the-art ZS-SBIR models.

Keywords: Zero-shot Sketch-based Image Retrieval · Hybrid Information Fusion · Multi-level Feature Representation · Dual-sample Relationships

1 Introduction

In real-world scenarios, the increasing sophistication of modern technology has made it easier to create hand-drawn sketches on ubiquitous touchscreen devices,

Supported by the National Natural Science Foundation of China under Grant 62173186 and 62076134. Email: lijuncst@njnu.edu.cn.

which allows human to more effectively express the shape and pose of the target image in a visually concise manner. Therefore, sketch-based image retrieval (SBIR) has received more attention than traditional cross-modal text-image retrieval or classic content-based image retrieval. In particular, it significantly facilitates the situations where it may be challenging to provide a textual description or an appropriate image for the required query, while users can easily and spontaneously draw sketches of the desired objects on a touchscreen. Compared to the traditional SBIR task, zero-shot sketch-based image retrieval (ZS-SBIR), which combines zero-shot learning and SBIR, is more challenging, since it requires the model to retrieve natural images of the same category from a large image database given a sketch of a category that is not present in the training set.

Although dramatic progress has been made, two main issues pose great challenges to ZS-SBIR. On the one hand, there are significant visual differences between sketches and images. Sketches are typically simplified figures represented by abstract lines and irregular strokes, whereas images contain rich colors, abundant textures, and detailed visual clues. This substantial domain gap between sketches and images leads to inconsistent inter-class similarity relationships, making the training of jointly embedding models more challenging. In addition, since sketches drawn by different artists exhibit various styles, there are considerable variations in the sketches of the same category. Therefore, ZS-SBIR models need to overcome both the sketch-image domain gap but also the intra-class variations. On the other hand, due to the uneven distribution of different object categories, sketch samples for some categories are exceedingly scarce, limiting the model's capability of generalizing to these categories. Since it is difficult to incorporate all potential sketch categories into the training set, new categories in open-world scenarios may not have sufficiently discriminative feature representations learned from the training data. In this context, it is crucial for exploring how to fully utilize the information from each sample and help the model to learn discriminative feature representations from limited data.

To address the aforementioned issues, we have proposed a novel ZS-SBIR model with hybrid information fusion and sample relationship modeling in this study. More specifically, we integrate a hybrid information fusion module (HIFM) that comprehensively combines the structured information (e.g., shapes, contours) and sequential information (e.g., stroke order) of sketches and images. Moreover, this module employs multi-level feature representations that can capture both low-level feature details and textures, as well as high-level geometric relationships and semantic structures. By merging these two types of visual clues, it significantly enhances the representation power of feature embeddings, reduces the domain gap between sketches and images, and improves the model's adaptability to different types of sketches. To explore intra-sample and inter-sample relationships in multi-modal information extraction for sketch retrieval, the dual-sample relationship modeling (DSRM) is introduced to model relationships within and between samples. Based on the Vision Transformer (ViT) architecture, attention mechanisms are employed in both visual and semantic spaces to embed features from sketches and images into the common space. Besides, DSRM deeply explores the relationships among samples within mini-batches to learn more robust feature representations. Meanwhile, the cross-modal attention mechanism is utilized to learn the relationships between the image and sketch features, allowing the alignment of these two modalities in the feature space and thereby reducing the modality gap. To our knowledge, this is the first attempt to investigate intra-sample and inter-sample relationships in multi-modal information extraction for ZS-SBIR. To summarize, the contributions of this paper are threefold as follows:

- We propose a Transformer-based ZS-SBIR model in which a hybrid information fusion module (HIFM) is integrated such that multi-level features of sketches and images can be characterized by combining both structured and serialized information, thereby enhancing the comprehensiveness and robustness of feature representations.
- We introduce the dual-sample relationship modeling method (DSRM) to capture inter-sample and intra-sample relationships. This not only enhances the robustness of feature representations but also maximizes the utilization of each sample's information in data-scarce scenarios, thereby improving the discrimination of unseen class features. In addition, the representations of the two modalities are well aligned within the feature space via the cross-modal attention module, reducing the inter-modal gap and capturing richer semantic information.
- Extensive experiments are conducted on three benchmarks, i.e., Sketchy, TU-Berlin and QuickDraw, demonstrating the superiority of our proposed approach for zero-shot sketch-based image retrieval.

2 The Proposed Method

The overall framework of our proposed model is shown in Fig. 1. After providing the problem formulation, we will elaborate on our proposed model in which the hybrid information fusion module and intra-sample relationship modeling module are discussed in detail.

2.1 Problem Formulation

In ZS-SBIR, the complete dataset is partitioned into two parts: the training set and the testing set. The training set, denoted as $D_{seen} = \{I_{seen}, S_{seen}\}$, is used to train the retrieval model. In this context, I_{seen} and S_{seen} correspond to images and sketches from the seen categories, respectively. Similarly, the testing set, denoted as $D_{unseen} = \{I_{unseen}, S_{unseen}\}$, is used to validate the performance of the retrieval model. A ZS-SBIR model which is trained with images from seen classes is expected to retrieve corresponding photos by sketches from the unseen set. To satisfy the zero-shot setting, $D_{seen} \cap D_{unseen} = \varnothing$.

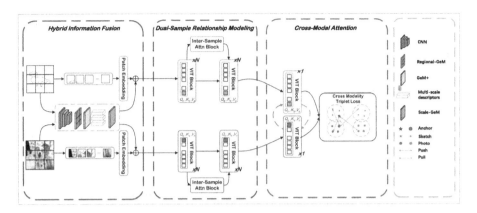

Fig. 1. The architecture of our proposed model which consists of two important components, namely hybrid information fusion module (HIFM) and dual-sample relationship modeling module (DSRM). The former can encode both structured and sequential information into comprehensive multi-level features of sketches and images, while the latter can simultaneously capture inter-sample and intra-sample relationships to fully take advantage of the sample information. Thus, the cross-modal representations can be well aligned via the cross-attention module, which can effectively reduce the inter-modal gap.

2.2 Hybrid Information Fusion Module

Conventional ZS-SBIR methods primarily rely on structured information from sketches, such as shapes, contours, and edges. These methods either fine-tune pre-trained Convolutional Neural Networks (CNNs) to extract features and then establish projection models to learn a shared embedding space [7], or train the entire model in an end-to-end manner [14]. While these approaches excel at capturing the geometric shapes of sketches, they often struggle to sufficiently express the semantic information of sketches. With the development of sequence-modeling techniques, sequential information has been introduced into the sketch retrieval field. In sketches, this information typically consists of the order and trajectory of strokes, which can better represent the detailed variations in sketches. Recent research demonstrates that ViT [6] can be used as a desirable alternative to CNN by serializing image patches and utilizing self-attention mechanisms to capture global dependencies within images. However, since sketches are mostly composed of simple and irregular lines, only serializing sketches is not helpful for those made up of sparse strokes, leading to a loss of outline information and a large number of blank patches. On the other hand, only considering sequential information falls short in capturing details and textures without providing a comprehensive encoding.

To address the above challenges, our approach integrates both structured and sequential clues, leveraging the complementarity between them to produce a more robust and comprehensive feature representation. In detail, given a sketch S or its corresponding image I, respectively, we initially use a backbone model to generate their structured feature representations $D \in \mathbb{R}^{H \times W \times C}$. Afterwards, we adopt the GeM pooling strategy [17] which introduces the parameter p to allow flexible adjustments during feature aggregation. In particular, the improved GeM pooling technique GeM+ [19] helps the model to better capture and represent the common features of the sketches within the same class, thus reducing the intra-class variance. In GeM+, a parameter tuning phase is incorporated during training process, dynamically adjusting the p to optimize the pooling effect.

When applying GeM+ to global pooling in ZS-SBIR, our goal is to strengthen feature discrimination when aggregated into the final embedding. However, in addition to discriminative information at the global level, local information such as object shape and contour is also important to distinguish between different instances. Such fine-grained details may not be captured robustly when simply performing feature pooling at the global level. Therefore, in addition to global pooling at each scale, we perform regional aggregation by applying the L_p pooling method [9] parameterized by p_r to our network. This can be considered a variant of GeM pooling operating at the regional level. In this configuration, activations of the feature map D are aggregated in a convolutional way, producing a new feature map $M \in \mathbb{R}^{H \times W \times C}$. Subsequently, we combine M and D to generate a more robust feature map, resulting in enhanced features as follows:

$$G_r = W \left(\frac{1}{HW} \sum_{h,w} \left(\frac{M_{h,w} + D_{h,w}}{2} \right)^p \right)^{1/p} + b, \qquad (1)$$

where p denotes the generalized mean power.

In addition, we further enhance the feature representations by applying scale-GeM pooling scheme to aggregate feature maps at different scales. More specifically, we extract features at each scale according to Eq. (1) and then aggregate the multi-scale features using the parameter p_{ms}, enhancing the model's adaptability to images of varying scales. Integrating multi-scale information allows the model to capture structured features at different scales. Mathematically, it can be expressed as:

$$G_{ms} = \left(\frac{1}{N} \sum_{s=1}^{N} (g_s + \zeta_s)^{p_{ms}} \right)^{1/p_{ms}} - \zeta_s, \qquad (2)$$

where $\zeta_s = -min(g_s)$ denotes a shift of scale-specific global feature g_s, N denotes the number of scales and p_{ms} is the multi-scale power parameter used in feature aggregation.

In addition to the above structured features, sequential information is also encoded within our framework for generating comprehensive representations. Specifically, we utilize the method proposed in ViT [6] to represent S and I as a sequence of visual tokens. Specifically, we uniformly divide the image or the sketch into non-overlapping patches of equal size at the pixel level, and thus $S, I \in \mathbb{R}^{H \times W \times C}$ are transformed into a sequence of visual embeddings $V_{S,I} \in \mathbb{R}^{n \times d}$. Subsequently, a residual connection is introduced to combine both structured and serialized features, resulting in the final embedding formulated as:

$$F_{S,I} = G_{ms} + V_{S,I}, \quad (3)$$

which are integrated into the subsequent Transformer encoder module to generate comprehensive representations.

2.3 Dual-Sample Relationship Modeling

The introduced DSRM explores the semantic relationships among a batch of image samples and utilizes the learned complementary semantic information to enhance the features of the input images or sketches. DSRM mainly includes three steps:

Intra-sample Relationship Modeling. The output embeddings derived from the hybrid information fusion module are used as input for ViT. However, instead of employing the MLP head provided by the vanilla ViT [6], an additional learnable fusion token [Mix] $\in \mathbb{R}^d$ is introduced into the multi-head self-attention and MLP blocks to learn a global representation of the image or sketch [12]. For both sketches and images, the global token after multi-head attention can be obtained as:

$$G_0 = [\text{Mix}; G^1; G^2; ...; G^n], \quad (4)$$

$$G_l = \text{MSA}(\text{LN}(G_{l-1})) + G_{l-1}, l = 1...L, \quad (5)$$

$$G_l = \text{MLP}(\text{LN}(G_l)) + G_l, l = 1...L, \quad (6)$$

where $G \in \{S, I\}$, $LN(\cdot)$ represents the layer norm, while L is the number of layers. Meanwhile, the residual connection is introduced in Eq. (5) and (6). $MSA(\cdot)$ denotes multi-head self-attention, which is formulated as:

$$\text{SelfAttn}(Q^g, K^g, V^g) = \text{softmax}(\frac{Q^g K^{gT}}{\sqrt{d}})V^g, \quad (7)$$

where Q, K, and V represent the query, key, and value, respectively, which are obtained by mapping the same token through three different linear projection heads $[W_q, W_k, W_v]$.

Inter-sample Relationship Modeling. As indicated in MBGNN [15], the input images exhibit varying levels of similarity within a given mini-batch. In this sense, the model performance can be significantly improved by effectively extracting complementary semantic information among different images in the mini-batch to enhance the feature representation power of the input images.

In practice, the training set of ZS-SBIR may be noisy, whilst the number of samples in different categories varies greatly. To overcome these drawbacks, we also utilize the multi-head attention mechanism to facilitate knowledge transfer among sample embeddings [11]. To be specific, for the minibatch sample embeddings $H \in \mathbb{R}^{B \times N \times C} = \left\{ G_l^{(i)} \in \mathbb{R}^{N \times C}, i = 1, 2, ..., B \right\}$, H is transformed into:

$$r_{ij} = \mathbf{v}^T \tanh(\mathbf{W} \cdot H + \mathbf{b}), \tag{8}$$

$$\alpha_{ij} = \text{softmax}\,(r_{ij}) = \frac{\exp{(r_{ij})}}{\sum_{k=1}^{C} \exp{(r_{ik})}}, \tag{9}$$

$$H' = \sum_{j=1}^{C} \alpha_{ij} * H, \tag{10}$$

where $\mathbf{W}$, $\mathbf{v}$, and $\mathbf{b}$ are parameters to be learned. Then, the multi-head mechanism is employed to capture features from various aspects, allowing the model to learn invariant features among samples:

$$H_0 = [H_1'; H_2'; \ldots; H_k'], \tag{11}$$

where k is the number of heads. Subsequently, a residual connection followed by layer normalization is applied to achieve the normalized mini-batch embedding $H^* \in \mathbb{R}^{B \times C}$ as follows:

$$H^* = \text{LN}(H_0 \cdot W^d + H), \tag{12}$$

where W^d denotes the weight to align the dimensionality.

The embeddings of test samples can be directly utilized for prediction without undergoing the aforementioned process, which implies that generating transformed or synthetic samples is unnecessary. This approach allows us to avoid the layer normalization based on batch statistics as described in Eq. (12).

2.4 Cross-Modal Attention Module

In our method, a cross-attention mechanism is employed to estimate the local visual correspondences between sketch and photo tokens. By exchanging sketch queries Q_S and image queries Q_I, new Q, K, V representations are generated, identifying pairwise connections between visual tokens of different modalities (i.e., sketches and photos). The cross-modal attention is calculated as follows:

$$\text{CrossAttn}(Q_I, K_S, V_S) = \text{softmax}(\frac{Q_I K_S^T}{\sqrt{D}}) V_S \tag{13}$$

In this manner, we can update the sketch token embeddings with information from the image tokens, and similarly update the image token embeddings with information from the sketch tokens.

Next, we align the tokens after cross-modal attention. The sketch and image tokens after applying cross-modal attention are represented as H_S and H_I. Similar to [3], we exploit a triplet loss applied on the [Mix] to train our framework.

Given a triplet $< \text{Mix}^a(H_S), \text{Mix}^p(H_I), \text{Mix}^n(H_I) >$, where $\text{Mix}^a(H_S)$ is an anchor sketch feature, $\text{Mix}^p(H_I)$ is a positive image feature with the same class label, $\text{Mix}^n(H_I)$ is a negative image feature with a different class label. During the training process, it is expected that for any anchor sketch image, its positive photo image should be closer than its negative photo image in the embedding space. To achieve the above goal, the triplet loss is formulated as follows:

$$\mathcal{L}_{tri} = \frac{1}{T} \sum_{i=1}^{T} \max \{ \|\text{Mix}^a(H_S) - \text{Mix}^p(H_I)\| - \|\text{Mix}^a(H_S) - \text{Mix}^n(H_I)\| + m, 0 \}, \quad (14)$$

where T is the total number of triplets, and m is the margin.

3 Experiments

To validate the effectiveness of our model, we have conducted experiments in different public benchmarking datasets for category-level ZS-SBIR. Additionally, we have carried out cross-dataset ZS-SBIR experiments to assess the generalization of learned visual correspondences in zero-shot settings across different datasets.

3.1 Datasets and Settings

Three large public datasets used for category-level ZS-SBIR include TU Berlin Ext [25], Sketchy Ext [13], and QuickDraw Ext [5]. The TU Berlin Ext dataset comprises 250 categories, each of which contains 80 sketches and a total of 204,489 photos sourced from ImageNet [4] and web images, paired with the sketches. Sketchy Ext dataset is an enlarged photo repository comprising roughly 73k images. It contains 125 categories, each of which has 100 photos and each photo has 5-8 corresponding sketches. In particular, Sketchy-25 refers to a partition of 100 training classes and 25 testing classes, while Sketchy-21 [24] refers to the version of 104/21 train/test classes, which selects classes that do not overlap with ImageNet categories as unseen classes. QuickDraw Ext is the largest SBIR dataset, featuring over 330K sketches across 110 categories and 204K photos.

In accordance with the standard evaluation protocol, mean average precision (mAP), Prec@100, and Prec@200 are used for the performance measure.

3.2 Implementation Details

In implementation. all the experiments are conducted using a NVIDIA RTX3090 GPU under PyTorch framework. For uniformity, either Sketch or image is scaled to 224×224. In our model, ViT-B/16 is pre-trained on ImageNet-1K, while AdamW optimizer is used with learning rate set as 10^{-5} in the training process. The total model training time is approximately 7h. For HIFM, we follow the same parameters as in [19] in which $p = 4.6$, $p_r = 2.5$, and $p_{ms} = 10$.

Table 1. Comparison of our proposed method with the other competitors. "ESI" denotes external semantic information. The best and the second best scores are highlighted **in bold** and underlined, respectively. The results of the competitors are directly derived from the original papers.

Method	ESI	$\mathbb{R}^D$	TU-Berlin Ext mAP	Prec@100	Sketchy-25 mAP	Prec@100	Sketch-21 mAP@200	Prec@200	QuickDraw Ext mAP	Prec@200
ZSIH [20]	✓	64	0.220	0.291	0.254	0.340	-	-	-	-
CC-DG [16]	✗	256	0.247	0.392	0.311	0.468	-	-	-	-
DOODLE [5]	✓	256	0.109	-	0.369	-	-	-	0.075	0.068
SEM-PCYC [7]	✓	64	0.297	0.426	0.349	0.463	-	-	-	-
SAKE [14]	✓	512	0.475	0.599	0.547	0.692	0.497	0.598	0.130	0.179
SketchGCN [10]	✓	300	0.324	0.505	0.382	0.538	-	-	-	-
StyleGuide [8]	✗	200	0.254	0.355	0.376	0.484	0.358	0.400	-	-
BDA-SketRet [2]	✓	128	0.375	0.504	0.437	0.514	**0.556**	0.458	**0.154**	**0.355**
SBTKNet [22]	✓	512	0.480	0.608	0.553	0.698	0.502	0.596	-	-
DSN [23]	✓	512	0.484	0.591	0.583	0.704	-	-	-	-
Sketch3T [18]	✓	512	0.507	-	0.575	-	-	-	-	-
TVT [21]	✓	384	0.484	<u>0.662</u>	0.648	0.796	0.531	<u>0.618</u>	<u>0.149</u>	0.149
ZSE-SBIR-RN [12]	✓	512	<u>0.542</u>	0.657	<u>0.698</u>	<u>0.797</u>	0.525	**0.624**	0.145	0.216
Ours	✗	512	**0.612**	**0.672**	**0.766**	**0.830**	<u>0.547</u>	0.616	0.147	<u>0.223</u>

3.3 Competitors

In our comparative studies, we compare our model against several state-of-the-art ZS-SBIR approaches, including ZSIH [20], CC-DG [16], DOODLE [5], SEM-PCYC [7], SAKE [14], StyleGuide [8], DSN [23], SketchGCN [10], BDA-SketRet [2], SBTKNet [22], TVT [21], Sketch3T [18], ZSE-SBIRZRN [12]. In particular, external semantic information is used in all competing methods except CC-DG [16] and StyleGuide [8]. In contrast, our method relies only on the learned visual correspondences between sketch-photo pairs.

3.4 Results

As shown in Table 1, our proposed method achieves superior results compared to the other methods in most experiments, and even outperforms numerous approaches using text-assisted semantic information. For example, our approach beats the state-of-the-art ZSE-SBIR-RN method in all the datasets across different metrics except that a slightly inferior Prec@200 score is achieved in Sketch-21. In particular, dramatic performance advantages are exhibited in both the TU-Berlin Ext and Sketchy-25 datasets, demonstrating that our model surpasses ZSE-SBIR-RN by 8.0% and 6.8% mAP accuracies, respectively. These results fully suggest that our method enjoys better generalization capability for unseen classes. Despite showing slight inferiority in Sketchy-21 and QuickDraw Ext datasets, our model still achieves top-three results among all the ZS-SBIR approaches. Notably, BDA-SketRet [2] and TVT [21] report better results than our model in the QuickDraw dataset, possibly due to the lower sketch quality

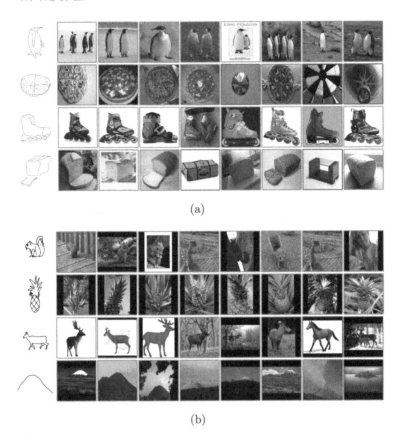

Fig. 2. Some representative query examples in TU-Berlin Ext (a) and Sketchy-25 datasets (b). The query sketch is shown on the left while the top-returned eights images are displayed on the right. The ground-truth images and the false positive results are highlighted in a green and red box, respectively. (Color figure online)

and the significant diversity of the sketches in Quickdraw. This will result in considerable domain differences between sketches and images, which poses greater challenges to our method. Since both BDA-SketRet and TVT incorporate specialized domain adaptation mechanisms (such as bi-level domain adaptation and multi-modal hypersphere learning), these mechanisms are more helpful in handling domain discrepancies between sketches and images.

3.5 Qualitative Results

In addition to the above quantitative evaluations, we also qualitatively evaluate our method and present some query examples, as illustrated in Fig. 2. It can be observed that our model successfully retrieves the correct candidates in most cases except for some incorrect candidates with similar structures. For example, the query sketch "pizza" in the second row is similar to hot air balloons in structure and outline. Neither complex backgrounds nor the presence of multiple objects significantly affect the retrieval performance of our method.

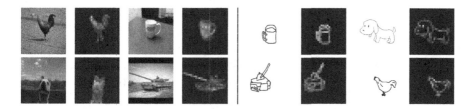

Fig. 3. Illustrative attention maps of some examples obtained by our network for unseen categories.

3.6 Ablation Study

In this section, we conduct extensive ablation studies and explore the effect of individual components on the performance of the model. To be specific, we remove each individual component while keeping the other parts intact, and the results are presented in Table 2. It can be observed that our complete model achieves the best results in both Sketchy-25 and TU-Berlin Ext datasets. More specifically, removing HIFM leads to performance decline of 1.9% mAP and 0.7% mAP in respective datasets, suggesting the beneficial effects of multi-level feature representations. When exploring the DSRM module, different groups of ablation experiments are conducted. For example, when the inter-sample relation modeling is not involved in our model, the mAP scores decrease by approximately 2.0% and 0.9%, respectively in the experiment of **w/o DSRM (inter)**. Since inter-sample relation modeling within DSRM can extract complementary semantic information among different images in the mini-batch, it primarily identifies invariant features within images of the same class and significantly contributes to improving ZS-SBIR performance.

To further examine the important role of DSRM within our model, the attention maps obtained by employing the fusion token as a query and measuring its correlation with each visual token by a vector dot product similar to [1]. As illustrated in Fig. 3, the attention maps focus on the key features of the target objects in the images (e.g., the comb of a rooster) and the line textures in the sketches, which is conducive to capturing correlation among different samples and benefits the model generalization ability of transferring to unseen classes.

3.7 Cross-Dataset Category-Level ZS-SBIR

To assess the generalization ability of our method across different datasets, additional evaluations are carried out in a zero-shot setting where the model is trained on a dataset and tested on another unseen dataset. Beyond the conventional within-dataset ZS-SBIR benchmarks, cross-dataset ZS-SBIR is more challenging

Table 2. Ablation results in Sketchy-25 and TU-Berlin Ext datasets. The best results are highlighted in bold.

Model	Sketchy-25		TU-Berlin Ext	
	mAP	Prec@100	mAP	Prec@100
w/o C-attn	0.293	0.356	0.301	0.367
w/o DSRM	0.235	0.341	0.212	0.318
w/o DSRM(Inter) and HIFM	0.734	0.794	0.593	0.654
w/o DSRM(Inter)	0.746	0.809	0.603	0.659
w/o HIFM	0.747	0.818	0.605	0.664
Complete (Ours)	**0.766**	**0.830**	**0.612**	**0.672**

Table 3. Comparison of different methods for cross-dataset ZS-SBIR. "S" and "T" denote the Sketchy-25 and TU-Berlin Ext datasets, respectively. The numbers in brackets denote the number of test categories unseen in the zero-shot setting. The best and the second best scores are highlighted in bold and underlined, respectively.

Method	S→T(21)		T→S(8)	
	mAP	Prec@100	mAP	Prec@100
CC-DG [16]	0.252	0.403	0.570	0.660
DSN [23]	0.384	0.480	0.646	0.673
SAKE [14]	0.421	0.549	0.657	<u>0.722</u>
ZSE-SBIR-RN [12]	<u>0.476</u>	<u>0.590</u>	**0.746**	**0.816**
Ours	**0.580**	**0.632**	<u>0.677</u>	0.705

in practical applications, as sketches from various datasets are typically drawn in different styles with significant inter-class variances. As illustrated in Table 3, our method reports the highest results in both metrics when transferring from Sketchy-25 to TU-Berlin Ext and exhibits competitive performance in a converse setting, indicating that our method can capture visual correspondences across different datasets and thus is conducive to zero-shot task.

4 Conclusion

In this study, we have proposed a ZS-SBIR model with hybrid information fusion and sample relationship modeling. More specifically, the hybrid information fusion can capture multi-level features of sketches and images, which is capable of enhancing the representation power of feature representations while improving the model's adaptability to different classes of sketches. In addition, the dual-sample relationship modeling method is incorporated to strengthen integrated representations and improve model generalization. With the help of the cross-modal attention mechanism, the resulting representations of two different modalities are well aligned within the feature space. Extensive experiments on three benchmark datasets demonstrate the promise of our method with desirable zero-shot performance.

References

1. Caron, M., et al.: Emerging properties in self-supervised vision transformers. In: Proceedings of the IEEE/CVF International Conference on Computer Vision (ICCV) (2021)
2. Chaudhuri, U., Chavan, R., Banerjee, B., Dutta, A., Akata, Z.: BDA-Sketret: bi-level domain adaptation for zero-shot SBIR. arXiv preprint arXiv:2201.06570 (2022)
3. Chechik, G., Sharma, V., Shalit, U., Bengio, S.: Large scale online learning of image similarity through ranking. J. Mach. Learn. Res. **11**, 1109–1135 (2010)
4. Deng, J., Dong, W., Socher, R., Li, L.J., Li, K., Fei-Fei, L.: ImageNet: a large-scale hierarchical image database. In: Proceedings of the 2009 IEEE Conference on Computer Vision and Pattern Recognition (CVPR), pp. 248–255. IEEE (2009)
5. Dey, S., Riba, P., Dutta, A., Llados, J., Song, Y.Z.: Doodle to search: practical zero-shot sketch-based image retrieval. In: Proceedings of the IEEE/CVF Conference on Computer Vision and Pattern Recognition (CVPR), pp. 2179–2188 (2019)
6. Dosovitskiy, A., et al.: An image is worth 16x16 words: transformers for image recognition at scale. arXiv preprint arXiv:2010.11929 (2020)
7. Dutta, A., Akata, Z.: Semantically tied paired cycle consistency for zero-shot sketch-based image retrieval. In: Proceedings of the IEEE/CVF Conference on Computer Vision and Pattern Recognition (CVPR), pp. 5089–5098 (2019)
8. Dutta, T., Singh, A., Biswas, S.: StyleGuide: zero-shot sketch-based image retrieval using style-guided image generation. IEEE Trans. Multimedia **23**, 2833–2842 (2020)
9. Gulcehre, C., Cho, K., Pascanu, R., Bengio, Y.: Learned-Norm pooling for deep feedforward and recurrent neural networks. In: Proceedings of the European Conference on Machine Learning and Principles and Practice of Knowledge Discovery in Databases (ECML-PKDD) (2014)
10. Gupta, S., et al.: Zero-shot sketch based image retrieval using graph transformer. In: Proceedings of the 2022 26th International Conference on Pattern Recognition (ICPR), pp. 1685–1691. IEEE (2022)
11. Hou, Z., Yu, B., Tao, D.: BatchFormer: learning to explore sample relationships for robust representation learning. In: Proceedings of the IEEE/CVF Conference on Computer Vision and Pattern Recognition (CVPR), pp. 7256–7266 (2022)
12. Lin, F., Pang, K., Bhunia, A.K., Song, Y.Z., Xiang, T.: Zero-shot everything: sketch-based image retrieval, and in explainable style. In: Proceedings of the IEEE/CVF Conference on Computer Vision and Pattern Recognition (CVPR) (2023)
13. Liu, L., Shen, F., Shen, Y., Liu, X., Shao, L.: Deep sketch hashing: fast free-hand sketch-based image retrieval. In: Proceedings of the IEEE Conference on Computer Vision and Pattern Recognition (CVPR) (2017)
14. Liu, Q., Xie, L., Wang, H., Yuille, A.L.: Semantic-aware knowledge preservation for zero-shot sketch-based image retrieval. In: Proceedings of the IEEE/CVF International Conference on Computer Vision (ICCV), pp. 3662–3671 (2019)
15. Mondal, A.K., Jain, V., Siddiqi, K.: Mini-batch graphs for robust image classification. arXiv preprint arXiv:2105.03237 (2021)

16. Pang, K., Siddiquie, B., Zhang, Y., Qi, Y., Song, Y.Z., Xiang, T.: Generalising fine-grained sketch-based image retrieval. In: Proceedings of the IEEE/CVF Conference on Computer Vision and Pattern Recognition (CVPR), pp. 677–686 (2019)
17. Radenovic, F., Tolias, G., Chum, O.: Fine-tuning CNN image retrieval with no human annotation. IEEE Trans. Pattern Anal. Mach. Intell. **41**, 1655–1668 (2019)
18. Sain, A., Bhunia, A.K., Potlapalli, V., Chowdhury, P.N., Xiang, T., Song, Y.Z.: Sketch3T: test-time training for zero-shot SBIR. In: Proceedings of the IEEE/CVF Conference on Computer Vision and Pattern Recognition (CVPR), pp. 7462–7471 (2022)
19. Shao, S., Chen, K., Karpur, A., Cui, Q., Araujo, A., Cao, B.: Global features are all you need for image retrieval and reranking. In: Proceedings of the IEEE/CVF International Conference on Computer Vision (ICCV) (2023)
20. Shen, Y., Liu, L., Shen, F., Shao, L.: Zero-shot sketch-image hashing. In: Proceedings of the IEEE Conference on Computer Vision and Pattern Recognition, pp. 3598–3607 (2018)
21. Tian, J., Song, Y.Z., Xiang, T., Zhang, Z., Cao, J.: TVT: three-way vision transformer through multi-modal hypersphere learning for zero-shot sketch-based image retrieval. In: Proceedings of the AAAI Conference on Artificial Intelligence, vol. 36, pp. 2082–2090 (2022)
22. Tursun, O., Denman, S., Sridharan, S., Goan, E., Fookes, C.: An efficient framework for zero-shot sketch-based image retrieval. Pattern Recogn. **126**, 108528 (2022)
23. Wang, Z., Xu, X., Xu, B., Wang, Y., Zhou, Y., Wang, J.: Domain-smoothing network for zero-shot sketch-based image retrieval. arXiv preprint arXiv:2106.11841 (2021)
24. Yelamarthi, S.K., Reddy, S.K., Mishra, A., Mittal, A.: A zero-shot framework for sketch based image retrieval. In: Proceedings of the European Conference on Computer Vision (ECCV) (2018)
25. Zhang, H., Liu, S., Zhang, C., Ren, W., Wang, R., Cao, X.: SketchNet: sketch classification with web images. In: Proceedings of the IEEE Conference on Computer Vision and Pattern Recognition (CVPR) (2016)

Special Session: ExpertSUM: Special Session on Expert-Level Text Summarization from Fine-Grained Multimedia Analytics

CalorieVoL: Integrating Volumetric Context Into Multimodal Large Language Models for Image-Based Calorie Estimation

Hikaru Tanabe[ID] and Keiji Yanai[✉][ID]

The University of Electro-Communications, Chofu, Tokyo, Japan
{tanabe-h,yanai}@mm.inf.uec.ac.jp

Abstract. Multimodal Large Language Models (MLLMs) can perform various food-related tasks with high quality. Notably, high-performance MLLMs, such as GPT-4V, can even estimate caloric content from food images. However, these MLLMs often struggle to accurately recognize volume information, which often leads to errors in calorie estimation. To address this issue, we propose a new MLLM framework called CalorieVoL, designed to enhance the recognition of volume information in food items. By integrating this framework into MLLMs like GPT-4V, we achieved higher scores in terms of MAE and correlation coefficients on Nutrition5k compared to simple MLLMs. Our experiments also showed that the volume-aware recognition improved responses in scenarios where accurate volume estimation is critical.

Keywords: Multimodal Large Language Models · Image-based Calorie Estimation · Volume Estimation

1 Introduction

By recording daily food intake, we can obtain valuable information that helps achieve health-related goals such as dieting and bodybuilding. For this purpose, manual dietary survey frameworks like food diary method, 24-hour recall method, and food frequency questionnaire are widely used among nutrition experts. However, these methods are time-consuming and require participants to weigh or recall their food intake, which poses a challenge to them. Moreover, caloric content of daily food is a critical metric for maintaining a healthy lifestyle. Therefore, quickly and easily estimating the caloric content of food can have a significant impact on the healthcare field.

Taking pictures of food using smartphones or camera-equipped AR devices is an easier approach for people to keep a food diary. However, daily food varies widely in type and quantity, which leads to variance in the caloric content. Thus, it is important to build models that can correctly recognize the type and quantity of food from images for calorie estimation.

Large Language Models (LLMs) have acquired a wide range of commonsense knowledge about the world. Mainly, LLMs that have undergone recent instruction tuning can reason tasks based on the commonsense they possess when provided with prompts designed for the tasks [1]. Multimodal Large Language Models (MLLMs) can solve image recognition tasks while retaining the reasoning capabilities of LLMs [2]. Some models have acquired specific knowledge in fields such as bio-medicine, achieving human-level performance [3]. Some advanced models can also perform reasoning food-related tasks by recognizing various types of food [4]. However, these models cannot recognize the volumetric amount of food accurately.

In this study, we propose a framework called CalorieVoL that utilizes food recognition capabilities of MLLMs for image-based calorie estimation. While leveraging MLLMs to cover the diversity of food that conventional calorie estimation methods could not achieve, we introduce a novel volume estimator to complement the volume estimation capabilities that MLLMs struggle with. This approach allows us to estimate the caloric content of food images with high quality, even for food images that were not explicitly trained for calorie estimation tasks.

The main contributions of this study are as follows:

- We introduce CalorieVoL, a framework that enables volume-aware recognition for image-based calorie estimation using MLLMs.
- We introduce a new plug-in volume estimator by utilizing off-the-shelf SOTA models to integrate volume information into MLLMs.
- We evaluate the performance of CalorieVoL on the Nutrition5k [5] dataset and discuss the effectiveness and challenges of this method.

2 Related Work

2.1 Image-Based Calorie Estimation

Estimating the caloric content of food items shown in images has been attempted through various methods due to its applicability [6]. There are primary approaches called size-based methods, where a pipeline is constructed that combines multiple image recognition modules to estimate caloric content. The basic procedure involves first segmenting the food regions from the meal image, then estimating the food category, followed by estimating the volume or mass of the food region. Subsequently, the caloric content is estimated based on these results. By taking these steps before estimating the caloric content, these methods can particularly consider the quantity of food.

To determine the metric size of the food region, some methods estimate the actual size of objects included in the food image. Okamoto et al. [7] used a credit card or a long wallet as a reference, Akpa et al. [8] used chopsticks as a reference, and Ege et al. [9] used rice grains as a reference. Furthermore, Tanno et al. [10] employed a method using anchors placed in an AR space, obtaining the actual size through interaction with the user. DepthCalorieCam [11] significantly reduced the error in calorie estimation by estimating the food volume using a depth camera and a segmentation model. Naritomi et al. [12] reconstructed high-quality 3D meshes of the dish and food using an implicit function representation.

However, these size-based methods lack variety in food. For example, DepthCalorieCam is limited to estimating only three categories of food, which causes the lack of applicability.

In this study, inspired by size-based methods, we create a food volume estimator by combining an open-set segmentation model, a promptable segmentation model, and a monocular depth estimation model. We aim to achieve zero-shot calorie estimation with high quality, without the need for training on the target dataset.

2.2 Multimodal Large Language Models

In recent years, Large Language Models (LLMs), which are language models trained under large-scale conditions with a substantial number of model parameters, data, and computational resources, have achieved high performance across various language tasks. These models exhibit a power-law improvement in performance as the scale of learning conditions increases [13]. they also demonstrate emergent abilities where their performance improves dramatically at a certain stage as the learning conditions are scaled up [14]. These new aspects, which were not observed in conventional language models, are attracting significant attention.

Multimodal Large Language Models (MLLMs) are constructed by extending the ability of LLMs to other modalities. Flamingo [15] acquired the ability to answer various vision-language questions by fusing the visual features with text features using gated cross-attention dense blocks. LLaVA [2] employed a linear or MLP layer to transform the visual features into the shape of the language tokens. Additionally, it adopted a training framework called Visual Instruction Tuning, which resulted in high-quality instruction-following ability for vision-language tasks. MiniGPT-4 [16] and InstructBLIP [17] also acquired abilities to solve a wide range of tasks by using Q-Former [18] as the vision-language connector and applied training framework similar to Visual Instruction Tuning.

In food domain, FoodLMM [19] achieved high performance in various food-related tasks, including image-based calorie estimation. we particularly focuses on improving the performance of calorie estimation from various food images without the need to train.

3 Methods

The overview of CalorieVoL is shown in Fig. 1. CalorieVoL consists of two main components: a part that uses MLLMs as a calorie estimator and a part that estimates the volume of food (Sect. 3.1). By combining these components, we construct CalorieVoL (Sect. 3.2).

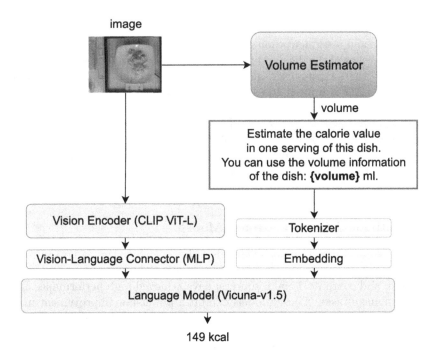

Fig. 1. Overview of CalorieVoL (the case of LLaVA-v1.5)

3.1 Volume Estimator

We construct a model that recognizes the food portion from a meal image and estimates its volume. The process from inputting the image to outputting the volume value is shown in Fig. 2.

The process of volume estimation is as follows (Fig. 2). First, the bounding box of the dish is obtained using Grounding DINO [20], an open-set object detection model. Next, the process is divided into three parts using this region of interest. First, the dish's region mask is obtained by applying Segment Anything (SAM) [21] to the region of interest. Second, the bounding box of the food portion is obtained by applying Grounding DINO to the region of interest, followed by applying SAM to obtain the region mask. Third, the depth map is obtained by applying Marigold [22], a monocular depth estimation model, to the region of interest.

Based on these two obtained masks and the depth map, the actual volume is estimated. First, the element-wise Hadamard product between the depth map and each mask is taken to extract the regions of each mask. Next, the maximum value within the dish's region in the depth map is obtained and used as the depth of the dish's reference plane. Then, the difference is taken between each value in the depth map of the food region and the depth of the dish's top surface. This provides the ratio of the height from the dish's reference plane to the top surface of the food to the height from the shooting position. Furthermore, the actual

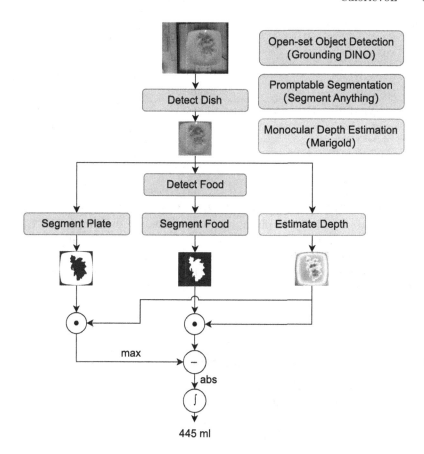

Fig. 2. Structure of the food volume estimator

height from the dish's reference plane to the top surface of the food for each pixel is obtained based on the actual height from the dish's reference plane to the shooting position and this ratio. Finally, the volume integration is performed for all pixels in the food region's depth map based on this height and the actual area per pixel. The volume integration value at this time is calculated using (1). Here, V represents the actual volume of the food, D_{ij} represents the actual height of the pixel at row i and column j, and A_{ij} represents the actual area of the pixel at row i and column j.

$$V = \sum_{i=1}^{n} \sum_{j=1}^{m} D_{ij} A_{ij} \quad (1)$$

Table 1. Results of zero-shot calorie estimation by MLLMs on the Nutrition5k dataset

Model	MAE / kcal ↓	MAPE / % ↓	r ↑
LLaVA-13B	109.6	92.8	0.656
GPT-4V	106.6	**54.8**	0.688
LLaVA-13B + CalorieVoL (Ours)	6122.7	6591.4	−0.041
GPT-4V + CalorieVoL (Ours)	**101.7**	56.8	**0.708**

3.2 CalorieVoL

We use GPT-4V [4] and LLaVA-v1.5 [2] as the MLLMs for the reasoning of calorie estimation. The text prompt is constructed to encourage the LLM to estimate the calorie value in one serving of the dish. Additionally, it includes a placeholder {volume}, where the volume value estimated from the volume estimator will be substituted. This allows the LLM to estimate the calorie value based on the volume of the food.

4 Experiments

4.1 Evaluation of Calorie Estimation

We conducted evaluations using the test split of the Nutrition5k dataset. It should be noted that none of the models were trained on Nutrition5k dataset, making this an evaluation of zero-shot calorie estimation. The temperature is set at 0 through the evaluation.

Table 1 shows the results of zero-shot calorie estimation. If the output text of a model did not include the calorie value, the result was obtained by repeating the same question up to five times with a temperature parameter set to 0.2. As a result, 79 data points were excluded from the evaluation for the model that combined GPT-4V with the food volume estimator. It can be observed that the model combining the proposed food volume estimator with the base model, GPT-4V, achieved better scores on MAE and correlation coefficient.

Figure 3 and Fig. 4 show scatter plots of the estimated values and the ground truth for zero-shot calorie estimation. Although there is not a significant difference overall, the correlation coefficient is higher for the model combined with the food volume estimator.

Additionally, Fig. 5 and Fig. 6 present examples of the model responses in zero-shot calorie estimation. In Fig. 5, it can be seen that the estimation results are improved significantly as the volume estimation results from the proposed method are considered in the calorie estimation process. On the other hand,

Fig. 6 shows an example of overestimation when LLaVA-v1.5 is used as the MLLM. Observing the reasoning process, it can be seen that the calorie value of the dish is initially estimated relatively accurately. However, in the latter part, the calorie value seems to have been incorrectly multiplied by the volume value, leading to the final overestimated result. This suggests that the overestimation occurred due to the MLLM's inability to properly recognize units and perform calculations accurately.

4.2 Evaluation of Volume Estimation

Figure 7 shows the estimated volume values. It can be observed that the shape of the distribution resembles the distribution of the true calorie values (Fig. 3).

Additionally, Fig. 8 presents the results of object detection, segmentation, and depth estimation during the volume estimation process. For object detection and segmentation, appropriate regions for both the dish and food were successfully extracted, demonstrating overall high-quality estimation. For depth estimation, variations in the uneven surfaces within the image are well represented. Furthermore, in the image containing multiple types of food, the depth values in areas with different heights are noticeably distinct from the surrounding values.

5 Direction of Improvement in Volume Estimation

The food volume estimator proposed in this study has two main characteristics that may lead to an overestimation of volume (Fig. 9). First, the volume from the bottom of the food to the reference plane of the dish may be overestimated. Second, when the lowest part of the dish is covered by food, an incorrect reference plane of the dish may be selected.

In the research of DepthCalorieCam [11], after the volume is calculated using a method similar to the one in this study, the volume value is input into a food mass regression model. This model adjusts the estimate by accounting for the overestimated volume, leading to an accurate calorie estimation. Another approach, as suggested by Naritomi et al. [12], is to reconstruct high-quality 3D shapes of the dish and food, ensuring that the area between the bottom of the food and the reference plane of the dish does not affect the food volume.

However, creating models to implement these methods would require a large amount of food data, which would be burdensome to prepare. For the method that estimates the excess volume, a large amount of annotated data on volume would be needed to train the model. Additionally, for the method of reconstructing 3D shapes, the existing performance of 3D shape reconstruction methods is not sufficient, especially in the food domain, necessitating a large amount of 3D shape data for improvement.

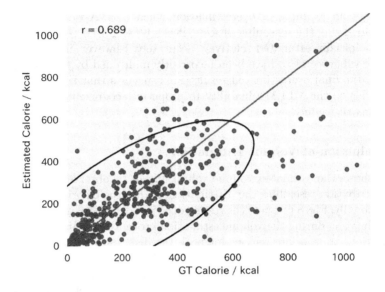

Fig. 3. Scatter plot of estimated calorie values by GPT-4V

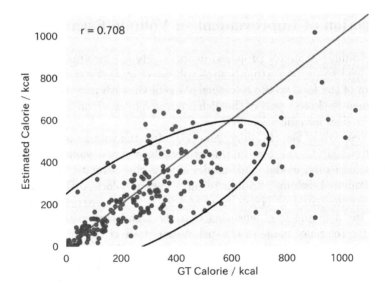

Fig. 4. Scatter plot of estimated calorie values by GPT-4V and the food volume estimator

CalorieVoL   361

Answer: **687 kcal**

User: Estimate the calorie value in one serving of the dish.
Return the single calorie value in the following format: "[[x]] calories".
This information is essential and must be provided in every response.

Example: "[[100]] calories"

GPT-4V: This is a bowl of almonds. A typical serving size of almonds is about 1 ounce (28 grams), which is roughly 23 almonds and contains approximately [[160]] calories. However, without a reference for the exact amount in this bowl, it's difficult to provide an accurate calorie count. If this bowl contains about one serving, then the calorie estimate would be around **[[160]] calories**. If it contains more, the calorie count would be higher accordingly.

User: Estimate the calorie value in one serving of the dish.
Return the single calorie value in the following format: "[[x]] calories".
This information is essential and must be provided in every response.

Example: "[[100]] calories"

You can use the volume information of the dish: 527 ml.

GPT-4V+Vol (Ours): The image shows a bowl of almonds. Almonds are a nutrient-dense food, and a typical serving size of almonds is about 1 ounce (28 grams), which is approximately 23 almonds. This serving size generally contains about 160 to 170 calories. However, the bowl in the image appears to contain more than a single serving.

Without knowing the exact weight or number of almonds in the bowl, it's challenging to provide an accurate calorie count. If we assume the bowl contains around 3 servings (which is a rough estimate based on visual assessment), the calorie content would be approximately 480 to 510 calories.

Given the volume information of the dish (527 ml), and assuming the bowl is filled mostly to the top, it could potentially hold around 4 to 5 ounces of almonds (since almonds are not a liquid and their volume would not directly convert to weight). This would mean the bowl might contain around 640 to 850 calories, depending on the exact amount and whether the almonds are raw or roasted, which can slightly affect the calorie content.

For the purpose of providing a single calorie value, I will estimate the calorie content of the bowl of almonds as follows:

[[700]] calories

Please note that this is a rough estimate and the actual calorie content could be higher or lower depending on the precise amount of almonds in the bowl.

Fig. 5. Example of calorie estimation by the model combining GPT-4V and the food volume estimator

Answer: **149 kcal**

User: Estimate the calorie value in one serving of the dish.
Return the single calorie value in the following format: "[[x]] calories".
This information is essential and must be provided in every response.

Example: "[[100]] calories"

You can use the volume information of the dish: 445 ml.

LLaVA-13B+Vol (Ours): To estimate the calorie value in one serving of the dish, we need to know the ingredients and their respective calorie counts. Since the image shows a plate of scrambled eggs, we can assume that the dish is primarily composed of eggs.

The calorie value of one serving of scrambled eggs can vary depending on the number of eggs used and the cooking method. However, as a rough estimate, we can assume that one serving of scrambled eggs contains around 200-250 calories.

To calculate the calorie value for the entire dish, we need to know the volume of the dish, which is 445 ml. To convert the volume to weight, we can assume that the dish is made up of 445 grams of scrambled eggs.

So, the calorie value of the dish would be approximately **98,000 to 112,000 calories**. However, this is a rough estimate and may not be accurate, as the actual calorie count can vary depending on the specific ingredients and cooking method used.

Fig. 6. Example of overestimation by the model combining LLaVA-v1.5 and the food volume estimator

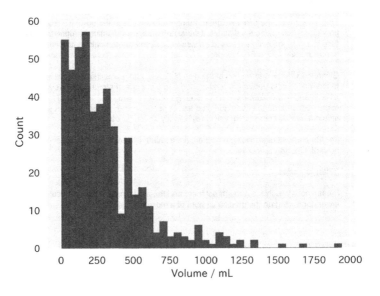

Fig. 7. Distribution of estimated food volumes

On the other hand, recent efforts have been made to achieve high-quality 3D shape reconstruction methods based on techniques such as NeRF [23] and 3D Gaussian Splatting [24]. By leveraging these approaches, it is expected that the challenges in volume estimation will be comprehensively addressed.

In terms of accurately capturing the spatial information of input images in MLLMs, methods such as SpatialVLM [25] exist. This method extracts various information from the input image using expert models like depth estimation and segmentation, integrates these results, and trains the MLLM based on the reconstructed 3D information. It is a promising direction to utilize off-the-shelf models to construct 3D information and train MLLMs to recognize food more spatially.

Furthermore, to prevent overestimation of calorie values under zero-shot conditions, it seems promising to prompt the language model to revise and correct the reasoning process when an output with an unreasonable calorie value is generated, based on the commonsense knowledge on food in MLLMs.

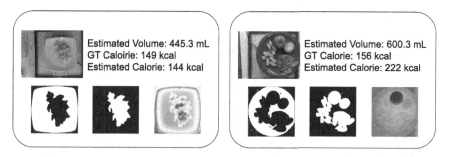

Fig. 8. Results of object detection, segmentation, and depth estimation. Top left: original image, bottom: dish region mask, food region mask, and depth map.

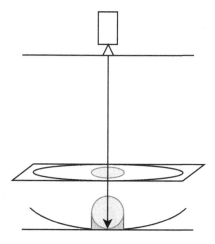

Fig. 9. Overestimation of volume by the food volume estimator when the food is assumed to be spherical. Blue: food region, Red: excess region. (Color figure online)

6 Conclusion

In this study, we proposed CalorieVoL, a novel framework that enhances the calorie estimation from food images by incorporating volumetric context into MLLMs. Our evaluations on the Nutrition5k dataset under zero-shot conditions demonstrate that CalorieVoL with GPT-4V improves the estimation accuracy, particularly in scenarios where the quantity of food plays a critical role.

We also identified challenges related to the volume estimation and the overestimation of food volume. We discussed potential approaches to address these issues, including the incorporation of mass regression models and advanced 3D shape reconstruction techniques. Looking forward, integrating the methods such as NeRF and 3D Gaussian Splatting, as well as adopting spatial-aware MLLMs training frameworks, could further enhance the accuracy and applicability of calorie estimation models.

Overall, CalorieVoL represents a promising step towards more accurate and reliable calorie estimation from food images, with potential applications in personalized nutrition and healthcare. Future work will focus on refining the volume estimation process and correcting the reasoning procedure with MLLMs.

Acknowledgments. This work was supported by JSPS KAKENHI Grant Numbers, 22H00540 and 22H00548.

References

1. Wei, J., et al.: Finetuned language models are zero-shot learners. In: Proceedings of International Conference on Learning Representations (2021)
2. Liu, H., Li, C., Wu, Q., Lee, Y.J.: Visual instruction tuning. In: Advances in Neural Information Processing Systems (2023)
3. Li, C., et al.: LLaVA-Med: training a large language-and-vision assistant for biomedicine in one day. Adv. Neural Info. Process. Syst. **36** (2024)
4. Yang, Z., et al.: The dawn of LMMs: preliminary explorations with GPT-4V (ision). arXiv preprintarXiv:2309.17421 (2023)
5. Quin Thames, Arjun Karpur, Wade Norris, Fangting Xia, Liviu Panait, Tobias Weyand, and Jack Sim. Nutrition5k: towards automatic nutritional understanding of generic food. In: Proceedings of IEEE Computer Vision and Pattern Recognition, pp. 8903–8911 (2021)
6. Sultana, J., Ahmed, B.M., Masud, M.M., Huq, A.O., Ali, M.E., Naznin, M.: A study on food value estimation from images: Taxonomies, datasets, and techniques. IEEE Access **11** , 45910–45935 (2023)
7. Okamoto, K., Yanai, K.: An automatic calorie estimation system of food images on a smartphone. In: Proceedings of the 2nd International Workshop on Multimedia Assisted Dietary Management (2016)
8. Akpa, Y.A.E.A.H., Suwa, H., Yasumoto, K.: Smartphone-based food weight and calorie estimation method for effective food journaling. SICE J. Control Meas. Syst. Integr. **10**(5), 360–369 (2017)
9. Ege, T., Shimoda, W., Yanai, K.: A new large-scale food image segmentation dataset and its application to food calorie estimation based on grains of rice. In: Proceedings of ACMMM Workshop on Multimedia Assisted Dietary Management (2019)

10. Ryosuke Tanno, Takumi Ege, and Keiji Yanai. AR DeepCalorieCam V2: food calorie estimation with CNN and AR-based actual size estimation. In: Proc. of the 24th ACM Symposium on Virtual Reality Software and Technology (2018)
11. Yoshikazu Ando, Takumi Ege, Jaehyeong Cho, and Keiji Yanai. DepthCalorieCam: a mobile application for volume-based foodcalorie estimation using depth cameras. In: Proceedings of the 5th International Workshop on Multimedia Assisted Dietary Management, pp. 76–81 (2019)
12. Naritomi, S., Yanai, K.: Hungry Networks: 3D mesh reconstruction of a dish and a plate from a single dish image for estimating food volume. In: Proceedings of the 2nd ACM International Conference on Multimedia in Asia (2021)
13. Kaplan, J., et al.: Scaling laws for neural language models. arXiv preprint arXiv:2001.08361 (2020)
14. Wei, J., et al.: Emergent abilities of large language models. arXiv preprint arXiv:2206.07682 (2022)
15. Alayrac, J.-B., et al.: Flamingo: a visual language model for few-shot learning. Adv. Neural Info. Process. Syst. **35**, 23716–23736 (2022)
16. Zhu, D., Chen, J., Shen, X., Li, X., Elhoseiny, M.: MiniGPT-4: enhancing vision-language understanding with advanced large language models. arXiv preprint arXiv:2304.10592 (2023)
17. Dai, W., et al.: InstructBLIP: towards general-purpose vision-language models with instruction tuning. arXiv preprintarXiv:2305.06500 (2023)
18. Li, J., Li, D., Savarese, S., Hoi, S.: BLIP-2: bootstrapping language-image pre-training with frozen image encoders and large language models. In: Proceedings of International Conference on Machine Learning (2023)
19. Yin, Y., Qi, H., Zhu, B., Chen, J., Jiang, Y.G., Ngo, C.W.: FoodLMM: a versatile food assistant using large multi-modal model. arXiv preprintarXiv:2312.14991 (2023)
20. Liu, S., et al.: Grounding DINO: marrying DINO with grounded pre-training for open-set object detection. arXiv preprint arXiv:2303.05499 (2023)
21. Kirillov, A., et al.: Segment anything. arXiv preprint arXiv:2304.02643 (2023)
22. Ke, B., Obukhov, A., Huang, S., Metzger, N., Daudt, R.C., Schindler, K.: Repurposing diffusion-based image generators for monocular depth estimation. arXiv preprintarXiv:2312.02145 (2023)
23. Mildenhall, B., Srinivasan, P.P., Tancik, M., Barron, J.T., Ramamoorthi, R., Ng, R.: NeRF: representing scenes as neural radiance fields for view synthesis. In: Proceedings of European Conference on Computer Vision (2020)
24. Kerbl, B., Kopanas, G., Leimkühler, T., Drettakis, G.: 3D Gaussian splatting for real-time radiance field rendering. In: Proceedings of ACM Transactions on Graphics, vol. 42 (2023)
25. Chen, B., et al.: SpatialVLM: endowing vision-language models with spatial reasoning capabilities. arXiv preprint arXiv:2401.12168 (2024)

Can Masking Background and Object Reduce Static Bias for Zero-Shot Action Recognition?

Takumi Fukuzawa[1](✉), Kensho Hara[2], Hirokatsu Kataoka[2], and Toru Tamaki[1]

[1] Nagoya Institute of Technology, Nagoya, Japan
t.fukuzawa.986@nitech.jp, tamaki.toru@nitech.ac.jp
[2] National Institute of Advanced Industrial Science and Technology (AIST), Tsukuba, Japan
{kensho.hara,hirokatsu.kataoka}@aist.go.jp

Abstract. In this paper, we address the issue of static bias in zero-shot action recognition. Action recognition models need to represent the action itself, not the appearance. However, some fully-supervised works show that models often rely on static appearances, such as the background and objects, rather than human actions. This issue, known as static bias, has not been investigated for zero-shot. Although CLIP-based zero-shot models are now common, it remains unclear if they sufficiently focus on human actions, as CLIP primarily captures appearance features related to languages. In this paper, we investigate the influence of static bias in zero-shot action recognition with CLIP-based models. Our approach involves masking backgrounds, objects, and people differently during training and validation. Experiments with masking background show that models depend on background bias as their performance decreases for Kinetics400. However, for Mimetics, which has a weak background bias, masking the background leads to improved performance even if the background is masked during validation. Furthermore, masking both the background and objects in different colors improves performance for SSv2, which has a strong object bias. These results suggest that masking the background or objects during training prevents models from overly depending on static bias and makes them focus more on human action.

Keywords: zero-shot action recognition · VLM models · background bias · object bias · static bias · masking

1 Introduction

Action recognition is a task for recognizing the actions of a person in a video and is expected to have various applications [8,14,29]. Supervised learning of action recognition models requires a large dataset [11,15,27], however, collecting such a dataset can be costly, and the amount of collectable training data is often limited in real-world applications. For this reason, zero-shot action recognition models have been proposed [5]. Zero-shot learning involves predicting unseen categories during testing, and a current popular approach is to use VLM models such as CLIP [23] for semantic proximity between visual elements and category name texts in latent space [17,21,24,32].

Action recognition requires a representation of the action itself. However, as reported by [4], many existing methods tend to rely on static appearances, such as the background of the scene and objects in the scene, rather than human actions. This issue is called *static bias*, often referred to as *background bias* or *object bias* in action recognition tasks, commonly known as representation bias [9,16]. For practical applications, an action recognition system should be able to correctly predict actions even with a strong background bias; for example, dancing on a basketball court should be classified as "dancing", not "playing basketball". This issue can cause incorrect predictions, especially in zero-shot learning where training and validation categories differ, and is more critical than in fully-supervised action recognition.

Datasets commonly used for action recognition, such as Kinetics400 [11] and UCF101 [27], are known to have a strong *background bias*. For example, many videos in the category "Hitting a baseball" include substantial background information such as the baseball ground. A single frame of videos often contains sufficient information to predict the action category [33], which leads to models becoming overly dependent on background bias. Chung et al. [4] demonstrated that experiments involving training on "background-only" videos of Kinetics400, with the person inpainted out, showed that many action recognition models perform about 50% in accuracy. Even when the person in the scene is masked with a single-colored rectangle, models are reported to still be able to predict categories reasonably [3,10,31]. This background bias has been extensively analyzed in the context of fully-supervised action recognition [3,4,10,20,31]. However, it is still uncertain for zero-shot action recognition with VLM models that extend CLIP, which is proficient at representing appearance features related to languages.

In contrast, Something-Something v2 (SSv2) [7] exhibits a weak background bias. The categories are represented as sentences with a placeholder [something] and the background is so simple that there are hardly any clues. As a result, it needs to focus on temporal changes such as human posture, hand shape, and object placement across multiple frames. However, SSv2 has a significant *object bias* and models tend to rely on objects that appear frequently. This bias is particularly evident in categories like "Tearing [something] into two pieces"; tearable objects such as paper rarely appear in other categories, so having just paper in the scene leads to predicting that category. This object bias might cause poorer performance for unseen categories [17,24]. Therefore, it is necessary to evaluate the behavior of zero-shot models using CLIP for object bias.

Previous *fully-supervised* studies have addressed static bias by masking the background [4] or objects [20]. Some methods use a bounding box (bbox) to mask the person [3] or the object [20]. Others mask and replace the background with another video background [4,28]. Inspired by these masking approaches, we use two masking approaches to investigate background and object biases for *zero-shot* action recognition, as shown in Fig. 1. Our approach for background bias masks the background with random colors to remove any potential clues. This makes the model focus solely on the person in action, similar to [4]. To address object bias, we mask objects using either the object's bounding box, its shape, or the entire background. In either case, human regions are excluded. Masking objects with bounding boxes removes the object's shape information, whereas masking the object's shape eliminates its texture but retains its shape. Masking the whole background can effectively remove object bias; however, we opt to keep the object's shape, as completely removing it may not always be effective during training.

In the experiment, we introduce new evaluation metrics, B-top1 and P-top1, to assess whether the model's predictions are based on the background or person regions. The experimental results confirmed that using the masking approaches improved P-top1, a performance indicator focusing on person regions, against both background and object bias.

2 Related Work

2.1 Action Recognition

Action recognition involves predicting human actions, and various methods have been proposed [1,2,6,8,14,25,26]. Famous datasets used in this task include HMDB51 [15], UCF101 [27], and Kinetics400 [11]. These datasets contain videos that last from a few seconds to several dozen seconds, and their action categories often comprise a few words, which makes them suitable for zero-shot setting as well as fully-supervised settings. SSv2 [7] is another commonly used dataset, and its categories are given as template sentences. The background is relatively simple, so it is necessary to observe the temporal changes in the video.

2.2 Zero-Shot Action Recognition

Zero-shot action recognition has also been studied extensively, often using CLIP [23] with the ability to compute the similarity between images and texts in the embedding space. X-CLIP [21] and ActionCLIP [30] introduced temporal interaction between frames by using interframe attention or temporal transformers, while ViFi-CLIP [24] simply averages CLIP embeddings of each frame with prompt learning to preserve the pre-trained CLIP features. These VLM-based models work well for zero-shot settings with datasets, such as Kinetics400, which have label texts that include names of objects that appear in the scene. However, the performance of SSv2 is not as good as that of Kinetics400 because the labels are given as templates without object names. In addition, no zero-shot methods have explicitly considered background and object biases.

2.3 Analyzing Static Bias

There is prior work that aims to analyze and counteract background and object biases for fully-supervised action recognition. Action-Swap [4] removes background bias by extracting the person region from a video and pasting it onto different backgrounds from videos of different actions. It also proposes a metric to measure how much a model focuses on actions by evaluating performance with these background-swapped videos. S3Aug [28] generates diverse backgrounds using a generative model while preserving the person region. Choi et al. [3] propose scene-independent action features by masking a person with bounding boxes and separating the background and person information. DARK [20] masks objects with bounding boxes and predicts action verbs and nouns in separate branches.

In contrast, we address the open question for zero-shot action recognition: How are VLM models that utilize language affected by static bias? This has not been considered yet.

3 Masking Approaches

In this section, we introduce two masking approaches: one for addressing background bias and the other for addressing object bias. In the following, masking by a binary mask $M_i \in \{0,1\}^{H \times W}$ is represented as:

$$X_i = \overline{M}_i \odot V_i + \mathbf{c} \cdot M_i, \quad i = 1, \ldots, T, \quad (1)$$

where $V_i \in R^{3 \times H \times W}$ is an RGB frame of a given video clip with T frames of height H and width W, and X_i is the masked frame. The complement $\overline{M}_i = \mathbf{1} - M_i$ denotes the flipping of the mask values, where $\mathbf{1} \in \{1\}^{H \times W}$ is a mask filled with ones. In the following, we omit the frame index i because there is no confusion. The 3d vector $\mathbf{c} \in R^3$ represents the masking color, $\odot$ denotes an element-wise multiplication with dimension alignment between the mask and the frame, and $\cdot$ represents a tensor product of the color vector and the mask to generate an RGB image.

If performance decreases when models are trained with the background or objects masked compared to when they are not masked, it suggests that they depend on background or object biases [4,9,16].

3.1 Masking for Background Bias

To mitigate background bias, we mask the background of each frame to eliminate any potential clues from the background. This approach forces the model to focus only on the foreground, that is, the person performing the action. Instance segmentation is used to create a mask $M_{\text{person}}^{\text{shape}} \in \{0,1\}^{H \times W}$, assigning 1 to pixels in the human region and 0 otherwise. The human region is preserved while the background is masked as follows:

$$M_{\text{bg}} = \overline{M_{\text{person}}^{\text{shape}}}, \quad X = \overline{M_{\text{bg}}} \odot V + \mathbf{c} \cdot M_{\text{bg}}. \quad (2)$$

This approach is similar to the background masking in [4], however, we do not use a fixed color for masking because the same masking color for all videos might not be useful for learning features. For each video, we sample an RGB color $\mathbf{c} \sim \mathcal{N}(\mathbf{0}, I)$ from a standard normal distribution $\mathcal{N}$ with mean $\mathbf{0}$ and unit standard deviation I, assuming that the pixel values are normalized. This results in different videos having their backgrounds masked in different colors, as shown in Fig. 1(a).

(a) (b)

Fig. 1. Examples of masking. (a) Masking background. (left) Original frames, and (right) masking with sampled colors. (b) Masking objects. From left to right, original frames, masking the object with bbox and its shape, and masking the background and object. (Color figure online)

3.2 Masking for Object Bias

To eliminate object bias, we mask objects in three ways: the object's bounding box, its shape, or the entire background (Fig. 1(b)).

Masking Object Bounding Box. The first is "object bounding box masking" which masks objects with bounding boxes. This removes the shape information of the objects while also hiding a small part of the background. Note that no masking is applied to human regions within the bounding box.

Let $M^{\text{bbox}}_{\text{object}} \in \{0,1\}^{H \times W}$ be a mask of the bounding boxes of the object, assigning 1 to the pixels inside the bounding boxes and 0 otherwise. Then the masking is done by;

$$M^{\text{bbox}'}_{\text{object}} = \overline{M^{\text{bbox}}_{\text{object}}} \cup M^{\text{shape}}_{\text{person}}, \quad X = M^{\text{bbox}'}_{\text{object}} \odot V + \mathbf{c} \cdot \overline{M^{\text{bbox}'}_{\text{object}}}, \quad (3)$$

where $\cup$ is element-wise logical OR.

Masking Object Shape. The second is "object shape masking". This masks the object regions, not the bounding box. While it does not hide non-object areas, the shape of objects can still be visible through the mask's shape.

Let $M^{\text{shape}}_{\text{object}} \in \{0,1\}^{H \times W}$ be a mask of the regions of the objects, assigning 1 to pixels inside the regions and 0 otherwise. Then the masking is done by;

$$X = \overline{M^{\text{shape}}_{\text{object}}} \odot V + \mathbf{c} \cdot M^{\text{shape}}_{\text{object}}, \quad (4)$$

assuming $M^{\text{shape}}_{\text{object}} \cap M^{\text{shape}}_{\text{person}} = \mathbf{0}$, where $\cap$ is an element-wise logical AND.

Masking Background. The third is "background masking", similar to background masking discussed in Sect. 3.1. However, masking background (*i.e.*, all non-human regions) could be problematic because it removes both objects and the background from the scene. For example, in videos of the categories "Pretending to put [something] onto [something]" and "Putting [something] onto [something]", the temporal change is caused by the presence and movement of objects, not by human actions. To address this issue, we mask the background and objects in different colors. This eliminates bias from the object's texture while preserving the information required for category prediction.

This background masking is represented by

$$X = M_{\text{person}}^{\text{shape}} \odot V + \mathbf{c}_{\text{bg}} \cdot (\overline{M_{\text{person}}^{\text{shape}}} \cap \overline{M_{\text{object}}^{\text{shape}}}) + \mathbf{c}_{\text{obj}} \cdot (\overline{M_{\text{person}}^{\text{shape}}} \cap M_{\text{object}}^{\text{shape}}), \quad (5)$$

where $\mathbf{c}_{\text{obj}}$ and $\mathbf{c}_{\text{bg}}$ are colors for background and objects, respectively.

3.3 Training Procedure

We implement the masking approaches through data augmentation, in a manner similar to [12,28]. The training procedure is illustrated in Fig. 2. For each sample pair of video frames V_i and category text T, we randomly apply one of the masking types with a predefined ratio (we call *masking ratio*). For background bias, the masking ratio is represented as a number pair for choosing "no masking" or "masking background". For object bias, it is a quadruplet for "no masking", "masking object bounding box", "masking object shape", or "masking background and object with different colors". Next, masked frames X_i are fed into a video encoder f_v to produce video embeddings v_i, while category text T are passed to a text encoder f_t to generate text embedding t. Then the infoNCE [22] loss is used to learn both embeddings.

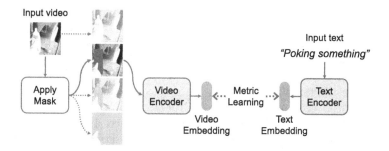

Fig. 2. Training procedure with the masking approaches.

4 Experiments

4.1 Settings

Datasets. For our experiments, we used three datasets: Kinetics400 [11] and Mimetics [31] to assess background bias, and SSv2 [7] to assess object bias. In Kinetics400 [11], each video is sourced from YouTube and assigned to one of 400 action categories. The maximum video length is 10 s. The training set consists of about 240,000 videos, with each category containing between 250 and 1,000 video clips. The evaluation set includes 20,000 videos, and each category contains 50 video clips. Mimetics [31] is an evaluation-only dataset. It consists of 713 videos across 50 categories that are a subset of Kinetics400. It includes video clips that deviate from typical scenarios, such as indoor surfing or pantomiming tennis. Therefore, these videos have less background bias and hence are challenging for models relying on the bias. In SSv2 [7], each video is recorded by crowd workers and assigned to one of the 174 template sentence categories. The average video length is 4.03 s and we excluded videos with shorter than 16 frames, resulting in 168,108 training videos and 24,679 validation videos.
Model. We used ViFi-CLIP [24] and ActionCLIP [30] as zero-shot action recognition models. ViFi-CLIP is a straightforward extension of CLIP [23] for video, where each frame is fed to CLIP to extract features followed by aggregation with temporal average. ActionCLIP is also an extension of CLIP for video, but it performs temporal aggregation with a 6-layer transformer. We used a pretrained CLIP model of ViT-B/32 for both models and fine-tuned all parameters of both the image and text encoders. For training, we create a video clip consisting of 8 frames uniformly sampled from each video. Unless noted otherwise, we used a batch size of 32 and AdamW [19]. For ViFi-CLIP, the learning rate was set to 2e-6 and the number of epochs to 12. For ActionCLIP, we used 15 epochs and the other settings of the original paper [30].
Masking. We used GroundingDINO [18] and Segment Anything Model (SAM) [13], which are open-vocabulary object detection and segmentation models, to create bounding boxes and regions of objects and persons, each applied independently to every frame of a given video. For person detection and segmentation, we used the words "person" and "hand" as prompts. When masking objects of SSv2, we used object names as prompts, which are provided to replace placeholders [something] in the template sentence. We applied the masking frame by frame. A frame and the prompts were fed to GroundingDINO to detect the bounding boxes of the person and objects. These boxes were then fed into SAM as prompts for person and object segmentation. The masking ratio for background bias varied from 1:0 to 0:1 for scenarios without and with masking. Similarly, the masking ratio for object bias starts from a quadruplet of 1:0:0:0. A ratio of 1:0 (or 1:0:0:0) implies that no masking was applied during training, which is essentially the same as training the original ViFi-CLIP [24] or ActionCLIP [30], although in different training settings with our re-implementation.
Base-to-Novel Zero-Shot Setting. We adopted the base-to-novel split used by [24] which divides the category set into base (seen) and novel (unseen) cat-

egories. This includes three few-shot splits, with each base category containing 16 samples. The base categories are made up of the top 50% most frequent categories in the training set, while the novel categories include the bottom 50% least frequent categories. We used a learning rate of 8e-6 with 40 epochs for this setting.

Cross-Dataset Fully-Supervised Setting. This setting is not zero-shot, but the datasets differ between training and validation. We trained models on the 50 categories of the Kinetics400 training set that are present in Mimetics, and then evaluated the performance on Mimetics, as in [31].

Cross-Dataset Zero-Shot Setting. This setting assesses the zero-shot performance of models trained on one dataset and evaluated on another dataset containing categories not seen during training. We trained on 350 categories of Kinetics400 which are not present in Mimetics, using a learning rate of 1e-5, and evaluated the zero-shot performance on Mimetics.

(a) (b)

Fig. 3. Examples of masking for validation videos. Video frames from (a) Kinetics400 and (b) SSv2. From left to right, (1st) original images, masking (2nd) person bounding boxes and (3rd) background, respectively.

4.2 Evaluation Metrics

We use the top1 performance on the validation set for performance evaluation and report the best result over the training epochs (or the average across the three splits for the base-to-novel zero-shot setting). In this case, no masking is applied to the videos in the validation set; masking is used only during training.

While this is commonly used in the action recognition literature, it does not reveal any potential model bias towards the background and objects. To assess how much the model relies on the background, we introduce the following two masking approaches for videos in the validation set (Fig. 3).

- "Masking background" is the same with Eq. (2) for background bias and Eq. (5) for object bias. We use the masked validation videos to compute the top1 performance, which we refer to as *Person top1* or simply *P-top1*. This term is used because, in this case, models only see the person.
- "Masking person bounding box" is intended to remove any persons from the scene. Let $M_{\text{person}}^{\text{bbox}} \in \{0,1\}^{H \times W}$ be a mask of the person bounding boxes,

assigning 1 to pixels inside the bounding boxes and 0 otherwise, and the masking is represented by;

$$M_{\text{person}}^{\text{bbox}'} = \overline{M_{\text{person}}^{\text{bbox}}} \cup M_{\text{object}}^{\text{shape}}, \quad X_i = M_{\text{person}}^{\text{bbox}'} \odot V_i + \mathbf{c} \cdot \overline{M_{\text{person}}^{\text{bbox}'}}. \quad (6)$$

We refer to the top1 performance of this masking as *Background top1* or *B-top1*. This is because, similar to P-top1, models can only see the background. Note that a smaller B-top1 is better because a large B-top1 indicates the model depends on the background bias.

These metrics offer advantages over existing ones such as SHAcc and SBErr [4] because our P-top1 and B-top1 can be calculated with lower computational cost.

4.3 Results of Masking for Background Bias in Kinetics

Base-To-Novel Zero-Shot. Table 1 shows the results of the base-to-novel zero-shot setting. The decrease in B-top1 as the mask ratio increases indicates that the masking approach suppresses the models from learning background bias. A decrease in top1 performance is expected because models use background bias, while masking the background during training likely prevents models from using this bias. P-top1 was significantly lower than B-top1 without masking; however, it improved significantly with masking. This suggests that the ability to make inferences without depending on the background information, and based solely on the information of the person, has improved.

The two models differ in the mask ratio that results in the highest P-top1, but both show that masking background of all videos does not significantly improve P-top1. ViFi-CLIP outperformed ActionCLIP, probably because ActionCLIP trains the temporal aggregation transformer from scratch, making few-shot learning more difficult.

Cross-Dataset Full-Supervision. Table 2 shows the results of fully-supervised setting evaluated on Mimetics. In addition to P-top1, the top1 performance improved with masking, in particular by more than 5% for Action-CLIP. Unlike in the experiments above, top1 is improved even when masking the background of all videos. This shows that preventing the model from using background information does not negatively impact performance when the background bias of validation videos is weak.

We compared our approach with S3Aug [28], a data augmentation technique for fully-supervised action recognition that generates diverse backgrounds using segmentation and generative models, while leaving human regions intact. Our masking approach outperformed S3Aug while requiring less computational cost, suggesting that masking is effective for background bias.

Cross-Dataset Zero-Shot. Table 3 shows the results of the cross-dataset zero-shot setting evaluated on Mimetics. As expected, without masking, relying on

Table 1. Performances of base-to-novel zero-shot setting on Kinetics400 with background masking. The masking ratio represents a ratio between "no masking" (—) and "masking background" (mask). *our re-implementation

model	ratio	base (seen)			novel (unseen)		
	— : mask	top1	B-top1	P-top1	top1	B-top1	P-top1
ViFi-CLIP [24]*	1.0 : 0.0	**68.87**	50.27	12.08	**51.57**	38.60	6.91
	0.5 : 0.5	68.24	49.24	26.08	50.97	38.44	13.98
	0.33 : 0.67	67.42	48.40	**26.61**	50.58	37.88	**14.08**
	0.0 : 1.0	59.68	42.34	25.40	47.99	36.21	13.46
ActionCLIP [30]*	1.0 : 0.0	**59.11**	42.42	10.00	**30.23**	22.67	5.68
	0.5 : 0.5	52.92	37.21	**18.40**	25.92	19.66	**8.31**
	0.33 : 0.67	48.55	33.24	17.70	23.78	18.15	8.01
	0.0 : 1.0	28.99	17.30	13.83	14.68	10.33	6.27

Table 2. Performances of cross-dataset fully-supervised setting where the model is trained on 50 classes of Kinetics400 and evaluated on the same 50 classes of Mimetics.

Model	ratio	Mimetics		
	— : mask	top1	B-top1	P-top1
ViFi-CLIP [24]*	1.0 : 0.0	19.23	5.45	13.30
	0.5 : 0.5	22.57	7.81	21.35
	0.33 : 0.67	**23.78**	8.51	**23.26**
	0.0 : 1.0	20.67	7.05	22.44
ActionCLIP [30]*	1.0 : 0.0	20.31	7.47	16.32
	0.5 : 0.5	**25.35**	8.85	23.61
	0.33 : 0.67	24.31	9.38	**24.13**
	0.0 : 1.0	21.53	5.90	23.26
S3Aug [28]	—	22.40	—	—

background bias results in a higher top1 performance on Kinetics400 and poorer on Mimetics. In contrast, the masking approaches increased top1 on Mimetics by 1.56% for ViFi-CLIP and 3.47% for ActionCLIP, indicating again that the masking is effective on videos with a weak background bias. While P-top1 improved in both datasets, B-top1 decreased in Kinetics400, indicating that the model's use of background information is being suppressed also in this setting.

4.4 Results of Masking for Object Bias in SSv2

Table 4 shows the results of the base-to-novel zero-shot setting on SSv2 with masking objects and background. The first row of each model shows the baseline performance without masking, and the next four rows show the performance with masking at different mask ratios.

Table 3. Performances of cross-dataset zero-shot setting where the model is trained on 350 classes of Kinetics400 and evaluated on both the rest 50 classes of Kinetics400 and Mimetics.

Model	Ratio		Mimetics			Kinetics400		
	—	: mask	top1	B-top1	P-top1	top1	B-top1	P-top1
ViFi-CLIP [24]*	1.0	0.0	16.67	6.77	11.46	**74.42**	61.55	19.33
	0.5	0.5	17.71	5.38	15.97	73.60	61.39	37.99
	0.33	0.67	**19.27**	6.60	16.50	72.41	61.23	**42.52**
	0.0	1.0	18.92	6.25	**16.67**	67.64	55.59	41.69
ActionCLIP [30]*	1.0	0.0	16.84	5.90	13.71	63.73	52.30	18.38
	0.5	0.5	**20.31**	8.16	**19.44**	**63.77**	52.38	36.23
	0.33	0.67	20.14	7.64	17.36	63.32	50.66	**36.60**
	0.0	1.0	16.15	5.38	15.62	52.67	40.87	34.17

Table 4. Performances of base-to-novel zero-shot setting on SSv2. The masking ratio is for "no masking" (—), "masking object bounding box" (bbox), "masking object shape" (shape), and "masking background and object with different colors" (bg).

Model	Ratio				Base (seen)			Novel (unseen)		
	—	: bbox	: shape	: bg	top1	B-top1	P-top1	top1	B-top1	P-top1
ViFi-CLIP [24]*	1.0	0.0	0.0	0.0	12.89	9.36	5.20	9.62	7.26	4.40
	0.5	0.5	0.0	0.0	11.78	8.21	5.77	8.65	6.11	4.22
	0.5	0.0	0.5	0.0	12.47	9.03	7.95	9.45	6.37	5.55
	0.5	0.0	0.0	0.5	**13.93**	9.96	11.01	9.13	6.50	6.65
	0.33	0.0	0.33	0.33	13.76	9.45	**11.09**	**9.64**	6.29	**6.89**
ActionCLIP [30]*	1.0	0.0	0.0	0.0	8.67	6.80	4.10	5.80	4.90	3.27
	0.5	0.5	0.0	0.0	7.86	5.94	5.73	6.09	4.44	4.00
	0.5	0.0	0.5	0.0	8.84	6.74	7.04	6.34	4.80	4.91
	0.5	0.0	0.0	0.5	**11.34**	8.35	**10.96**	**8.06**	6.09	**7.03**
	0.33	0.0	0.33	0.33	10.74	7.73	10.77	7.50	5.36	6.61

Masking the Object Bounding Box and the Object Shape. The second (0.5:0.5:0:0) and third (0.5:0:0.5:0) rows show performances with masking the object bounding box and the object shape, respectively. In both cases, background information is available during the training. B-top1, which is the performance when the background (and also the objects) are available during evaluation, decreases in all cases. This suggests the following. First, the difference in B-top1 with and without masking is marginal (± 1%), even when background information is available during both training and evaluation. This means that there is no clue in the background (i.e., weak background bias). Second, masking objects during training leads to poorer performance, and the object appearances available during evaluation do not help. Therefore, object appearance may serve as an important clue (i.e., a strong object bias). Third, masking the object bounding box is worse than masking object shapes. Hence, the shape of the object is a cue for prediction, and hiding it may be useful for object bias.

Masking the Object and Background in Different Colors. The fourth (0.5:0:0:0.5) row shows the performances with masking the object and background in different colors. Since the background is masked but the object shape is available, the performance is expected to remain the same as when only the object shape is masked. However, the performance is higher than expected. This might be due to the masking approach. By removing unnecessary background information, models can focus directly on the shape of the objects.

A Combination of Masking. The fifth row (0.33:0:0.33:0.33) shows the performance of a combination of two masking approaches, which is a similar or higher performance to the fourth row. During training, we use both masking object shapes (but the background is available) and masking background and objects (but object shape is available). This combination of masking approaches might provide useful information to the models and lead to better performance.

Model Architecture Difference. The fourth row shows the best top-1 performance for the seen categories in most cases. However, the best performance for unseen categories is achieved at different mask ratios for different models. In the fourth row, ActionCLIP performance improved by 2%, while ViFi-CLIP performance decreased by 0.5% from baseline. Therefore, mask ratios or masking approaches need to be tuned differently for different model architectures.

5 Conclusion

In this study, we addressed the issue of static bias in zero-shot action recognition by investigating the impact of background and object biases on model performance. We tackled this issue by masking backgrounds and objects and evaluated their effectiveness on Kinetics400, Mimetics, and SSv2. Our experimental results showed that zero-shot models depend too much on background bias. Masking the background during training prevents the model from focusing excessively on this bias, as evidenced by the improvement of P-top1. Models also rely on object bias. Masking objects effectively reduced the model's reliance on this bias, demonstrated by the improved performance when both backgrounds and objects were masked with different colors. Future work may include devising a model based on the results that incorporates a structure to mask the background or objects, thereby focusing more on human actions.

Acknowledgements. This work was supported in part by JSPS KAKENHI Grant Number JP22K12090.

References

1. Arnab, A., Dehghani, M., Heigold, G., Sun, C., Lučić, M., Schmid, C.: Vivit: a video vision transformer. In: Proceedings of the IEEE/CVF International Conference on Computer Vision (ICCV), pp. 6836–6846, October 2021

2. Carreira, J., Zisserman, A.: Quo vadis, action recognition? a new model and the kinetics dataset. In: Proceedings of the IEEE Conference on Computer Vision and Pattern Recognition (CVPR), July 2017
3. Choi, J., Gao, C., Messou, J.C.E., Huang, J.B.: Why can't i dance in the mall? learning to mitigate scene bias in action recognition. In: Advances in Neural Information Processing Systems, vol. 32. Curran Associates, Inc. (2019)
4. Chung, J., Wu, Y., Russakovsky, O.: Enabling detailed action recognition evaluation through video dataset augmentation. In: Thirty-sixth Conference on Neural Information Processing Systems Datasets and Benchmarks Track (2022)
5. Estevam, V., Pedrini, H., Menotti, D.: Zero-shot action recognition in videos: a survey. Neurocomputing **439**, 159–175 (2021)
6. Feichtenhofer, C.: X3d: expanding architectures for efficient video recognition. In: Proceedings of the IEEE/CVF Conference on Computer Vision and Pattern Recognition (CVPR), June 2020
7. Goyal, R., et al.: The "something something" video database for learning and evaluating visual common sense. In: Proceedings of the IEEE International Conference on Computer Vision (ICCV), October 2017
8. Hara, K.: Recent advances in video action recognition with 3d convolutions. IEICE Trans. Fundamentals Electron. Commun. Comput. Sci. **E104.A**(6), 846–856 (2021)
9. Hara, K., Ishikawa, Y., Kataoka, H.: Rethinking training data for mitigating representation biases in action recognition. In: Proceedings of the IEEE/CVF Conference on Computer Vision and Pattern Recognition (CVPR) Workshops, pp. 3349–3353, June 2021
10. He, Y., Shirakabe, S., Satoh, Y., Kataoka, H.: Human action recognition without human. In: Hua, G., Jégou, H. (eds.) ECCV2016, pp. 11–17 (2016)
11. Kay, W., et al.: The kinetics human action video dataset. CoRR abs/1705.06950 (2017)
12. Kimata, J., Nitta, T., Tamaki, T.: Objectmix: data augmentation by copy-pasting objects in videos for action recognition. In: Proceedings of the 4th ACM International Conference on Multimedia in Asia. MMAsia '22. Association for Computing Machinery, New York (2022)
13. Kirillov, A., et al.: Segment anything. In: Proceedings of the IEEE/CVF International Conference on Computer Vision (ICCV), pp. 4015–4026, October 2023
14. Kong, Y., Fu, Y.: Human action recognition and prediction: a survey. Int. J. Comput. Vis. **130**(5), 1366–1401 (2022)
15. Kuehne, H., Jhuang, H., Garrote, E., Poggio, T.A., Serre, T.: HMDB: a large video database for human motion recognition. In: Metaxas, D.N., Quan, L., Sanfeliu, A., Gool, L.V. (eds.) IEEE International Conference on Computer Vision, ICCV 2011, Barcelona, Spain, November 6-13, 2011, pp. 2556–2563. IEEE Computer Society (2011)
16. Li, Y., Li, Y., Vasconcelos, N.: Resound: towards action recognition without representation bias. In: Proceedings of the European Conference on Computer Vision (ECCV), September 2018
17. Lin, W., et al.: Match, expand and improve: Unsupervised finetuning for zero-shot action recognition with language knowledge. In: Proceedings of the IEEE/CVF International Conference on Computer Vision (ICCV), pp. 2851–2862, October 2023
18. Liu, S., et al.: Grounding dino: marrying dino with grounded pre-training for open-set object detection. arXiv preprint arXiv:2303.05499 (2023)

19. Loshchilov, I., Hutter, F.: Decoupled weight decay regularization. In: International Conference on Learning Representations (2019)
20. Luo, Z., Ghosh, S., Guillory, D., Kato, K., Darrell, T., Xu, H.: Disentangled action recognition with knowledge bases. In: Proceedings of the 2022 Conference of the North American Chapter of the Association for Computational Linguistics: Human Language Technologies (Jul 2022)
21. Ma, Y., Xu, G., Sun, X., Yan, M., Zhang, J., Ji, R.: X-CLIP: end-to-end multi-grained contrastive learning for video-text retrieval. arXiv preprint arXiv:2207.07285 (2022)
22. van den Oord, A., Li, Y., Vinyals, O.: Representation learning with contrastive predictive coding. CoRR abs/1807.03748 (2018)
23. Radford, A., et al.: Learning transferable visual models from natural language supervision. CoRR abs/2103.00020 (2021)
24. Rasheed, H., khattak, M.U., Maaz, M., Khan, S., Khan, F.S.: Finetuned clip models are efficient video learners. In: The IEEE/CVF Conference on Computer Vision and Pattern Recognition (2023)
25. Selva, J., Johansen, A.S., Escalera, S., Nasrollahi, K., Moeslund, T.B., Clapés, A.: Video transformers: a survey. CoRR abs/2201.05991 (2022)
26. Simonyan, K., Zisserman, A.: Two-stream convolutional networks for action recognition in videos. In: Ghahramani, Z., Welling, M., Cortes, C., Lawrence, N., Weinberger, K. (eds.) Advances in Neural Information Processing Systems, vol. 27. Curran Associates, Inc. (2014)
27. Soomro, K., Zamir, A.R., Shah, M.: UCF101: a dataset of 101 human actions classes from videos in the wild. CoRR abs/1212.0402 (2012)
28. Sugiura, T., Tamaki, T.: S3aug: segmentation, sampling, and shift for action recognition. In: Proceedings of the 19th International Joint Conference on Computer Vision, Imaging and Computer Graphics Theory and Applications - Volume 2: VISAPP, pp. 71–79. INSTICC, SciTePress (2024)
29. Ulhaq, A., Akhtar, N., Pogrebna, G., Mian, A.: Vision transformers for action recognition: a survey. CoRR abs/2209.05700 (2022)
30. Wang, M., Xing, J., Liu, Y.: Actionclip: a new paradigm for video action recognition. CoRR abs/2109.08472 (2021)
31. Weinzaepfel, P., Rogez, G.: Mimetics: towards understanding human actions out of context. Int. J. Comput. Vision **129**(5), 1675–1690 (2021)
32. Wu, W., Wang, X., Luo, H., Wang, J., Yang, Y., Ouyang, W.: Bidirectional cross-modal knowledge exploration for video recognition with pre-trained vision-language models. In: Proceedings of the IEEE/CVF Conference on Computer Vision and Pattern Recognition (CVPR), pp. 6620–6630 (June 2023)
33. Zhu, Y., et al.: A comprehensive study of deep video action recognition. CoRR abs/2012.06567 (2020)

Special Session: MLLMA: Special Session on Multimodal Large Language Models and Applications

Enhanced Anomaly Detection in 3D Motion Through Language-Inspired Occlusion-Aware Modeling

Su Li[1,2], Liang Wang[2(✉)], Jianye Wang[2], Ziheng Zhang[2], Junjun Zhang[3], and Lei Zhang[4]

[1] University of Science and Technology of China, Hefei, China
[2] Hefei Institutes of Physical Science, Chinese Academy of Sciences, Hefei, China
liang.wang@inest.cas.cn
[3] Anhui University of Science and Technology, Heifei, China
[4] Laboratory of Vision Engineering (LoVE), School of Computer Science, University of Lincoln, Lincoln, UK

Abstract. In the field of 3D motion analysis, enhancing anomaly detection through language-inspired occlusion-aware modeling is crucial. It can be effective in interpreting human motion under occlusion and aids the model in comprehending complex movement patterns. This paper introduces the OAD2D framework, which detects motion abnormalities by reconstructing 3D coordinates of mesh vertices and human joints from monocular videos, with a specific focus on occluded scenes. OAD2D utilizes optical flow to capture motion prior information from video streams, thereby enriching the data on occluded human movements and ensuring temporal-spatial alignment of poses. Furthermore, we innovate in abnormal behavior detection by integrating it with the Motion-to-Text model, which employs VQVAE to quantize 3D motion features. This method maps motion tokens to text tokens, facilitating a semantically interpretable analysis of motion and boosting the generalization of abnormal behavior detection through the use of a language model. Our approach demonstrates the robustness of anomaly detection against severe and self-occlusions, as it reconstructs human motion trajectories in global coordinates to effectively mitigate occlusion issues. We validate the effectiveness of our proposed method on the Human3.6M, 3DPW, and NTU RGB+D datasets, Outperforms other state-of-the-art methods by a substantial margin, with a 5.1% improvement in Accuracy and a 0.12 increase in F_1-Score. The code and data will be made publicly available upon publication.

Keywords: Anomaly Detection · Lanuage Model · Occlusion-Aware 3D Motion Modeling · Interpretable Motion

1 Introduction

Abnormal behavior detection is critical for enhancing security, healthcare, and sports applications. In this task, accurately identifying abnormal movements depends on precise, which involves reconstructing 3D human poses motions from monocular datasets. Nevertheless, significant challenges remain, including the ambiguous nature of defining an anomaly, the scarcity of abnormal data, difficulties in reconstructing 3D poses from 2D frames, and particularly, occlusions that obscure movements. To some extent, this task relies on advancements in 3D Human Pose Estimation (3DHPE) research.

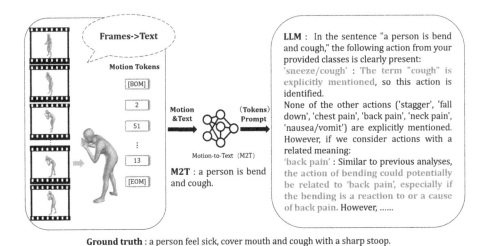

Fig. 1. The M2T model captures motion semantics, and LLM can use specifical prompts identify behavior classes in medical conditions.

3DHPE using a monocular camera directly regresses from RGB images to 3D human poses, utilizing single frames [15,19] or video sequences [16,17], employs deep Convolutional Neural Networks (CNNs) to predict 3D pose joints directly [18,42] or predict human pose parameters like those of the Skinned Multi-Person Linear model (SMPL) [24]. While achieving high accuracy on various indoor datasets, challenges like severe occlusions can disrupt model performance due to incorrect parameter regression and this misalignment may result in erroneous model outputs. To counteract occlusion, several studies [36,43] suggest incorporating priori physical knowledge into the regression process, enhancing robustness. Current studies are focused on spatio-temporal modeling [2,23], which utilize motion data to accurately depict human movements and handle occlusions effectively. These efforts focus on creating a robust human motion model that handles occlusions in current frames by learning implicit cues from contextual sequences. However, these models heavily rely on static features to ensure temporally consistent and smooth representations of 3D human body motion across video sequences [3]. Techniques like optical flow estimation [7,32] infer motion in occluded areas, aiding accuracy but still struggling to balance overfitting in typical scenes against underfitting in complex scenarios.

However, existing methods inadequately capture the semantic information of abnormal behaviors, resulting in a limited understanding of the implementation of behavior detection. Specifically, relying on feature clustering derived from simplistic action classification falls short in accurately identifying abnormal behaviors that carry unique significance. When confronted with a scarcity of samples for abnormal behaviors, comprehending the direction of model learning and the manifestation of feature meaning within the model's perspective becomes challenging. Nevertheless, the advent and rapid advancement of large language models (LLMs) [1] have yielded remarkable strides in semantic comprehension. Consequently, it becomes imperative to introduce semantic bridges in the realm of behavior detection, thereby establishing a solid groundwork for the integration of LLMs. By integrating with LLMs for few-shot learning, we can achieve better performance than purely visual models, providing alternative direction for interpretable, as shown in Fig. 1.

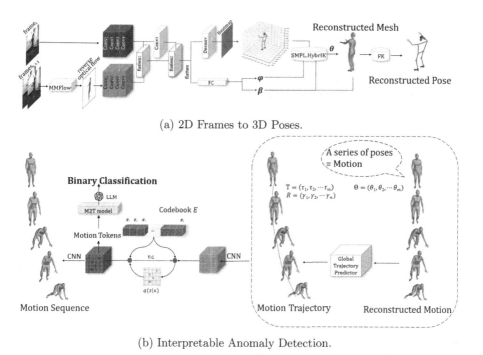

(a) 2D Frames to 3D Poses.

(b) Interpretable Anomaly Detection.

Fig. 2. Overview of our method (OAD2D). The entire pipeline consists of pose estimation using image and optical flow, followed by trajectory optimization and motion quantization, along with LM classification.

Building on the insights from previous research, in this article, we propose OAD2D (**O**cclusion-**A**ware **A**bnormality **D**etection in **2D** Frames), a novel method for detecting abnormal actions in occlusion. The overall pipeline is shown in Fig. 2. Initially, we employ optical flow [5,14] for temporal and spatial alignment, as well as for compensating occlusion, to facilitate the regression of keypoint heatmaps. Subsequently, to accurately reconstruct human behavior understanding in 3D pose from monocular video, recovering the motion trajectory under global coordinates [45] is crucial. To this end, we enhance motion information using physical constraints [20] and a global coordinate transformation, thereby achieving a more precise motion representation under occlusion. Finally, based on the reconstructed 3D pose, we reconceptualize abnormal behavior detection by treating motion as a language, utilizing the Motion-to-Text (M2T) model [28] to translate motion tokens into text tokens for semantic interpretation. Specifically, we employ a Vector Quantized Variational Autoencoder (VQVAE) [38] to compress 3D motion data, encapsulating motion features as discrete tokens within a codebook. The experiments demonstrate that the effectiveness of detecting pose anomalies in occlusion scenarios using our method and it also allows for semantic interpretation of abnormal behaviors.

The contributions of this paper are as follows: (1) We propose a new approach named OAD2D, which sets a new benchmark for accurately detecting abnormal actions in scenarios with occlusion. (2) We employ a novel 3D pose regressor to facilitate 2D-to-3D pose regression under occlusion along with a global trajectory predictor to enhance global trajectory prediction for the generated poses. (3) We pioneer the use of LLM in conjunction with our occlusion-aware detection system to provide semantic interpretation of detected motions.

2 Related Work

The challenge of estimating 3D human pose and shape from monocular video [17,21,41] has propelled the development of several advanced methods. These approaches, grounded in model-based techniques [15,19], leverage temporal information and advanced neural network architectures [3,37] to enhance pose estimation. Choi et al. introduced the Time Consistent Mesh Restoration System (TCMR) [3], which employs bidirectional GRUs to focus on temporal features, while others like PoseNet3D [37] utilize knowledge distillation from 2D skeletons to estimate SMPL [24] parameters without requiring 3D training data. Attention mechanisms and hybrid inverse kinematics [20] solutions further improve the accuracy of pose reconstruction, accounting for kinematic constraints and depth ambiguities. In parallel, advancements in optical flow estimation [5,25], such as the FlowNet [8] series and PWC Net [35], have deepened our understanding of pixel motion, contributing to more stable and accurate motion predictions. These innovations in optical flow techniques are enriched by the integration of semantic segmentation and motion propagation modules, enhancing the robustness and consistency of flow estimation in dynamically complex environments. Representing motion features in videos has been another focus, which involves capturing motion dynamics such as trajectory [45] and velocity using methods like dense optical flow and vector fields. The pre-training and fine-tuning stages of MotionBERT [46], along with modular designs like FrankMocap [29], help in creating more sophisticated and detailed 3D motion feature representations. Moreover, recent studies advocate for learning discrete motion representations using VQVAE [38], enabling the generation and mapping of complex human motion patterns to textual descriptions.

For the abnormal behavior detection methods [11,13,34], they principally rely on classifiers of visual cue and features, powered by deep learning networks. However, these works, anchored in 2D features for human behavior anomaly detection, lack interpretability and heavily depend on precise anomaly label information. Furthermore, semantic ambiguity significantly hampers abnormal behavior discrimination, particularly in medical conditions.

3 Proposed Method

The pipeline of our proposed method is illustrated in Fig. 2. In Fig. 2a, the two-stream network is trained on both video frames and their optical flow to generate 3D poses even in the presence of occlusion (Sect. 3.1). Subsequently, based on the generated poses shown in Fig. 2b, global trajectory prediction (Sect. 3.2) is completed via CVAE and fed into VQVAE and reference to generate motion tokens. These tokens are then utilized to derive semantic information using the M2T model. Finally, abnormal behaviors are discriminated by LLMs (Sect. 3.3).

3.1 3D Pose Regressor with Occlusion

Some prior researches [18,30] often experience missing 2D joints resulting due to occlusion, subsequently hindering accurate human mesh inference. Building upon previous works in kinesiology [20,43,44], we have developed a 3D human pose generator that fuses optical flow with kinematics. This integration not only alleviates the impact of occlusions on human reconstruction but also enhances regression of human shape parameters. Inspired by HybrIK [20], in our method, the pose is estimated through direct joint regression, complemented by the inverse kinematics with SMPL [24].

3D Keypoint Regression. We first utilize MMFlow [5] to infer the corresponding optical flow images for video sequences. After generating a 3D heatmap through deconvolution layers followed by a 1×1 convolution kernel, we apply soft-argmax to extract the 3D poses from the heatmap. The keypoint locations are then predicted under the supervision of $L1$ loss:

$$L_{keypoints} = \frac{1}{K} \sum_{k=1}^{K} \|p_k - \hat{p}_k\|_1 , \qquad (1)$$

where K is the total number of keypoints and $\hat{p}_k$ is the ground-truth 3D keypoint.

SMPL Parameters Regression. Upon generating the twist angles φ and shape parameters β through the fully-connected layer, hybrid analytical-neural inverse kinematics is then utilized to determine the relative rotation (i.e. the pose parameters θ) of the 3D poses. The twist angles φ are supervised using $L2$ loss:

$$L_{twist} = \frac{1}{K}\sum_{k=1}^{K}\|(\cos\varphi_k,\cos\varphi_k)-(\cos\hat{\varphi}_k,\sin\hat{\varphi}_k)\|_2,\qquad(2)$$

where $\hat{\varphi}_k$ is the ground-truth twist angle.

We use these parameters put into the SMPL model to reconstruct the human mesh. During the training phase, the shape parameters β and pose parameters θ of SMPL are supervised as follows:

$$L_{shape}=\|\beta-\hat{\beta}\|_2,\quad L_{pose}=\|\theta-\hat{\theta}\|_2,\qquad(3)$$

3.2 Global Trajectory Predictor

When the 3D pose regressor is used to estimate the pose θ with occlusion, it struggles to accurately depict the direction, trajectory, and global coordinates of human motion due to the absence of root information. In this case, we utilize the Global Trajectory Predictor $\mathcal{T}$ to take the pose $\Theta = (\theta_1,\ldots,\theta_t)$ as input and forecast the corresponding root translations $T = (\tau_1,\ldots,\tau_t)$ and rotations $R = (\gamma_1,\ldots,\gamma_t)$. To reduce the trajectory ambiguity caused by occlusion, the Global Trajectory Predictor operates within the CVAE framework:

$$\Psi = \mathcal{T}(\Theta,v),\qquad(4)$$

$$(T,R) = EgoToGlobal(\Psi),\qquad(5)$$

where $\mathcal{T}$ is the decoder of CVAE and v is the latent code obtained by the pose passing through the encoder of CVAE, the output $\Psi = (\psi_1,\ldots,\psi_t)$ is the self-centered trajectory and converted to a global trajectory (T,R) via the conversion function.

3.3 Interpretable Motion Expressor

As shown on the left in Fig. 2b, our approach leverages LLMs to detect abnormal behaviors, the powerful textual performance of LLMs derives from the Transformers [39] framework. The detection of abnormal behavior can be framed as a binary classification problem (abnormal or normal), then by achieving an accurate semantic representation of human motion and using the right prompts, we can avoid relying on traditional visual classification methods [9,33]. This not only enriches the motion representations but also enhances their interpretability.

Guided by the aforementioned motivation, our methodology facilitates the transformation from 2D frames to 3D poses under occlusion as detailed in Sect. 3.1 and Sect. 3.2. We leverage trajectory information, including the root node, and enhance it by incorporating detailed motion data-such as velocity and acceleration of the joints-referenced from the HumanML3D [10] data format. This approach provides a more comprehensive representation of motion information. We utilize the VQVAE [38] to convert motion into tokens by building a codebook through motion quantization and modeling motion-to-text translation.

Motion Quantization. To reduce data dimensionality and facilitate the replacement of complex motion data with representative features, we employ a VQVAE to learn latent codes for 3D human motion. A series of motion sequences $m \in \mathbb{R}^{T \times D_p}$ as input, where T is the number of poses contained in all frames and D_p is pose dimension, is processed along the temporal dimension using $1 \times 1D$ convolution to generate latent vectors $\hat{e} \in \mathbb{R}^{t \times d}$, where d represents the number of convolution kernels. This transformation is expressed as $\hat{e} = Encoder(m)$. The vectors $\hat{e}$ are then subjected to discrete quantization Q, converting them into a set of codebook entries e_k. The learnable codebook $E = \{e\}_{k=1}^{K}$ consists of K potential embedded d-dimensional vectors. The quantization process Q involves mapping each vector $\hat{e}_k$ into the nearest vector in the codebook E, defined as follows:

$$e_k = Q(\hat{e}) := \left(\text{argmin}_{e_i \in E} \|\hat{e}_i - e_k\|\right) \in \mathbb{R}^{t \times d}. \tag{6}$$

And then, projecting e_k as a sequence of poses $\hat{m}$ back into the space of 3D motion through a decoder D, the process is expressed as:

$$\hat{m} = D(e_k) = D(Q(E(m))). \tag{7}$$

The loss function of VQVAE consists of a reconstruction loss, a codebook loss, and an encoder output loss:

$$L_{vqvae} = \|\hat{m} - m\|_1 + \|\text{sg}[E(m)] - e_k\|_2^2 + \beta \|E(m) - \text{sg}[e_k]\|_2^2, \tag{8}$$

where β is the empirical hyperparameter and sg is the stop gradient.

Motion-to-Text Model Learning. Following the discrete quantization of motion, we can convert motion into tokens, which can then be directly used with the Motion-to-Text (M2T) model. This allows us to learn the mapping relationship between human motions and textual descriptions. The target of this mapping is the text tokens $c \in \{1, \ldots, |V|\}^N$ following encoding methods like Glove [27], where V represents the vocabulary and N is the number of words. In the M2T model, motion tokens are fed into the transformer encoder, and subsequently, the decoder predicts the probability distribution of potential discrete text tokens at each step $p_\theta(c \mid s) = \prod_i p_\theta(c_i \mid c_{<i}, s)$. This process defines the model's loss function as follows:

$$L_{m2t} = -\sum_{i=0}^{N-1} \log p_\theta(c_i \mid c_{<i}, s). \tag{9}$$

Abnormal Behavior Detection with LLMs. As previously mentioned, we utilize motion tokens to map text tokens. After inference using the decoder of the transformer, we obtain the semantic representation of the motion. Subsequently, we enable LLMs to directly perform the task of detecting abnormal behavior by using a specific prompt through in context learning.

In the sphere of abnormal behavior detection, the direct involvement of LLMs is a pioneering strategy. By guiding LLMs to engage directly in the task of abnormal behavior detection through the integration of specific prompts within an in-context learning framework, we empower these models to utilize their extensive linguistic understanding to address the challenges posed by anomaly detection. This strategic fusion of multimodal data and language processing enhances the interpretability of the model's outputs and signifies a significant step toward achieving more comprehensive and nuanced anomaly detection capabilities. This concerted effort is meticulously depicted in Fig. 1, illustrating the interconnectedness of language prompts and behavior detection in a visually compelling manner.

4 Experiments

4.1 Experimental Setup

Datasets. For the 3D pose regressor, we utilize the **Human3.6M** [12] and **3DPW** [26] datasets for both training and evaluation. Given that our approach is based on an optical flow driven two-stream network, we also incorporated the corresponding optical flow datas from Human3.6M and 3DPW datasets. Regarding the detection of abnormal behaviors, we select the **NTU RGB+D 120** [22,31] dataset, which features explicit movement classification. We focus on abnormal behaviors included movements associated with medical conditions (NTU-MC) as they relate to everyday life activities, so we also add **HumanML3D** [10] dataset as a supplement to the daily normal behavior data.

Implementation Details. All experiments are conducted using PyTorch. On our two-stream model phase, building upon prior research, the pre-trained weight of image pathway are used to continue learning from 2D source data. The optical flow data inferred from Human3.6M via MMflow is utilized as the input, with all images standardized to a size of 256 × 192. Due to the small frame count of NTU-MC and HumanML3D, the motion window size is set to 32 to capture motion features. The codebook size is 1024. During the motion to text phase, a transformer is trained to map motion tokens to text tokens, utilizing a single GPU with a batch size of 1200 for 400 epochs, at a learning rate of 1×10^{-4}.

Evaluation Metric. In the 3D pose preprocessing approach, we deploy three established metrics to evaluate pose estimation: **MPJPE** calculates average discrepancy between predicted and real joint positions in 3D human pose estimations. **PA-MPJPE** is an improved MPJPE applying Procrustes analysis to minimize translation, rotation, and scale differences before error computing, enhancing joint localization accuracy. **MPVPE** measures the average difference in 3D position between predicted and true model vertices, assessing 3D model accuracy. In binary classification, **Accuracy** measures overall correctness, while F_1-**Score** is the harmonic mean of precision and recall, crucial for imbalanced datasets. **Recall** gauges the model's detection of positives, and precision the accuracy of these detections. **Macro and Weighted Averages** offer insights into performance across classes, with the latter accounting for class frequency. These metrics together provide a nuanced picture of predictive performance.

Table 1. 3D Human Pose Estimation: Evaluation on 3DPW and Human3.6M.

Methods	3DPW-TEST			Human3.6M-TEST	
	PA-MPJPE (↓)	MPJPE (↓)	MPVPE(↓)	PA-MPJPE (↓)	MPJPE (↓)
HMR [15]	76.7	130.0	-	56.8	88.0
SPIN [19]	59.2	96.9	116.4	41.1	62.5
Pose2Mesh [4]	58.3	88.9	106.3	46.3	88.9
TCMR [3]	52.7	86.5	103.2	52.0	86.5
VIBE [17]	51.9	82.9	99.1	41.4	65.6
Ours	**49.7**	**80.3**	**97.6**	**37.3**	**58.8**

4.2 Main Results

Quantitative Results. In order to verify the validity of our model in different settings, we tested our model on different datasets, i.e., 3DPW [26], Human3.6M [12], NTU-MC [22], and HumanML3D [10].

We initiate our evaluation by comparing the performance of our 3D pose regressor against various baseline models as detailed in Table 1. Reference to prior works [17,20,45], we fine-tuning our model using data from Human3.6M-TRAIN and 3DPW-TRAIN. The results on the validation sets of both datasets demonstrate our model's superior performance, evidenced by the following results: on Human3.6M-TEST, we achieved a PA-MPJPE of 37.3 mm and an MPJPE of 58.8 mm; on 3DPW-TEST, we recorded a PA-MPJPE of 49.7 mm, an MPJPE of 80.3 mm, and an MPVPE of 97.6 mm.

In order to evaluate the performance of the abnormal behavior detection method through interpretable 3D motion, we evaluate the precision of detecting 12 distinct disease-related behaviors in the NTU-MC dataset, as depicted in Table 3. Since we are more interested in disease behaviors that express body language, we derive a subset named MC-Half (See bolded classes in Table 3) to validate the results of our classification. The detailed evaluation results in Table 2.

To further validate the generalization of our method, we crafte a hybrid dataset by merging HumanML3D with NTU-MC, conscientiously maintaining a low-probability distribution for abnormal behaviors, as typically observed in everyday life. As illustrated in Table 4, using this approach, we achieved an accuracy of 85% in the identification of abnormal behaviors. Moreover, the F_1-score for the recognition of normal behaviors reached a high mark of 0.90.

Last but not least, we compare our method and the state-of-the-art method [6] with two other baseline methods [34,40] on the MC-Half and HumanML3D mixed dataset, and we keep the model training settings the same for a fair comparison. As the final stage of classification in our approach involves deploying the LLM to directly discern abnormal behavior, namely, through binary classification, the utilization of the ROC curve and AUC for model performance validation becomes unfeasible. Consequently, we opt to directly employ

Table 2. Abnormal Behavior Detection Performance on NTU-MC and half classes of NTU-MC(MC-Half).

Datasets	Action Classes	Accuracy	F_1-Score
NTU-MC	12	87.3%	0.93
MC-Half	6	89.3%	0.94

Table 3. Medical Conditions.

A41: sneeze/cough	A42: staggering
A43: falling down	A44: headache
A45: chest pain	A46: back pain
A47: neck pain	A48: vomiting
A49: fan self	A103: yawn
A104: stretch oneself	A105: blow nose

Table 4. LLM Classification Performance on HumanML3D and MC-Half.

	Precision	Recall	F_1-Score	Support
Normal	0.88	0.92	0.90	17264
Abnormal	0.78	0.89	0.94	6638
Accuracy	-	-	0.85	23902
Macro Avg	0.82	0.79	0.81	23902
Weighted Avg	0.85	0.85	0.85	23902

Table 5. Performance comparison on the mixed dataset.

Method	Accuracy	F_1-Score
Hasan et al. [34]	72.6%	0.68
Wang et al. [40]	75.1%	0.70
Degardin et al. [6]	79.3%	0.73
OAD2D(Ours)	84.4% (↑ 5.1%)	0.85

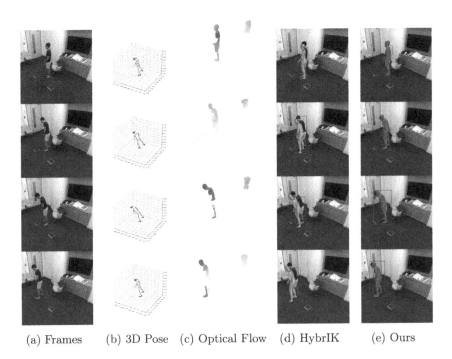

(a) Frames (b) 3D Pose (c) Optical Flow (d) HybrIK (e) Ours

Fig. 3. Qualitative Results on NTU-MC.

Accuracy and F_1-Score to conduct model comparisons. As shown in Tab. 5, it proves that our model has high performance on the abnormal behavior detection task.

Qualitative Results. We have conducted a qualitative evaluation using the Human3.6M, as shown in Fig. 3. We observe that, in contrast to the HybrIK [20], our 3D pose regressor effectively captures information about occluded motions, as highlighted in the red rectangles. Furthermore, we enrich the dataset by incorporating optical flow data, which significantly supplements shape information. Additionally, predicting global trajectories enhances motion information, thereby improving overall robustness. This comparison underlines the strengths of our pose regression method, especially in scenarios involving occlusions.

4.3 Ablation Studies

In order to enhance the performance of the baseline 3D pose regressor under occlusion, we integrate the optical flow and the global trajectory. Therefore, we perform ablation studies to evaluate the contribution of them to the overall pipeline performance. The results of its quantification are shown in Table 6.

Effect on Optical Flow. Building upon prior research, optical flow data is widely acknowledged as a representation of spatio-temporal feature, demonstrating its utility in 3D pose estimation. Optical flow proves particularly valuable in a multitude of occlusion scenarios-in instances of self-occlusion which can hinder the model's accurate inference of joint positions. Only by utilizing the optical flow information from the initial frames, or from the preceding frame alone, we effectively counteract the disruption caused by self-occlusion, and delivering data augmentation.

In response to this insight, we have designed a two-stream network dedicated to the regression of 3D poses alongside their corresponding 3D mesh parameters, employing no more than the reverse optical flow from the current frame.

Effect on Global Trajectory. Recognizing that many approaches to 3D pose estimation overlook the computational significance of the root node, thereby neglecting its critical role in human locomotion analysis, we adopt a global trajectory predictor. It facilitates the calculation of the motion direction, velocity, and acceleration pertaining to the joints, enabling us to capture more nuanced and representative motion.

Table 6. Quantitative results showing the impact of optical flow and trajectory prediction on pose estimation.

Optical flow	Traj pre	Human3.6M-TEST		3DPW-TEST	
		PA-MPJPE ($\downarrow$)	MPJPE ($\downarrow$)	PA-MPJPE ($\downarrow$)	MPJPE ($\downarrow$)
✓		43.9	65.2	56.5	85.5
	✓	47.6	67.7	62.3	93.0
✓	✓	**37.3**	**58.8**	**49.7**	**80.3**

5 Conclusion

In this paper, we present the OAD2D, an innovative approach for detecting abnormal behaviors (e.g. medical conditions) from 3D motion derived from 2D frames. We reconceptualize abnormal behavior detection by viewing motion as a language, using the Motion-to-Text (M2T) model to transform motion tokens into text tokens for semantic interpretation. A critical step adopted is the use of optical flow to capture 3D motion and pose, which are supplemented by the CVAE and global trajectory optimization to mitigate the occlusion. The integration of a large language model (LLM) enhances our model's generalization in abnormal behavior detection and ability to interpret motion in occluded settings. Our experiments on the NTU-MC and HumanML3D datasets demonstrate state of the art results. Supported by the Instrument and Equipment Function Development Project of Chinese Academy of Sciences under Grant E2ADCV112T1.

References

1. Achiam, J., et al.: Gpt-4 technical report. arXiv preprint arXiv:2303.08774 (2023)
2. Arnab, A., Doersch, C., Zisserman, A.: Exploiting temporal context for 3d human pose estimation in the wild. In: Proceedings of the IEEE/CVF Conference on Computer Vision and Pattern Recognition, pp. 3395–3404 (2019)
3. Choi, H., Moon, G., Chang, J.Y., Lee, K.M.: Beyond static features for temporally consistent 3d human pose and shape from a video. In: Proceedings of the IEEE/CVF Conference on Computer Vision and Pattern Recognition, pp. 1964–1973 (2021)
4. Choi, H., Moon, G., Lee, K.M.: Pose2mesh: graph convolutional network for 3d human pose and mesh recovery from a 2d human pose. In: Computer Vision–ECCV 2020: 16th European Conference, Glasgow, UK, August 23–28, 2020, Proceedings, Part VII 16, pp. 769–787. Springer (2020)
5. Contributors, M.: MMFlow: Openmmlab optical flow toolbox and benchmark (2021). https://github.com/open-mmlab/mmflow
6. Degardin, B.M.: Weakly and partially supervised learning frameworks for anomaly detection. Master's thesis, Universidade da Beira Interior (Portugal) (2020)
7. Ding, M., Wang, Z., Zhou, B., Shi, J., Lu, Z., Luo, P.: Every frame counts: joint learning of video segmentation and optical flow. In: Proceedings of the AAAI Conference on Artificial Intelligence, vol. 34, pp. 10713–10720 (2020)
8. Dosovitskiy, A., et al.: Flownet: learning optical flow with convolutional networks. In: Proceedings of the IEEE International Conference on Computer Vision, pp. 2758–2766 (2015)
9. Goodfellow, I., et al.: Generative adversarial nets. Advances in neural information processing systems **27** (2014)
10. Guo, C., Zou, S., Zuo, X., Wang, S., Ji, W., Li, X., Cheng, L.: Generating diverse and natural 3d human motions from text. In: Proceedings of the IEEE/CVF Conference on Computer Vision and Pattern Recognition (CVPR), pp. 5152–5161, June 2022
11. Hao, Y., Tang, Z., Alzahrani, B., Alotaibi, R., Alharthi, R., Zhao, M., Mahmood, A.: An end-to-end human abnormal behavior recognition framework for crowds with mentally disordered individuals. IEEE J. Biomed. Health Inform. **26**(8), 3618–3625 (2021)

12. Ionescu, C., Papava, D., Olaru, V., Sminchisescu, C.: Human3.6m: large scale datasets and predictive methods for 3d human sensing in natural environments. IEEE Trans. Pattern Anal. Mach. Intell. **36**(7), 1325–1339 (jul 2014)
13. Ji, X., Zhao, S., Li, J.: An algorithm for abnormal behavior recognition based on sharing human target tracking features. International Journal of Intelligent Robotics and Applications, pp. 1–13 (2024)
14. Jiang, S., Campbell, D., Lu, Y., Li, H., Hartley, R.: Learning to estimate hidden motions with global motion aggregation. In: Proceedings of the IEEE/CVF International Conference on Computer Vision, pp. 9772–9781 (2021)
15. Kanazawa, A., Black, M.J., Jacobs, D.W., Malik, J.: End-to-end recovery of human shape and pose. In: Proceedings of the IEEE Conference on Computer Vision and Pattern Recognition, pp. 7122–7131 (2018)
16. Kanazawa, A., Zhang, J.Y., Felsen, P., Malik, J.: Learning 3d human dynamics from video. In: Proceedings of the IEEE/CVF Conference on Computer Vision and Pattern Recognition, pp. 5614–5623 (2019)
17. Kocabas, M., Athanasiou, N., Black, M.J.: Vibe: video inference for human body pose and shape estimation. In: Proceedings of the IEEE/CVF Conference on Computer Vision and Pattern Recognition, pp. 5253–5263 (2020)
18. Kocabas, M., Huang, C.H.P., Hilliges, O., Black, M.J.: Pare: part attention regressor for 3d human body estimation. In: Proceedings of the IEEE/CVF International Conference on Computer Vision, pp. 11127–11137 (2021)
19. Kolotouros, N., Pavlakos, G., Black, M.J., Daniilidis, K.: Learning to reconstruct 3d human pose and shape via model-fitting in the loop. In: Proceedings of the IEEE/CVF International Conference on Computer Vision, pp. 2252–2261 (2019)
20. Li, J., Xu, C., Chen, Z., Bian, S., Yang, L., Lu, C.: Hybrik: a hybrid analytical-neural inverse kinematics solution for 3d human pose and shape estimation. In: Proceedings of the IEEE/CVF Conference on Computer Vision and Pattern Recognition, pp. 3383–3393 (2021)
21. Ling, L., et al.: Dl3dv-10k: a large-scale scene dataset for deep learning-based 3d vision. arXiv preprint arXiv:2312.16256 (2023)
22. Liu, J., Shahroudy, A., Perez, M., Wang, G., Duan, L.Y., Kot, A.C.: Ntu rgb+d 120: a large-scale benchmark for 3d human activity understanding. IEEE Trans. Pattern Anal. Mach. Intell. **42**(10), 2684–2701 (2020)
23. Liu, J., Rojas, J., Li, Y., Liang, Z., Guan, Y., Xi, N., Zhu, H.: A graph attention spatio-temporal convolutional network for 3d human pose estimation in video. In: 2021 IEEE International Conference on Robotics and Automation (ICRA), pp. 3374–3380. IEEE (2021)
24. Loper, M., Mahmood, N., Romero, J., Pons-Moll, G., Black, M.J.: Smpl: a skinned multi-person linear model. In: Seminal Graphics Papers: Pushing the Boundaries, vol. 2, pp. 851–866 (2023)
25. Lu, Y., et al.: Transflow: transformer as flow learner. In: Proceedings of the IEEE/CVF Conference on Computer Vision and Pattern Recognition, pp. 18063–18073 (2023)
26. von Marcard, T., Henschel, R., Black, M., Rosenhahn, B., Pons-Moll, G.: Recovering accurate 3d human pose in the wild using imus and a moving camera. In: European Conference on Computer Vision (ECCV), September 2018
27. Pennington, J., Socher, R., Manning, C.D.: Glove: Global vectors for word representation. In: Proceedings of the 2014 Conference on Empirical Methods in Natural Language Processing (EMNLP), pp. 1532–1543 (2014)

28. Radouane, K., Tchechmedjiev, A., Lagarde, J., Ranwez, S.: Motion2language, unsupervised learning of synchronized semantic motion segmentation. Neural Comput. Appl. **36**(8), 4401–4420 (2024)
29. Rong, Y., Shiratori, T., Joo, H.: Frankmocap: fast monocular 3d hand and body motion capture by regression and integration. arXiv preprint arXiv:2008.08324 (2020)
30. Rong, Y., Shiratori, T., Joo, H.: Frankmocap: A monocular 3d whole-body pose estimation system via regression and integration. In: Proceedings of the IEEE/CVF International Conference on Computer Vision, pp. 1749–1759 (2021)
31. Shahroudy, A., Liu, J., Ng, T.T., Wang, G.: Ntu rgb+d: a large scale dataset for 3d human activity analysis. In: Proceedings of the IEEE Conference on Computer Vision and Pattern Recognition, pp. 1010–1019 (2016)
32. Shi, X., et al.: Videoflow: exploiting temporal cues for multi-frame optical flow estimation. In: Proceedings of the IEEE/CVF International Conference on Computer Vision, pp. 12469–12480 (2023)
33. Simonyan, K., Zisserman, A.: Very deep convolutional networks for large-scale image recognition. arXiv preprint arXiv:1409.1556 (2014)
34. Sultani, W., Chen, C., Shah, M.: Real-world anomaly detection in surveillance videos. In: Proceedings of the IEEE Conference on Computer Vision and Pattern Recognition, pp. 6479–6488 (2018)
35. Sun, D., Yang, X., Liu, M.Y., Kautz, J.: Pwc-net: Cnns for optical flow using pyramid, warping, and cost volume. In: Proceedings of the IEEE Conference on Computer Vision and Pattern Recognition, pp. 8934–8943 (2018)
36. Tripathi, S., Müller, L., Huang, C.H.P., Taheri, O., Black, M.J., Tzionas, D.: 3d human pose estimation via intuitive physics. In: Proceedings of the IEEE/CVF Conference on Computer Vision and Pattern Recognition, pp. 4713–4725 (2023)
37. Tripathi, S., Ranade, S., Tyagi, A., Agrawal, A.: Posenet3d: learning temporally consistent 3d human pose via knowledge distillation. In: 2020 International Conference on 3D Vision (3DV), pp. 311–321 (2020). https://doi.org/10.1109/3DV50981.2020.00041
38. Van Den Oord, A., Vinyals, O., et al.: Neural discrete representation learning. Advances in neural information processing systems **30** (2017)
39. Vaswani, A., et al.: Attention is all you need. Advances in neural information processing systems **30** (2017)
40. Wang, T., Qiao, M., Lin, Z., Li, C., Snoussi, H., Liu, Z., Choi, C.: Generative neural networks for anomaly detection in crowded scenes. IEEE Trans. Inf. Forensics Secur. **14**(5), 1390–1399 (2018)
41. Wang, Y., Lu, Y., Xie, Z., Lu, G.: Deep unsupervised 3d sfm face reconstruction based on massive landmark bundle adjustment. In: Proceedings of the 29th ACM International Conference on Multimedia, pp. 1350–1358 (2021)
42. Wehrbein, T., Rudolph, M., Rosenhahn, B., Wandt, B.: Probabilistic monocular 3d human pose estimation with normalizing flows. In: Proceedings of the IEEE/CVF International Conference on Computer Vision, pp. 11199–11208 (2021)
43. Xie, K., Wang, T., Iqbal, U., Guo, Y., Fidler, S., Shkurti, F.: Physics-based human motion estimation and synthesis from videos. In: Proceedings of the IEEE/CVF International Conference on Computer Vision, pp. 11532–11541 (2021)
44. Xu, J., Yu, Z., Ni, B., Yang, J., Yang, X., Zhang, W.: Deep kinematics analysis for monocular 3d human pose estimation. In: Proceedings of the IEEE/CVF Conference on computer vision and Pattern recognition, pp. 899–908 (2020)

45. Yuan, Y., Iqbal, U., Molchanov, P., Kitani, K., Kautz, J.: Glamr: global occlusion-aware human mesh recovery with dynamic cameras. In: Proceedings of the IEEE/CVF Conference on Computer Vision and Pattern Recognition, pp. 11038–11049 (2022)
46. Zhu, W., Ma, X., Liu, Z., Liu, L., Wu, W., Wang, Y.: Motionbert: a unified perspective on learning human motion representations. In: Proceedings of the IEEE/CVF International Conference on Computer Vision, pp. 15085–15099 (2023)

Evaluating VQA Models' Consistency in the Scientific Domain

Khanh-An C. Quan[1,2(✉)], Camille Guinaudeau[3], and Shin'ichi Satoh[4]

[1] University of Information Technology, VNU-HCM, Ho Chi Minh City, Vietnam
anqck@uit.edu.vn
[2] Vietnam National University, Ho Chi Minh City, Vietnam
[3] Université Paris-Saclay/Japanese French Laboratory for Informatics, CNRS, Tokyo, Japan
[4] National Institute of Informatics, Tokyo, Japan

Abstract. Visual Question Answering (VQA) in the scientific domain is a challenging task that requires a high-level understanding of the given image to answer a given question. Although having impressive results on the ScienceQA dataset, both LLaVA and MM-CoT models exhibit inconsistent answers when a simple modification is applied to the textual input of the question (e.g., choices re-ordering). In this paper, we propose two approaches that slightly modify the image-question pair without changing the question's meaning to gain a deeper comprehension of VQA models' question understanding: choices permutation and question rephrasing. Along with these two proposed approaches, we introduce two metrics, namely Consistency across Choice Variations (CaCV) and Consistency across Question Variations (CaQV), to measure the consistency of the VQA models. The experimental results show that both LLaVA and MM-CoT give inconsistent answers regardless of the accuracy. We further conducted a comparison between the proposed metrics and the Accuracy metric, demonstrating that relying solely on the Accuracy is inadequate. By revealing the limitations of existing VQA models and the Accuracy metric through evaluation results in the scientific domain, we aim to provide insights for motivating future research.

Keywords: Visual Question Answering · LLMs evaluation

1 Introduction

Visual Question Answering (VQA) is a challenging task that requires a high-level understanding of the given image to provide the answer to a given question. In particular, the model must understand various visual elements in this task, including recognizing instances, reading text, comprehending the visual characteristics of objects, or reasoning based on visual data to provide a response. On the other hand, integrating various forms of data, such as images and text, adds complexity to this task as the model needs to comprehend and leverage the relationships between these modalities.

This work was conducted during Khanh-An C. Quan's internship at the National Institute of Informatics, Tokyo, Japan.

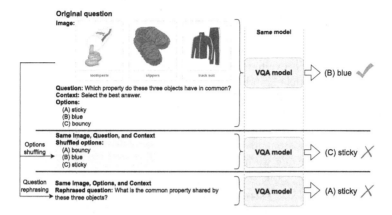

Fig. 1. Illustrate the inconsistent results predicted by the current VQA model using the same image-question pair but altering the order of options or rephrasing the question.

Scientific problem-solving benchmarks [8,14,17] have been employed to evaluate the multi-hop reasoning skills and the interpretability of AI systems. To address the questions in this field, the model must not only comprehend multimodal content but also retrieve external information to determine the correct answer. Among these recently proposed benchmarks for a scientific domain, the ScienceQA dataset [14] is a large-scale multichoice dataset with multimodal science questions along with explanations and has a wide range of domains.

Large Language Models (LLMs) have recently demonstrated impressive performance across a range of Natural Language Processing tasks [16,18]. Additionally, LLMs have shown the capability to address complex reasoning problems through chain-of-thought (CoT) processes by leveraging a small number of demonstration examples [2]. When integrated with input image data, Multimodal Large Language Models (MLLMs) achieve promising visual question answering (VQA) results, both in general contexts and specifically within scientific domains [12,23]. Despite MLLMs achieving notable results in the scientific domain, MLLMs require high computational costs due to their nature. On the other hand, another recent approach to this problem is VLM. In this direction, Multimodal-CoT (MM-CoT) is a starting point and achieves a comparative result to MLLMs. Specifically, MM-CoT combines textual and visual data within a two-stage approach, separating the rationale generation process from the answers inference stage. In comparison to the MLLMs approach, the VLM approach has significantly fewer parameters and a much faster computational speed.

Although both LLMs and VLMs obtain remarkable results on the ScienceQA dataset, these models can provide inconsistent output. Specifically, by simply altering the order of choices, these models can produce different answers to the same question and image. Figure 1 demonstrates the inconsistency in answer predictions of current VQA models when given the same image-question pair but with different choice orders or rephrased questions.

In this paper, we propose two approaches that modify the image-question pair without changing the question's meaning to have a better comprehension of the VQA model's question understanding. In the first one, we assess each question in the dataset using all possible permutations of the choices instead of their original order. In the second approach, we rephrase the question into various forms and then evaluate VQA models using these rephrased questions. Essentially, VQA models should produce the same rationales and answers for the same question, regardless of the order of choices or the form of the question.

We demonstrate that relying solely on Accuracy metrics to evaluate VQA models is insufficient. Specifically, despite high-accuracy examples, the models still provide inconsistent answers with two proposed evaluation approaches. To address this limitation, we introduce two metrics: Consistency across Choice Variations (CaCV) and Consistency across Question Variations (CaQV) to measure the consistency of the VQA models. We assess the performance of two recent VQA models, LLaVA [12] and MM-CoT [24], using the ScienceQA dataset [14].

Our contributions can be summarized in four folds as follows:

- we introduce two approaches that make minor adjustments to the image-question pair without altering the question's meaning to gain a better understanding of VQA models: choices permutation evaluation and rephrasing question evaluation;
- we propose two metrics to measure the consistency of VQA models: CaCV and CaQV. we further compare these metrics with Accuracy, highlighting the limitations of using Accuracy as a sole measure;
- we conduct experiments on two current VQA models, LLaVA [12] and MM-CoT [24], and show that they achieve 89.07% and 94.12% respectively of CaCV and 87.48% and 91.77% of CaQV on the ScienceQA dataset [14];
- Finally, we draw insights into these inconsistent sample characteristics.

2 Related Works

Large Language Models (LLMs). Recently, the advancement of LLMs has demonstrated remarkable performance across various natural language tasks [18]. Various methods have been suggested to enhance multimodal understanding by leveraging the robust generality of LLMs, especially when integrated with other modalities like images [12,16,23]. In the vision-language field, Multimodal Large Language Model (MMLMs) yields remarkable results in various downstream tasks, especially in multimodal reasoning and visual question-answering (VQA) [12,16]. However, one of the main difficulties with MLLMS is its high computational cost and the requirement for large-scale, high-quality training data.

MLLMs Evaluation. As MLLMs have advanced, many benchmarks have been proposed to evaluate comprehension abilities, such as [3,11,13,21,22]. Recent benchmarks, e.g. MME [3], MMBench [13], and SEED-Bench [11], assess MLLMs' comprehension abilities by creating multiple choice questions that span a range of ability dimensions. Li et al. [11] show that most MLLMs still exhibit limited performance on tasks that require fine-grained instance-level comprehension.

Chain-of-Though Reasoning. LLMs have recently demonstrated impressive results by utilizing Chain-of-Thought (CoT) prompting techniques [9,20]. Specifically, CoT methods prompt the LLM to produce a step-by-step reasoning chain to address a problem. There are two primary mechanisms to perform CoT reasoning on LLMs: Zero-Shot-CoT and Few-Shot-CoT. Kojima et al. [9] show that LLMs can perform Zero Shot-CoT by simply appending a prompt such as "Let's think step by step" to the question can trigger CoT reasoning. In Few-Shot-CoT, language models learn reasoning through a few examples demonstrating the step-by-step reasoning process. Recent research indicates that fine-tuned language models can evoke CoT reasoning in smaller models [4,5,15].

Visual Question Answering for Scientific Domain. Solving science problems is a difficult task requiring an AI system to not only grasp multimodal information within the scientific domain but also require the model to explain how to address the questions. There are many proposed benchmarks for VQA in the scientific domain, such as AI2D [7], DVQA [6], VLQA [17], FOODWEDS [10], and ScienceQA [14]. Among these datasets, ScienceQA [14] incorporated reasoning into the VQA task, establishing a standard for multimodal chain-of-thought analysis. The ScienceQA dataset includes approximately 21,000 multimodal multiple-choice questions covering a wide range of science topics, along with annotated answers, related lectures, and explanations.

There are many recent works researching solving this problem, but in general, there are two main directions: utilizing LLM's capabilities and training a Vision-Language Model. Using chain-of-thought prompting, Lu et al. [14] demonstrate that a few-shot GPT-3 model can enhance reasoning performance on the ScienceQA dataset and produce reasonable explanations. However, since GPT-3 is an unimodal model that processes only language, captioning models are required to convert visual information into language modality. Employing caption generation models can lead to considerable information loss when dealing with highly complicated images. To overcome this issue, LLaVA [12] proposes a mechanism to embed visual information into LLM and achieve remarkable results on the ScienceQA dataset [14]. On the other hand, Multimodal-CoT (MM-CoT) [24] implements a two-stage framework, which separates the rationale step from the answer step and training with annotated CoT rationales. Compared to LLaVA, although having the same overall result, the computational cost of MM-CoT is significantly lower than LLaVA. Recently, T-SciQ [19] has shown that by combining the MM-CoT architecture and LLM's reasoning, the VQA performance can be further improved.

3 Methodology

3.1 Preliminaries

In this study, we concentrate on the task of Visual Question Answering [1], which requires the model to deliver the answer by utilizing the information given in the question along with the associated image. Specifically, considering a VQA dataset consisting of k $\{X, Y\}$ samples, where X represents multimodal inputs and Y indicates the corresponding ground-truth answers. The multimodal input X can be denoted as $X = \langle T, I \rangle$, where T refers to the text content and I represents the image content associated with the given question. Text content

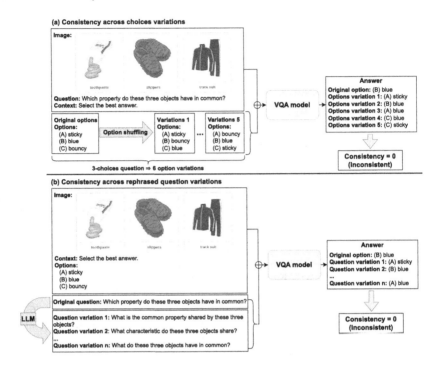

Fig. 2. Illustration of the workflow of our two proposed approaches for assessing the consistency of the current VQA model in the scientific domain.

T can be decomposed into $T = \langle Q, C, M \rangle$, where Q represents the question, C denotes the context, and $M = (m_1, \ldots, m_k)$ is the list of possible options, and k is the number of choices of the given question. It is important to note that the list of choices M in the input for current VQA models is an ordered list. In simple form, the visual question answering can be described as follows:

$$Y = \arg\max_{Y'} p\left(Y' \mid T, I\right) \tag{1}$$

where $p(Y' \mid T, I)$ represents the likelihood of the answer Y' given the textual content T and the visual content I. Based on the basic VQA model, the visual question-answering reasoning model (such as GPT-3 [14], LLaVA [12]) involves generating a rationale R, which explains the chain-of-though that supports the answer Y. This can be mathematically described as follows:

$$Y, R = \arg\max_{Y', R'} p\left(Y', R' \mid T, I\right) \tag{2}$$

where $p(Y', R' \mid T, I)$ is the probability of the answer Y' and the rationale R' given the text content T and the image content I. In MM-CoT [24], rationale generation and answer inference are divided into two distinct stages as follows:

$$\begin{aligned} R &= \arg\max_{R'} p\left(R' \mid T, I\right) \\ Y &= \arg\max_{Y'} p\left(Y' \mid R, T, I\right) \end{aligned} \tag{3}$$

In the following section, we will use Eq. 1 to simplify the description of the VQA model, which takes the image-text input and predicts the answer only.

3.2 Method 1: Choices Permutation Evaluation

Choices Permutation. In this kind of evaluation, rather than using the predefined order of choices in the dataset, we attempt to assess the VQA models using all possible permutations of the choices. Generally, VQA systems are expected to produce the same answer when presented with the same question and a list of choices, regardless of the order in which the choices are arranged. Specifically, given a list of choices $M = (m_1, \ldots, m_k)$, we construct a set M^*, which includes every possible permutation of the list C. M^* can be denoted as:

$$M^* = \{(m_p, m_q, \ldots, m_r) \mid 1 \leq p, q, \ldots, r \leq k, \\ \text{and } (p, q, \ldots, r) \text{ is a permutation of } (1, 2, \ldots, k)\} \quad (4)$$

Typically, a question with k-choices results in $k!$ different permutations of the choices (e.g., with a 3-choices question, we will have $3! = 6$ choices variations.) Subsequently, we create a text input set $T^* = \langle Q, C, M^* \rangle$ derived from all permutations of choices M^*. All variations of choices are input into the VQA models to yield $k!$ answers as follows:

$$Y^* = \{\arg\max_{Y'} p(Y' \mid T^*, I)\} \quad (5)$$

Finally, based on a list of predicted answers Y^*, we propose a metric - CaCV - to measure the stability of VQA models with the same question content and all permutations of choices.

Propose Metric: Consistency (%) Across All Choices Variations (CaCV). The Consistency across all Choice Variations defines whether the model yields the same answer in all different choice variations or not. With an answers' list Y^* predicted by the VQA model based on different choice variations, CaCV for each sample in the dataset is measured as:

$$CaCV = \begin{cases} 1, & \text{if } y_i = y_j \ \forall \ y_i, y_j \in Y^* \\ 0, & \text{otherwise} \end{cases} \quad (6)$$

The overall CaCV for the dataset is derived by taking the mean of the CaCV values for each individual sample. We consider samples with a CaCV of zero as inconsistent and those with a CaCV of one as consistent.

3.3 Method 2: Question Rephrasing Evaluation

Question Rephrasing. Along with assessing the models by altering the order of choices in the questions, we also evaluated them based on rephrased questions. The aim of this evaluation is to determine whether the models genuinely understand the questions and provide answers based on their content. In this type of evaluation, we use a large language model (LLM) to rephrase the original question Q into various forms, denoted as $Q_{rephrase} = (Q_1, \ldots, Q_n)$, where n is the number of rephrased questions. In this work, we use ChatGPT-3.5 to rephrase the question with the prompt "Rephrase this question into n different form." In this paper, we rephrase the original question in 5 different forms. The illustration of the rephrased question is presented in Fig. 2. We also manually verified the correctness of the rephrasing question generated by the LLMs. Next, we concatenate the rephrased questions $Q_{rephrase}$ with the original question Q to create Q^*. We then construct a text input set $T^{**} = \langle Q^*, C, M \rangle$ from all variations of Q^* and input this set into the VQA models.

$$Y^{**} = \{\arg\max_{Y'} p(Y' \mid T^{**}, I)\} \tag{7}$$

Using a list of predicted answers Y^{**}, we introduce CaQV metric to evaluate the robustness of VQA models when faced with different variations of questions.

Propose Metric: Consistency (%) Across Questions Variations (CaQV). Similar to CaCV presented in Sect. 3.2, this metric aims to measure the consistency of the VQA models with different variations of questions. In particular, given a list of answers predicted by the VQA model from different question variations, CaQV is defined as:

$$CaQV = \begin{cases} 1, & \text{if } y_i = y_j \ \forall \ y_i, y_j \in Y^{**} \\ 0, & \text{otherwise} \end{cases} \tag{8}$$

The overall CaQV for the dataset is obtained by averaging the CaQV values of each individual sample. Samples with a CaQV of zero are categorized as inconsistent, while those with a CaQV of one are categorized as consistent as the same CaCV.

4 Experiment

4.1 Experiment Setup

Dataset. We use the ScienceQA [14] dataset for evaluation and analysis. This multimodal multiple-choice science question dataset includes $21,000$ questions across three subjects, covering 26 topics, 127 categories, and 379 unique skills. The dataset is divided into training, validation, and test sets, with $12,726, 4,241$, and $4,241$ samples respectively. In this paper, we focus on the questions with 2, 3, or 4 choices from the ScienceQA test set, which includes $2,228$ questions with 2 choices, 971 questions with 3 choices, and $1,004$ questions with 4 choices.

Metrics. For evaluation, we use the *Accuracy* metric, which compares the answer predicted by the model with the ground-truth from the dataset. In the evaluation of choice permutation, the accuracy of each sample is calculated by averaging the accuracy of all choice variations generated by the evaluated model. Similarly, in question rephrasing evaluation, the accuracy of each sample is measured by averaging the accuracy across all rephrased questions. We also use the proposed metrics $CaCV$ (See Sect. 3.2) and $CaQV$ (See Sect. 3.3) for choice permutation evaluation and question rephrasing evaluation, respectively.

Competing VQA Methods. In this paper, we utilize two VQA models for benchmarking: LLaVA [12] and MM-CoT [24]. For the LLaVA model, we use the ScienceQA version pre-trained with 13 billion parameters and set its temperature to 0 for reproducibility. For the MM-CoT model, we employ the largest ScienceQA pre-trained model with 768 million parameters that achieved the highest performance.

4.2 Results and Analysis

Overall. The overall result of both choice shuffling and question rephrasing evaluations is shown in Table 1. Despite having the same accuracy, MM-CoT shows higher Consistency compared to LLaVA, which are 94.12% and 89.07% for choice shuffling and 87.48% and 91.77%, respectively. In comparison to the choice shuffling evaluation, question rephrasing exhibited lower Consistency for both LLaVA and MM-CoT. This might be due to the challenges posed by comprehending rephrased questions in various forms. It can be seen that Consistency is not related to the number of choices. In particular, the lowest Consistency is

Table 1. Overall percentage of Consistency (CaCV and CaQV) and Accuracy (Acc.) for different types of questions for LLaVA [12] and MM-CoT [24] models with choice shuffling (on the left, in white) and question rephrasing (on the right in blue) approaches on ScienceQA dataset [14]. The best results between the two models are highlighted in bold.

	Choice shuffling				Question rephrasing			
	LLaVA		MM-CoT		LLaVA		MM-CoT	
Question type	CaCV	Acc.	CaCV	Acc.	CaQV	Acc.	CaQV	Acc.
2-choices	95.37	92.93	**99.64**	92.14	85.18	**91.15**	**89.99**	90.78
3-choices	74.15	85.35	**88.36**	**86.06**	87.12	84.74	**92.79**	**86.03**
4-choices	89.54	94.33	**91.53**	92.80	92.92	93.71	**94.72**	**97.94**
Overall	89.07	91.53	**94.12**	90.96	87.48	90.28	**91.77**	**91.39**

Table 2. Results obtained in two types of situations. On the left, accuracy for consistent (Con.) vs. inconsistent (Inc.) examples. On the right, consistency (CaCV or CaQV) for questions with (Img.) and without images (W/o).

	Consistent / Inconsistent				With / Without Image			
	LLaVA		MM-CoT		LLaVA		MM-CoT	
	Inc.	Con.	Inc.	Con.	Img.	W/o	Img.	W/o
Choice shuffling	51.42	96.45	45.72	93.78	86.48	91.93	92.87	95.33
Question rephrasing	56.56	95.10	60.16	94.19	87.95	88.97	92.59	90.04

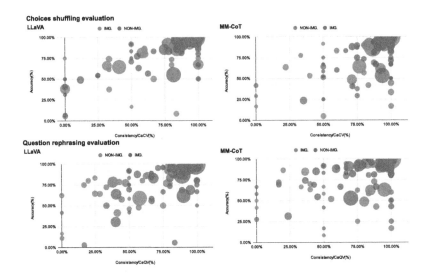

Fig. 3. Scatter plot of Consistency and Accuracy for each question in the ScienceQA dataset (circle sizes represents the number of question examples).

the 3-choices question, while 2-choices have the highest Consistency for both the LLaVA and MM-CoT models.

Table 2 (right) illustrates the Consistency comparison between questions with images and text-only questions within the ScienceQA dataset [14]. When evaluated with choice shuffling, questions that included images showed lower Consistency in comparison to those that were text-only. Although the Consistency for questions with images remained unchanged during the question rephrasing evaluation, there was a slight decrease in Consistency for text-only questions.

Comparison with Accuracy. Table 2 (left) highlights the accuracy of both inconsistent and consistent samples using two evaluation methods. While consistent samples have a high accuracy of around 95% for both models, inconsistent samples still have an accuracy of around 50%. It can be seen that the model somehow can predict the correct answer by chance; however, the accuracy of inconsistent samples still contributes significantly to the overall accuracy across all choice variations and the overall accuracy across rephrased questions.

Figure 3 shows the relationship between the Consistency and Accuracy of each question in the dataset by using the two proposed approaches. The plot reveals that there are many cases where Consistency and Accuracy disagree. With the high-accuracy and low-consistency cases, the VQA model does not perform well and might give the correct answer by chance. On the other hand, with low-accuracy and high-consistency cases, the VQA model does not fully comprehend the question and predicts incorrect answers. Thus, the proposed Consistency metrics effectively complements Accuracy, providing a deeper understanding of the VQA model's behaviors.

We also measure the impact of Consistency on Accuracy in Table 4 in 4 cases: original choices, all choices variations, best cases, and worst cases. In the case of original choices, we assess the accuracy of VQA models exclusively on the original choice provided by the ScienceQA dataset, whereas, in the all-choice variations scenario, the accuracy is evaluated across all possible choice variations. For the best cases, a final answer is deemed correct if any of the choice variations are correct. In contrast, in the worst cases, a final answer is considered incorrect if any of the choice variations is incorrect. In all choice variations, accuracy remains consistent with the original choice; however, in the best-case/worst-case scenarios, there is a notable increase/decrease in this figure.

Choice Shuffling Evaluation. We found that the majority of inconsistent samples in the choice shuffling evaluation were questions containing images, accounting for 67.32% and 64.37% for LLaVA and MM-CoT, respectively. Table 2 (right) also highlights that the Consistency of questions with images is also lower than text-only questions with choice shuffling evaluation. We also illustrate the top-5 questions having the most inconsistent samples through choice shuffling evaluation as presented in Table 3. We also group all questions that require understanding map-image in 'geography' topic into 'Question related to map'. By analyzing the inconsistent samples, we notice that there are some characteristics of the question that can affect the Consistency as follows:

- *Question requires fine-grained instance-level comprehension of the image.* We observed that many inconsistent samples are related to questions needing a detailed understanding of the provided image. For example, with the question 'Think about the magnetic force between the magnets in each pair. ...' (First row of Table 3, Fig. 4.(1)) requires VQA models to identify two pairs of magnets in the image, as well as the relative size, orientation of each magnet, or distance between magnets in each pair. In general, this demands that the model have particular capabilities such as text recognition, instance identification, and instance interaction. With this type of question, both LLaVA and MM-CoT show remarkably poor Consistency and Accuracy. In addition, these models also output inconsistent rationales about understanding the image. With questions related to map (Second row of Table 3), we noticed that inconsistent samples require fine-grained understanding, as shown in Fig. 4.(4).
- *Require specific knowledge/logical reasoning.* With questions require logical reasoning such as using guide words skill question (Fig. 4.(3)), which requires comparing words in alphabetical order. The words given in these questions in the test set are completely different from the training set. Thus, these questions show low Consistency and Accuracy, and the VQA's reasoning for these questions is entirely incorrect. With compare properties of objects question (Forth row of Table 3), although having higher accuracy than other inconsistent questions, the attributes of each object in the VQA's rationales remain incorrect, even inconsistent samples (Fig. 4.(2)).

Table 3. Top-5 question with most inconsistent samples of both LLaVA [12] and MM-CoT [24] models with choices shuffling and question rephrasing approaches. (Has img: whether the question has an image, #Total: the number of samples with the question listed in the dataset, #Inc: number of inconsistent samples predicted by VQA models, Acc.: average accuracy across choice variations/rephrased question with choices shuffling/rephrasing approach.)

Question	Num. of choices	Has image	#Total	LLaVA #Inc.	LLaVA Acc.	MM-CoT #Inc.	MM-CoT Acc.
Top-5 questions with the most inconsistent samples with *choices shuffling* approach							
Think about the magnetic force between the magnets in each pair. Which of the following statements is true?	3	✓	120	71	64.44	28	64.16
Question related to map.	4	✓	266	46	89.09	38	87.29
Which solution has a higher concentration of {} particles?	3	✓	45	45	38.15	13	48.53
Which property do these three/four objects have in common?	3	✓	70	29	80.95	11	91.66
Use guide words skill question.	2		140	26	55.00	9	50.71
Top-5 questions with most inconsistent sample with *question rephrasing* approach							
Use guide words skill question.	2		140	59	58.81	42	52.86
Suppose {} decides to {}. Which result would be a cost?	2		45	28	77.78	14	91.11
Will these magnets attract or repel each other?	2	✓	52	31	63.78	8	62.50
Question related to map.	4	✓	266	28	88.29	19	97.43
Which solution has a higher concentration of {} particles?	3	✓	45	27	30.37	9	48.52

– *Missing image (data issue).* We noticed that some questions require understanding the provided image; However, no image was supplied. Consequently, this question exhibited poor Consistency and Accuracy.

Question Rephrasing Evaluation. Compared to the choice-shuffling evaluation, the question rephrasing evaluation shows the same analysis with inconsistent samples as described above. Along with the reduction in the Consistency of text-only questions, as shown in Table 2 (right), there are also new text-only inconsistent samples compared to the choice shuffling evaluation. Most of the new text-only inconsistent samples require the logical reasoning of the model.

Consistent Samples' Characteristic. By analyzing the remaining consistent samples of both the choices shuffling and question rephrasing evaluations, we found that the majority of these are questions that require an understanding of the provided text to deduce the answer from this understanding Fig. 4.(9). Some of these samples include images, which serve merely as illustrations and are not essential for answering the question (Fig. 4.(7)). Questions related to maps, which necessitate a global understanding of the image, also exhibit high consistency Fig. 4.(8). Although there are some low-consistency samples that require detailed discrimination of features within the image, the other still shows good Consistency and Accuracy.

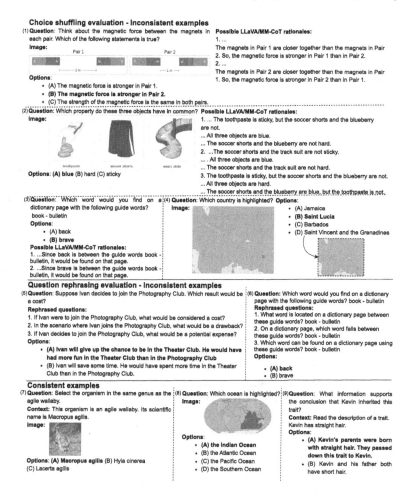

Fig. 4. Illustrate the rationales and answers predicted by the VQA models with our two proposed evaluations.

Table 4. Comparison of Overall Accuracy(%) of LLaVA [12] and MM-CoT [24] with different scenerios.

Methods	Original choices (Baseline)	All choices variations	Best cases	Worst cases
LLaVA [12]	91.60%	91.53% (-0.07%)	96.17%	85.91% (-5.63%)
MM-CoT [24]	90.91%	90.96%	93.58%	88.27% (-2.64%)

5 Conclusion

In this paper, we introduce two new kinds of evaluation, namely choice shuffling and question rephrasing, in order to understand the behavior of VQA models in the scientific domain. Along with two new approaches, we also proposed two metrics, which are Consistency across all Choice Variations (CaCV) and Consistency across Question Variations (CaQV), to evaluate the consistency of the VQA models. We show that depending solely on Accuracy metrics to assess VQA models is inadequate, and combining both Accuracy and Consistency metrics provides a more comprehensive understanding of VQA models. Our experimental results on the ScienceQA dataset [14] show that current VQA models might have inconsistent results with the same question-image pairs regardless of the accuracy. By understanding and objectively evaluating LLaVA [12] and MM-CoT [24], we have observed the following findings.

1. Number of choices is not correlated with Consistency. Table 1 illustrates that increasing the number of choices does not decrease the consistency. In particular, the 3-choice questions exhibit the lowest Consistency among the cases with three different choices. When compared to 3-choice questions, the 4-choice questions demonstrate higher Consistency. Instead, Consistency is more related to the question's content.
2. In both kinds of evaluations, LLaVA and MM-CoT show **poor consistency** for the question that requires fine-grained instance-level comprehension of the image, specific knowledge/logical reasoning, or data issues.
3. In contrast, LLaVA and MM-CoT show **good consistency** for questions that require an understanding of the provided text to provide the answer or a global understanding of the given image.
4. Despite having the same accuracy, MM-CoT demonstrates higher consistency than LLaVA.

These findings aim to offer valuable insights that can inspire future research on VQA models within the scientific domain. Future endeavors may explore the integration of stronger image representations into MLLMs to further enhance performance.

Acknowledgments. The first author of this paper is supported by the NII International Internship Program and stayed in NII from March to September 2024.

References

1. Antol, S., Agrawal, A., Lu, J., Mitchell, M., Batra, D., Zitnick, C.L., Parikh, D.: VQA: visual question answering. In: ICCV (2015)
2. Brown, T., et al.: Language models are few-shot learners. In: Larochelle, H., Ranzato, M., Hadsell, R., Balcan, M., Lin, H. (eds.) Advances in Neural Information Processing Systems, vol. 33, pp. 1877–1901. Curran Associates, Inc. (2020)

3. Fu, C., et al.: Mme: a comprehensive evaluation benchmark for multimodal large language models (2024)
4. Ho, N., Schmid, L., Yun, S.Y.: Large language models are reasoning teachers. In: Rogers, A., Boyd-Graber, J., Okazaki, N. (eds.) ACL, pp. 14852–14882 (Jul 2023)
5. Hsieh, C.Y., et al.: Distilling step-by-step! outperforming larger language models with less training data and smaller model sizes. In: Rogers, A., Boyd-Graber, J., Okazaki, N. (eds.) ACL, pp. 8003–8017, Jul 2023
6. Kafle, K., Price, B., Cohen, S., Kanan, C.: Dvqa: understanding data visualizations via question answering. In: 2018 IEEE/CVF Conference on Computer Vision and Pattern Recognition, pp. 5648–5656 (2018)
7. Kembhavi, A., Salvato, M., Kolve, E., Seo, M., Hajishirzi, H., Farhadi, A.: A diagram is worth a dozen images. In: Leibe, B., Matas, J., Sebe, N., Welling, M. (eds.) ECCV, pp. 235–251. Springer International Publishing (2016)
8. Kembhavi, A., Seo, M., Schwenk, D., Choi, J., Farhadi, A., Hajishirzi, H.: Are you smarter than a sixth grader? textbook question answering for multimodal machine comprehension. In: 2017 IEEE Conference on Computer Vision and Pattern Recognition (CVPR), pp. 5376–5384 (2017)
9. Kojima, T., Gu, S.S., Reid, M., Matsuo, Y., Iwasawa, Y.: Large language models are zero-shot reasoners. NeurIPS **35**, 22199–22213 (2022)
10. Krishnamurthy, J., Tafjord, O., Kembhavi, A.: Semantic parsing to probabilistic programs for situated question answering. In: Su, J., Duh, K., Carreras, X. (eds.) EMNLP, pp. 160–170 (Nov 2016)
11. Li, B., Wang, R., Wang, G., Ge, Y., Ge, Y., Shan, Y.: Seed-bench: Benchmarking multimodal llms with generative comprehension. arXiv preprint arXiv:2307.16125 (2023)
12. Liu, H., Li, C., Wu, Q., Lee, Y.J.: Visual instruction tuning. In: NeurIPS (2023)
13. Liu, Y., et al.: Mmbench: Is your multi-modal model an all-around player? (2024)
14. Lu, P., et al.: Learn to explain: multimodal reasoning via thought chains for science question answering. In: NeurIPS (2022)
15. Magister, L.C., Mallinson, J., Adamek, J., Malmi, E., Severyn, A.: Teaching small language models to reason. In: Rogers, A., Boyd-Graber, J., Okazaki, N. (eds.) Proceedings of the 61st Annual Meeting of the Association for Computational Linguistics (Volume 2: Short Papers), pp. 1773–1781. ACL (Jul 2023)
16. OpenAI: GPT-4 technical report. CoRR abs/2303.08774 (2023)
17. Sampat, S.K., Yang, Y., Baral, C.: Visuo-linguistic question answering (VLQA) challenge. In: Cohn, T., He, Y., Liu, Y. (eds.) EMNLP, pp. 4606–4616, November 2020
18. Touvron, H., et al.: Llama: open and efficient foundation language models. arXiv preprint arXiv:2302.13971 (2023)
19. Wang, L., et al.: T-sciq: teaching multimodal chain-of-thought reasoning via large language model signals for science question answering. AAAI **38**, 19162–19170 (03 2024)
20. Wei, J., Wang, X., Schuurmans, D., Bosma, M., Xia, F., Chi, E., Le, Q.V., Zhou, D., et al.: Chain-of-thought prompting elicits reasoning in large language models. NeurIPS **35**, 24824–24837 (2022)
21. Xu, P., et al.: Lvlm-ehub: a comprehensive evaluation benchmark for large vision-language models. arXiv preprint arXiv:2306.09265 (2023)
22. Yin, Z., et al.: Lamm: language-assisted multi-modal instruction-tuning dataset, framework, and benchmark. Advances in Neural Information Processing Systems **36** (2024)

23. Zhang, R., et al.: Llama-adapter: efficient fine-tuning of language models with zero-init attention. arXiv preprint arXiv:2303.16199 (2023)
24. Zhang, Z., Zhang, A., Li, M., Zhao, H., Karypis, G., Smola, A.: Multimodal chain-of-thought reasoning in language models. Transactions on Machine Learning Research (2024)

Image2Text2Image: A Novel Framework for Label-Free Evaluation of Image-to-Text Generation with Text-to-Image Diffusion Models

Jia-Hong Huang, Hongyi Zhu[✉], Yixian Shen, Stevan Rudinac, and Evangelos Kanoulas

University of Amsterdam, Amsterdam, The Netherlands
h.zhu@uva.n

Abstract. Evaluating the quality of automatically generated image descriptions is a complex task that requires metrics capturing various dimensions, such as grammaticality, coverage, accuracy, and truthfulness. Although human evaluation provides valuable insights, its cost and time-consuming nature pose limitations. Existing automated metrics like BLEU, ROUGE, METEOR, and CIDEr attempt to fill this gap, but they often exhibit weak correlations with human judgment. To address this challenge, we propose a novel evaluation framework called Image2Text2Image, which leverages diffusion models, such as Stable Diffusion or DALL-E, for text-to-image generation. In the Image2Text2Image framework, an input image is first processed by a selected image captioning model, chosen for evaluation, to generate a textual description. Using this generated description, a diffusion model then creates a new image. By comparing features extracted from the original and generated images, we measure their similarity using a designated similarity metric. A high similarity score suggests that the model has produced a faithful textual description, while a low score highlights discrepancies, revealing potential weaknesses in the model's performance. Notably, our framework does not rely on human-annotated reference captions, making it a valuable tool for assessing image captioning models. Extensive experiments and human evaluations validate the efficacy of our proposed Image2Text2Image evaluation framework.

Keywords: Image Captioning · Metrics for Automated Evaluation · Text to Image Generation Models

1 Introduction

Evaluating sentences generated by automated methods remains a significant challenge in image captioning. Existing metrics for assessing image descriptions

---

J.-H. Huang—Equal contribution.

aim to capture various desirable qualities, including grammaticality, coverage, correctness, truthfulness, and more. While human evaluation is crucial for accurately quantifying these attributes—often using Likert scales or pairwise comparisons [33,39]—it is expensive, difficult to replicate, and time-consuming. This has led to an increasing demand for automated evaluation measures that closely align with human judgment. The primary challenge in developing such metrics lies in integrating these diverse evaluation criteria into a cohesive measure of sentence quality.

Several automated metrics, such as BLEU [35], ROUGE [30], METEOR [11], and CIDEr [46], have been developed to evaluate image descriptions generated by automated methods. BLEU, originally designed for machine translation, focuses on precision, while ROUGE, from the summarization domain, emphasizes recall. METEOR aims to assess the overall quality of image descriptions. However, studies have shown that these metrics often exhibit a weak correlation with human judgment [11,40,41]. In contrast, the consensus-based metric CIDEr measures the similarity between a generated sentence and a set of human-authored ground truth sentences, and it generally aligns well with human consensus. However, CIDEr's effectiveness depends on having a sufficiently large and diverse set of ground truth sentences, which can be a limitation when such data is scarce [46]. This limitation also applies to other methods like CLAIR [4] and the previously mentioned metrics. Additionally, some approaches involve caption ranking [16], but they often fall short in evaluating novel image descriptions.

To address the challenge of evaluating image descriptions, we introduce a novel framework that leverages modern vision large language models (VLLMs), such as GPT-4 [3] or Gemini [45], capable of creating images. The advancements in VLLMs, exemplified by models like GPT-4V, enable the creation of textual prompts that generate images closely aligned with the semantic meaning of the input text. The core design philosophy of our proposed framework is that if an image captioning model is effective, its generated description should be accurate enough to reconstruct the original image, or a highly similar one, using VLLM-based image generation. The continuous evolution of VLLM technology underpins and strengthens the foundation of this framework.

Our proposed framework for evaluating image captioning models starts by defining the task of taking an image as input and generating a corresponding textual description. The input image is processed by the image captioning model under evaluation, producing a descriptive text. This description is then fed into a text-to-image generation model, such as Stable Diffusion, to generate a new image based on the text. Next, we extract features from both the original input image and the generated image, comparing them using the cosine similarity metric. Importantly, our framework does not require human-annotated reference captions for this evaluation. In this framework, a high cosine similarity score indicates that the generated description is of high quality, allowing the text-to-image generation to accurately recreate an image that closely resembles the original. Conversely, a low cosine similarity score suggests that the text description lacks accuracy, leading to a generated image that diverges from the original. This discrepancy signals the suboptimal performance of the image captioning model. Moreover, we apply our framework to detect multi-modal hallucinations in VLLMs, particularly in VQA and image description responses. After extensive experiments, our method shows a significant advantage in detecting hallucinations generated by VLLMs compared to other baseline models. Consequently, the proposed framework proves valuable for assessing the effectiveness of a given image captioning model.

The key contributions of this work can be summarized as follows:

- **Novel Framework for Evaluating Image-to-Text Models:** We present a novel framework that uses the text-to-image generation model, such as Stable Diffusion or DALL-E, to evaluate the quality of image descriptions generated by an image captioning model. The proposed evaluation framework does not necessitate human-annotated reference captions.
- **Human Assessment of the Framework:** To verify the effectiveness of our evaluation framework and facilitate large-scale comparison with human judgment, we introduce a new dataset by modifying and enhancing well-known benchmarks, originally made for conventional computer vision tasks.
- **Comprehensive Experiments on Established Datasets:** We perform extensive experiments to demonstrate the efficacy of the proposed evaluation framework using widely used image captioning datasets. We also test the capability of our evaluation framework to make the human likert scale judgment and detect the hallucinations contained in the VLLM-generated image captions.

2 Related Work

In this section, we begin by reviewing existing related literature, covering topics such as the image captioning methods and the evolution of automated metrics.

2.1 Image Captioning Methods

The encoder-decoder network architecture has become fundamental in image captioning, as demonstrated by various studies [17,47,49,55]. Typically, these networks use a CNN as the encoder to extract global image features and an RNN as the decoder to generate word sequences. Mao et al. [32] introduce a method for generating referring expressions, which describe specific objects or regions within an image. In [48], a bidirectional LSTM-based method for image captioning leverages both past and future information to learn long-term visual-language interactions. Attention mechanisms have significantly improved the performance of image captioning models. Pedersoli et al. [36] introduce an area-based attention model that predicts the next word and the corresponding image regions at each RNN timestep. While these advancements are significant, they mainly focus on single-image-based description generation. However, certain abstract concepts or descriptions might not be fully captured using only image data [26]. [18,19] have explored using expert-defined keyword sequences to enhance model capabilities in generating more accurate and contextually relevant descriptions. Recent advancements have also explored transformer-based architectures, such as Vision Transformers (ViT), which have shown promise in capturing finer details and global context in images for caption generation [10]. Furthermore, integrating multimodal learning approaches, where models are trained on both visual and textual data, has led to significant improvements in generating contextually richer and more nuanced image descriptions [31].

The domain of medical image captioning has seen significant advancements, particularly through methods that combine human expertise with algorithmic capabilities. [29] developed a Hybrid Retrieval-Generation Reinforced Agent, which integrates human prior knowledge with AI-based caption generation for medical images. This agent alternates between a generative module and a retrieval mechanism that uses a template database reflecting human expertise, producing multi-faceted, sequential sentences. [22] contributed to this field with a multi-task learning framework that simultaneously predicts tags and generates captions. Their method, which focuses on abnormal areas in chest radiology images using an attention mechanism and a hierarchical LSTM, offers detailed descriptions. These methods primarily focus on generating reports for chest radiology images, which differ in object size and detail compared to retinal images [19,26]. Additionally, the color features in chest radiology and retinal images differ significantly, with the former being predominantly grey-scale and the latter being colorful [19,26]. Most existing methods rely primarily on the image input for caption generation. Recent advancements also include enhancing the CNN-RNN framework with the TransFuser model [18]. This model adeptly combines features from different modalities and addresses the challenge of incorporating unordered keyword sequences with visual inputs, minimizing information loss [18]. This development represents a significant stride in medical image captioning, reflecting the growing complexity and capability of these methods. Further progress in deep learning, particularly the application of ViTs, has offered promising results in medical imaging [6]. ViTs excel in capturing intricate details and providing a broader context for more accurate medical image analysis and caption generation. The evaluation framework proposed in this paper is versatile and capable of assessing any existing image captioning approaches.

2.2 Automatic Metrics for Image Captioning

The evolution of image captioning has been significantly influenced by the development and application of automatic metrics for evaluating caption quality [2,20,25,30,35,46]. These metrics guide the training of captioning models and provide a scalable means for performance assessment. The BLEU score, a pioneering metric by [35], measures n-gram precision in the generated text against a reference. ROUGE [30] emphasizes recall through the overlap of n-grams and longest common subsequences. Subsequent innovations introduced refined approaches. METEOR [2] aligns more closely with human judgment by incorporating synonym matching and stemming. In [46], the CIDEr metric, specifically designed for image captioning, assesses the similarity of generated captions to a set of reference captions. The SPICE metric [1] evaluates semantic content and the depiction of objects, attributes, and relationships. Additionally, the NLG-Eval toolkit [42] provides a comprehensive suite of metrics for a more holistic evaluation of natural language generation. However, these metrics have limitations. Metrics like BLEU and ROUGE often fail to capture the contextual nuances of captions [30,35]. The challenge of evaluating creativity and novelty in caption generation is also evident, as automated metrics may penalize deviations from standard references [1,46]. Recently, advancements like BERTScore [52] and CLIPScore [15], which utilize contextual embeddings and visual-textual alignment, respectively, have been proposed to address these challenges.

In this study, human evaluation is employed to validate the effectiveness of the proposed evaluation framework.

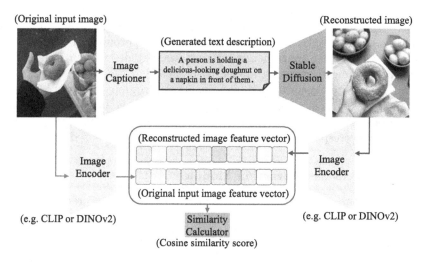

Fig. 1. Flowchart of the proposed evaluation framework. The proposed framework consists of four main components: an image captioning module, an image encoder, a text-to-image generation model (Stable Diffusion), and a similarity calculator. The image captioning module employs a chosen model to process an input image and generate textual descriptions. The image encoder is tasked with extracting features from the input image. The text-to-image generation model utilizes the text descriptions produced by the image captioning model to generate the corresponding image. Finally, the similarity calculator computes the similarity between the features of the input image and the image generated by the text-to-image generation model.

3 Methodology

The evaluation framework consists of four main components: an image captioning module, a Stable Diffusion-based text-to-image generator, an image feature extraction module, and a similarity calculator, as illustrated in Fig. 1. The following subsections will provide a detailed introduction to each component.

3.1 Image Captioning Module

The module includes an image captioning model that will be evaluated using the proposed framework. This model generates a text description from an input image. To help users understand the evaluation framework, we use the Instruct-BLIP model [9] as an example in Sect. 4. This example demonstrates the entire process of using the framework to evaluate an image captioning model, making it clear and accessible for users.

3.2 Stable Diffusion-Based Text-to-Image Generator

Numerous studies [3,45] have shown that text-to-image generators, such as GPT-4V and Stable Diffusion, can produce high-quality images that closely match the semantic meaning of given text prompts. Within the proposed framework, the Diffusion model-based image generator uses the text description generated by the preceding image captioning model. If the image captioning model performs well and produces an accurate and high-quality description, the Diffusion model-based image generator will create an image similar to the original input image. This demonstrates the connection between effective image captioning and the generation of corresponding images by the Diffusion model-based approach.

3.3 Image Feature Extraction Module

The image feature extraction module primarily consists of a pre-trained image encoder. This module takes an image as input and produces a feature vector representing the input image. To enhance user understanding of the proposed evaluation framework, we use DINOv2 [34] as an example for image feature extraction in Sect. 4. DINOv2 is a vision-only self-supervised model achieved by pre-training a ViT [10] on 142 million images without labels or annotations. Through this pre-training process, the model is distilled into smaller models with millions of parameters, achieving notable performance across various vision downstream tasks, including image recognition, video action recognition, object detection, instance segmentation, and semantic segmentation, all without extensive supervised fine-tuning. The demonstration in Sect. 4 highlights the complete process, including image feature extraction for calculating similarity scores between the input and generated images. It illustrates how the proposed framework can be used to assess an image captioning model, providing users with a clear understanding. It is worth noting that the image feature extractor can be substituted with other pre-trained CNNs, such as VGG-16 [44] or ResNet-52 [14]. Unlike vision language models such as CLIP [37] and Coca [51], which pre-train the image encoder by aligning text-image pairs, self-supervised vision-only models focus solely on vision information extraction, making them more suitable for image-to-image comparison.

3.4 Similarity Calculator

Cosine similarity, as defined in Eq. (1), serves as a metric for quantifying the similarity between two vectors in a multi-dimensional space. It evaluates the cosine of the angle between these vectors, offering insight into their degree of similarity or dissimilarity. The advantage of cosine similarity lies in its ability to assess directional similarity rather than magnitude, rendering it robust against variations in scale and orientation. This characteristic makes it a widely adopted metric in diverse domains, including image processing and NLP. In these fields, cosine similarity is frequently employed to assess the similarity between images, documents, or sentences represented as vectors in high-dimensional spaces. The cosine similarity value $\text{CosSim}(\cdot,\cdot) \in [-1,1]$, where a value of 1 signifies that the vectors are identical, 0 indicates orthogonality (i.e., no similarity), and -1 indicates complete dissimilarity or opposition.

$$\text{CosSim}(\mathbf{i_o}, \mathbf{i_g}) = \frac{\mathbf{i_o} \cdot \mathbf{i_g}}{\|\mathbf{i_o}\|\|\mathbf{i_g}\|}, \qquad (1)$$

where $\mathbf{i_o} \cdot \mathbf{i_g}$ denotes the dot product (also known as the inner product) of the original input image feature vector $\mathbf{i_o}$ and the LLM-generated image feature vector $\mathbf{i_g}$. $\|\mathbf{i_o}\|$ and $\|\mathbf{i_g}\|$ represent the Euclidean norms (also known as the magnitudes or lengths) of vectors $\mathbf{i_o}$ and $\mathbf{i_g}$, respectively. In other words, cosine similarity measures the cosine of the angle between two vectors, which represents their similarity in direction and magnitude.

4 Experiments and Analysis

In this section, we aim to evaluate the effectiveness of the proposed evaluation framework for image captioning models. To do this, we will validate our framework using our newly introduced image caption dataset, widely adopted human-annotated image captioning judgment datasets, and multi-modal hallucination detection datasets. For the image captioning datasets, since all datasets have undergone human annotation, our primary objective is to determine whether the evaluation results from our framework align with human consensus or judgment. Specifically, a correct caption-matching the human-annotated counterpart-should yield a high cosine similarity score between the generated and original images, as measured by our framework. Conversely, an incorrect caption-deviating from the human-annotated version-should result in a lower cosine similarity score. For the hallucination detection datasets, we aim to test whether our evaluation framework assigns a higher score to the accurate image description compared to an image caption containing textual descriptions not present in the image. This approach allows us to empirically validate the effectiveness of our proposed evaluation framework in aligning with human judgment and detecting patterns of hallucinations produced by the VLLMs. The code and models are released at https://github.com/s04240051/Image2Text2Image.git

4.1 Experimental Settings

To demonstrate the application of the proposed framework for evaluating an image captioning model, we use the InstructBLIP [9] model in our image captioning module. This model incorporates the pre-trained language model Vicuna-7B [53] to generate image descriptions. Image captions are generated using the prompt "<Image> A short image caption:", guiding the model to produce sentences with fewer than 100 tokens, excluding special symbols. For the text-to-image generation, we employ Stable Diffusion 3 medium [12]. This model uses three fixed, pre-trained encoders: CLIP L/14 [37], OpenCLIP bigG/14 [7], and T5-v1.1-XXL [38]. The entire diffusion model is pre-trained on the CC12M dataset [5]. Both original and generated images are encoded by DINOv2, which is distilled from ViT-B/14 on the LVD-142M dataset. For the proposed image caption dataset, human evaluation serves as the validation method for the framework. Each image in the dataset is accompanied by five human-annotated captions, and performance is measured using the average cosine similarity score.

4.2 Human Likert Scale Judgment

Evaluation on Proposed Dataset. We aim to test the capability of our method to measure the model's image caption quality without ground-truth annotation. To ensure the validity of the evaluation results based on our framework-specifically, their alignment with human judgment-we propose a dataset enhanced from the open-domain image caption dataset MSCOCO and an evaluation framework to test the feasibility of our method. In our study, we enhance the existing MSCOCO Caption dataset by incorporating an additional 30,000 human-annotated image-description pairs from Flickr30k. This augmented dataset serves as the basis for evaluating the alignment of our proposed evaluation method with human-annotated image descriptions.

The dataset introduced in this work consists of pairs of images and captions, each accompanied by five distinct human-generated captions. The details of our human evaluation process are outlined below. In Step 1, we use the human-annotated ground truth caption to generate an image through a text-to-image generation model, such as Stable Diffusion or DALL-E. In Step 2, we extract the image features of both the ground truth caption's corresponding image and the image generated by the text-to-image model. In Step 3, we apply the cosine similarity formula from Sect. 3.4 to compute the cosine similarity scores between these two sets of image features. Given that the caption is a human-annotated ground truth description, accurately portraying the corresponding image, we expect the similarity score from Step 3 to be high. Conversely, if a caption inaccurately describes a given image, the cosine similarity score from Step 3 should be low. Consistency between the experimental result and these expectations indicates the effectiveness of the proposed evaluation framework in aligning with human consensus.

The evaluation results depicted in Fig. 2 reveal notable insights. The blue lines in Fig. 2 illustrate the impact of the provided captions on the cosine similarity scores. Specifically, when the provided caption matches the correct human-annotated description (upper blue line), the average cosine similarity score reaches approximately 0.67. Conversely, when the caption is incorrect (lower blue line), the average cosine similarity score drops to around 0.47. This discrepancy results in a similarity gap of approximately 0.2. These findings underscore the effectiveness of the proposed evaluation framework, as it closely aligns with human judgment. The robustness of this human evaluation method is attributed to the remarkable text-to-image generation capabilities of modern VLLMs. Widely recognized models such as GPT-4V and Gemini have been extensively acclaimed in various studies and by the broader community [3,45,54].

Figure 2 reveals consistent trends in the evaluation results across MSCOCO, Flickr30k, and our dataset. Similar patterns are observed in MSCOCO and Flickr30k, where there is a notable decrease in the average cosine similarity when the model-generated image caption differs from the human-annotated ground truth caption. These findings affirm the effectiveness and reliability of the proposed evaluation framework for assessing image captioning models.

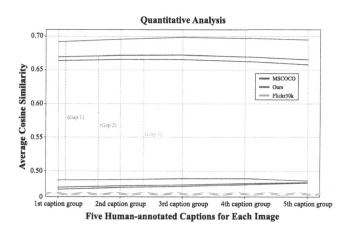

Fig. 2. Human evaluation results. The top three lines represent scenarios where the provided caption aligns with the correct human-annotated description, while the bottom three lines represent scenarios where the caption is incorrect. "Gap 1", "Gap 2", and "Gap 3" signify the disparities in average cosine similarity scores.

Flickr8K-Expert and Flickr8K-CF [16]. The Flickr8K dataset consists of three main components: images, paired captions (or captions from other images), and human judgment to evaluate the correctness of the captions. Flickr8K-Expert contains $5,664$ image-caption pairs, each annotated by three human experts who provide judgment scores ranging from 1 to 4. A score of 1 indicates that the caption does not describe the image at all, while a score of 4 indicates that the caption accurately describes the image. Following the CLIPScore setting, we flatten all human judgment scores into a list of $16,992$ ($5,664 \times 3$) data

Table 1. Correlations with human judgment for the Flickr8K-Expert.

	τ_c
BLEU-1 [35]	32.3
BLEU-4 [35]	30.8
ROUGE-L [30]	32.3
BERT-S (RoBERTa-F)	39.2
METEOR [2]	41.8
CIDEr [46]	43.9
SPICE [1]	44.9
LEIC(τ_b) [8]	46.9
BERT-S++ [50]	46.7
TIGEr [21]	49.3
NUBIA [23]	49.5
ViLBERTScore-F [27]	50.1
CLIP-S [15]	51.2
RefCLIP-S [15]	53.0
Image2Text2Image (Ours)	**53.5**

Table 2. Correlation with human judgment for Flickr8k-CF, a version of Flickr-8k dataset annotated through crowdsourcing.

	τ_b
BLEU-4 [35]	16.9
ROUGE-L [30]	19.9
BERT-S (RoBERTa-F)	22.8
METEOR [2]	22.2
CIDEr [46]	24.6
SPICE [1]	24.4
LEIC [8]	29.5
CLIP-S [15]	34.4
RefCLIP-S [15]	**36.4**
Image2Text2Image (Ours)	35.2

samples. Flickr-CF [16] is another human-annotated dataset with 145K binary quality judgments collected from CrowdFlower for 48K image-caption pairs and 1K unique images. Each pair has at least three binary human judgments, and we use the percentage of "yes" responses as the judgment score. For both Flickr-Expert and Flickr-CF judgments, we use Kendall's τ coefficient [24] to evaluate the correlation between the metric and human judgment, with τ_c and τ_b used for Flickr8K-Expert and Flickr8K-CF, respectively.

The experimental results for Flickr8K-Expert are shown in Table 1, and for Flickr8K-CF in Table 2. Our method achieves higher correlations with human judgments than CLIP-S (without references) and other previous methods that rely on references. Additionally, our method is only slightly inferior to the best reference-based method RefCLIP-S in the Flickr8K-CF dataset and outperforms it in the Flickr8K-Expert dataset. This provides strong evidence that our method can produce judgments similar to human evaluations for image captions without ground truth captions. Furthermore, unlike the CLIPScore method, which uses cross-modality (text-to-image) comparison, our method maintains the comparison within the same modality. The image generated from the caption can reconstruct semantic information into spatial information. This comparison within the spatial domain is more contrastive and improves image caption evaluation quality while reducing hallucination.

4.3 Evaluation on the Hallucination Sensitivity

Hallicination Detection on Foil. In the context of image captioning for VLMs, hallucinations can manifest as responses containing incorrect references or descriptions of the input image [28]. To mitigate these hallucinations and enhance the reliability and accuracy of image captioning models in real-life use cases, a pragmatic and efficient detection system is required. We first test our method on the FOIL dataset [43] to verify its ability to detect potential misdescriptions in image captions. FOIL reformulates the COCO image caption dataset by swapping a single noun phrase, e.g., substituting "bike" for "motorcycle". Following previous work, we select sample pairs with FOIL words and compute the accuracy of each evaluation metric in assigning a higher score to the true candidate versus the FOIL. Table 3 shows the results. The results reveal that when the metrics are exposed to more references, the evaluation improves significantly. This is likely because the reference captions contain the original words, leading to better measurement scores. Our method outperforms all reference-free metrics, including CLIP-S, but is slightly inferior to the best reference-based model. This demonstrates its outstanding ability to detect obvious incorrect information within image descriptions without the assistance of references.

Hallicination Detection on M-Hallucination. The FOIL dataset is designed to have only one mismatched word between the image caption and the original image. However, the FOIL setting cannot simulate the hallucinations that occur in real-world applications. Hallucination text often contains multiple and more complex descriptions of non-existent objects. To assess this aspect, we use the M-HalDetect dataset [13], which includes fine-grained annotations at a sub-sentence level over detailed image descriptions. This dataset consists of image descriptions generated by the recent VLLM InstructBLIP, prompted by randomly selected questions from a pool of instructions for describing an image. For hallucination annotation, each sentence of a generated image description is manually labeled as either "Accurate" or "Inaccurate". According to the original work setting, we conduct sentence-level binary hallucination classification. Following the previous experiment, we compute the accuracy of each evaluation metric in assigning a higher score to the ground truth image caption compared to the hallucinated parts of the caption sentences generated by a VLLM-based image captioning model. As shown in Table 4, our method outperformed all baseline models, including the best reference-free model CLIP-S, and the best reference-based model RefCLIP. These results indicate that image caption hallucinations from real-world applications pose a challenge to existing evaluation methods. Our method is capable of effectively evaluating under this scenario without exposure to reference annotations. We can observe from Fig. 3 that the baseline models like CLIP-s [15] and RefCLIP-s mostly focused on the correct part of the description but ignored the hallucinated part. Our method can reconstruct both correct and hallucinated parts of the description and makes more effective single-modal comparisons with the original image. while some ambiguous object feature descriptions such as whether the "table" is wooden can still confuse our method.

424   J.-H. Huang et al.

Table 3. Pairwise accuracy results on the FOIL hallucination detection. The baseline models use either one or four references.

	1-ref	4-ref
length	50.2	50.2
BLEU-4 [35]	66.5	82.6
BERT-S	88.6	92.1
METEOR [2]	78.8	85.4
CIDEr [46]	82.5	90.6
SPICE [1]	75.5	86.1
CLIP-S [15]	87.2	87.2
RefCLIP-S [15]	91.0	92.6
Image2Text2Image (Ours)	87.86	87.86

Table 4. Pairwise accuracy results on the M-HalDetection. The baseline models use one reference because this dataset lacks multiple references.

	1-ref
length	15.3
BLEU-1 [35]	20.1
BERT-S	34.8
METEOR [2]	28.4
CIDEr [46]	32.3
SPICE [1]	23.6
CLIP-S [15]	35.2
RefCLIP-S [15]	38.5
Image2Text2Image (Ours)	**57.3**

GT: A little league baseball team in orange uniforms posing for a group picture.

Hall: In the image, there is a group of young baseball players posing for a team photo on a grassy field and several cars parked in the background, adding to the outdoor setting.

Score: CLIP-S/RefCLIP-S/Ours = (0.77/0.68/**0.55**)

GT: A couple of men trying to lasso a cow.

Hall: In the image, there are two cowboys riding horses in a rodeo arena and one of the cowboys is wearing a white shirt, while the other is wearing a brown shirt.

Score: CLIP-S/RefCLIP-S/Ours = (0.83/0.72/**0.59**)

GT: A lunch box of some sorts is placed on the table.

Hall: In the image, there is a black suitcase placed on top of a wooden table.

Score: CLIP-S/RefCLIP-S/Ours = (0.91/0.87/**0.86**)

Fig. 3. The visualization of the hallucination detection. **GT** and **Hall** denote ground truth and hallucinated captions, respectively. The hallucinated descriptions are highlighted in red. In each row, the first image represents the original image, and the second image represents the image generated from the hallucinated caption. For each caption, we report the scores of CLIP-S [15], RefCLIP-S [15], and our method. The first two rows show the corrected examples, while the third row shows an example that all three methods failed to detect. A smaller score is better for detecting the hallucinated captions.

5   Conclusion

In this study, we have introduced a novel framework called Image2Text2Image for evaluating automatically generated image descriptions, aiming to overcome the limitations of existing evaluation metrics like BLEU, ROUGE, METEOR, and CIDEr. Our framework leverages advancements in text-to-image generation models such as Stable Diffusion or DALL-E to utilize image descriptions generated by an image captioning model for creating corresponding images. By quantifying the cosine similarity between the representation of the original input image in the image captioning model and the representation of the generated image, we can effectively assess the image captioning model's performance without relying on human-annotated reference captions. Through extensive experiments on established datasets like Flickr8k-Expert, Flickr8k-CF, and our proposed dataset, we have demonstrated the effectiveness of the proposed evaluation framework. Our experimental results suggest that the proposed framework's performance closely correlates with human judgment, offering a valuable method for evaluating the effectiveness of image captioning models. Additionally, human evaluations conducted on our introduced dataset validate the framework's efficacy in capturing various aspects such as grammaticality, coverage, correctness, and truthfulness in automatically generated image descriptions. Moving forward, the proposed framework presents new opportunities for evaluating image captioning models, offering a more efficient and reliable alternative to traditional human evaluations and existing automated evaluation metrics. It is designed to complement, rather than replace, human judgment. In summary, our work contributes to the ongoing development of robust evaluation frameworks for image captioning models, bridging the gap between automated metrics and human judgment, and driving advancements in this field.

References

1. Anderson, P., Fernando, B., Johnson, M., Gould, S.: Spice: semantic propositional image caption evaluation. In: ECCV. Springer (2016)
2. Banerjee, S., Lavie, A.: Meteor: an automatic metric for mt evaluation with improved correlation with human judgments. In: ACL Workshop (2005)
3. Brown, T., Mann, B., Ryder, N., et al.: Language models are few-shot learners. Advances in neural information processing systems (2020)
4. Chan, D., Petryk, S., Gonzalez, J.E., Darrell, T., Canny, J.: Clair: evaluating image captions with large language models. arXiv preprint arXiv:2310.12971 (2023)
5. Changpinyo, S., Sharma, P., et al.: Conceptual 12m: pushing web-scale image-text pre-training to recognize long-tail visual concepts. In: CVPR (2021)
6. Chen, J., He, Y., Frey, E.C., Li, Y., Du, Y.: Vit-v-net: vision transformer for unsupervised volumetric medical image registration. arXiv:2104.06468 (2021)
7. Cherti, M., Beaumont, R., Wightman, R., Wortsman, M., et al.: Reproducible scaling laws for contrastive language-image learning. In: CVPR (2023)
8. Cui, Y., Yang, G., Veit, A., Huang, X., Belongie, S.: Learning to evaluate image captioning. In: CVPR (2018)
9. Dai, W., Li, J., Li, D., et al.: Instructblip: towards general-purpose vision-language models with instruction tuning. arxiv 2023. arXiv preprint arXiv:2305.06500
10. Dosovitskiy, A., Beyer, L., Kolesnikov, A., Weissenborn, D., Zhai, X., Unterthiner, T., et al.: An image is worth 16 × 16 words: Transformers for image recognition at scale. arXiv preprint arXiv:2010.11929 (2020)

11. Elliott, D., Keller, F.: Image description using visual dependency representations. In: EMNLP (2013)
12. Esser, P., Kulal, S., Blattmann, A., et al.: Scaling rectified flow transformers for high-resolution image synthesis. In: ICML (2024)
13. Gunjal, A., Yin, J., Bas, E.: Detecting and preventing hallucinations in large vision language models. In: AAAI (2024)
14. He, K., et al.: Deep residual learning for image recognition. In: CVPR (2016)
15. Hessel, J., Holtzman, A., Forbes, M., Bras, R.L., Choi, Y.: Clipscore: a reference-free evaluation metric for image captioning. arXiv preprint arXiv:2104.08718 (2021)
16. Hodosh, M., Young, P., Hockenmaier, J.: Framing image description as a ranking task: Data, models and evaluation metrics. JAIR (2013)
17. Huang, J.H., Alfadly, M., Ghanem, B.: Vqabq: Visual question answering by basic questions. arXiv preprint arXiv:1703.06492 (2017)
18. Huang, J.H., Wu, T.W., Yang, Lin, I., Tegner, J., Worring, M., et al.: Non-local attention improves description generation for retinal images. In: WACV (2022)
19. Huang, J.H., Yang, C.H.H., Liu, F., et al.: Deepopht: medical report generation for retinal images via deep models and visual explanation. In: WACV (2021)
20. Huang, J.H., Zhu, H., et al.: A novel evaluation framework for image2text generation. In: ACM SIGIR workshop (2024)
21. Jiang, M., Huang, Q., Zhang, L., Wang, X., Zhang, P., et al.: Tiger: text-to-image grounding for image caption evaluation. arXiv preprint arXiv:1909.02050 (2019)
22. Jing, B., Xie, P., Xing, E.: On the automatic generation of medical imaging reports. arXiv preprint arXiv:1711.08195 (2017)
23. Kane, H., Kocyigit, M.Y., Abdalla, A., et al.: Nubia: neural based interchangeability assessor for text generation. arXiv preprint arXiv:2004.14667 (2020)
24. Kendall, M.G.: A new measure of rank correlation. Biometrika (1938)
25. Khan, O.S., Sharma, U., Zhu, H., et al.: Exquisitor at the lifelog search challenge 2024: Blending conversational search with user relevance feedback. In: ICMR Lifelog Search Challenge (2024)
26. Laserson, J., Lantsman, C.D., Cohen-Sfady, M., et al.: Textray: mining clinical reports to gain a broad understanding of chest x-rays. In: MICCAI. Springer (2018)
27. Lee, H., Yoon, S., Dernoncourt, F., Kim, D.S., Bui, T., Jung, K.: Vilbertscore: evaluating image caption using vision-and-language bert. In: Eval4NLP (2020)
28. Li, Y., Du, Y., Zhou, K., Wang, J., Zhao, W.X., Wen, J.R.: Evaluating object hallucination in large vision-language models. arXiv preprint arXiv:2305.10355 (2023)
29. Li, Y., Liang, X., et al.: Hybrid retrieval-generation reinforced agent for medical image report generation. Advances in neural information processing systems (2018)
30. Lin, C.Y.: Rouge: a package for automatic evaluation of summaries. In: Text Summarization Branches Out (2004)
31. Lu, J., Batra, D., Parikh, D., Lee, S.: Vilbert: pretraining task-agnostic visiolinguistic representations for vision-and-language tasks. NeurIPS (2019)
32. Mao, J., Huang, J., Toshev, A., Camburu, O., Yuille, A.L., Murphy, K.: Generation and comprehension of unambiguous object descriptions. In: CVPR (2016)
33. Mitchell, M., et al.: Midge: generating descriptions of images. In: INLG (2012)
34. Oquab, M., Darcet, T., Moutakanni, T., Vo, H., et al.: Dinov2: learning robust visual features without supervision. arXiv preprint arXiv:2304.07193 (2023)
35. Papineni, K., Roukos, S., Ward, T., Zhu, W.J.: Bleu: a method for automatic evaluation of machine translation. In: ACL (2002)
36. Pedersoli, M., et al.: Areas of attention for image captioning. In: ICCV (2017)
37. Radford, A., Kim, J.W., Hallacy, C., Ramesh, A., Goh, G., et al.: Learning transferable visual models from natural language supervision. In: ICML (2021)

38. Raffel, C., Shazeer, N., Roberts, A., Lee, K., Narang, S., et al.: Exploring the limits of transfer learning with a unified text-to-text transformer. JMLR (2020)
39. Rohrbach, M., Qiu, W., Titov, I., Thater, S., Pinkal, M., Schiele, B.: Translating video content to natural language descriptions. In: ICCV (2013)
40. Rudinac, S., Larson, M., et al.: Learning crowdsourced user preferences for visual summarization of image collections. IEEE Trans. Multimed. (2013)
41. by Saheel, S.: Baby talk: understanding and generating image descriptions
42. Sharma, S., et al.: Relevance of unsupervised metrics in task-oriented dialogue for evaluating natural language generation. arXiv preprint arXiv:1706.09799 (2017)
43. Shekhar, R., Pezzelle, S., Klimovich, Y., Herbelot, A., Nabi, E., et al.: FOIL it! find one mismatch between image and language caption. In: ACL (2017)
44. Simonyan, K., Zisserman, A.: Very deep convolutional networks for large-scale image recognition. arXiv preprint arXiv:1409.1556 (2014)
45. Team, G., Anil, R., Borgeaud, S., Wu, Y., Alayrac, J.B., Yu, J., et al.: Gemini: a family of highly capable multimodal models. arXiv:2312.11805 (2023)
46. Vedantam, R., Lawrence Zitnick, C., Parikh, D.: Cider: consensus-based image description evaluation. In: CVPR (2015)
47. Vinyals, O., Toshev, A., Bengio, S., Erhan, D.: Show and tell: a neural image caption generator. In: CVPR (2015)
48. Wang, C., Yang, H., et al.: Image captioning with deep bidirectional lstms. In: ACMMM (2016)
49. Xiao, X., Wang, L., Ding, K., Xiang, S., Pan, C.: Deep hierarchical encoder–decoder network for image captioning. IEEE Transactions on Multimedia (2019)
50. Yi, Y., Deng, H., Hu, J.: Improving image captioning evaluation by considering inter references variance. In: ACL (2020)
51. Yu, J., Wang, Z., Vasudevan, V., et al.: Coca: Contrastive captioners are image-text foundation models. arXiv preprint arXiv:2205.01917 (2022)
52. Zhang, T., Kishore, V., Wu, F., Weinberger, K.Q., Artzi, Y.: Bertscore: evaluating text generation with bert. arXiv preprint arXiv:1904.09675 (2019)
53. Zheng, L., Chiang, W.L., Sheng, Y., Zhuang, S., Wu, Z., Zhuang, Y., et al.: Judging llm-as-a-judge with mt-bench and chatbot arena. ArXiv abs/2306.05685 (2023)
54. Zhu, H., Huang, J.H., et al.: Enhancing interactive image retrieval with query rewriting using large language models and vision language models. In: ICMR (2024)
55. Zhu, H., Salah, A.A., et al.: Video-based estimation of pain indicators in dogs. In: ACII (2023)

Quantifying Image-Adjective Associations by Leveraging Large-Scale Pretrained Models

Chihaya Matsuhira[1]([✉]), Marc A. Kastner[1,2], Takahiro Komamizu[1], Takatsugu Hirayama[1,3], and Ichiro Ide[1]

[1] Nagoya University, Nagoya, Aichi, Japan
matsuhirac@cs.is.i.nagoya-u.ac.jp
[2] Hiroshima City University, Hiroshima, Japan
[3] University of Human Environments, Okazaki, Aichi, Japan

Abstract. Quantifying the associations between images and adjectives, i.e., how much the visual characteristics of an image are connected with a certain adjective, is important for better image understanding. For instance, the appearance of a *kitten* can be associated with adjectives such as "soft", "small", and "cute" rather than the opposite "hard", "large", and "scary". Thus, giving scores for a kitten photo considering the degree of its association with each antonym adjective pair (termed *adjective axis*, e.g., "round" vs. "sharp") aids in understanding the image content and its atmosphere. Existing methods rely on subjective human engagement, making it difficult to estimate the association of images with arbitrary adjective axes in a single framework. To enable the extension to arbitrary axes, we explore the use of large-scale pretrained models, including Large Language Models (LLMs) and Vision Language Models (VLMs). In the proposed training-free framework, users only need to specify a pair of antonym nouns that negatively and positively describe the target axis (e.g., "roundness" and "sharpness"). Evaluation confirms that the proposed framework can predict negative and positive associations between adjectives and images as correctly as the manually-assisted comparative. The result also highlights the pros and cons of utilizing the VLM's textual or visual embedding for specific types of adjective axes. Furthermore, computing the similarities among four adjective axes unveils how the proposed framework connects them with each other, such as its tendency to regard a sharp object as being small, hard, and quick in motion.

Keywords: Large Language Model · Vision Language Model · Adjective-Image Association · Training-Free Framework

1 Introduction

Quantifying visual characteristics in multiple aspects is important for better image understanding. To characterize such characteristics from various perspectives, linguistic properties such as adjectives are useful [10]. For instance, the appearance of a *kitten* can be described by adjectives such as "soft", "small", and

Quantifying Image-Adjective Associations by Leveraging LLMs and VLMs

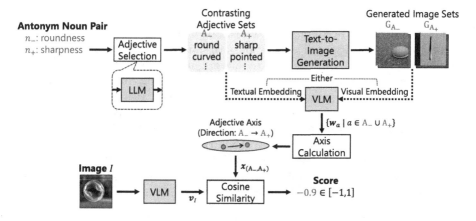

Fig. 1. Proposed framework for predicting the degree of association of an image with arbitrary antonym adjective pairs. Given an antonym noun pair denoting the negativity and positivity of a certain adjective axis, the framework first identifies such an axis in the Vision Language Model (VLM)'s embedding space by comprehending its semantics using a Large Language Model (LLM). It then computes a score for an input image regarding how much the image is associated with the adjective axis.

even more subjectively, "cute", rather than their opposite counterparts: "hard", "large", and "scary". Estimating how much an image is associated with each antonym adjective pair (termed *adjective axis*) would enhance the understanding of image content and its atmosphere, which can facilitate many vision-language-related tasks, such as image retrieval, image tagging, and image captioning.

Machine-learning-based automatic estimation requires data acquisition that often necessitates subjective human engagement. Traditional approaches utilize labeled data, e.g., annotated by users on social media, for training a classification or regression model [5,17]. An alternative approach is to scrape data from the Web via search engines, as taken in some related studies [16,20]. However, such data contain noise, i.e., many irrelevant data, requiring manual checks before being adopted as a reliable data source. Therefore, these approaches are not truly automatic, inhibiting the expansion of the estimation to various adjectives.

To construct a versatile framework for arbitrary adjective axes, this paper explores the use of large-scale pretrained models. In recent ages, many such models, including Large Language Models (LLMs) [3,6,9,18] and Vision Language Models (VLMs) [4,8,13,14], have shown their plausible capabilities in various tasks. One of their notable features is zero-shot performance, meaning that they can process data from a wide range of domains without additional training.

Figure 1 illustrates the proposed framework to measure the association between an image and an adjective axis by leveraging a generative LLM and a VLM. This framework is based on the existing method proposed by Alper and Averbuch-Elor [1] which outputs a score regarding the object's roundness and sharpness in an image. Their method needed to manually devise a set of adjectives to rule an axis directing from roundness to sharpness on a VLM's embedding space. To automate this manual adjective selection, we exploit a generative LLM. Our framework takes two inputs; a pair of antonym nouns and an image, to output a score denoting the degree of association of the image with the adjective axis specified by the antonym nouns. In this framework, users only need to specify a pair of antonym nouns describing the negativity and positivity of a target axis (e.g., "roundness" and "sharpness" for the round-sharp axis). Based on this input, an LLM in the proposed framework lists multiple adjectives describing each polarity of the axis. With these adjective sets, an adjective axis is defined on its vision-language feature space that points to the direction from negativity to positivity. Lastly, for a target image, a score is computed by plotting the image onto the axis. This framework supplies flexibility for expansion to arbitrary axes while keeping it not fully black-boxed. This enables an analysis of how the proposed framework associates visual characteristics with each adjective axis.

2 Related Work on Estimating Image-Adjective Associations

Some existing studies have estimated the visual characteristics in images that can be described using adjectives. Conventionally, these tasks start by collecting manual annotations for training data [5,12,15,17]. As an alternative approach, some studies use search engines such as Google[1] and Bing[2] to obtain similar data [16,20]. However, these methods cannot predict the association of images with arbitrary adjectives, since their models are trained on datasets collected for limited sets of adjectives. Also, the data provided by search engines are often noisy and contain irrelevant data, hindering a fully automatic way to extend the model to other adjective axes. More recent studies have used VLMs for estimating the visual sharpness of objects and aesthetics of images [1,7,19]. However, in all these studies, the prediction target is limited to one specific axis and thus cannot be easily extended to others.

One way to construct a versatile estimator for arbitrary adjective axes is to use pretrained models guaranteeing zero-shot performance. Lazaridou et al. [10] have shown that the representation space of pretrained language models, such as Word2Vec [11], can be useful for image adjective tagging. In line with this finding, the proposed framework in this paper explores the use of large-scale pretrained models, LLMs and VLMs, for predicting arbitrary adjective axes.

3 Proposed Framework

Figure 1 illustrates the proposed training-free framework to estimate the degree of association between an image and any adjective axis. Given an image I and a pair of antonym nouns n_- and n_+ denoting the negativity and positivity on a target adjective axis (e.g., "roundness" and "sharpness" for the round-sharp axis), the framework yields a score reflecting how negatively or positively the image is associated with the target adjective axis.

---
[1] https://images.google.com/ (Accessed on October 23, 2024).
[2] https://www.bing.com/images/feed/ (Accessed on October 23, 2024).

Table 1. Prompt used to instruct an LLM to list up K adjectives. [K], [N-], and [N+] are replaced with the values of K, n_-, and n_+, respectively.

```
Pick up [K] adjectives describing the [N+] of an object.
Then, as antonyms, please pick up another [K] adjectives describing the [N-] of an
object.\n
During the word selection, please be careful about the following five points:\n
- All words must be commonly used adjectives. Also, they must not be nouns.\n
- They must not contain hyphens.\n
- Their meanings must be clear. They must not be interpreted to mean other things.\n
- Their word stems must be diverse. That is, no two words must have the same stem.\n
- They can be used in the attributive position. That is, they can be used in the noun
phrase "the <WORD> thing" where <WORD> represents each selected adjective.\n
```

3.1 Automatic Adjective Selection

In most studies, the first step to defining the target axis is to choose adjectives that negatively and positively describe the visual characteristics [1,7,19]. This paper represents these two sets as A_- and A_+, respectively. The method by Alper and Averbuch-Elor [1], on which the proposed framework is based, requires manual labor to obtain them. Specifically, they need to select 10 adjectives describing n_- (roundness) and n_+ (sharpness) of objects, respectively, to predict the sharpness of an object in an image.

This paper proposes an automatic way to expand the method to arbitrary adjective axes by exploiting a generative LLM. Specifically, using the prompt shown in Table 1, an LLM is instructed to provide two sets of K adjectives that negatively and positively describe the target axis indicated by the two contrasting nouns n_- and n_+, respectively.

Here, since the output of an LLM is non-deterministic, the output adjectives differ in every trial. Thus, using only the output of the first trial does not yield a reliable selection. To extract adjectives having the highest likelihood of being selected inside the LLM and ensure the highest relevance of those adjectives to the input prompt, this paper conducts the trial 50 times to take the most frequently selected K adjectives as the components of each of A_- and A_+.

This procedure sometimes yields cases where words chosen by the LLM violate the rules embedded in the prompt. For instance, a selected word may contain a hyphen, or it is even possible that the word is not an adjective. To ensure precision regarding this perspective, in the evaluation described in Sect. 4, we manually filtered and removed them from the set of selected words. This process can be automated using tools such as Natural Language ToolKit [2] when a larger number of words needs to be scanned. Besides, there may be cases where the resultant A_- and A_+ do not contain the root adjectives of the specified nouns n_- and n_+ (e.g., the adjective "sharp" for the noun "sharpness"). To maximize the consistency between the results and the user's intention, if the root adjective is not included in A_-, we define a new A_- as a concatenation of the root adjective and the top $K-1$ adjectives in the original A_- (the same procedure is applied to A_+, too).

3.2 Adjective Axis and Score Calculation

The proposed framework uses a VLM, such as Contrastive Language Image Pretraining (CLIP) [13], to compute embeddings for each adjective which rule an adjective axis on the VLM's embedding space. Based on Alper and Averbuch-Elor's method [1], we implement two embedding calculation strategies. Both strategies, which will be explained in the last part of this section, yield an embedding for each adjective a, which is denoted as w_a.

After obtaining the embeddings w_a, an adjective axis $x_{(A_-,A_+)}$ is computed on the VLM's embedding space. Our adjective axis is realized as a probe vector proposed by Alper and Averbuch-Elor [1]. The adjective axis, which points to the direction from A_- to A_+, is computed according to the following formula:

$$x_{(A_-,A_+)} = \sum_{a \in A_+} \hat{w}_a - \sum_{a \in A_-} \hat{w}_a, \qquad (1)$$

where $\hat{w}_a$ denotes a unit-normalized vector of w_a. This becomes a vector pointing to the direction in the VLM's embedding space which distinguishes between the two adjective sets A_- and A_+ [1].

Lastly, given an input image I, the framework computes the degree of its association with the adjective axis as a single score. The score is computed as:

$$\text{score}(I, A_-, A_+) = \text{cossim}(v_I, x_{(A_-,A_+)}), \qquad (2)$$

where v_I denotes the VLM's visual embedding for the input image I and $\text{cossim}(\cdot, \cdot)$ is a cosine similarity between two vectors.

Below, we explain the two embedding calculation strategies implemented in this paper.

Textual Embedding: Textual embedding is computed by directly inputting an adjective a into the VLM as a text with prompt engineering. Textual embeddings result in an adjective axis that reflects the VLM's ability to match an adjective specified in texts with the corresponding visual appearance. This strategy does not explicitly utilize the visual information of the image domain.

Visual Embedding: Visual embedding is computed by combining a VLM and a text-to-image generation model such as Stable Diffusion [4,14]. For each adjective, M images are first generated to obtain the visual appearance described by the adjective. This results in two sets of generated images, G_{A_-} and G_{A_+}. Then, the VLM computes the visual embedding for each image. Lastly, the visual embedding for the adjective a is computed by averaging the M visual embeddings of a. The axis calculated based on visual embeddings becomes aware of the visual appearance of adjectives represented in generated images.

4 Experiment

This experiment evaluates, in the first place, whether LLMs can substitute the manual adjective selection required by the existing method [1]. Second, we are interested in how the proposed framework computes scores for each image, since it does not rely on explicit training data. To shed light on its behavior, two axis calculation strategies are compared to discuss their pros and cons.

4.1 Targeted Adjective Axes

Four adjective axes are targeted in this evaluation: round-sharp, small-large, slow-quick, and firm-soft axes. The round-sharp axis is chosen according to the existing study [1], and the other axes are selected to cover various types of visual characteristics. The characteristics of the former two out of the four axes can be measured directly from the visual properties of objects. In contrast, the latter two would require understanding more salient correspondence between appearance and characteristics. For instance, although it is hard to infer the softness of a *marshmallow* without touching it, it is possible if we know the textures of other soft objects. Note that as an antonym of "soft", "firm" is preferred over "hard" due to its ambiguity in meaning.

4.2 Implementation

For each axis, the following antonym noun pairs of (n_-, n_+) are used to be embedded in the prompt on Table 1: (roundness, sharpness), (smallness, largeness), (slowness, quickness), and (firmness, softness).

Generative Pretrained Transformer (GPT) [3], CLIP [13], and Stable Diffusion [4,14] are adopted as the LLM, the VLM and the text-to-image generation model in the proposed framework, respectively[3]. The number of selected adjectives K is set as 10, and the number of images generated for each adjective M is set as 50 in line with the implementation of Alper and Averbuch-Elor's work [1]. Table 2(b) shows $K = 10$ selected adjectives for each adjective axis under this implementation. For the textual and visual embedding calculation, the prompt 'a photo of a <ADJ> object' is used throughout the experiment where <ADJ> is replaced with each adjective. This differs from the existing study [1] that used the prompt 'a 3D rendering of a <ADJ> object' since they focused only on measuring the sharpness of the object's shape in an image.

To assess how appropriate the LLM's adjective selection is, a comparative method is implemented with manually selected adjectives. As the source of manually selected sets for the round-sharp axis, we adopt those proposed by Alper and Averbuch-Elor [1], thus resulting in an identical implementation to their method. For the other three axes, since there are no known criteria for selection, we select them manually by ourselves, as shown in Table 2(b), based on

---

[3] gpt-3.5-turbo-0125, CLIP ViT-L/14, and stable-diffusion-v1.4, listed on their respective model cards, are used throughout this evaluation.

Table 2. Antonym noun pair and adjective sets used in the proposed framework for each of the four adjective axes targeted in the experiment.

(a) Sets of adjectives automatically selected by the LLM. In each set, more frequently selected adjectives are shown earlier

Antonym pair	Automatically selected set of adjectives
n_- = roundness	A_- = {round, spherical, circular, smooth, curved, oval, bulbous, rotund, blunt, plump}
n_+ = sharpness	A_+ = {sharp, piercing, jagged, pointed, cutting, serrated, keen, acute, edgy, prickly}
n_- = smallness	A_- = {small, tiny, petite, miniature, diminutive, microscopic, lilliputian, compact, minuscule, mini}
n_+ = largeness	A_+ = {large, massive, immense, enormous, gigantic, huge, vast, colossal, mammoth, tremendous}
n_- = slowness	A_- = {slow, leisurely, gradual, languid, tardy, sluggish, plodding, delayed, lethargic, ponderous}
n_+ = quickness	A_+ = {quick, swift, speedy, brisk, hasty, rapid, fast, fleet, prompt, agile}
n_- = firmness	A_- = {firm, tough, rigid, solid, resilient, robust, hard, sturdy, steady, strong}
n_+ = softness	A_+ = {soft, tender, smooth, fluffy, silky, pliable, gentle, delicate, velvety, plush}

(b) Sets of adjectives manually selected by humans. Underlined adjectives indicate those that overlap with the automatically selected sets shown in Table 2(a)

Antonym pair	Manually selected set of adjectives
n_- = roundness	A_- = {round, circular, smooth, curved, rotund, plump, soft, fat, chubby, plush}
n_+ = sharpness	A_+ = {sharp, jagged, pointed, edgy, prickly, spiky, angular, hard, rugged, uneven}
n_- = smallness	A_- = {small, tiny, petite, miniature, microscopic, little, slender, slight, teeny, weeny}
n_+ = largeness	A_+ = {large, massive, enormous, gigantic, huge, big, sizable, astronomical, jumbo, fat}
n_- = slowness	A_- = {slow, leisurely, languid, tardy, sluggish, idle, dilatory, dull, lazy, slothful}
n_+ = quickness	A_+ = {quick, swift, speedy, brisk, hasty, rapid, fast, flying, sudden, rushed}
n_- = firmness	A_- = {firm, rigid, solid, robust, hardened, sturdy, rough, heavy, stiff, dense}
n_+ = softness	A_+ = {soft, fluffy, silky, gentle, spongy, cottony, flabby, mild, fragile, fluent}

the same five criteria as listed in Table 1. Note that they were selected before the experimentation using an LLM, without observing the adjectives listed in Table 2(a). The comparison between the two tables indicates that more than half of the elements overlap between the automatically and manually selected adjective sets.

Furthermore, we also compare the proposed framework with a method that only uses a single pair of the root adjectives of the noun pair (n_-, n_+). This corresponds to the implementation that only uses the first element of each set shown in Table 2(a). This comparison would highlight the effect of the adjective augmentation by the LLM used in the proposed framework.

4.3 Evaluation Data

We prepare images that possess visual characteristics negatively and positively associated with each of the four adjective axes. To collect such images, we first list concepts that should possess such characteristics. By feeding the prompt

Table 3. Prompt used to instruct an LLM to list 10 concepts for the evaluation. [N-] and [N+] are replaced with the values of n_- and n_+, respectively.

Pick up 10 nouns denoting an object with high [N+].
Then, pick up another 10 nouns denoting an object with high [N-].\n
During the word selection, please be careful about the following four points:\n
- All words must be commonly used one-word nouns.\n
- They must not contain hyphens.\n
- Their meanings must be clear. They must not be interpreted to mean other things.\n
- Their word stems must be diverse. That is, no two words must have the same stem.\n
---

Table 4. Sets of 10 nouns of concepts that are likely to have negative and positive associations with each polarity of each adjective axis.

Antonym pair	Set of 10 nouns of concepts
n_- = roundness	{ball, orange, globe, peach, sphere, melon, circle, plum, cherry, bubble}
n_+ = sharpness	{needle, sword, razor, dagger, knife, scissors, axe, arrow, saw, chisel}
n_- = smallness	{ant, atom, pebble, dust, grain, molecule, insect, microbe, seed, drop}
n_+ = largeness	{mountain, whale, elephant, building, canyon, ship, glacier, sequoia, airplane, galaxy}
n_- = slowness	{sloth, snail, tortoise, glacier, slug, loris, koala, turtle, iceberg, panda}
n_+ = quickness	{cheetah, rocket, bullet, ferrari, lightning, speedboat, jet, falcon, arrow, greyhound}
n_- = firmness	{brick, concrete, steel, rock, wood, plastic, glass, marble, diamond, metal}
n_+ = softness	{marshmallow, cloud, pillow, cotton, feather, silk, sponge, blanket, velvet, mousse}

shown in Table 3, which is similar to the one in Table 1, into the same version of LLM (GPT) 50 times, we select 10 nouns denoting 10 different concepts with the highest likelihood for each polarity of the adjective axes.

Next, we collect real images showing each concept. Bing image search‡ results are used to collect the top-20 images for each concept. Here, a single-noun query is used to collect images. The number of images is restricted to 20 to avoid less relevant images being included in the evaluation data, since images scraped from the Web contain much noise. This results in $10 \times 20 = 200$ images for negative and positive polarities of each adjective axis, respectively.

10 nouns of concepts selected for each polarity of each adjective axis are shown in Table 4. Regarding this selection, we manually removed concepts whose Bing images did not contain them due to the overlap with existing brand, product, and character names (e.g., *apple* and *thunderbolt*).

4.4 Task and Metrics

The proposed framework is evaluated in an image retrieval task. We evaluate whether images negatively associated with the adjective axis (e.g., *ball* for the round-sharp axis) are scored lower than those positively associated with the axis (e.g., *needle* for the same axis).

Table 5. Evaluation of image retrieval results with different implementations of the proposed framework. Bold indicates the best score among the three methods within each embedding strategy.

Adjective selection	Embedding	Recall@200 ↑ (Total 400 images)			
		round-sharp	small-large	slow-quick	firm-soft
Single	Textual	0.940	0.885	0.810	0.915
Manual	Textual	0.975	0.935	0.890	**0.930**
LLM (GPT [3])	Textual	0.955	0.935	**0.910**	0.920
Single	Visual	0.955	0.945	0.800	0.910
Manual	Visual	0.910	0.950	0.795	0.825
LLM (GPT [3])	Visual	**0.990**	**0.955**	0.780	0.830

An image pool is constructed as $200 \times 2 = 400$ real images collected for both the negative and positive polarities of each adjective axis. Then, images in the pool are ranked according to the score assigned to each image, and the top-k images are retrieved. By regarding the 200 images positively associated with each axis as the set of positive samples, Recall@k is computed as the metric where k is set as the number of positive samples, i.e., 200.

4.5 Results

Results with different implementations are reported in Table 5. First, regarding the adjective selection, compared with the manually selected adjective sets, we can see that the sets selected by an LLM contributed to the performance very comparably. The comparison to the single adjective pair setting, denoted as "Single" in the table, also indicates the effectiveness of the LLM's adjective selection. Notably, in the setting with textual embeddings, the method adopting the LLM better predicted the scores in all adjective axes than the one adopting a single noun pair.

When contrasting the two implementations of the embedding calculation, we can observe that the textual embedding was effective in quickness and softness prediction, whereas the visual embedding had gain for the other two. The decisive factor is whether the image-adjective association can be measured directly from images or predicting it requires understanding more salient correspondence between the appearance and the adjectives. As mentioned earlier, it is not easy to measure the softness of a *marshmallow* solely from its appearance. Similarly, predicting the speed of a *slug* may require observing its motion, which is not visible in a one-frame picture. The textual embedding could predict it since the word "slug" often co-occurs with words like "slow" in text corpora. On the other hand, since the textual embedding is relatively blind to the visual appearance, the visual embedding better predicts object properties such as shape and size.

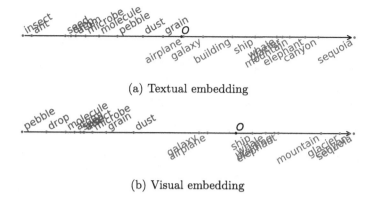

Fig. 2. Average scores for each concept on the small-large axis, computed by the proposed framework with each axis calculation strategy.

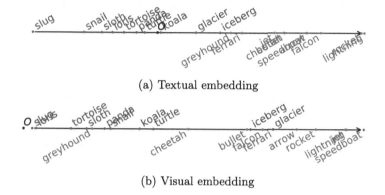

Fig. 3. Average scores for each concept on the slow-quick axis, computed by the proposed framework with each axis calculation strategy.

4.6 Qualitative Study

Figures 2 and 3 show scores for each concept used in the evaluation averaged over 20 images, plotted on the small-large and slow-quick axes, respectively. The scores are computed by the proposed framework with the LLM's adjective selection. O on each axis indicates the point of the score 0.

Between Figs. 2(a) and 3(b), we can observe that, as expected, the visual embedding makes a better distinction between visually small and large concepts snapped in images than the textual embedding. In contrast, a comparison between Figs. 3(a) and 3(b) reveals that the visual embedding is weak at predicting the quickness of concepts, e.g. a *greyhound*, which is hard to be predicted relying solely on the appearance. The relative position of O in Fig. 3(b) also shows that the scores of all concepts in this setting are positive, suggesting that even *slugs* are not regarded as an absolutely slow concept.

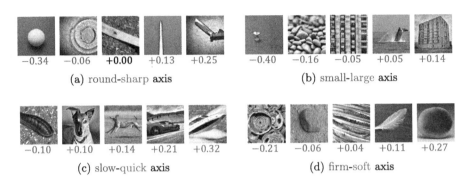

Fig. 4. Predicted scores for single generated images. Scores are computed by the proposed framework with the image embedding strategy.

Figure 4 showcases scores for single images on all four axes computed by the proposed framework with the image embedding strategy. Considering the copyrights, as an alternative to the Bing image pool, we created a new image pool for this visualization by generating 50 images for the 10 concept nouns in Table 4 using `stable-diffusion-v1.4` and recalculated scores for each generated image. For each axis, we picked five images representing the 0, 25, 50, 75, and 100% data samples in the pool when sorted by score. The figure confirms that, even for the axes that are hard to be measured directly from images, the scores reasonably correspond to the visual characteristic of the image. For instance, in Fig. 4(c), the image of a *running greyhound* scores higher than that of a *still greyhound*.

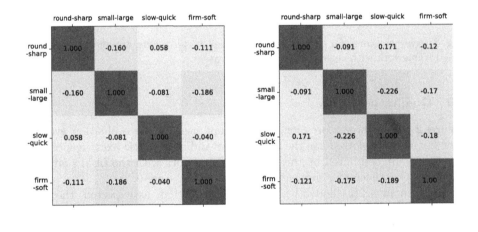

Fig. 5. Cosine similarity matrix of the four adjective axes in the experiment for two cases: Computing the axis vectors either with textual or visual embedding.

To further analyze the proposed framework, we measure the cosine similarity among the four target adjective axes, i.e., $x_{(A_-, A_+)}$ in Eq. (1). Figure 5 showcases the similarity matrix where each cell shows a cosine similarity between two axes. This matrix unveils how the proposed framework interprets each adjective axis. For instance, a negative similarity between the round-sharp and small-large axes suggests that the proposed framework tends to predict low sharpness scores for large concepts. Likewise, a positive similarity between the round-sharp and slow-quick axes indicates that concepts with a sharp contour tend to move quickly. By these observations, we can see the grounding behind the prediction by the proposed framework.

5 Conclusion

This paper proposed a training-free framework to estimate the degree of association of the visual characteristics of images with any adjective axis. We realized this by leveraging large-scale pretrained models, i.e., a generative LLM and a VLM. Experimental results indicated that the proposed framework predicts the negative and positive associations between adjectives and images as correctly as the method adopting manually selected sets of adjectives. The results also highlighted the strengths of textual and visual embeddings in predicting specific types of adjective axes. Lastly, the similarities among adjective axes indicated how the proposed framework associated them with each other.

One limitation of the proposed framework is that its behavior reflects the bias in the adopted LLMs and VLMs. Especially, we observed that GPT was less likely to output potentially harmful words. For instance, in the adjective selection, GPT did not output words like *fat* or *chubby* in describing the roundness of concepts, whereas manually selected adjective sets include them [1]. This might be degrading the intuitiveness of the predicted scores, since humans might associate visual characteristics even with such derogatory adjectives.

Our future work includes extending the framework to adjectives that cannot be described by an antonym pair. Since the current framework requires an antonym pair that rules the adjective axis on the VLM's embedding space, it becomes an obstacle when considering an extension to more diverse adjectives, such as emotions, due to the absence of an obvious antonym for each emotion.

Acknowledgments. This work was partly supported by Microsoft Research CORE16 program and JSPS Grant-in-aid for Scientific Research (22H03612 and 23K24868). The computation was carried out using the General Projects on supercomputer "Flow" at Information Technology Center, Nagoya University. The first author was financially supported by JST SPRING's "THERS Make New Standards Program for the Next Generation Researchers" (Grant Number JPMJSP2125).

References

1. Alper, M., Averbuch-Elor, H.: Kiki or Bouba? sound symbolism in vision-and-language models. Adv. Neural Inform. Process. Syst. **36**, 78347–78359 (2023)

2. Bird, S., Klein, E., Loper, E.: Natural language processing with Python: analyzing text with the Natural Language Toolkit. O'Reilly Media Inc, Sebastopol, CA, USA (2009)
3. Brown, T., et al.: Language models are few-shot learners. Adv. Neural Inform. Process. Syst. **33**, 1877–1901 (2020)
4. Computer Vision and Learning Research Group at Ludwig Maximilian University of Munich: Stable diffusion (2022). https://github.com/CompVis/stable-diffusion/ (Accessed 22 July 2024)
5. Datta, R., Joshi, D., Li, J., Wang, J.Z.: Studying aesthetics in photographic images using a computational approach. In: Leonardis, A., Bischof, H., Pinz, A. (eds.) ECCV 2006. LNCS, vol. 3953, pp. 288–301. Springer, Heidelberg (2006). https://doi.org/10.1007/11744078_23
6. Devlin, J., Chang, M.W., Lee, K., Toutanova, K.: BERT: pre-training of deep bidirectional transformers for language understanding. In: Proceedings of the 2019 Conference of the North American Chapter of the Association for Computational Linguistics: Human Language Technologies, Volume 1 (Long and Short Papers), pp. 4171–4186 (2019). https://doi.org/10.18653/v1/N19-1423
7. Hentschel, S., Kobs, K., Hotho, A.: CLIP knows image aesthetics. Front. Artif. Intell. **5**(976235), 11p (2022). https://doi.org/10.3389/frai.2022.976235
8. Jia, C., et al.: Scaling up visual and vision-language representation learning with noisy text supervision. In: Proceedings of the 38th International Conference on Machine Learning, vol. 139, pp. 4904–4916 (2021)
9. Jiang, A.Q., et al.: Mistral 7B. Comput. Res. Reposit., arXiv Preprints, arXiv:2310.06825 (2023). https://doi.org/10.48550/arxiv.2310.06825
10. Lazaridou, A., Dinu, G., Liska, A., Baroni, M.: From visual attributes to adjectives through decompositional distributional semantics. Trans. Assoc. Comput. Linguist. **3**, 183–196 (2015). https://doi.org/10.1162/tacl_a_00132
11. Mikolov, T., Chen, K., Corrado, G.S., Dean, J.: Efficient estimation of word representations in vector space. Comput. Res. Reposit., arXiv Preprints, arXiv:1301.3781 (2013)
12. Murray, N., Marchesotti, L., Perronnin, F.: AVA: a large-scale database for aesthetic visual analysis. In: Proceedings of 2012 IEEE Conference on Computer Vision and Pattern Recognition, pp. 2408–2415 (2012). https://doi.org/10.1109/CVPR.2012.6247954
13. Radford, A., et al.: Learning transferable visual models from natural language supervision. In: Proceedings of the 38th International Conference on Machine Learning Research, vol. 139, pp. 8748–8763 (2021)
14. Rombach, R., Blattmann, A., Lorenz, D., Esser, P., Ommer, B.: High-resolution image synthesis with latent diffusion models. In: Proceedings of 2022 IEEE/CVF Conference on Computer Vision and Pattern Recognition, pp. 10684–10695 (2022). https://doi.org/10.1109/CVPR52688.2022.01042
15. Russakovsky, O., Fei-Fei, L.: Attribute learning in large-scale datasets. In: Kutulakos, K.N. (ed.) ECCV 2010. LNCS, vol. 6553, pp. 1–14. Springer, Heidelberg (2012). https://doi.org/10.1007/978-3-642-35749-7_1
16. Shimoda, W., Yanai, K.: A visual analysis on recognizability and discriminability of onomatopoeia words with DCNN features. In: Proceedings of 2015 IEEE International Conference on Multimedia and Expo, 6 p. (2015). https://doi.org/10.1109/ICME.2015.7177453
17. Tong, H., Li, M., Zhang, H.-J., He, J., Zhang, C.: Classification of digital photos taken by photographers or home users. In: Aizawa, K., Nakamura, Y., Satoh,

S. (eds.) PCM 2004. LNCS, vol. 3331, pp. 198–205. Springer, Heidelberg (2004). https://doi.org/10.1007/978-3-540-30541-5_25
18. Touvron, H., et al.: LLaMA: open and efficient foundation language models. Comput. Res. Reposit., arXiv Preprints, arXiv:2302.13971 (2023)
19. Wang, J., Chan, K.C., Loy, C.C.: Exploring CLIP for assessing the look and feel of images. In: Proceedings of AAAI Conference on Artificial Intelligence, vol. 37(2), pp. 2555–2563 (2023). https://doi.org/10.1609/aaai.v37i2.25353
20. Zhao, N., Cao, Y., Lau, R.W.: What characterizes personalities of graphic designs? ACM Trans. Graph. **37**(4), 1–15 (2018). https://doi.org/10.1145/3197517.3201355

TACST: Time-Aware Transformer for Robust Speech Emotion Recognition

Wei Wei[1,2], Bingkun Zhang[1,2(✉)], and Yibing Wang[1,2]

[1] Dalian Minzu University, Dalian 116650, Liaoning, China
immersions@foxmail.com
[2] National Ethnic Affairs Commission of the Peoples Republic of China Key Laboratory of Big Data Applied Technology, Dalian Minzu University, Dalian, China

Abstract. Speech Emotion Recognition (SER) is an important research direction in the fields of human-computer interaction and affective computing. Effectively extracting emotional features from complex speech signals has always been a challenging task. This paper proposes a Time-Aware Convolutional Speech Transformer (TACST), which combines a Hybrid Convolutional Extractor (HCE) and a Time-Attention Module (TAM) to effectively improve emotion recognition performance. The HCE, by integrating Convolutional Neural Networks (CNN) and Multi-Head Attention, is able to capture both local and global features. The TAM introduces an adaptive attention mechanism along the time dimension, further enhancing the model's ability to capture the global temporal dynamics of emotional features. Experimental results show that this method achieves significant improvements in various metrics, including Weighted Accuracy, Unweighted Accuracy, and F1 scores on datasets such as IEMOCAP, DAIC-WOZ, and MELD, outperforming existing SOTA models.

Keywords: Speech Emotion Recognition · Time-Aware Transformer · Convolutional Extractor

1 Introduction

Speech is a primary means of communication for humans and an important medium for conveying emotions. With the rapid development of artificial intelligence technology, Speech Emotion Recognition (SER) [1-3] has gained widespread attention and application due to its deep understanding and accurate interpretation of human emotional expression. For example, Catania et al. [4] used an emotion-aware chatbot to help patients with difficulties in emotional expression; Han et al. [5] improved SER performance in customer care calls by converting the classification SER task into a temporal SER task.

Traditional methods of speech emotion recognition typically use techniques such as Support Vector Machines (SVM) [6] and Hidden Markov Models (HMM)

[7]. However, these methods often have low recognition accuracy. With the introduction of deep learning models like Convolutional Neural Networks (CNN) [8] and Long Short-Term Memory networks (LSTM) [9], the accuracy of speech emotion recognition has significantly improved. Deep learning-based models can effectively capture complex features in speech signals, thereby improving emotion classification performance. For instance, Zhao et al. [10] combined CNN and LSTM to learn local and global emotion-related features from speech and log-Mel spectrograms, greatly improving recognition accuracy.

The introduction of the Attention Mechanism [11] has provided new technical solutions for speech emotion recognition and has been widely applied in models such as Transformer [12]. For example, Clement et al. [13] designed a Self-Speaker Attention (SSA) mechanism to modulate speaker identity by calculating self and cross-attribute attention scores, focusing on emotion-related parts of the speech. Chen et al. [14] proposed a Deformable Speech Transformer (DST) model that dynamically adjusts window flexibility to capture more fine-grained emotional information embedded in speech.

Although the above methods have demonstrated significant performance in Speech Emotion Recognition (SER), they still struggle to capture the rich information in complex, highly nonlinear, and dynamic speech signals. This is partly because current sequential models process speech signals at different time steps equally, leading to a loss of important information. Additionally, existing attention mechanisms often focus only on local features of speech while neglecting global contextual information.

To address the limitations of current speech feature extraction methods in handling complex, dynamic speech signals, the contributions of this paper can be summarized as follows:

1) We design and propose a Hybrid Convolutional Extractor (HCE). By integrating Convolutional Neural Networks (CNN) and Multi-Head Attention modules, the HCE effectively captures both local and global features, overcoming the limitations of traditional feature extraction methods.

2) This paper proposes a Time-Aware Convolutional Speech Transformer (TACST), which leverages a Time-Attention Module (TAM) to allow the model to dynamically focus on important information at any position in the sequence and weigh it according to contextual information. This enables the model to adaptively attend to features at different time scales, automatically identifying and focusing on key features for emotion recognition, thereby improving model performance.

3) Extensive experiments on the IEMOCAP dataset [15], DAIC-WOZ dataset [16], and MELD dataset [17] demonstrate that TACST outperforms SOTA methods.

2 Method

As shown in Fig. 1, TACST is composed of an HCE (as illustrated in Step 1) and multiple stacked TAM blocks (as illustrated in Step 2). The HCE module improves the model's computational efficiency and generalization ability by integrating local and global features and incorporating Multi-Head Attention. Each TAM module mainly consists of Time-Attention and a Feedforward Neural Network (FFN). By capturing temporal features at different levels and dynamically adjusting attention weights, the model's performance is significantly enhanced, improving the extraction of key emotional information and making the model more robust and adaptable.

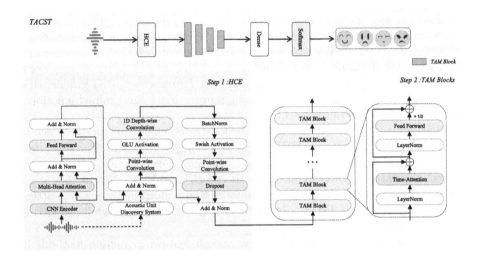

Fig. 1. Structure of the TACST model. The top part shows the data flow diagram of TACST, Step 1 illustrates the HCE structure, and Step 2 depicts the TAM Block structure

2.1 Hybrid Convolutional Extractor

The Hybrid Convolutional Extractor (HCE) (as shown in Fig. 1) is a feature extraction method that combines Convolutional Neural Networks (CNN) and self-supervised learning, aiming to extract effective features from complex speech signals. Let the input speech data be X, and after preprocessing, its shape is $X \in \mathbb{R}^{T \times F}$, where T is the time step and F is the frequency dimension. Local temporal and frequency features are extracted using CNN with a convolution kernel size of $K \times F$ and an output channel number of C. The convolution output is then given by:

$$X_{CNN} = Conv2D(X, K, C) \in \mathbb{R}^{T' \times 1 \times C} \qquad (1)$$

where T' represents the time dimension after convolution, C is the number of output channels, and X_{CNN} is the feature processed by the CNN Encoder.

Next, multiple parallel attention mechanisms enhance the model's expressive capability in capturing dependencies between different time steps or positions.

This can be expressed as follows:

$$Attention(Q, K, V) = softmax\left(\frac{QK^T}{\sqrt{d_k}}\right) V \quad (2)$$

$$X_{Attention} = Concat\,(head_1, head_2, \ldots, head_h)\,W^O \quad (3)$$

where Q, K, V represent the Query, Key, and Value, respectively; d_k denotes the dimension of the keys, typically used for scaling to avoid gradient vanishing. W^O is the parameter to be learned.

To fuse and compress feature representations from different frequencies or time segments, Point-wise Convolution enhances feature expressiveness by establishing associations between different feature channels. The Gated Linear Unit (GLU) introduces a gating mechanism that helps the model adaptively select important features, further strengthening the model's focus on key information. These can be expressed as follows:

$$X_{PW} = X_{LN} W_{pw} + b_{pw} \quad (4)$$
$$X_{GLU} = X_{PW} \odot \sigma(X_{PW} W_g + b_g) \quad (5)$$

where W_{pw} represents the weight matrix of the convolution, b_{pw} is the bias term of the convolution, and X_{LN} is the feature after layer normalization, aimed at improving the model's stability and convergence speed. X_{PW} is the feature processed by point-wise convolution, W_g denotes the gating weight matrix, and b_g is the gating bias term. X_{GLU} is the feature after GLU activation, which enhances the feature's expressive capability.

After passing through the GLU, the data undergoes local convolution for each channel using 1D depth-wise convolution to capture fine-grained features. BatchNorm and LayerNorm improve the stability of the training process through normalization techniques, reducing internal covariate shift and accelerating convergence. Furthermore, the Swish activation function is chosen to enhance the model's nonlinear expressive capability, allowing negative-valued features to be effectively utilized, thereby improving overall performance. The design of HCE enhances the model's ability to extract features from complex speech signals. This architectural design aims to address the diversity and complexity within speech signals, ensuring that the extracted features possess rich information content and can be efficiently used for downstream emotion classification tasks.

2.2 TAM Block

Time-Attention Module. The Time-Attention Module (TAM) is the core of TACST, and its main structure is illustrated in Fig. 2. Unlike traditional attention mechanisms, which focus on global interactions among queries, keys, and values across the entire sequence to extract features, TAM emphasizes the dependencies between time steps. This makes it particularly suitable for tasks like speech emotion recognition, which require capturing temporal dynamics and changes.

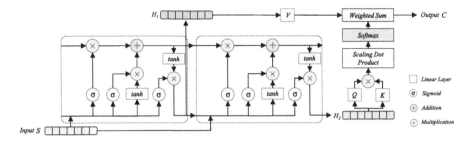

Fig. 2. Time-Attention Module

The speech feature sequence $S \in \mathbb{R}^{T \times d}$ captured by HCE is first passed to the TAM, where d represents the feature dimension. The first LSTM then models the features extracted by HCE over time, generating a hidden state sequence H_1, which serves as the V matrix in the attention mechanism. The LSTM effectively captures long-term dependencies in sequential data, addressing the issue of purely convolutional features being unable to capture long-term dependencies. H_1 provides a sequence modeled for temporal dependencies, better reflecting the emotional information in the input data.

The second LSTM further processes the sequential data, generating the hidden state H_2, which is used to compute Q and K in the attention mechanism. The specific equations are as follows:

$$H_1 = LSTM_1(S) \tag{6}$$

$$V = H_1(H_1 \in \mathbb{R}^{T \times d_{h1}}) \tag{7}$$

$$H_2 = LSTM_2(S) \tag{8}$$

$$Q = H_2 W_Q (W_Q \in \mathbb{R}^{d_{h2} \times d_k}) \tag{9}$$

$$K = H_2 W_K (W_K \in \mathbb{R}^{d_{h2} \times d_k}) \tag{10}$$

The first LSTM generates the Value for context representation, while the second LSTM generates higher-level Q and K, allowing for more flexible control of feature interactions in the attention mechanism. By stacking the two LSTMs, richer temporal information can be captured, and features can be hierarchically organized, enhancing the model's ability to express complex emotional features.

By performing scaled dot-product and softmax operations on Q and K, attention weights are generated. These weights then summarize the important information in V. Through dynamic selection, the model can flexibly determine which time steps are most critical to the current output, thereby avoiding the limitations of traditional methods that treat all time steps uniformly. The specific operations are as follows:

$$E = \frac{QK^T}{\sqrt{d_k}} (E \in \mathbb{R}^{T \times T}) \tag{11}$$

$$\alpha = Softmax(E) \tag{12}$$

$$C = \alpha V (C \in \mathbb{R}^{d_v}) \tag{13}$$

Here, E is the energy score used to measure the similarity between each time step. α represents the attention weights, and C is the context vector. The final context vector C is the weighted sequence information that not only contains complete temporal information but is also weighted according to the importance of the emotional features. By dynamically adjusting the weights, the model addresses the issues of information overload and ambiguity within the sequence features, allowing for more precise capture of emotion-related features.

The design of TAM, by combining convolutional layers and LSTMs with the introduction of an attention mechanism, effectively captures both local and global features while dynamically focusing on key time steps. This architecture not only resolves dependency issues in processing long sequences but also enhances the model's sensitivity to key emotional features, improving the performance and generalization ability of speech emotion recognition.

Feed Forward Module. In the TACST model, the feedforward module follows the TAM module to further enhance the model's representation capabilities. This feedforward module consists of two linear layers, combined with a nonlinear activation function to introduce nonlinear feature mapping.

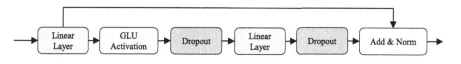

Fig. 3. Feed Forward Module

To improve the model's performance, a residual connection is designed in the feedforward layer, allowing information to flow more effectively through the network. The output is then subjected to layer normalization to stabilize the training process.

Additionally, both the residual units within the module and the signals entering the first linear layer undergo layer normalization, effectively reducing the gradient vanishing problem during training. To further enhance the network's regularization capability, the module employs the Gated Linear Unit (GLU) activation function and Dropout mechanism to alleviate overfitting and improve the model's generalization ability. The detailed structure of this feedforward module is illustrated in Fig. 3.

3 Experiments

3.1 Datasets

1) IEMOCAP is a widely used speech emotion recognition dataset from the SAIL laboratory at the University of Southern California. It contains 12 h of audio,

video, and text data, with 5,531 expressions labeled as anger, neutral, happiness, and sadness. For this dataset, this study employs 5-fold cross-validation for model training and evaluation.

2) DAIC-WOZ dataset focuses on depression analysis, containing 189 clinical interviews with durations ranging from 7 to 33 min. These interviews are labeled as "depressed" or "non-depressed" based on the patients' performances, providing valuable resources for research in emotion recognition and mental health assessment.

3) MELD dataset encompasses 13,708 utterances of different emotions, including happiness, anger, neutral, sadness, fear, surprise, and disgust. This dataset is divided into training, validation, and test sets in a 7:2:1 ratio to ensure the comprehensiveness and reliability of model evaluation.

3.2 Experimental Configuration and Evaluation Metrics

In HCE feature extraction, each frame lasts 25 milliseconds, with a step size of 20 milliseconds to generate overlapping frames, resulting in a 5-millisecond overlap between adjacent frames. This method extracts 1,024-dimensional frame-level features for each utterance sample. Moreover, experimental data indicate that approximately 80% of the sample lengths are less than 326. Therefore, to set a reasonable sequence length, this study sets the maximum sequence length for the IEMOCAP dataset to 326, 426 for the DAIC-WOZ dataset, and 224 for the MELD dataset.

A total of 120 training iterations are conducted, with an initial learning rate of 0.0005, gradually decreasing to the initial 1% through cosine annealing. The batch size is set to 32, and model parameters are updated using stochastic gradient descent (SGD) with momentum of 0.9. The hardware platform used for the experiments is the NVIDIA RTX3090.

This study employs four evaluation metrics to measure the performance variations of the model across different datasets: Weighted Accuracy (WA), Unweighted Accuracy (UA), Weighted F1 Score (WF1), and Macro-F1 Score.

3.3 Experimental Results and Analysis

Table 1 presents a performance comparison of TACST with other advanced methods on the IEMOCAP dataset. TACST demonstrates outstanding performance in Weighted Accuracy (WA), Unweighted Accuracy (UA), and Weighted F1 Score (WF1), achieving WA of 0.732, UA of 0.741, and WF1 of 0.719, all of which surpass other comparative methods. In contrast, although ShiftSER and DST show similar performance, they do not exceed TACST. The performance of STC and LSTM-GIN is significantly lower than that of TACST, indicating that TACST not only outperforms traditional methods in computational efficiency but also excels in emotion recognition tasks.

Table 2 presents a performance comparison of TACST with other advanced methods on the DAIC-WOZ dataset. TACST leads in Weighted Accuracy (WA),

TACST: Time-Aware Transformer for Robust Speech Emotion Recognition 449

Table 1. Performance Comparison of TACST with Other Advanced Methods on the IEMOCAP Dataset

Method	WA	UA	WF1
STC [18]	0.613	0.604	0.617
LSTM-GIN [19]	0.647	0.655	-
CA-MSER [20]	0.698	0.711	-
SpeechFormer [21]	0.629	0.645	-
ISNet [22]	0.704	0.650	-
DST [14]	0.718	0.736	-
ShiftSER [23]	0.721	0.727	-
TACST (Ours)	**0.732**	**0.741**	**0.719**

Table 2. Performance Comparison of TACST with Other Advanced Methods on the DAIC-WOZ Dataset

Method	WA	UA	MF1
FVTC-CNN [24]	0.735	0.656	0.640
Saidi [25]	0.680	0.680	0.680
EmoAudioNet [26]	0.732	0.649	0.653
Solieman [27]	0.660	0.615	0.610
SIMSIAM-S [28]	0.703	-	-
TOAT [29]	0.717	0.429	0.480
SpeechFormer [21]	0.686	0.650	0.694
TACST (Ours)	**0.748**	**0.692**	**0.70**

Unweighted Accuracy (UA), and Macro F1 Score (MF1), achieving WA of 0.748, UA of 0.692, and MF1 of 0.70. In contrast, the performance of other methods lags behind the TACST method proposed in this paper, indicating that TACST demonstrates stronger overall performance and balanced recognition capabilities among emotional categories in the depression recognition task.

Table 3 shows the performance comparison of TACST with other advanced methods in terms of Weighted F1 Score (WF1) on the MELD dataset. As can be seen from the table, TACST achieves a WF1 score of 0.428, which is slightly higher than that of other methods, demonstrating TACST's consistent advantage in emotion recognition tasks.

3.4 Ablation Experiments

This paper conducts ablation experiments to analyze the HCE module in depth, further validating its effectiveness and contribution to the overall model.

In the ablation experiments, the effectiveness and contribution of the HCE module within the overall model were thoroughly analyzed. By comparing the

Table 3. Performance Comparison of TACST with Other Advanced Methods on the MELD Dataset

Method	WF1
ConGCN [30]	0.422
MMFA-RANN [31]	0.423
MM-DFN [32]	0.427
CTNet [33]	0.382
SpeechFormer [21]	0.419
TACST (Ours)	**0.428**

Table 4. Ablation Experiment. Here, (+) indicates an improvement, while (-) indicates a decline

WE	IEMOCAP			DAIC-WOZ			MELD
	WA	UA	WF1	WA	UA	MF1	WF1
w/o	0.723	0.726	0.71	0.729	0.672	0.668	0.413
w/	0.732	0.741	0.719	0.748	0.692	0.70	0.428
Percentage	+1.22%	+2.08%	+1.4%	+2.61%	+3.01%	+4.73%	+3.58%
FLOPs							
w/o	7.05G			8.9G			5.14G
w/	7.64G			9.62G			5.60G
Percentage	+8.52%			+8.14%			+9.05%

performance metrics of two configurations in Table 4-"w/o" (without HCE module) and "w/" (with HCE module)-it can be observed that the model's WA, UA, and WF1 significantly improved on the IEMOCAP, DAIC-WOZ, and MELD datasets. In the IEMOCAP dataset, WA increased from 0.723 to 0.732, an improvement of 1.22%; in the DAIC-WOZ dataset, UA increased by 2.08

Although the introduction of the HCE module resulted in an increase in FLOPs and exhibited a rising trend across all datasets, the increase in computational complexity reflects an effective balance between efficiency and performance.

The results of the ablation experiments demonstrate that the inclusion of the HCE module significantly enhances the model's performance across various datasets, further proving its importance and effectiveness in the task of speech emotion recognition.

4 Conclusion

In this paper, we proposed the Time-Aware Convolutional Speech Transformer (TACST) and designed the Hybrid Convolutional Extractor (HCE) and Time

Attention Mechanism Module (TAM) to enhance the performance of speech emotion recognition. Experimental results on multiple datasets (IEMOCAP, DAIC-WOZ, and MELD) demonstrate that TACST outperforms existing state-of-the-art methods in metrics such as weighted accuracy, unweighted accuracy, and F1 score, showcasing its robust emotion recognition capabilities. The HCE module effectively integrates local and global features, while the TAM dynamically adjusts the model's attention to different time steps, enabling a deep understanding of complex speech signals. The experimental results also validate the effectiveness of HCE within the overall model, further indicating the potential of TACST in handling complex and dynamic speech signals. These findings provide new insights and methodologies for future research in speech emotion recognition, promising to advance the development and application of this field.

Acknowledgments. The authors would like to extend their sincere thanks to the School of Computer Science and Engineering at Dalian Minzu University for its support. This research was jointly supported by the Basic Research Projects for Universities of Liaoning Provincial Department of Education of China (JYTMS20231802) and the Fundamental Research Funds for the Central Universities (0919-140248).

Disclosure of Interests. The authors have no competing interests to declare that are relevant to the content of this article.

References

1. Koolagudi, S.G., Sreenivasa Rao, K.: Emotion recognition from speech: a review. Int. J. Speech Technol. **15**(2), 99–117 (2012)
2. Schuller, D.M., Schuller, B.W.: A review on five recent and near-future developments in computational processing of emotion in the human voice. Emot. Rev. **13**(1), 44–50 (2021)
3. Singh, Y.B., Goel, S.: A systematic literature review of speech emotion recognition approaches. Neurocomputing **492**, 245–263 (2022)
4. Catania, F., Garzotto, F.: A conversational agent for emotion expression stimulation in persons with neurodevelopmental disorders. Multimed. Tools Appl. **82**(9), 12797–12828 (2022)
5. Han, W., et al.: Ordinal learning for emotion recognition in customer service calls. In: IEEE International Conference on Acoustics, Speech, and Signal Processing, pp. 6494–6498 (2020)
6. Hsu, J.-H., et al.: Speech emotion recognition considering nonverbal vocalization in affective conversations. IEEE-ACM Trans. Audio Speech Lang. Process. **29**, 1675–1686 (2021)
7. Ke, X., et al.: Speech emotion recognition based on PCA and CHMM. In: IEEE Joint International Information Technology and Artificial Intelligence Conference, pp. 667–671 (2019)
8. Othmani, A., Kadoch, D., Bentounes, K., Rejaibi, E., Alfred, R., Hadid, A.: Towards robust deep neural networks for affect and depression recognition from speech. In: Proc. Int. Conf. Pattern Recognit., Int. Workshops Challenges, pp. 5–19 (2021)

9. Sundermeyer, M., Ney, H., Schlüter, R.: From feedforward to recurrent LSTM neural networks for language modeling. IEEE-ACM Trans. Audio Speech Lang. Process., 517–529 (2015)
10. Zhao, J., Mao, X., Chen, L.: Speech emotion recognition using deep 1D & 2D CNN LSTM networks. Biomed. Signal Process. Control **47**, 312–323 (2019)
11. Subakan, C., et al.: Attention is all you need in speech separation. In: ICASSP 2021 - 2021 IEEE International Conference on Acoustics, Speech and Signal Processing (ICASSP) abs/2010.13154, pp. 21–25 (2021)
12. Vaswani, A., et al.: Attention Is All You Need. Adv. Neural. Inf. Process. Syst. **30**, 5998–6008 (2017)
13. Moine, C.L., et al.: Speaker attentive speech emotion recognition. In: Conference of the International Speech Communication Association, pp. 2866–2870 (2021)
14. Chen, W., Xing, X., Xu, X., Pang, J., Du, L.: DST: deformable speech transformer for emotion recognition. In: ICASSP 2023 - 2023 IEEE International Conference on Acoustics, Speech and Signal Processing, pp. 1–5 (2023)
15. Busso, C., et al.: IEMOCAP: interactive emotional dyadic motion capture database. Language resources and evaluation, pp. 335–359 (2008)
16. Hsu, W.-N., et al.: HuBERT: self-supervised speech representation learning by masked prediction of hidden units. IEEE-ACM Trans. Audio Speech Lang. Process., 3451–3460 (2021)
17. Poria, S., Hazarika, D., Majumder, N., Naik, G., Cambria, E., Mihalcea, R.: MELD: a multimodal multi-party dataset for emotion recognition in conversations. arXiv preprint arXiv:1810.02508 (2019)
18. Guo, L., et al.: Representation learning with spectro-temporal-channel attention for speech emotion recognition. In: IEEE International Conference on Acoustics, Speech, and Signal Processing, pp. 6304–6308 (2021)
19. Liu, J., Wang, H.: Graph isomorphism network for speech emotion recognition. In: Conference of the International Speech Communication Association, pp. 3405–3409 (2021)
20. Zou, H., Si, Y., Chen, C., Rajan, D., Chng, E.S.: Speech emotion recognition with co-attention based multi-level acoustic information. In: IEEE International Conference on Acoustics, Speech, and Signal Processing, pp. 7367–7371 (2022)
21. Chen, W., Xing, X., Xu, X., Pang, J., Du, L.: SpeechFormer: a hierarchical efficient framework incorporating the characteristics of speech. In: Conference of the International Speech Communication Association, pp. 346–350 (2022)
22. Fan, W., Xu, X., Cai, B., Xing, X.: ISNet: individual standardization network for speech emotion recognition. IEEE-ACM Trans. Audio Speech Lang. Process., 1803–1814 (2022)
23. Shen, S., Liu, F., Zhou, A.: Mingling or misalignment? temporal shift for speech emotion recognition with pre-trained representations. In: ICASSP 2023 - 2023 IEEE International Conference on Acoustics, Speech, and Signal Processing, pp. 1–5 (2023)
24. Huang, Z., Epps, J., Joachim, D.: Exploiting vocal tract coordination using dilated CNNs for depression detection in naturalistic environments. In: IEEE International Conference on Acoustics, Speech, and Signal Processing, pp. 6549–6553 (2020)
25. Saidi, A., Ben Othman, S., Ben Saoud, S.: Hybrid CNN-SVM classifier for efficient depression detection system. In: 2020 4th International Conference on Advanced Systems and Emergent Technologies (IC_ASET), pp. 229–234 (2020)
26. Othmani, A., Kadoch, D., Bentounes, K., Rejaibi, E., Alfred, R., Hadid, A.: Towards robust deep neural networks for affect and depression recognition from speech. In: International Conference on Pattern Recognition, pp. 5–19 (2020)

27. Solieman, H., Pustozerov, E.A.: The detection of depression using multimodal models based on text and voice quality features. In:IEEE Conference of Russian Young Researchers in Electrical and Electronic Engineering, pp. 1843–1848 (2021)
28. Dumpala, S.H., Sastry, C.S., Uher, R., Oore, S.: On combining global and localized self-supervised models of speech. In: Conference of the International Speech Communication Association, pp. 3593–3597 (2022)
29. Yanrong, G., Chenyang, Z., Shijie, H., Richang, H.: A topic-attentive transformer-based model for multimodal depression detection (2022)
30. Zhang, D., Wu, L., Sun, C., Li, S., Zhu, Q., Zhou, G.: Modeling both context- and speaker-sensitive dependence for emotion detection in multi-speaker conversations. In: Proc. Int. Joint Conf. Artif. Intell., pp. 5415–5421 (2019)
31. Ho, N.-H., Yang, H.-J., Kim, S.-H., Lee, G.: Multimodal approach of speech emotion recognition using multi-level multi-head fusion attention-based recurrent neural network. IEEE Access **8**, 61672–61686 (2020)
32. Hu, D., Hou, X., Wei, L., Jiang, L., Mo, Y.: MM-DFN: multimodal dynamic fusion network for emotion recognition in conversations. In: Proc. IEEE Int. Conf. Acoust., Speech, Signal Process., pp. 7037–7041 (2022)
33. Lian, Z., Liu, B., Tao, J.: CTNet: conversational transformer network for emotion recognition. IEEE/ACM Trans. Audio, Speech, Lang. Process. **29**, 985–1000 (2021)

TS-MEFM: A New Multimodal Speech Emotion Recognition Network Based on Speech and Text Fusion

Wei Wei[1,2], Bingkun Zhang[1,2(✉)], and Yibing Wang[1,2]

[1] Dalian Minzu University, Dalian 116650, Liaoning, China
immersions@foxmail.com
[2] National Ethnic Affairs Commission of the People's Republic of China Key Laboratory of Big Data Applied Technology, Dalian Minzu University, Dalian, China

Abstract. Speech Emotion Recognition (SER) holds a significant position in the fields of Natural Language Processing (NLP) and Affective Computing. Traditional unimodal approaches are constrained by the quality of speech signals and variations in speaker identity, thereby affecting the accuracy and comprehensiveness of emotion recognition. This paper proposes a novel multimodal speech emotion recognition model that fuses speech and text, named the Text and Speech Multimodal Emotion Fusion Model (TS-MEFM). The model introduces relative positional encoding and residual units through the Text Multimodal Attention SFKAN Encoder (TMAK), improving the encoder's adaptability and robustness when processing utterances of varying lengths. By incorporating the Speech Multimodal Temporal SFKAN Encoder (SMTK) and the Speech Multimodal Attention Module (SMAM), the model more effectively captures subtle changes and transient features in speech signals while reducing computational overhead. Additionally, the Super Fast KAN (SFKAN) module enhances the model's nonlinear modeling capacity and reduces parameter size. Experimental comparisons on the IEMOCAP and MELD datasets demonstrate that the proposed TS-MEFM model significantly outperforms the current state-of-the-art models in speech emotion recognition performance. Ablation studies further verify the contributions of each module to the overall performance of the model, proving the effectiveness of the proposed approach.

Keywords: Speech Emotion Recognition · Multimodal · Transformer · Fusion Model

1 Introduction

Speech Emotion Recognition (SER) is an important topic in the fields of Natural Language Processing and Affective Computing. SER aims to identify the emotional state of a speaker, such as happiness, anger, sadness, etc., by analyzing speech signals [1]. With the development of human-computer interaction technologies, systems that enable computers to accurately recognize and understand human emotions are becoming increasingly important, particularly in applications such as intelligent voice assistants, voice-based customer service, and speech-driven healthcare. SER not only helps improve human-computer interaction but also provides a deeper understanding of the complexity of emotional expression and subtle variations in speech. Progress in this field will drive advancements in affective computing technologies, helping us better understand and simulate human emotions.

In traditional SER, emotion recognition is typically based solely on unimodal speech signals. This approach is susceptible to factors such as environmental noise, recording equipment, and speaker state (e.g., health condition), leading to reduced accuracy in emotion recognition [2]. In reality, emotions are multidimensional, and it is difficult to capture them comprehensively using only speech signals. Additionally, different speakers may exhibit significant individual differences when expressing the same emotion, further increasing the difficulty of recognizing emotions using unimodal approaches [3]. Latif et al. [4] overcame the impact of noise on SER accuracy in real environments by integrating DenseNet, Long Short-Term Memory (LSTM) networks, and highway networks to learn noise-robust representations. However, this method is still insufficient when dealing with unimodal data in high-noise environments, as the model's performance significantly deteriorates when the noise level exceeds a certain threshold.

To overcome the limitations of unimodal approaches, Shah et al. [5] proposed a method that uses an improved Restricted Boltzmann Machine (RBM) to distinguish emotional signals from three modalities: acoustic signals, text information, and facial expressions. They then fused these modalities for decision-making. Multimodal methods, which combine various sources of information such as speech, body language, tone, vocabulary, and physiological signals, provide a more comprehensive and accurate way to recognize emotions. These methods not only offer higher accuracy and robustness but also leverage the complementarity between different modalities to improve recognition performance. Cho et al. [6] used LSTM networks and multi-resolution convolutional networks (MCNN) to enhance recognition performance by fusing acoustic and word features, but the model still had issues with the mutual promotion of the two modalities.

Currently, with the application of Transformer models in SER, it is feasible to use multimodal and different temporal information fusion methods to recognize emotions in speech. For example, Huang et al. [7] used Transformer models to fuse audio and video modalities, achieving better results than feature fusion alone. However, the Multilayer Perceptron (MLP) in Transformer has a large number of parameters and suboptimal accuracy. In this paper, we designed the Super Fast KAN (SFKAN) module and replaced the MLP in the Transformer, reducing the number of parameters while improving SER performance by eliminating natural prosody variations in speech.

To better capture emotional information from speech, we modelled the complementarity between speech and text and proposed a novel fusion model for multimodal emotion recognition that combines both-TS-MEFM (Text and

Speech Multimodal Emotion Fusion Model)-to promote diversified complementary learning of emotional features. The main contributions of this paper are:

(1) A novel multimodal emotion fusion neural network model is proposed to capture fine-grained emotional information from both acoustic and textual modalities. By utilizing speech and its corresponding transcriptions, this model captures subtle interactions between the two modalities, thereby reducing the impact of unimodal bias.

(2) The proposed SFKAN module replaces the traditional MLP to reduce the model's parameter size and improve accuracy. In the encoder, we designed a new Speech Multimodal Attention Module (SMAM) to replace the original Multi-Head Self-Attention (MSA). SMAM limits the global attention computation to a smaller range of tokens, significantly reducing the computational load.

(3) To ensure mutual interaction between the text and speech modalities, this paper uses the RoBERTa model [8] and HuBERT model [9] for feature extraction of text and speech, respectively. On this basis, the TMAK Encoder (Text Multimodal Attention SFKAN Encoder) and SMTK Encoder (Speech Multimodal Temporal SFKAN Encoder) are designed to implicitly align the two modalities. Through the Bi-LSTM Dense Module (BLDM), efficient fusion and processing of multimodal features are achieved, ensuring that the model captures fine-grained emotional information. The results of the proposed model on the IEMOCAP [10] and MELD [11] datasets surpass those of other state-of-the-art methods.

2 Related Work

2.1 Unimodal Speech Emotion Recognition

Unimodal speech emotion recognition primarily relies on audio features from speech signals, such as pitch, intensity, rhythm, and speech rate, to identify the emotional state of the speaker. Early studies focused on Low-Level Descriptors (LLDs) [12–14], such as frequency-domain, time-domain, and time-frequency domain features. Using machine learning methods like Support Vector Machines (SVM) [15] and Hidden Markov Models (HMM) [16,17], researchers achieved decent results in classifying these extracted features. However, the heavy reliance on feature extraction limited the generalization ability of these models.

In recent years, with the rapid development of deep learning technology, models such as Convolutional Neural Networks (CNN) [18,19] and Recurrent Neural Networks (RNN) [20,21] have been widely applied to speech emotion recognition. Deep learning methods significantly improved the performance of emotion recognition by automatically learning and extracting features. For example, Long Short-Term Memory (LSTM) networks [22,23] effectively capture the temporal dependencies in speech signals, resulting in more accurate emotion recognition. The current state-of-the-art in speech emotion recognition uses HuBERT as the speech encoder, which can significantly improve recognition accuracy by extracting high-quality speech features.

2.2 Multimodal Speech Emotion Recognition

Multimodal speech emotion recognition combines various signal sources, such as speech, facial expressions, gestures, and physiological signals, to identify a speaker's emotional state. Among these, the "speech-text" combination is the most common, leveraging both speech signals and their transcribed text to enhance accuracy by utilizing the strengths of linguistic and paralinguistic information [24]. Speech signals offer audio features like pitch, volume, and speech rate that convey emotional cues, while the text modality provides deeper insights through semantic and syntactic analysis. For example, anger may be indicated by a high volume and fast speech rate in audio, while the corresponding text may use intense vocabulary and abrupt sentence structures. Merging these modalities allows for a more comprehensive understanding of the speaker's emotions.

The rapid advancement of deep learning has propelled multimodal emotion recognition, leading to the creation of more sophisticated models through feature fusion. For instance, Tripathi et al. [25] integrated self-attention with a bidirectional LSTM (BiLSTM) to extract multimodal features from speech and text. Similarly, Li et al. [26] employed a Bi-LSTM model with an attention mechanism to automatically learn optimal temporal features, enhancing emotion recognition performance. However, challenges remain, such as information loss with long sequential data and the potential for attention mechanisms to overly focus on specific features, overlooking others and thereby reducing accuracy. Further improvements in multimodal emotion recognition methods are necessary.

3 Method

The proposed TS-MEFM model, as shown in Fig. 1, consists of two main components: text feature learning and speech feature learning. The text feature learning component combines the RoBERTa [8] text encoder with the proposed TMAK encoder. RoBERTa, pre-trained on large-scale text data, captures subtle semantic and emotional cues from the text. The speech feature learning component combines the HuBERT [9] speech encoder with the proposed SMTK encoder, which captures both acoustic features and semantic information at multiple levels, resulting in more comprehensive and detailed feature extraction. Afterward, the extracted text and speech features are implicitly aligned and fused to obtain richer and more integrated emotional features. Since emotions may manifest across different time sequences, the model utilizes BLDM to capture temporal dependencies between contexts, enhancing the understanding of relationships among features and improving the overall performance and robustness of emotion recognition.

3.1 TMAK Encoder

The TMAK encoder consists of the Text Multimodal Attention Module (TMAM) and SFKAN, as illustrated in Fig. 1(a). TMAM applies a relative sinusoidal positional encoding scheme from the Transformer architecture. This scheme enhances

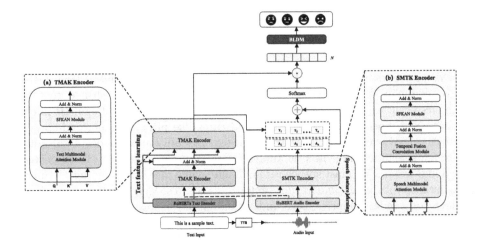

Fig. 1. TS-MEFM model. The left part represents the text feature learning structure, and the right part represents the speech feature learning structure. Together, they form the TS-MEFM through joint learning

the model's understanding of local and global dependencies in a sequence by capturing the relative positions between words. Unlike traditional absolute positional encoding, which relies on the absolute positions of words in a sequence, the relative positional encoding avoids issues related to poor generalization across sequences of varying lengths. This approach allows TMAM to adapt more flexibly to different input lengths and improves the encoder's robustness in handling varying utterance lengths.

Additionally, TMAM incorporates residual units with Dropout, which plays a crucial role in training and regularizing deep models. In the forward propagation of the residual unit, input data is first normalized, followed by a linear transformation and nonlinear activation. Finally, the output is added back to the input through a residual connection. This design helps stabilize gradients, alleviating the vanishing gradient problem in deep models, and, with Dropout regularization, prevents overfitting and improves the model's generalization ability.

3.2 SMTK Encoder

The SMTK encoder consists of the Speech Multimodal Attention Module (SMAM) and SFKAN, as shown in Fig. 1(b).

Speech Multimodal Attention Module. Figure 2 illustrates the structure of the Speech Multimodal Attention Module (SMAM). In this module, an attention window is employed to limit all attention computations to a small range of neighboring tokens, significantly reducing the computational burden. The computational process is as follows:

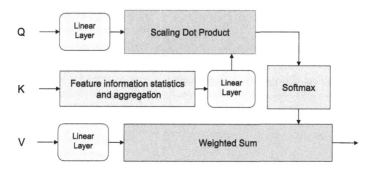

Fig. 2. SMAM. In the SMAM, scaled dot product and feature statistics are used to limit token counting, thereby reducing the computational burden

(1) **Attention Calculation:** Let H_w be the attention window size, and t be the time step. From the key sequence K, a window of size H_w is selected, centered around the current token at position t, i.e., from $t - \frac{H_w}{2}$ to $t + \frac{H_w}{2}$. The tokens within this window are then used to perform a scaled dot-product operation with the query sequence Q at position t. These dot products are normalized using the Softmax function, resulting in the attention weights $\alpha(t)$. The formula is as follows:

$$scores(t) = \left(\frac{Q(t) \cdot K_{t-\frac{H_w}{2}}}{\sqrt{d_k}}, \frac{Q(t) \cdot K_{t-\frac{H_w}{2}+1}}{\sqrt{d_k}}, \ldots, \frac{Q(t) \cdot K_{t+\frac{H_w}{2}}}{\sqrt{d_k}} \right) \quad (1)$$

$$\alpha(t) = Softmax(scores(t)) \quad (2)$$

Here, $\sqrt{d_k}$ represents a scaling factor that stabilizes the values of the dot product to avoid overly large results. In attention mechanisms, the dot product between Q and K is used to compute relevance scores. If scaling is not applied, the dot product can become large as d_k increases, which could result in very small gradients for the Softmax function, negatively affecting model training. To mitigate this issue, the dot-product results are scaled by $\sqrt{d_k}$.

(2) **Output Calculation:** A window of size H_w is selected from the value sequence V, corresponding to the same range as in the key sequence. These tokens are multiplied by the respective attention weights, and the results are summed to obtain the output at position t:

$$output(t) = \sum_{i=t-\frac{H_w}{2}}^{t+\frac{H_w}{2}} \alpha_i(t) \cdot V(i) \quad (3)$$

(3) **Repeat:** Steps (1) and (2) are repeated for each time step t, from 1 to T.

3.3 SFKAN Module

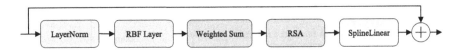

Fig. 3. SFKAN Module

Figure 3 illustrates the forward propagation process of each layer in the SFKAN model. The SFKAN module enhances the model's nonlinear modeling capability by using a Radial Basis Functions Layer (RBF Layer) and Reflectional Switch Activation (RSA), improving its ability to represent complex data while reducing the number of parameters.

First, SFKAN normalizes the input data to ensure that subsequent operations are more stable. It then employs the RBF to replace traditional fully connected layers, performing localized response processing on the input to capture the nonlinear features of the data. The reason for using RBF instead of fully connected layers (MLP) is that fully connected layers typically perform global operations on all inputs, resulting in a large number of parameters. In contrast, the localized response of RBF reduces global dependence and better represents complex nonlinear relationships. The specific formulas are as follows:

$$\hat{x}_i = \frac{x_i - \mu}{\sigma} \tag{4}$$

$$\phi_j(\hat{x}_i) = \exp\left(-\frac{\|\hat{x}_i - c_j\|^2}{2\sigma_j^2}\right) \tag{5}$$

where x_i is the input data, μ and σ are the mean and standard deviation, $\hat{x}_i$ is the normalized data, c_j is the center of the RBF, σ_j is the width parameter, and ϕ_j is the output of the j-th Gaussian radial basis function.

To address the issue that the output of the RBF alone may not fully express global patterns in the data, the output of the RBF layer is multiplied by weights and summed to obtain intermediate representations of the hidden layer, allowing the model to integrate different local responses. Subsequently, the RSA performs a nonlinear transformation on the hidden layer output, reflecting negative values as positive values to retain more information and enhance the model's ability to represent complex signals. The formulas are as follows:

$$z_i = \sum_{j=1}^{M} w_j \cdot \phi_j(\hat{x}_i) \tag{6}$$

$$RSA(z_i) = \begin{cases} z_i, & z_i \geq 0 \\ -z_i, & z_i < 0 \end{cases} \tag{7}$$

where w_j is the weight of the output from the j-th RBF, and M is the number of RBF functions.

The SFKAN module retains nonlinear expressive capabilities while reducing the number of parameters through an efficient structure, thereby improving the speed and performance of the model. Compared to traditional MLPs, this design can handle complex multimodal data more flexibly, and the reduction in parameters enhances the training and inference speed of the model.

3.4 Bi-LSTM Dense Module

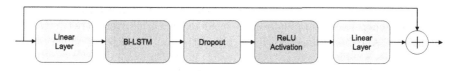

Fig. 4. BLDM

The structure of the BLDM model is shown in Fig. 4. This model combines the advantages of Bi-LSTM [27,28] and dense layers, further processing and optimizing feature representations after feature fusion, significantly enhancing the model's ability to understand emotional and semantic information. It effectively handles and merges complex features in sequential data, improving the extraction and analysis of emotional information. By capturing both local and global temporal dependencies, the BLDM model enhances the transfer and utilization of features, providing richer and more accurate emotional feature representations in multimodal emotion analysis tasks.

4 Experiments

4.1 Dataset

(1) IEMOCAP is a popular dataset for speech emotion recognition, collected by the Speech and Emotion Research Laboratory at the University of Southern California. It includes 12 h of multimodal data-audio, video, and text-from five sessions, with 5,531 expressions labeled as anger, neutral, happiness, and sadness. A five-fold cross-validation strategy is used for training, with one session reserved as the test set. Model performance is measured using Weighted Accuracy (WA) and Unweighted Accuracy (UA).

(2) MELD is another key dataset for emotion recognition, containing 13,708 utterances that represent seven emotions: happiness, anger, neutral, sadness, fear, surprise, and disgust. It is split into training (9,989 samples), validation (1,109 samples), and test sets (2,610 samples). The Weighted F1 Score (WF1) is used to evaluate model performance in the experiments.

4.2 Experimental Setup

In text processing, the model is first used to back-translate the text. During the back-translation process, five translations are generated, and the best-performing one is selected as the augmented text. Then, RoBERTa-BASE is utilized as the text encoder to extract features from both the original and augmented text, and the processed text features are input into TMAK for subsequent operations.

In speech processing, a TTS model is used to generate the corresponding speech signals based on the augmented text. Then, the pre-trained HuBERT-Base model, which has been trained on 960 h of LibriSpeech data [29], is used to process the speech. The duration of each frame is set to 25 ms, with a hop length of 20 ms to generate overlapping frames, resulting in a 5-ms overlap between adjacent frames. A 1024-dimensional feature is extracted for each utterance sample and passed to SMTK.

Experiments are conducted on an NVIDIA RTX A4000 GPU with 120 training epochs, a learning rate of 1×10^{-5}, and a batch size of 2. The IEMOCAP and MELD datasets are set to a dimension of 1024. Each training and inference step takes about 8 min, and TS-MEFM has approximately 221.3 million parameters.

4.3 Experimental Results and Analysis

Table 1. Performance Comparison of TS-MEFM on the IEMOCAP Dataset. The best results for the metrics are highlighted in bold

Model	WA (%)	UA (%)
TAB [30]	77.57	78.41
ResNet + BERT [31]	76.07	75.76
Key-Sparse Transformers [32]	74.30	75.30
Modality Calibration [33]	75.60	77.60
MSE [34]	75.40	74.20
DMF [35]	76.80	77.30
Speechformer++ [36]	70.50	71.50
MMER [37]	80.18	78.69
MSER [38]	77.20	76.56
MF-AED-AEC [39]	-	79.30
TS-MEFM (Ours)	**84.21**	**82.20**

Table 1 and Table 2 present a performance comparison between TS-MEFM and other advanced methods on the IEMOCAP and MELD datasets, respectively. A specific comparison is made regarding WA, UA, and WF1. From the tables, it can be observed that the proposed TS-MEFM model outperforms other comparison methods on both datasets, achieving significant improvements. Specifically, the proposed model reaches a WA of 84.21% and an UA of 82.2% on the IEMOCAP dataset, while on the MELD dataset, the WF1 metric is 69.14%.

The experimental results demonstrate that incorporating a multimodal Transformer fusion model can effectively combine the information from speech and text modalities, utilizing their complementarity to enhance emotion recognition performance. In particular, the designed TMAK encoder and SMTK encoder significantly improve the accuracy of emotion recognition by capturing fine-grained emotional information in both text and speech.

Compared to other methods, the approach presented here is more effective in capturing complex emotional expressions. The experimental results further validate the superiority of this method, indicating that the TS-MEFM model has great potential and application value in the field of multimodal emotion recognition.

Table 2. Performance Comparison of TS-MEFM on the MELD Dataset. The best results for the metrics are highlighted in bold

Model	WF1(%)
SDT [40]	66.60
CFN-ESA [41]	66.70
M2FNet [42]	66.71
FacialMMT [43]	66.73
SACL-LSTM [44]	66.86
HiDialog [45]	66.96
DF-ERC [46]	67.03
EACL [47]	67.12
SPCL-CL-ERC [48]	67.25
TelME [49]	67.37
TS-MEFM (Ours)	**69.14**

4.4 Ablation Experiments

To verify the impact of each module on the overall performance of the model, ablation experiments were conducted on different datasets, with results shown in Table 3. The experiments evaluated the performance of the TMAK (T), SMTK (S), and SFKAN (K) modules under different modality combinations.

From the table, it can be seen that the recognition performance of using the text modality TMAK and the speech modality SMTK individually is relatively low. When combining both speech and text modalities without the SFKAN module and using traditional methods, the performance on the IEMOCAP dataset reaches 83.33%, while on the MELD dataset it is 67.24%, which is significantly

Table 3. Ablation Experiments. In the table, (T) represents TMAK, (S) represents SMTK, and (K) represents SFKAN

Model	Modality	FLOPs	IEMOCAP WA(%)	IEMOCAP UA(%)	MELD WF1(%)
T	Text	135.8M	76.63	75.36	62.89
S	Speech	156.4M	75.77	74.41	58.66
S(w/o K)+T(w/o K)	Speech+Text	256.8M	83.33	80.64	67.24
S+T	Speech+Text	**221.3M**	**84.21**	**82.20**	**69.14**

higher than that of a single modality. This indicates that multimodal fusion has a significant advantage in emotion recognition. After introducing the SFKAN module, the model's weighted accuracy (WA) improved by 0.88% and unweighted accuracy (UA) improved by 1.56% on the IEMOCAP dataset, while on the MELD dataset, the weighted F1 score (WF1) increased by 1.9%. At the same time, the parameter count was reduced by approximately 13.82%. This confirms that the SFKAN module enhances the model's accuracy and robustness while reducing the number of parameters.

4.5 Conclusion

This paper introduces TS-MEFM, a novel multimodal emotion fusion network that shows strong performance in emotion recognition. By integrating speech and text information with the RoBERTa and HuBERT encoders, the model effectively captures fine-grained emotional features. The SFKAN module reduces parameters while improving accuracy. Experimental results reveal that TS-MEFM outperforms existing methods on the IEMOCAP and MELD datasets, achieving WA, UA, and WF1 scores of 84.21%, 82.20%, and 69.14%, respectively. These results enhance emotion recognition accuracy and robustness, providing a solid foundation for future research in multimodal emotion analysis. Future work will explore the model's generalization ability and its application on broader datasets to tackle diverse emotional expressions.

Acknowledgments. The authors would like to extend their sincere thanks to the School of Computer Science and Engineering at Dalian Minzu University for its support. This research was jointly supported by the Basic Research Projects for Universities of Liaoning Provincial Department of Education of China (JYTMS20231802) and the Fundamental Research Funds for the Central Universities (0919-140248).

Disclosure of Interests. The authors have no competing interests to declare that are relevant to the content of this article.

References

1. Schuller, B.W.: Speech emotion recognition: Two decades in a nutshell, benchmarks, and ongoing trends. Commun. ACM **61**(5), 90–99 (2018)
2. Schuller, B., et al.: Recognising realistic emotions and affect in speech: state of the art and lessons learnt from the first challenge. Speech Commun. **53**(9–10), 1062–1087 (2011)
3. Shoumy, N.J., et al.: Multimodal big data affective analytics: a comprehensive survey using text, audio, visual and physiological signals. J. Network Comput. Appl. **149**, 102447–71 (2020)
4. Lian, Z., Tao, J., Liu, B., et al.: Context dependent domain adversarial neural network for multimodal emotion recognition. In: Interspeech. ISCA 2020, pp. 394–398 (2020)
5. Shah, M., et al.: A multi-modal approach to emotion recognition using undirected topic models. In: IEEE International Symposium on Circuits and Systems, pp. 754–757 (2014)
6. Cho, J., et al.: Deep neural networks for emotion recognition combining audio and transcripts. In: Conference of the International Speech Communication Association, pp. 247–251 (2019)
7. Huang, J., et al.: Multimodal transformer fusion for continuous emotion recognition. In: IEEE International Conference on Acoustics, Speech, and Signal Processing, pp. 3507–3511 (2020)
8. Liu, Y., et al.: Roberta: a robustly optimized BERT pretraining approach. arXiv preprint arXiv:1907.11692 (2019)
9. Hsu, W.-N., et al.: HuBERT: self-supervised speech representation learning by masked prediction of hidden units. IEEE-ACM Trans. Audio Speech Lang. Process., 3451–3460 (2021)
10. Busso, C., et al.: IEMOCAP: interactive emotional dyadic motion capture database. Language Resources and Evaluation, pp. 335–359 (2008)
11. Poria, S., et al.: MELD: a multimodal multi-party dataset for emotion recognition in conversations. arXiv preprint arXiv:1810.02508 (2019)
12. Rozgic, V., Ananthakrishnan, S., Saleem, S., et al.: Emotion recognition using acoustic and lexical features. In: Proceedings of the 13th Annual Conference of the International Speech Communication Association (INTERSPEECH), pp. 366–369 (2012)
13. Jin, Q., Li, C., Chen, S., et al.: Speech emotion recognition with acoustic and lexical features. In: 2015 IEEE International Conference on Acoustics, Speech and Signal Processing (ICASSP), pp. 4749–4753 (2015)
14. Gamage, K.W., Sethu, V., Ambikairajah, E.: Salience based lexical features for emotion recognition. In: 2017 IEEE International Conference on Acoustics, Speech and Signal Processing (ICASSP), pp. 5830–5834 (2017)
15. Shen, P., et al.: Automatic speech emotion recognition using support vector machine. In: Proc of International Conference on Electronic and Mechanical Engineering and Information Technology (EMEIT), Harbin, China, pp. 621–625 (2011)
16. Nwe, T.L., et al.: Speech emotion recognition using hidden Markov models. Speech Commun., 603–623 (2003)
17. Schuller, B., Rigoll, G., Lang, M.: Hidden Markov model-based speech emotion recognition. In: IEEE International Conference on Acoustics, Speech, and Signal Processing 1, 2003, pp. I-401-4 (2003)

18. Othmani, A., et al.: Towards robust deep neural networks for affect and depression recognition from speech. In: Proc. Int. Conf. Pattern Recognit., Int. Workshops Challenges, 2021, pp. 5–19 (2021)
19. Muppidi, A., Radfar, M.: Speech emotion recognition using quaternion convolutional neural networks. In: ICASSP 2021, Toronto, ON, Canada, June 6-11, 2021, pp. 6309–6313. IEEE (2021)
20. Bertini, F., et al.: An automatic Alzheimer's disease classifier based on spontaneous spoken English. Comput. Speech Lang. 72, Art. no. 101298 (2022)
21. Rajamani, S.T., Rajamani, K.T., Mallol-Ragolta, A., et al.: A novel attention-based gated recurrent unit and its efficacy in speech emotion recognition. In: ICASSP 2021, Toronto, ON, Canada, June 6-11, 2021, pp. 6294–6298. IEEE (2021)
22. Sundermeyer, M., Ney, H., Schlüter, R.: From Feedforward to Recurrent LSTM Neural Networks for Language Modeling, pp. 517–529. Speech and Language Processing, IEEE-ACM Transactions on Audio (2015)
23. Woellmer, M., Schuller, B., Eyben, F., Rigoll, G.: Combining long short-term memory and dynamic bayesian networks for incremental emotion-sensitive artificial listening. IEEE J. Sel. Top. Signal Process., 867–881 (2010)
24. Huang, J., Li, Y., Tao, J., et al.: Continuous multimodal emotion prediction based on long short term memory recurrent neural network. In: Proceedings of the 7th Annual Workshop on Audio/Visual Emotion Challenge, pp. 11–18 (2017)
25. Tripathi, S., Tripathi, S., Beigi, H.: Multi-modal emotion recognition on IEMOCAP dataset using deep learning, arXiv preprint arXiv:1804.05788 (2019)
26. Li, C., Bao, Z., Li, L., et al.: Exploring temporal representations by leveraging attention-based bidirectional LSTM-RNNs for multi-modal emotion recognition. Inform. Process. Manage. **57**(3), 102185 (2020)
27. Chen, M., Zhao, X.: A multi-scale fusion framework for bimodal speech emotion recognition. In: Conference of the International Speech Communication Association, pp. 374–378 (2020)
28. Su, B.-H., et al.: Self-assessed affect recognition using fusion of attentional BLSTM and static acoustic features. In: Conference of the International Speech Communication Association, pp. 536–540 (2018)
29. Panayotov, V., et al.: Librispeech: an ASR corpus based on public domain audio books. In: 2015 IEEE International Conference on Acoustics, Speech and Signal Processing (ICASSP), pp. 5206–5210 (2015)
30. Wu, W., et al.: Emotion recognition by fusing time synchronous and time asynchronous representations. In: IEEE International Conference on Acoustics, Speech, and Signal Processing abs/2010.14102, pp. 6269–6273 (2021)
31. Padi, S., et al.: Multimodal emotion recognition using transfer learning from speaker recognition and BERT-based models. In: The Speaker and Language Recognition Workshop, pp. 407–414 (2022)
32. Chen, W., et al.: Key-sparse transformer for multimodal speech emotion recognition. In: IEEE International Conference on Acoustics, Speech, and Signal Processing, pp. 6897–6901 (2022)
33. Hou, M., et al.: Multi-modal emotion recognition with self-guided modality calibration. In: IEEE International Conference on Acoustics, Speech, and Signal Processing, pp. 4688–4692 (2022)
34. Feng, L., et al.: Multimodal speech emotion recognition based on multi-scale MFCCs and multi-view attention mechanism. Multimed. Tools Appl. **82**(19), 28917–28935 (2023)

35. Prisayad, D., et al.: Dual memory fusion for multimodal speech emotion recognition. In: Conference of the International Speech Communication Association, pp. 4543–4547 (2023)
36. Chen, W., et al.: SpeechFormer++: a hierarchical efficient framework for paralinguistic speech processing. IEEE-ACM Trans. Audio Speech Lang. Process. **31**(1), 775–788 (2023)
37. Ghosh, S., et al.: MMER: multimodal multi-task learning for speech emotion recognition. In: Conference of the International Speech Communication Association, pp. 1209–1213 (2023)
38. Khan, M., et al.: MSER: multimodal speech emotion recognition using cross-attention with deep fusion. Expert Syst. Appl. 245 (2024)
39. He, J., et al.: MF-AED-AEC: speech emotion recognition by leveraging multimodal fusion, asr error detection, and asr error correction. In: ICASSP 2024 - 2024 IEEE International Conference on Acoustics, Speech and Signal Processing abs/2401.13260, 2024, pp. 11066–11070 (2024)
40. Ma, H., et al.: A transformer-based model with self-distillation for multimodal emotion recognition in conversations, CoRR abs/2310.20494, pp. 1–13 (2023)
41. Li, J., et al.: CFN-ESA: a cross-modal fusion network with emotion-shift awareness for dialogue emotion recognition. IEEE Trans. Affective Comput., 1–16 (2023)
42. Chudasama, V., et al.: M2FNet: multi-modal fusion network for emotion recognition in conversation. In: Computer Vision and Pattern Recognition, pp. 4651–4660 (2022)
43. Zheng, W., et al.: A facial expression-aware multimodal multi-task learning framework for emotion recognition in multi-party conversations. In: Annual Meeting of the Association for Computational Linguistics Proceedings of the 61st Annual Meeting of the Association for Computational Linguistics (Volume 1: Long Papers), pp. 15445–15459 (2023)
44. Hu, D., et al.: Supervised adversarial contrastive learning for emotion recognition in conversations. In: Annual Meeting of the Association for Computational Linguistics abs/2306.01505, pp. 10835–10852 (2023)
45. Liu, X., et al.: Hierarchical Dialogue Understanding with Special Tokens and Turn-level Attention. CoRR abs/2305.00262 (2023)
46. Li, B., et al.: Revisiting Disentanglement and Fusion on Modality and Context in Conversational Multimodal Emotion Recognition. CoRR abs/2308.04502, pp. 5923–5934 (2023)
47. Yu, F., et al.: Emotion-Anchored Contrastive Learning Framework for Emotion Recognition in Conversation. CoRR abs/2403.20289 (2024)
48. Song, X., et al.: Supervised prototypical contrastive learning for emotion recognition in conversation. In: Conference on Empirical Methods in Natural Language Processing, pp. 5197–5206 (2022)
49. Yun, T., et al.: TelME: teacher-leading multimodal fusion network for emotion recognition in conversation. CoRR abs/2401.12987 (2024)

Author Index

B
Bonatto, Daniele 128

C
Cao, Qian 268
Chang, Ding-Chi 58
Chen, Junjian 253
Chen, Nan 16
Chen, Xu 268
Chen, Zhuowei 16

D
Dai, Xiyue 100

F
Fachada, Sarah 128
Fang, Zhiyi 100
Fukuzawa, Takumi 366

G
Guðmundsson, Gylf Þór 198
Guinaudeau, Camille 212, 398

H
Hara, Kensho 366
Hirayama, Takatsugu 428
Huang, Jen-Wei 58
Huang, Jia-Hong 413
Huang, Mengqi 16
Huang, Qianni 240
Huron, Caroline 212

I
Ide, Ichiro 226, 428
Ioannidis, Konstantinos 283

J
Ju, Boyuan 30
Juarez, Eduardo 128
Juliussen, Bjørn Aslak 184

K
Kando, Noriko 170
Kanoulas, Evangelos 413
Kastner, Marc A. 226, 428
Kataoka, Hirokatsu 366
Kawanishi, Yasutomo 226
Komamizu, Takahiro 226, 428
Kompatsiaris, Ioannis 283
Kyas, Marcel 198

L
Lafruit, Gauthier 128
Latapie, Hugo 85
Li, Feng 325
Li, Jun 337
Li, Shiou-Chi 58
Li, Su 383
Li, Yizhou 155
Li, Yu 298
Lin, Xiumin 71
Lincker, Élise 212
Liu, Jing 44
Liu, Zhaoyang 3
Liu, Zihua 155
Lu, Congjian 3
Lu, Lingyi 114
Luo, Jing 30
Luo, Jiusong 325
Lv, Yishan 30

M
Mao, Zhendong 16
Martin, Stéphanie 212
Matsuhira, Chihaya 428
Monno, Yusuke 155

N
Ngu, Anne Hee Hiong 85

O
Okutomi, Masatoshi 155

P
Pan, Shanliang 141
Papadopoulos, Sotirios 283
Patras, Ioannis 283
Phueaksri, Itthisak 226

Q
Qi, Na 71
Qian, Yi 100
Quan, Khanh-An C. 398

R
Rudinac, Stevan 413

S
Sancho, Jaime 128
Satoh, Shin'ichi 212, 398
Shan, Ke 3
Sharma, Nikhil 85
Shen, Yixian 413
Shukla, Jainendra 212
Song, Ruihua 268
Springer, Joshua 198
Sun, Changchang 85
Suo, Zihao 141

T
Tamaki, Toru 366
Tanabe, Hikaru 353
Terada, Takamasa 311
Teratani, Mehrdad 128
Toyoura, Masahiro 311

V
Vrochidis, Stefanos 283

W
Wan, Yan 240
Wang, Jianye 383
Wang, Liang 383
Wang, Xiao 114
Wang, Yibing 442, 454
Wei, Wei 442, 454
Wu, Weijie 337
Wu, Zhijian 337

X
Xia, Wanjun 325
Xie, Zhenping 298
Xu, Jianhua 337
Xu, Xin 114

Y
Yadav, Saumya 212
Yamamoto, Shuhei 170
Yan, Yan 85
Yanai, Keiji 353
Yang, Xinyu 30
Yang, Xuan 253
Yao, Li 240

Z
Zhang, Bingkun 442, 454
Zhang, Hongkuan 3
Zhang, Junjun 383
Zhang, Lei 383
Zhang, Yongliang 44
Zhang, Ziheng 383
Zhao, Hui 71
Zhao, Zhenghao 85
Zhou, Shuwang 3
Zhu, Hongyi 413
Zhu, Qing 71

Printed in the United States
by Baker & Taylor Publisher Services